AF213617

NONLINEARITY, BIFURCATION AND CHAOS – THEORY AND APPLICATIONS

Edited by **Jan Awrejcewicz**
and **Peter Hagedorn**

Nonlinearity, Bifurcation and Chaos – Theory and Applications
http://dx.doi.org/10.5772/2888
Edited by Jan Awrejcewicz and Peter Hagedorn

Contributors

Igor Andrianov, Jan Awrejcewicz, Victor Olevs'kyy, John Alexander Taborda, Fabiola Angulo, Gerard Olivar, José Manoel Balthazar, Angelo Marcelo Tusset, Atila Madureira Bueno, Bento Rodrigues de Pontes Junior, A. O. Belyakov, A. P. Seyranian , A. Dumitrache, F. Frunzulica, T.C. Ionescu, Nikolai A. Magnitskii, Ulrich Fuellekrug, Tomasz Kopecki, Tomasz Kubiak, N.M. Evstigneev, N.A. Magnitskii, B. M. Podlevskyi, Miroslav Verbič

Published by InTech
Janeza Trdine 9, 51000 Rijeka, Croatia

Copyright © 2012 InTech
All chapters are distributed under the Creative Commons Attribution 3.0 license, which allows users to download, copy and build upon published articles even for commercial purposes, as long as the author and publisher are properly credited, which ensures maximum dissemination and a wider impact of our publications. However, users who aim to disseminate and distribute copies of this book as a whole must not seek monetary compensation for such service (excluded InTech representatives and agreed collaborations). After this work has been published by InTech, authors have the right to republish it, in whole or part, in any publication of which they are the author, and to make other personal use of the work. Any republication, referencing or personal use of the work must explicitly identify the original source.

Notice
Statements and opinions expressed in the chapters are these of the individual contributors and not necessarily those of the editors or publisher. No responsibility is accepted for the accuracy of information contained in the published chapters. The publisher assumes no responsibility for any damage or injury to persons or property arising out of the use of any materials, instructions, methods or ideas contained in the book.

Publishing Process Manager Iva Lipovic
Technical Editor InTech Prepress, Novi Sad
Cover InTech Design team

Additional hard copies can be obtained from orders@intechopen.com

Nonlinearity, Bifurcation and Chaos – Theory and Applications,
Edited by Jan Awrejcewicz and Peter Hagedorn
p. cm.
ISBN 978-953-51-0816-0

We are IntechOpen, the first native scientific publisher of Open Access books

3,450+
Open access books available

110,000+
International authors and editors

115M+
Downloads

151
Countries delivered to

Our authors are among the
Top 1%
most cited scientists

12.2%
Contributors from top 500 universities

WEB OF SCIENCE™

Selection of our books indexed in the Book Citation Index in Web of Science™ Core Collection (BKCI)

Interested in publishing with us?
Contact book.department@intechopen.com

Numbers displayed above are based on latest data collected.
For more information visit www.intechopen.com

For EU product safety concerns:
IN TECH d.o.o., Prolaz Marije Krucifikse Kozulić 3, 51000 Rijeka, Croatia,
info@intechopen.com or visit our website at intechopen.com.

Meet the editors

Jan Awrejcewicz graduated from the Lodz University of Technology in 1977 (Mechanics) and from the University of Lodz in 1978 (Philosophy), Poland. He obtained PhD (Habilitation) in 1981 (1990), and became a Full Professor in 1997. He authored and co-authored 42 monographs, 2 textbooks, 275 journal papers, 340 conference papers, 35 book chapters, 13 journal special issues, edited 5 books and 13 conference proceedings, and supervised 18 PhD theses. He served in editorial boards of 40 journals, gave 80 invited seminars at international universities, delivered 30 plenary/keynote lectures and served as a member of scientific committees to 120 conferences. He was a recipient of the Humboldt Research Award in 2011.

Peter Hagedorn was born in 1941 in Berlin, Germany. He graduated in mechanical engineering and earned a doctoral degree at EPUSP (Brazil). He completed his 'habilitation' at TU Karlsruhe, Germany. From 1973 to 1974 he was a visiting Research Fellow at the AA Department of Stanford University. In October 1974 he became full professor of mechanics at TU Darmstadt, Germany. Among other universities he served as visiting professor in Rio de Janeiro (Brazil), Berkeley, Paris, Irbid (Jordan) and Christchurch (New Zealand), where he also held an Adjunct Professorship at UCC. He served as Head of Department and Vice-President to TU Darmstadt and he is active in a number of professional and editorial committees. He is an author of over 200 papers and several books on dynamics, vibrations and analytical mechanics. He presently heads the Dynamics and Vibrations Group within the Institute of Numerical Methods in Mechanical Engineering at TU Darmstadt.

Contents

Preface

Nonlinearity, bifurcation and chaos are common features exhibited by both natural and technological processes. Therefore, in this book theoretical and applied studies are addressed with emphasis on engineering dynamical systems.

This book includes chapters of both theoretical and applied studies of nonlinearity, bifurcation and chaos not published yet.

Chapter 1 by I. Andrianow et al. presents a novel approach to the problem of inhomogeneous compression of shells with a zero Gaussian curvature. The authors provide a meromorphic continuation of the polynomial asymptotics of a solution of differential equations within the nonlinear shell theory. They apply Padé approximation of two variable functions (2D PAs) and then confirm the obtained results by experiments with stainless steel specimens using a holographic interferometry.

Digital pulse-width modulation (digital-PWM) controllers yield a novel alternative to control power converters. These controllers have many advantages such as the programmability, high reliability and easy implementation of the advanced control algorithms. They can be designed in the measured variables in order to guarantee the necessary computing time of the signal control. However, the performance of PWM controllers is affected by delays. Investigations of the incidence of delay in digital-PWM controllers based on two novel techniques, i.e. the so called Zero Average Dyamics (ZAD) and Fixed-Point Inducting Control (FPIC) are presented in **chapter 2** by F. Angulo and G. Olivar.

In **chapter 3** J. M. Balthazar et al. present research results of the behavior of Nano- and Micro-Electro-Mechanical Systems (NEMS and MEMS). The main interest of this chapter is focused on the relationship between structure and nanoscale materials properties and the molecular dynamics of the atomic force microscopy. In particular, micro-cantilever beams and interaction of their vibration modes are analyzed. In addition, various aspects of Atomic Force Microscopy (AFM) and modeling and dynamic behavior of Electro-Mechanical System (EMS) are presented with respect to their control by taking into account the nonlinear and chaotic dynamic behavior.

A. O. Belyakov and A. P. Seyranian in **chapter 4** study the dynamics of a pendulum of variable length, rotations of a pendulum with elliptically moving pivot and twirling of a hula-hoop. They assumed that the quasi-linearity of the analyzed systems permits the derivation of higher order approximations by the averaging method. To prove correctness of the method they compared all approximate solutions with results obtained by numerical simulation.

Jet flows or multi-jet flows are frequently observed to adhere around nearby solid boundaries. This general class of phenomena, which may be found in both liquid and gaseous jets, is known as the Conada effect. In **chapter 5** A. Dumitrache et al. present a number of approaches to study both external and internal flows with emphasis on the Conada effect.

N. M. Evstigneev and N. A. Magnitskii in **chapter 6** study the laminar-turbulent transition processes for some particular initial-boundary value problems. Using the 3D Navier-Stokes equations for incompressible fluids, they consider two problems: an advection problem for flow over a backward facing step and a free convection problem as the Oberbeck-Boussinesq approximation of the Rayleigh-Benard convection. Then, applying the previously confirmed and validated numerical methods, nonlinear dynamics phenomena within the phase space and its three and four dimensional subspaces are studied.

U. Fuellekrug in **chapter 7** illustrates the application of the non-linear identification by an analytical example and from measured data of a large aircraft. The applied analysis of technical structures is validated by computational models, which is usually done assuming a linear elasto-mechanical structures. However, in order to improve the accuracy of simulations, the author considers equations of structures taking into account the non-linearities and then compares the obtained results with the experimental dynamic identification and modal analysis.

T. Kopecki in **chapter 8** aims on research of the problem of obtaining credible results of nonlinear finite elements method analyses of thin-walled load-bearing structures subjected to post-critical loads. This chapter is highly recommended for designers working in the aircraft industry, since the post-critical loads decide about a structure's deformation state being the effect of a rapid (dangerous) change of the structure's shape occurring when the critical load levels are exceeded.

It is well known that thin-walled constructions are used in sports, industry, automotive, as well as the aerospace and civil engineering. As most structures exhaust carrying capacity not by exceeding the allowable stress but rather by the stability loss, not only critical load but also the postbuckling behavior of thin-walled structures subjected to static and dynamic load should be rigorously studied. **Chapter 9** written by T. Kubiak, devoted to buckling and postbuckling behavior of thin-walled ortotrophic structures subjected to static and dynamic loads, addresses this problem, and the results obtained there can directly be applied by engineers and designers.

Chapter 10 by N. A. Magnitskii is devoted to the theory of development of complexity in nonlinear systems through subharmonic and harmonic cascades of bifurcations of stable limit cycles or stable two-dimensional or many-dimensional invariant tori. It is illustrated and shown that this universal theory of dynamical chaos is applicable to all kinds of nonlinear differential equations including dissipative and conservative, non-autonomous and autonomous dynamical systems governed by ordinary and partial differential equations, as well as differential equations with delay.

In **chapter 11** B.M. Podlevskyi presents numerical examples for the numerical algorithms of finding the eigenvalue curves and the bifurcation points. The methods allow finding branching lines of solutions of nonlinear integral equations, obtained as a result of solution of a synthesis problem, where also problems associated with the linearization procedure of the integral operator are rigorously solved.

Heterogenous agent models are present in various fields of economic analysis, such as marker models, exchange rate models, monetary policy models, overlapping generations models, asset price models and models of socio-economic behavior. In **chapter 12** M. Verbič discusses the role of memory in an asset pricing model with heterogeneous beliefs. A particular emphasis is paid to how memory in the fitness measure affects the stability of evolutionary adaptive systems and the survival of technical trading.

We hope that the presented book will be useful to academic researchers, engineers as well as post-graduate students.

Jan Awrejcewicz
The Lodz University of Technology
Poland

Peter Hagedorn
Dynamics and Vibrations Group, FNB,
Mechanical Engineering, TU Darmstadt
Germany

Chapter 1

Applications of 2D Padé Approximants in Nonlinear Shell Theory: Stability Calculation and Experimental Justification

Igor Andrianov, Jan Awrejcewicz and Victor Olevs'kyy

Additional information is available at the end of the chapter

http://dx.doi.org/10.5772/48822

1. Introduction

The widest class of shells used in the civil and mechanical engineering is the class of shells with developable principal surface. The stress-strain state of shell structures under loads, which corresponds to buckling, is inhomogeneous, significantly bended, and nonlinear. Permanent interest of researchers in the problem of inhomogeneous compression of shells of zero Gaussian curvature has not led so far to a correct solution. Therefore, there is a need for the development and application of new methods that allow considering the problem in a complex setting, the most appropriate to study real behavior of structures.

The approximate analytic integration of nonlinear differential equations of the theory of flexible elastic shells in most practical cases is based on the method of continuation of solution on the artificially introduced parameter. They can be satisfactorily applied only with an effective method of summation. The most natural analytical continuation method is that using Padé approximants (PAs). PAs effectively solves the problem of analytical continuation of power series, and this is a basis of their successful application in the study of applied problems. Currently, the method of PAs is one of the most promising non-linear methods of summation of power series, and the localization of its singular points. Recently, the method of PAs for single-variable functions has been successfully extended to the approximation of two variable functions (2D PAs).

A method that provides polynomial asymptotics of the exact solution of the general form and its meromorphic continuation based on 2D Padé approximants is proposed in this work. Several examples of displacements, stability and vibration calculations for inhomogeneous loaded shells with developable principal surface are presented. The accuracy of 2D PAs theoretical results are confirmed by experiments with stainless steel

INTECH
open science | open minds

© 2012 The Author(s). Licensee InTech. This chapter is distributed under the terms of the Creative Commons Attribution License http://creativecommons.org/licenses/by/3.0), which permits unrestricted use, distribution, and reproduction in any medium, provided the original work is properly cited.

specimens based on holographic interferometry. It is shown that the application of PAs provides sufficient accuracy in the studied area that confirms the advantage of our proposed approach.

2. Padé approximants

Let us consider Padé approximants (1-D PAs) which allow us to perform somewhat the most natural continuation of the power series. Below, we are going to define the 1-D PAs [1] for a complex variable z:

$$F(z)=\sum_{i=0}^{\infty} f_i z^i,$$

$$F_{nm}(z)=\sum_{i=0}^{n} p_i z^i \Big/ \sum_{i=0}^{m} q_i z^i, \quad q_0 \equiv 1,$$

where coefficients p_i, q_i are determined from the following condition: the first $(m+n+1)$ components of the expansion of rational function $F_{nm}(z)$ in the McLaurin series coincide with the same components of $F(z)$ series. Then rational function F_{nm} is called [*n*/*m*] PAs or 1-D [*n*/*m*] PAs. The set of F_{nm} functions for different *m* and *n* forms the so called Padé table.

PAs make a meromorphic continuation of *F* due to the following theorem of Montessus de Ballore [10]:

Theorem. Let function $F(z)$ be meromorphic in a closed circle $|z| \le r$ with *m* different poles z_i of multiplicity μ_i in this circle

$$0<|z_1|\le|z_2|\le\ldots\le|z_m|<r$$

of total multiplicity M, $\sum_{i=1}^{m}\mu_i = M$.

Then sequence $F_{NM}(z)$ converges uniformly to $F(z)$ on compact subsets of this circle without poles, and z_i is attracting zeros of the Padé denominator according to its multiplicity:

$$\lim_{N\to\infty} F_{NM}(z)=F(z), \quad |z|\le r, z\ne z_i, i=\overline{1,m}. \qquad \blacksquare$$

The most common generalization of PAs are two-dimensional PAs (2-D PAs). For complex variables z_1, z_2 let

$$F(z_1,z_2)=\sum_{i=0}^{\infty} f_{ij} z_1^i z_2^j$$

be a holomorphic function near the origin. For any integer sets $n=(n_1,n_2)$ and $m=(m_1,m_2)$, i.e. for any $n,m \in Z_+^2$, let

$$R(n,m)=\left\{r=\frac{p}{q},p=\sum_{i=0}^{n_1}\sum_{j=0}^{n_2}p_{ij}z_1^iz_2^j,\quad q=\sum_{i=0}^{m_1}\sum_{j=0}^{m_2}q_{ij}z_1^iz_2^j,q_{00}\equiv 1\right\},$$

be the class of rational functions, i.e. the ratio of 2-D polynomials whose degrees do not exceed $n=(n_1,n_2)$ and $m=(m_1,m_2)$ for each variable. It may be written briefly as $\deg(p)\le n,\deg(q)\le m$.

Each rational function $r\in R(n,m)$ may be identified with its power series that converges in some neighborhood of the origin. It should be mentioned that $r=p/q\in R$ depends on $\tau_{nm}=(n_1+1)(n_2+1)+(m_1+1)(m_2+1)-1$ parameters (the coefficients of p and q).

The set of integer points $I(n,m)\subset Z_+^2$ for fixed $n=(n_1,n_2)$ and $m=(m_1,m_2)$ is called the determinative (interpolation) set, if it has the following properties:

1. $\dim I(n,m)=\tau_{nm}$,
2. $(n_1+m_1,0),(0,n_2+m_2)\in I(n,m)$ (this property guarantees that in the case when $z_1=0$ (or $z_2=0$) one would have the classical 1-D rational approximation of Padé type),
3. $n=(n_1,n_2)\in I(n,m)$,
4. if $(k_1,k_2)\in I(n,m)$ then $[0,k]\subset I(n,m)$, where $[0,k]=\{(s_1,s_2)\in Z_+^2:0\le s_j\le k_j,j=1,2\}$ – the rectangle rule,
5. $(n_1+m_1,m_2)\in I(n,m)$ or $(m_1,n_2+m_2)\in I(n,m)$.

a. Two and only two possible variants of these sets satisfying requirements are:

$$I_1(n,m)=\{(i,j):$$

$$[0\le i\le n_1,0\le j\le n_2]\cup[n_1+1\le i\le n_1+m_1,0\le j\le m_2]\cup[i=0,n_2+1\le j\le n_2+m_2]\},$$

$$I_2(n,m)=\{(i,j):$$

$$[0\le i\le n_1,0\le j\le n_2]\cup[0\le i\le m_1,n_2+1\le j\le n_2+m_2]\cup[n_1+1\le i\le n_1+m_1,j=0]\},$$

The generalized PAs for given $n=(n_1,n_2)$ and $m=(m_1,m_2)$ are defined as the rational function $F_{nm}\in R(n,m)$ for which $T_{ij}(F-F_{nm})=0$ for all $(i,j)\in I(n,m)$, where $T_{ij}(\phi)$ are Taylor's coefficients of the power series for function F. The rational function F_{nm} is called the 2-D PAs $[(n_1,m_1)/(n_2,m_2)]$ of $F(z_1,z_2)$ which corresponds to the determinative set $(i,j)\in I(n,m)$.

As in the 1-D case, the existence and uniqueness of PAs (in the sense of the above given definition) for C^2 require special type of analysis. It should be mentioned that PAs do not always exist in the sense of the given definition.

Let $m=(m_1,m_2)\in Z_+^2$ be fixed and let the class

$$M_m = M_m\left(C^2\right) = \left\{F : F(z_1,z_2) = \frac{P(z_1,z_2)}{Q_m(z_1,z_2)}\right\}$$

be defined as a class of functions with the properties:

- $P(z_1,z_2)$ is an entire function;
- $\deg Q_m = m$, i.e. $\deg Q_m(z_1,0)=m_1, \deg Q_m(0,z_2)=m_2$;
- $Q_m(0,0)=1$;
- functions $P(z_1,0), P(0,z_2)$ and polynomials $Q_m(z_1,0), Q_m(0,z_2)$ are not equal to zero simultaneously.

The most important theorem for using 2-D PAs for meromorphic continuation is the following Montessus de Ballore – type theorem [2]:

Theorem. Let $F(z_1,z_2)\in M_m$ be given by the power series, $m=(m_1,m_2)\in Z_+^2$ be fixed and $n=(n_1,n_2)\in Z_+^2$. Then:

1. For all $n'=\min(n_1,n_2)$ that are large enough, there is a unique Padé approximant $F_{nm}=P_n/q_n$ for each of the determinative sets $I_j(n,m), j=1,2$;
2. The sequence F_{nm} for $n'=\min(n_1,n_2)\to\infty$ converges uniformly to function $F(z_1,z_2)$ inside the compact subsets of $G=C^2\setminus\{Q_m=0\}$. For any compact $E\subset C^2$ the following relationships are true:

$$\lim_{n'\to\infty}\left\|Q_m - q_n\right\|_E^{1/n'} = 0,$$
$$\lim_{n'\to\infty}\left\|F - F_{nm}\right\|_E^{1/n'} = 0,$$

where j = 1, 2 and $\|*\|_E = \sup_{z\in E}|*|$.■

This is an analog of the classical Montessus de Ballore theorem for the convergence of the rows of Padé tables.

3. Modified method of parameter continuation

Let us introduce a formal definition of the proposed modified method of parameter continuation (MMPC) for systems of ODEs using the terminology of the perturbation method. It is known [3] that any ODE or system of ODEs may be represented by a normal system of ODEs of the first order in respect to unknown functions $\{u_i=u_i(\xi)\}_{i=1}^n$ in the vicinity of regular point in the interval $\Omega:\xi\in]0,1[$:

$$Lu_i + R_i(\xi, u_1, ..., u_n) + N_i(\xi, u_1, ..., u_n) = g_i(\xi),\ L = \frac{d}{d\xi}, \qquad i = \overline{1,n}, \tag{1}$$

with the BCs on the bounds $\partial\Omega : \xi = 0 \cup 1$

$$G_j(u_1, ..., u_n)\big|_{\partial\Omega} = 0,\ j = \overline{1,n} \tag{2}$$

Here L and R_i are the linear differential operators, whereas N_i and G_j are the non-linear differential operators. We assume also that point $\xi_0 = 0$ belongs to closure Ω, and R_i, N_i and G_j are the holomorphic functions for $\{u_i\}_{i=1}^{n}$.

Considering $\{u_i = u_i(\xi)\}_{i=1}^{n}$ and their derivatives as independent arguments, we introduce operators R_i, N_i, F and G_j as the multidimensional Taylor series

$$R_i + N_i = \sum_{j=1}^{n}\left(N_{ij}u_j + \frac{1}{2!}\sum_{p=1}^{n} N_{ijp}u_j u_p + ...\right),\quad i = \overline{1,n}, \tag{3}$$

$$F = \left(F_0 L u_1 + \frac{1}{2!}\sum_{p=1}^{n} F_{0p}u_p L u_1 + ...\right) + \sum_{j=1}^{n}\left(F_j u_j + \frac{1}{2!}\sum_{p=1}^{n} F_{jp}u_j u_p + ...\right),$$

$$G_j = \sum_{q=1}^{n}\left(G_{jq}\left(u_q - u_q\big|_{\partial\Omega}\right) + \frac{1}{2!}\sum_{p=1}^{n} G_{jqp}\left(u_q - u_q\big|_{\partial\Omega}\right)\left(u_p - u_p\big|_{\partial\Omega}\right) + ...\right), j = \overline{1,n}. \tag{4}$$

We also introduce the following power series

$$N_{ij} = \sum_{r=0}^{\infty} N_{ij}^{r}\xi^{r},\ N_{ijp} = \sum_{r=0}^{\infty} N_{ijp}^{r}\xi^{r}, ...\ F_j = \sum_{r=0}^{\infty} F_j^{r}\xi^{r},\ F_{jp} = \sum_{r=0}^{\infty} F_{jp}^{r}\xi^{r}, ...\ g_i = \sum_{j=0}^{\infty} g_{ij}\xi^{i}\ \ i, j, p = \overline{1,n}. \tag{5}$$

To implement the MMPC, we introduce parameter ε as follows

$$u_i = \sum_{j=0}^{\infty} u_{ij}^{M}\varepsilon^{j}, \tag{6}$$

$$Lu_i = \varepsilon\left(g_i - R_i(u_1, ..., u_n) - N_i(u_1, ..., u_n)\right), \quad i = \overline{1,n}, \tag{7}$$

$$G_j(u_1\big|_{\partial\Omega}, ..., u_n\big|_{\partial\Omega})\big|_{\partial\Omega} = 0,\ j = \overline{1,n}$$

where R_i and N_i are always the algebraic operators in this case.

Substituting power series (6) into (7) and splitting it with respect to the powers of ε, we get

$$\varepsilon^0: \quad Lu_{i0}^M = 0, \quad G_i(u_{10}^M,...,u_{n0}^M)\Big|_{\partial\Omega} = 0 \Rightarrow u_{i0}^M = u_i\Big|_{\partial\Omega}, i = \overline{1,n},$$

$$\varepsilon^1: \quad Lu_{i1}^M - g_i + \sum_{r=1}^{n}\left(N_{ir}u_{r0}^M + \frac{1}{2!}\sum_{p=1}^{n}N_{irp}u_{r0}^M u_{p0}^M + ...\right) = 0, \quad u_{i1}^H\Big|_{\partial\Omega} = 0 \Rightarrow$$

$$u_{i1}^M = \sum_{j=0}^{\infty}\frac{\xi^{j+1}}{(j+1)}\left(g_{ij} - \sum_{r=1}^{n}\left(N_{ir}^j u_r\Big|_{\partial\Omega} + \frac{1}{2!}\sum_{p=1}^{n}N_{irp}^j u_r\Big|_{\partial\Omega} u_p\Big|_{\partial\Omega} + ...\right)\right), \quad i = \overline{1,n},$$

$$\varepsilon^2: Lu_{i2}^M + \sum_{r=1}^{n}\left(N_{ir}u_{r1}^M + \frac{1}{2!}\sum_{p=1}^{n}N_{irp}\left(u_{r1}^M u_{p0}^M + u_{r0}^M u_{p1}^M\right) + ...\right) = 0, \quad u_{i2}^M\Big|_{\partial\Omega} = 0 \Rightarrow \tag{8}$$

$$u_{i2}^M = -\sum_{s=0}^{\infty}\sum_{r=1}^{n}\left(N_{ir}^s\left(\sum_{j=0}^{\infty}\frac{\xi^{s+j+2}}{(j+1)(s+j+2)}\left(g_{rj} - \sum_{l=1}^{n}\left(N_{rl}^j u_l\Big|_{\partial\Omega} + \frac{1}{2!}\sum_{q=1}^{n}N_{rlq}^j u_l\Big|_{\partial\Omega} u_q\Big|_{\partial\Omega} + ...\right)\right)\right)\right) +$$

$$+\frac{1}{2!}\sum_{p=1}^{n}N_{irp}^s\left(u_p\Big|_{\partial\Omega}\left(\sum_{j=0}^{\infty}\frac{\xi^{s+j+2}}{(j+1)(s+j+2)}\left(g_{rj} - \sum_{l=1}^{n}\left(N_{rl}^j u_l\Big|_{\partial\Omega} + \frac{1}{2!}\sum_{q=1}^{n}N_{rlq}^j u_l\Big|_{\partial\Omega} u_q\Big|_{\partial\Omega}\right)\right)\right)\right) +$$

$$+u_r\Big|_{\partial\Omega}\left(\sum_{j=0}^{\infty}\frac{\xi^{s+j+2}}{(j+1)(s+j+2)}\left(g_{pj} - \sum_{l=1}^{n}\left(N_{pl}^j u_l\Big|_{\partial\Omega} + \frac{1}{2!}\sum_{q=1}^{n}N_{plq}^j u_l\Big|_{\partial\Omega} u_q\Big|_{\partial\Omega}\right)\right)\right) + ...\Bigg)\Bigg) + ..., \; i = \overline{1,n}.$$

Summation in (8) of the coefficients with the same degrees of ξ for $\varepsilon = 1$ gives

$$u_i = \xi^0\left(\left[u_i\big|_{\partial\Omega}\right] + [0] + [0] + ...\right) +$$

$$+\xi^1\left([0] + \left[g_{i0} - \sum_{r=1}^{n}\left(N_{ir}^0 u_r\Big|_{\partial\Omega} + \frac{1}{2!}\sum_{p=1}^{n}N_{irp}^0 u_r\Big|_{\partial\Omega} u_p\Big|_{\partial\Omega} + ...\right)\right] + [0] + ...\right) +$$

$$+\xi^2\left([0] + \left[\frac{g_{i1}}{2} - \frac{1}{2}\sum_{r=1}^{n}\left(N_{ir}^1 u_r\Big|_{\partial\Omega} + \frac{1}{2!}\sum_{p=1}^{n}N_{irp}^1 u_r\Big|_{\partial\Omega} u_p\Big|_{\partial\Omega} + ...\right)\right] +\right.$$

$$+\left[-\sum_{r=1}^{n}\left(N_{ir}^0\left(\frac{g_{r0}}{2} - \frac{1}{2}\sum_{l=1}^{n}\left(N_{rl}^0 u_l\Big|_{\partial\Omega} + \frac{1}{2!}\sum_{q=1}^{n}N_{rlq}^0 u_l\Big|_{\partial\Omega} u_q\Big|_{\partial\Omega} + ...\right)\right)\right) + \right. \tag{9}$$

$$+\frac{1}{2!}\sum_{p=1}^{n} N_{irp}^0\left(u_p\Big|_{\partial\Omega}\left(\frac{g_{r0}}{2} - \frac{1}{2}\sum_{l=1}^{n}\left(N_{rl}^0 u_l\Big|_{\partial\Omega} + \frac{1}{2!}\sum_{q=1}^{n}N_{rlq}^0 u_l\Big|_{\partial\Omega} u_q\Big|_{\partial\Omega}\right)\right)\right) +$$

$$+u_r\big|_{\partial\Omega}\left(\frac{g_{p0}}{2}-\frac{1}{2}\sum_{l=1}^{n}\left(N_{pl}^{0}u_l\big|_{\partial\Omega}+\frac{1}{2!}\sum_{q=1}^{n}N_{plq}^{0}u_l\big|_{\partial\Omega}u_q\big|_{\partial\Omega}\right)\right)\Bigg)+\ldots\Bigg]+\ldots\Bigg)+\ldots,\quad i=\overline{1,n}.$$

Analysis of the obtained approximation suggests that it gives the exact value of coefficients in the power of the independent variable to the extent equal to the order of approximation (taking into account the expansion in power series of expressions in the equation). This guarantees stability of the computation with a limit-order approximation of the independent variable.

One of the possible fields for application of the proposed approach are the nonlinear problems of plates and shells theory. The equations of statics of the geometrically nonlinear thin-walled structures can be reduced to the resolving equations which contain the products and squares of the desired functions and their derivatives [4]. In this case the solution of equation (9) becomes

$$u_i=\xi^0 u_i\big|_{\partial\Omega}+\xi^1\left(g_{i0}-\sum_{r=1}^{n}\left(N_{ir}^{0}u_r\big|_{\partial\Omega}+\frac{1}{2!}\sum_{p=1}^{n}N_{irp}^{0}u_r\big|_{\partial\Omega}u_p\big|_{\partial\Omega}\right)\right)+$$
$$+\xi^2\left(\left(\frac{g_{i1}}{2}-\frac{1}{2}\sum_{r=1}^{n}\left(N_{ir}^{1}u_r\big|_{\partial\Omega}+\frac{1}{2!}\sum_{p=1}^{n}N_{irp}^{1}u_r\big|_{\partial\Omega}u_p\big|_{\partial\Omega}\right)\right)-\right.$$
$$-\sum_{r=1}^{n}\left(N_{ir}^{0}\left(\frac{g_{r0}}{2}-\frac{1}{2}\sum_{l=1}^{n}\left(N_{rl}^{0}u_l\big|_{\partial\Omega}+\frac{1}{2!}\sum_{q=1}^{n}N_{rlq}^{0}u_l\big|_{\partial\Omega}u_q\big|_{\partial\Omega}\right)\right)+\right. \tag{10}$$
$$+\frac{1}{2!}\sum_{p=1}^{n}N_{irp}^{0}\left(u_p\big|_{\partial\Omega}\left(\frac{g_{r0}}{2}-\frac{1}{2}\sum_{l=1}^{n}\left(N_{rl}^{0}u_l\big|_{\partial\Omega}+\frac{1}{2!}\sum_{q=1}^{n}N_{rlq}^{0}u_l\big|_{\partial\Omega}u_q\big|_{\partial\Omega}\right)\right)+\right.$$
$$\left.\left.\left.+u_r\big|_{\partial\Omega}\left(\frac{g_{p0}}{2}-\frac{1}{2}\sum_{l=1}^{n}\left(N_{pl}^{0}u_l\big|_{\partial\Omega}+\frac{1}{2!}\sum_{q=1}^{n}N_{plq}^{0}u_l\big|_{\partial\Omega}u_q\big|_{\partial\Omega}\right)\right)\right)\right)\right)+\ldots,\quad i=\overline{1,n}.$$

The approximation thus obtained is converted to 1-D PAs in respect to ξ or 2-D PAs. 2-D PAs in the form proposed by V. Vavilov [2] is very promising for the use as an analytical continuation. This technique allows us to choose the coefficients of 2-D Taylor series for the construction of unambiguous 2-D PAs with a given structure of the numerator and denominator. It also ensures optimal PAs features in the sense of the theorem of Montessus de Ballore-type. This means homogenous convergence of PAs to the approximated function with an increase of the degree of the numerator and denominator in all points of its meromorphic area. It should be noted that direct application of 2-D PAs does not lead to the anticipated merging of 1-D approximations. This is due to the initial requirements of the 2-D approximation to ensure its transition to 1-D in the case when the second variable is equal to zero [1]. At the same time it is necessary to ensure such a transition, when the parameter is equal to one. This can be achieved by combining this method with 2-D PAs from a converted parameter which maps the unit to zero.

4. Stability investigation

For the analytic continuation of meromorphic solutions to the region and acceleration of the convergence we use 2-D PAs. To do this, the resulting series are reconstructed in rational functions of the form

$$u_i = \sum_{j=0}^{m_1} \sum_{k=0}^{n_1} P_{ijk}(P)\varepsilon^j \xi^k \Big/ \sum_{j=0}^{m_2} \sum_{k=0}^{n_2} Q_{ijk}(P)\varepsilon^j \xi^k, \; Q_{i00} \equiv 1,$$

where P is the parameter of loading.

Since the proposed method is modified, all functions belonging to the boundary value problems can be expanded in powers of the independent variables and parameters. For $\varepsilon = 1$ we obtain

$$u_i = \sum_{j=0}^{m_3} \sum_{k=0}^{n_1} \overline{P}_{ijk} P^j \xi^k \Big/ \sum_{j=0}^{m_4} \sum_{k=0}^{n_2} \overline{Q}_{ijk} P^j \xi^k .$$

Rational functions have singular points which are determined by equating of the denominator to zero. According to the theory of bifurcation of solutions of ordinary differential equations, at these points either bifurcations or limit states are achieved. So we can get the estimation of critical point localization by solving equations of the form

$$\min_{i,\xi} P : \sum_{j=0}^{m_4} \sum_{k=0}^{n_2} \overline{Q}_{ijk} P^j \xi^k = 0 .$$

In practice this equation is transformed to its counterpart simpler form with respect to its characteristic point $\xi = \xi_0$ (which usually corresponds to the point of maximum transverse displacement for thin shells)

$$\sum_{j=0}^{m_4} \sum_{k=0}^{n_2} \overline{Q}_{ijk} P^j \xi_0^k = 0 .$$

Example

Let us consider the computational aspects of the proposed approach. We consider three types of PAs with respect to the independent variable, on the specified parameters, and 2-D.

A typical behavior of the approximations for the BVP is governed by the following problem

$$\begin{gathered}\varepsilon z'+z=1,\\ z(0)=0,\quad 0<\varepsilon<<1,\quad x\geq 0.\end{gathered} \tag{11}$$

where natural small parameter ε is the factor at the highest derivative, as shown in Fig. 1 for $\varepsilon = 0.1$. The exact solution of this BVP follows

$$z=1-\exp\left(-\frac{x}{\varepsilon}\right)=\frac{x}{\varepsilon}-\frac{x^2}{2\varepsilon}+\ldots+(-1)^{n+1}\frac{1}{n!}\left(\frac{x}{\varepsilon}\right)^n+\ldots. \tag{12}$$

Eq. (12) shows that the exact solution is regular for all real positive x for $\varepsilon \neq 0$. But the general term of power series (12) grows rapidly when $x > \varepsilon$, and we have to take into account many terms in (2) to obtain an acceptable and reliable approximation. Thus, the accuracy of the used truncated Taylor series is not uniform according to x value.

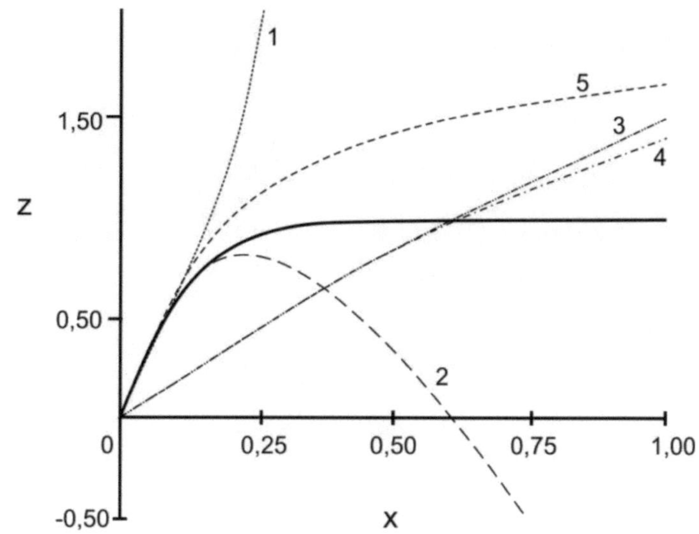

Figure 1. The exact solution (solid line) of Eq. (1) for $\varepsilon = 0.1$ and approximate solutions (1 – three terms ADM, 2 – $z^{(\varepsilon_1)}$ for ADM, 3 – three terms HAM, 4 – $z^{(x)}$ for HAM, 5 – 2-D PAs for MMPC, ADM and HAM).

Let us introduce parameter ε_1 as follows

$$z'=\frac{1}{\varepsilon}-\varepsilon_1\frac{z}{\varepsilon}, \tag{13}$$

and suppose

$$z=\sum_{i=0}^{\infty} z_i\varepsilon_1^i. \tag{14}$$

This way of introducing the parameter leads to a system of successive approximations of Adomian decomposition method (ADM, see [5,6]). Substituting power series (4) into Eq. (3) and splitting it with respect to the powers of ε_1, yields

$$\varepsilon_1^0: \ z_0' = \frac{1}{\varepsilon}, \ z_0(0) = 0 \ \Rightarrow \ z_0 = \frac{x}{\varepsilon};$$

$$\varepsilon_1^1: \ z_1' = -\frac{z_0}{\varepsilon} = -\frac{x}{\varepsilon^2}, \ z_1(0) = 0 \ \Rightarrow \ z_1 = -\frac{x^2}{2\varepsilon^2} = (-1)^1 \frac{1}{2!}\left(\frac{x}{\varepsilon}\right)^2;$$

$$\varepsilon_1^2: \ z_2' = -\frac{z_1}{\varepsilon} = -\frac{x^2}{\varepsilon^3}, \ z_2(0) = 0 \ \Rightarrow \ z_2 = \frac{x^3}{6\varepsilon^3} = (-1)^2 \frac{1}{3!}\left(\frac{x}{\varepsilon}\right)^3;$$

$$\varepsilon_1^n: \ z_n' = -\frac{z_{n-1}}{\varepsilon} = (-1)^n \frac{1}{(n-1)!}\left(\frac{x}{\varepsilon}\right)^n \left(\frac{1}{\varepsilon}\right), \ z_n(0) = 0 \ \Rightarrow \ z_n = (-1)^n \frac{1}{n!}\left(\frac{x}{\varepsilon}\right)^{n+1};$$

$$z = \frac{x}{\varepsilon} - \frac{x^2}{2\varepsilon^2}\varepsilon_1 + \frac{x^3}{6\varepsilon^3}\varepsilon_1^2 + \ldots + (-1)^n \frac{1}{n!}\left(\frac{x}{\varepsilon}\right)^{n+1} \varepsilon_1^n + \ldots.$$

For $\varepsilon_1 = 1$ one gets power series (2), which corresponds to the results reported in [5]. To accelerate the convergence we use 1-D and 2-D PAs. One can use 1-D PAs [1/1] ($z^{(\varepsilon_1)}$, when $x = const \neq 0$ and $z^{(x)}$, when $\varepsilon_1 = const \neq 0$) or 2-D PAs [(1,1)/(1,1)] ($z^{(\varepsilon_1, x)}$). Using PAs one obtains (for $\varepsilon = 1$)

$$z^{(\varepsilon_1)} = \frac{x}{\varepsilon}\left(1 - \frac{3x}{6\varepsilon + 2x}\right), \tag{15}$$

$$z^{(x)} = z^{(\varepsilon_1, x)} = \frac{2x}{2\varepsilon + x}. \tag{16}$$

R.h.s. of Eq. (15) contains singularity at point $\varepsilon = 0$ in contrast to Eqs. (16). Thus, Eqs. (16) give the approximation with uniform accuracy when x grows.

Let us rewrite Eq. (11) in the following form

$$z' = \varepsilon_1(1 - z + (1 - \varepsilon)z'). \tag{17}$$

After substitution of the power series (14) in Eq. (17) one obtains a successive approximation of the homotopy analysis method (HAM) in the so called homotopy perturbation form (HPM, see [7])

$$\varepsilon_1^0: \ z_0' = 0, \ z_0(0) = 0 \ \Rightarrow \ z_0 = 0;$$

$$\varepsilon_1^1: \; z_1' = 1 - z_0 + (1-\varepsilon)z_0' = 1, \quad z_1(0) = 0 \;\; \Rightarrow \;\; z_1 = x;$$

$$\varepsilon_1^2: \; z_2' = -z_1 + (1-\varepsilon)z_1' = -x + (1-\varepsilon), \quad z_2(0) = 0 \;\; \Rightarrow \;\; z_2 = -\frac{x^2}{2!} + (1-\varepsilon)x;$$

$$\varepsilon_1^3: \; z_3' = -z_2 + (1-\varepsilon)z_2' = \frac{x^2}{2} - (1-\varepsilon)x + (1-\varepsilon)(-x + (1-\varepsilon)), \quad z_3(0) = 0 \;\; \Rightarrow$$

$$\Rightarrow \;\; z_3 = \frac{x^3}{3!} - (1-\varepsilon)x^2 + (1-\varepsilon)^2 x;$$

$$\varepsilon_1^4: \; z_4' = -z_3 + (1-\varepsilon)z_3' = -\frac{x^3}{6} + (1-\varepsilon)x^2 - (1-\varepsilon)^2 x + (1-\varepsilon)\left(\frac{x^2}{2} - 2(1-\varepsilon)x + (1-\varepsilon)^2\right),$$

$$z_4(0) = 0 \;\; \Rightarrow \;\; z_4 = -\frac{x^4}{4!} + (1-\varepsilon)\frac{x^3}{2} - 3(1-\varepsilon)^2\frac{x^2}{2} + (1-\varepsilon)^3 x;$$

We obtain the HAM approximation in the following form

$$z = x\varepsilon_1 + \left(-\frac{x^2}{2!} + (1-\varepsilon)x\right)\varepsilon_1^2 + \left(\frac{x^3}{3!} - (1-\varepsilon)x^2 + (1-\varepsilon)^2 x\right)\varepsilon_1^3 + \dots .$$

For $\varepsilon_1 = 1$ one obtains

$$z = (1 + (1-\varepsilon) + (1-\varepsilon)^2 + (1-\varepsilon)^3 + \dots)x + \left(-\frac{1}{2!} + (1-\varepsilon) - 3(1-\varepsilon)^2\frac{1}{2} + \dots\right)x^2 + \\ + \left(\frac{1}{3!} + (1-\varepsilon)\frac{1}{2} + \dots\right)x^3 - \frac{x^4}{4!} + \dots . \qquad (18)$$

Eq. (18) coincides with Eq. (12) after expanding the coefficients of Eq. (12) in the vicinity of $\varepsilon = 1$.

For the obtained approximations we use PAs as described above and get the following results:

$$z^{(\varepsilon_1)} = z^{(\varepsilon_1, x)} = \frac{2x}{2\varepsilon + x},$$

$$z^{(x)} = \frac{2(2-\varepsilon)^2 x}{2(2-\varepsilon) + x}.$$

Let us introduce parameter ε_1 in such a way that

$$z' = \varepsilon_1 \frac{1-z}{\varepsilon}. \qquad (19)$$

After substitution of the power series (14) in Eq. (19), one obtains a new system of successive approximations:

$$\varepsilon_1^0: \; z_0' = 0, \quad z_0(0) = 0 \;\Rightarrow\; z_0 = 0;$$

$$\varepsilon_1^1: \; z_1' = \frac{1 - z_0}{\varepsilon} = \frac{1}{\varepsilon}, \quad z_1(0) = 0 \;\Rightarrow\; z_1 = \frac{x}{\varepsilon} = (-1)^2 \frac{1}{1!}\left(\frac{x}{\varepsilon}\right)^1;$$

$$\varepsilon_1^2: \; z_2' = -\frac{z_1}{\varepsilon} = -\frac{x}{\varepsilon^2}, \quad z_2(0) = 0 \;\Rightarrow\; z_2 = -\frac{x^2}{2\varepsilon^2} = (-1)^3 \frac{1}{2!}\left(\frac{x}{\varepsilon}\right)^2;$$

$$\varepsilon_1^n: \; z_n' = -\frac{z_{n-1}}{\varepsilon} = (-1)^{n+1} \frac{1}{(n-1)!}\left(\frac{x}{\varepsilon}\right)^{n-1}\left(\frac{1}{\varepsilon}\right), \quad z_2(0) = 0 \;\Rightarrow\; z_n = (-1)^{n+1} \frac{1}{n!}\left(\frac{x}{\varepsilon}\right)^n;$$

$$z = \frac{x}{\varepsilon}\varepsilon_1 - \frac{1}{2!}\left(\frac{x}{\varepsilon}\right)^2 \varepsilon_1^2 + \frac{1}{3!}\left(\frac{x}{\varepsilon}\right)^3 \varepsilon_1^3 + \dots.$$

For $\varepsilon_1 = 1$ one obtains PAs as follows:

$$z^{(\varepsilon_1)} = z^{(x)} = z^{(\varepsilon_1, x)} = \frac{2x}{2\varepsilon + x}.$$

The coincidence of these approximations demonstrates their proximity to a unique function representing the exact solution in the fractional-rational form.

The ADM approximation describes well the exact solution only for a distance which is comparable with the value of natural small parameter ε. Despite the fact that the error of solutions of HAM is substantially less than the ADM, HAM does not accurately reflect the nature of solutions, namely the phenomenon of boundary layer in the vicinity of zero. At the same time, PAs for the ADM approximations for independent variable and PAs for the MMPC (1-D and 2-D) give satisfactory qualitative and quantitative results.

Similar results are provided by the analysis of approximations of BVP of the following problem

$$\begin{gathered} \varepsilon z' + xz = x, \\ z(0) = 2, \quad 0 < \varepsilon << 1, \quad x \geq 0, \end{gathered} \tag{20}$$

whose coefficients are given depending on the variable for $\varepsilon = 0.2$. Its graphs are presented in Fig. 2.

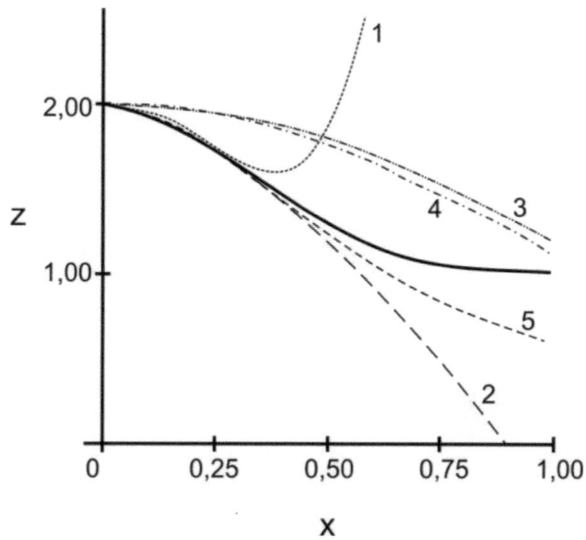

Figure 2. The exact solution (solid line) of Eq. (20) for $\varepsilon = 0.2$ and approximate solutions (1 – three terms ADM, 2 – $z^{(\varepsilon_1)}$ for ADM, 3 – three terms HAM, 4 – $z^{(x)}$ for HAM, 5 – 2-D PAs for MMPC, ADM and HAM).

Fig. 3 shows the graphs of approximations for strongly non-linear BVP of the form

$$\varepsilon z' = z^2 + x,$$
$$z(1) = 1, \quad 0 < \varepsilon << 1, \quad x \geq 0. \tag{21}$$

The graphs show that the solution is well described by the HAM approximation and MHAM-Padé «in average», and badly – in the boundary layer. The ADM approximation and MADM-Padé, on the contrary, is in good agreement with the behavior of solution in the vicinity of zero and in the bad one – on the stationary part. At the same time, 1-D and 2-D PAs, based on approximations of the MMPC, well describe the solution in the whole interval.

5. Calculation of nonlinear deformation and stability of shells

The proposed MMPC method has been applied to calculate the deformation and stability of a long flexible elastic circular cylindrical shell of radius R with half the central angle β_0 in the case of cylindrical bending under uniform external pressure with a simple support of the longitudinal edges. The corresponding system of resolving equations in the normal form is given in [8]. Dependences of "dimensionless intensity of pressure P – deflection w/R" for the top cross-section of the shell at different angles and dimensionless flexibility $C = 10^{-4}$ are shown in Fig. 4. The dependence of the dimensionless intensity of limit load P_* on the size of half angle β_0 is shown in Fig. 5.

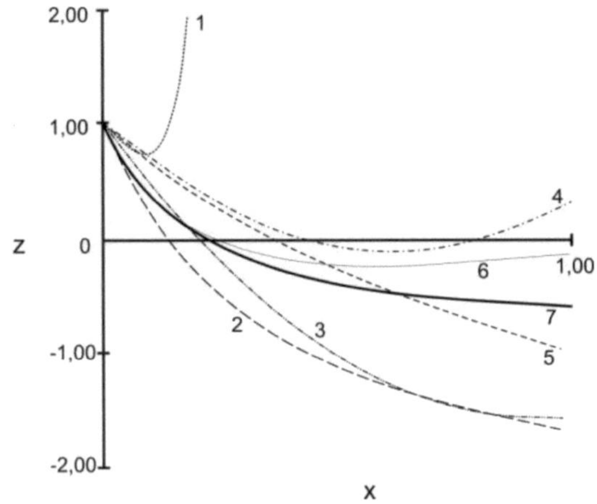

Figure 3. Approximate solutions of Eq. (18) for $\varepsilon = 0.2$ (1 – three terms ADM, 2 – $z^{(\varepsilon_1)}$ for ADM, 3 – $z^{(x)}$ for ADM, , 4 – three terms HAM, 5 – $z^{(x)}$ for HAM, 6 – $z^{(\varepsilon_1)}$ for HAM and MMPC, 7 – $z^{(x)}$ and 2-D Padé for MMPC).

For comparison, Fig. 4b also shows the dependence of the critical loads for inextensible shell obtained by S.P. Timoshenko [8]. We see that dependences are in good agreement, while consideration of deformation of the longitudinal axis substantially affects the value of critical loads of the construction.

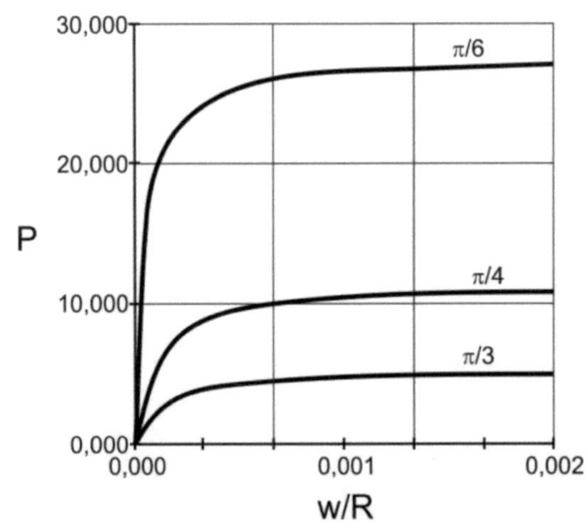

Figure 4. The dependence of the intensity of pressure P versus deflection w / R for different values of β_0 (the value of β_0 is indicated by curves).

The proposed method can be used in a combination with the known asymptotic method. Consider free vibrations of a flexible elastic circular cylindrical shell of radius R, thickness h and length L, backed by a set of uniformly distributed stringers having a simple support at the ends.

The calculation is based on mixed dynamical equations of the theory of shells after splitting them in powers of natural small parameters [9]. The shape of radial deflection w satisfies the boundary conditions given in the form

$$w = f_1(t)\sin s_1 x_1 \cos s_2 x_2 + f_2(t)\sin^2 s_1 x_1 .$$

Here f_1, f_2 functions depend on time and are related by the condition of continuity of displacements

$$f_2 = 0{,}25 R^{-1} s_2^2 f_1^2 ,$$

where $s_1 = \pi m l^{-1}$, $s_2 = n$ are the parameters characterizing the wave generation along the generator and directrix, respectively.

The governing equations can be reduced by the Bubnov – Galerkin method to the Cauchy problem with respect to $\xi = f_1 / R$ on $t_1 = t\sqrt{B_1/\rho R^2}$ (all symbols are taken in accordance with [9])

$$\begin{gathered} \ddot{\xi} + \alpha\xi\left[\left(\dot{\xi}\right)^2 + \xi\ddot{\xi}\right] + A_1\xi + A_2\xi^3 + A_3\xi^5 = 0, \\ \left(\dot{\ }\right) \equiv \frac{d(\ )}{dt_1}, t_1 = 0 : \xi = f, \dot{\xi} = 0 \end{gathered} \tag{22}$$

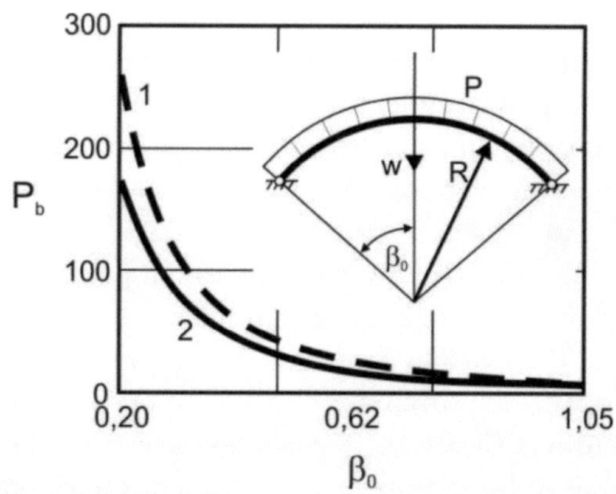

Figure 5. The dependence of limit loads P_b versus β_0 (1 - data [8], 2 - calculation).

The application of the proposed method of parameter continuation to the Cauchy problem (22) gives approximation of the second order for the artificial parameter for frequency Ω of nonlinear oscillations in the form

$$\Omega = \sqrt{\frac{1 + f^2\left(A_2 / A_1\right) + f^4\left(A_3 / A_1\right)}{\left(1 + \alpha f\right)}}..$$

It is seen that the oscillations are not isochronous. This agrees well with previous results reported in reference [9] (Fig. 6). However, our approach allows for a significant reduction of the computation time (in [9] to obtain similar results the approximation of the fourth order is taken).

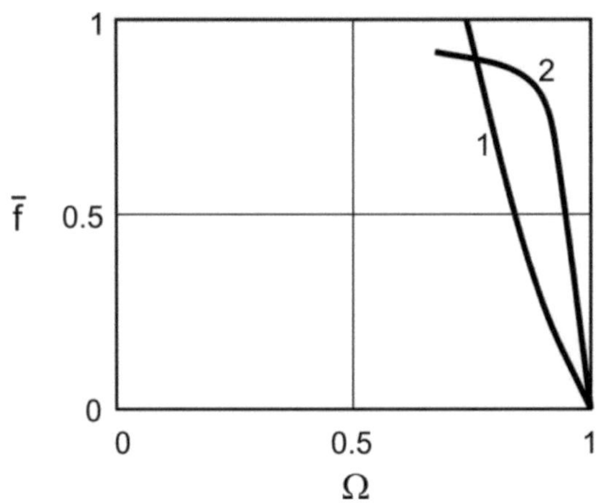

Figure 6. Amplitude of the initial disturbance versus oscillation frequency of stringer shell (1 – according the proposed method, 2 – data [9]).

6. Experimental technique

In solving various kinds of problems of modern development and improvement of thin-walled machine elements operating under the conditions of intensive manufacturing process, the use of a holographic interferometry method should be emphasized [10,11,14]. It allows for a more accurate and complete investigation of shell structures under complex stress-strain state. The accuracy of interpretation of holographic interferograms is mainly determined by the number of support points of the design used for the construction regarding displacements and stresses. Improvement of the accuracy requires a large amount of routine preparations for writing the coordinates of points and their corresponding numbers of lines when developing data on a computer, which is particularly important in the case of an experiment. The existing methods of automated data entry and processing of interferograms yield, as a rule, the specific configuration of the optical system and the types of strain state (flat, one-dimensional, etc.), making them difficult to use in this case. In addition, although several authors proposed methods of interpretation [10], they did not fully take into account the statistical nature of input data. For cylindrical shells a method for automated processing of the results of the holographic research has been proposed, which eliminates the above drawbacks [12]. Next, we have extended this technique to study the motion of shell structures of zero Gaussian curvature which is based on modern means of an interactive data processing. The surface of zero Gaussian curvature can be approximated with sufficient accuracy with respect to the system of flat rectangular panels whose sides are segments close to the case which occurred during the analysis of generators. To determine all components (points) of the displacement vector, three holograms of a circuit design interferometer based on a reference beam is used. The interferometer is shown schematically in Fig 7.

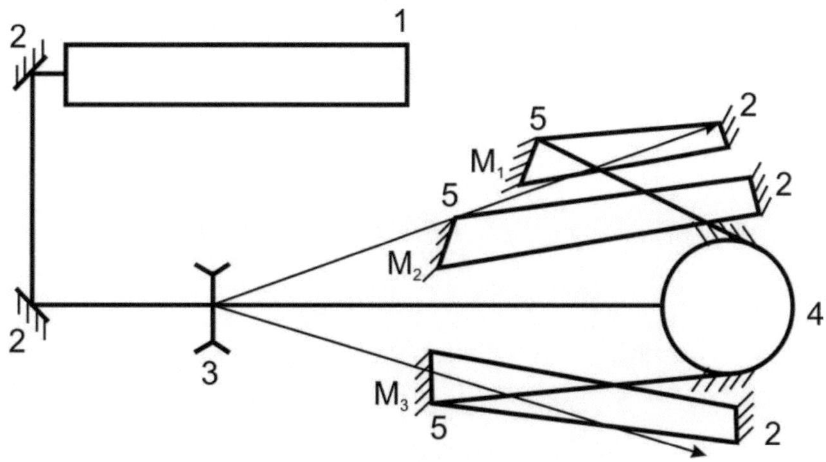

Figure 7. The scheme of the interferometer (1 – laser generator, 2 – mirror, 3 – expanding lens, 4 – studied object, 5 – camera)

After registering the two exposures, i.e. unloaded and loaded state of the object, we get a flat image of the interference pattern corresponding to the observation of points $M_i\left(x_{Hi}, y_{Hi}, z_{Hi}\right)$, $i=\overline{1,3}$. Let us enter the order line using a computer in the following manner. The photos of interferograms are scanned and entered into the computer memory in the form of graphic files with the extension, for example, jpg, which is the most popular choice of compression of graphic information on all platforms, or equivalently in other file formats. Next, the file is displayed on the screen in a specially designed box on the toolbar image processing. The information produced is removed by a successive mouse click on the corresponding image points at the request of a specially created database. Algorithms for further processing of the data are widely described in [12]. In the $X'O'Y'$ coordinate system (Fig. 8) associated with the imaging plate, base point $M_B\left(x'_B, y'_B\right)$ and a segment of the OY axis of the XOY coordinate system, whose direction coincides with the vertical axis of the projection, are given. Further calculations are performed in the XOY system in which the entered coordinates of the points of lines of equal order are transformed by the formulas

$$x = x'\cos\varphi + y'\sin\varphi - x_B,\ y = y'\cos\varphi - x'\sin\varphi - y_B,$$

where φ is the angle of rotation of the XOY system with respect to $X'O'Y'$

$$x_B = x'_B \cos\varphi + y'_B \sin\varphi, \; y_B = -x'_B \sin\varphi + y'_B \cos\varphi.$$

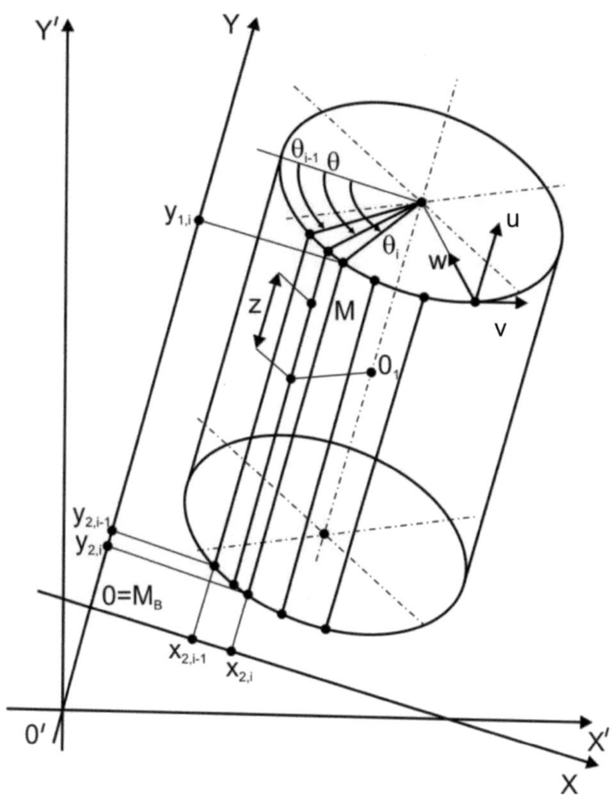

Figure 8. The scheme of approximation of the surface by the system of folds

While computing the physical coordinates, the approximation of the mentioned shell surface by the system of folds is applied (Fig. 8). In this case, the physical coordinates of point $M(r,z,\theta)$ in the i fold $(i=\overline{1,n})$ are defined by

$$\theta = \frac{\theta_i - \theta_{i-1}}{x_i - x_{i-1}}\left(x - x_{i-1}\right) + \theta_{i-1}$$

$$z = z_1 + \frac{y - \left[y_{2,i-1} + \left(y_{2,i} - y_{2,i-1}\right)\left(x - x_{2,i-1}\right) / \left(x_{2,i} - x_{2,i-1}\right)\right]}{y_{1,i} - y_{2,i}}\left(z_2 - z_1\right),$$

where r,θ,z are the coordinates of a point in the cylindrical coordinate system associated with the axis of the shell, and the shape of the surface is analytically given by equation $r = r(\theta,z)$; r_i, θ_j, z_k are the physical coordinates of corner points of the considered fold;

$(x_{1,i};y_{1,i})$, $(x_{2,i};y_{2,i})$ are the coordinates of corner points of the projection of the folds into the XOY system (Fig. 8).

Therefore, the so far obtained arrays of point coordinates of the lines of equal order corresponding to the three noncoplanar directions of observation allow us to approximate the surface bands. The most appropriate method to do this is the structural analysis of extrapolation (MSEA) [6] using step by step the best choice of the model. Indeed, the formalization of the input source data for inhomogeneous stress-strain state requires a large number of points.

The above method allows us to determine the coordinates of the centers of bands up to 0.1 mm without any additional devices. This procedure is used to significantly increase the number of input points (up to 200-400 for each direction of observation). The use of spline functions for smoothing requires the enumeration of all coordinates of control points for each calculation of the order of the band. This slows down the calculation and requires a significant memory space. In addition, these disadvantages are compounded by the increasing number of control points. The use of MSEA allows each step to obtain unbiased estimates of the effective coefficients of the model to ensure a maximum plausible value of the order of the reference points [11]. This eliminates the problem of choosing a smoothing parameter, with the number needed to calculate the coefficients one order of magnitude smaller than the number of coordinates of reference points. In addition, the incremental method allows us to formalize the process of selecting the optimal order of approximating polynomial based on the assessment of the significance of the model and the adequacy of its source data. Note that in this case the number of points is much larger than the number of estimated parameters, which suggests a considerable power of the statistical tests (like those of Student's, Fisher and Durbin-Watson), and indicates the validity of hypotheses taken in selecting the best model. An increase of the number of points improves a regression model, and a loss of accuracy in the summation can be successfully overcome by standardizing the original data according to the known methods.

Because the shape of the surface is analytically given by equation $r = r(\theta, z)$, it is possible to obtain two-dimensional regression models for the line of the i-order for the direction of observation in the form

$$N_i = \sum_{j=0}^{n(i)} \sum_{k=0}^{n(j)} b_{jki} \theta^j z^k.$$

Displacements are defined by the equation [4]:

$$MU = \lambda N,$$

where: M – optic matrix; $U = \{u, v, w\}$ – vector of displacement; λ – length of the laser wave; $N = \{N_i\}_{i=1}^{3}$ – vector of lines order.

Further transformation of movements, according to the Cauchy relations and equations of state of the environment, can also yield the stress state at the point. Performing the calculation of the stress-strain state parameters to form and direct the shell with a certain step, it is possible to obtain data for plotting the distribution of displacements and stresses.

7. Experimental justification of MMPC

Loading capacity of cylindrical shells is significantly affected by the unevenness of deformation caused by the ovality ends of the shell [13]. Imperfections in face of shells usually occur as a result of their deformation either under their own weight or during the mechanical handling, storage, as well as installation and assembly of the shells as individual elements. In the case of welded shells, the end face has the form of an oval with a and b axes and a/b compression ratio (or the actual ovality). At the same roundness of the upper end $(a/b)_B$ may be different from the roundness of the bottom one $(a/b)_H$ because of the conditions introduced by a collection with other elements of the design. In all cases the shape of each end should be within the required tolerances, and roundness introduced by the collection process should not be reduced by more than 0.8.

Another form of imperfections arising from the inaccuracy of the assembly is associated with a weak taper angle characterized by forming the membrane to its axle α. A number of studies [14] consider that a small taper with $\alpha \le 13^{\circ}$ has no significant effect on the magnitude of critical loads of axial compression. However, the results of stability studies of technologically imperfect cylindrical shells based on the multivariate approach [15,16] suggest that an increase of α to the value of 3° often leads to a significant change in carrying capacity, and in some cases the interaction with the oval and other factors yields an increase of the critical loads. These studies have shown the need for a more correct approach in establishing the correspondence between the magnitude of these abnormalities and the level of carrying capacity. As a consequence, it is necessary to study the nature of deformation of shells with different ratios of the parameters of roundness and taper.

In order to solve this problem, two-factor second-order experiments on two levels of α and a/b, and also on two levels of a/b at lower and upper ends when $\alpha^{\circ} = 1$ (when the taper is small the difference between the lower and upper end is missing) have been implemented. Welded specimens with radius $R = 71.5$ *mm* and length $L = 200$ *mm*, made from plate steel of the mark $H18N9-n$ and with thickness $\delta = 0.25$ *mm* have been tested. The use of a multi-factor approach allows one to solve correctly the problem of the nonlinear joint influence of defects on the loading capacity of the shell.

Tests on the stability of prototypes carried out on a UME-10TM machine showed that the exhaustion of loading capacity of the shell took place at one stage by reaching a limit point.

The loss of stability of a conical shell with the same low ovality ends (Fig. 9a) is in general related to a form close to its own form of stability loss of oval cylindrical shells under the action of uniform axial compression [17], but shifted to a larger shell butt. On one side of the shell there are two or three belt dents located at the larger end. They cover the smaller curvature of the plate and are shifted to the side of panel larger curvature. Local dents have a relatively large size and do not form a regular closed form buckling.

The increase of taper and roundness of the ends leads to a shift of the zone of wave generation into the longitudinal direction to a lower end (Fig. 9b) while maintaining the overall character of buckling.

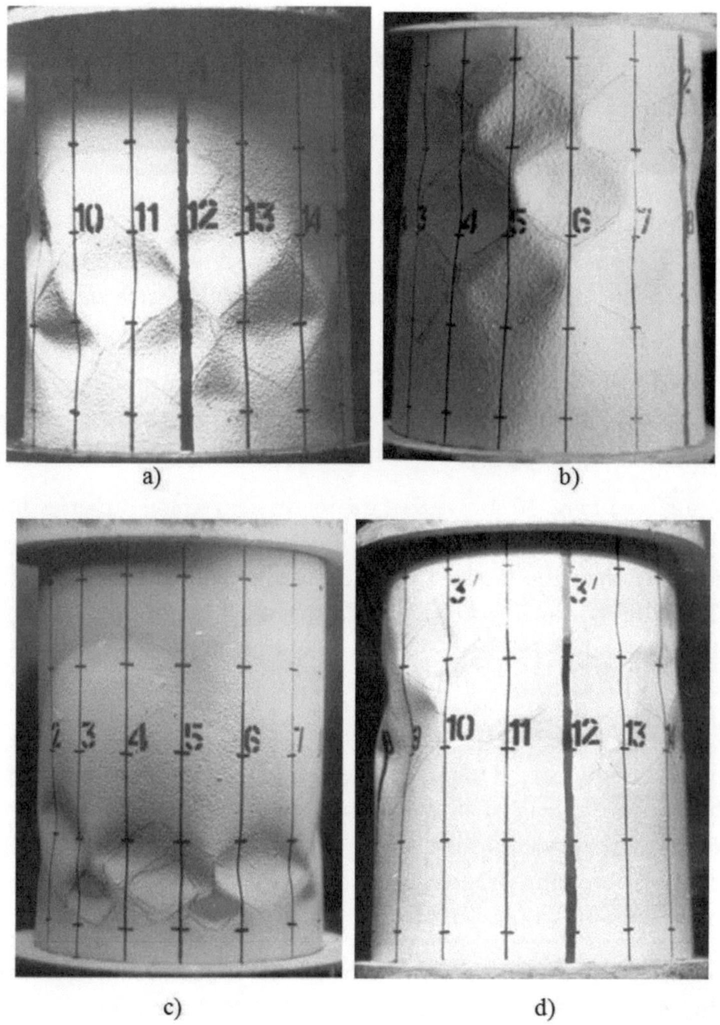

Figure 9. Forms of supercritical wave generation of the shell with a small taper and the same low ovality of ends (a); with a large taper, and the same large oval ends (b); with a large taper, and a large oval of the lower extremity (c); with a large taper, and a large oval upper end (d)

At high cone (within a given experiment) increased roundness of the lower end, while maintaining the shape of the upper longitudinal, increases the localization of buckling (Fig. 9c) shifting the dents closer to the lower end, while maintaining the variability in the circumferential direction. Conversely, the prevalence of high cone-roundness of the upper end leads to a significant shift of dents to the end of a large oval (Fig. 9d).

Results of the experiment allow us to derive mathematical models of the form

$$\left(a/b\right)_H^\circ = \left(a/b\right)_B^\circ = \left(a/b\right)^\circ : \; K = 0{,}379 + 0{,}0029\alpha^\circ - 0{,}012\left(a/b\right)^\circ - 0{,}014\alpha^\circ \left(a/b\right)^\circ ;$$

$$\alpha^\circ = 1: \; K = 0{,}346 - 0{,}017\left(a/b\right)_H^\circ - 0{,}0033\left(a/b\right)_B^\circ - 0{,}013\left(a/b\right)_H^\circ \left(a/b\right)_B^\circ ,$$

where $(...)^\circ$ is the standardized value; $K = \dfrac{T_{cr}}{2\pi E\delta^2}$ is the dimensionless ratio of the critical stress; T_{cr} is the critical compressive load; E is the Young's modulus.

The resulting models are adequate to the experimental data by Fisher criteria at the 5% significance level. The presence of significant second-order terms indicates a significant non-linearity of the relationship between the parameters, and, therefore, incorrect to separate consideration of the parameters and the placement of single-factor experiments.

Let us investigate the derived mathematical models. The corresponding surface of the pair interactions are shown in Fig. 10 and 11. They demonstrate good agreement between calculation results of MMPC and experimental data.

Analysis of the surface in Fig. 10 shows that the increase in single imperfections significantly reduces the carrying capacity of the shell. In addition, in these limits roundness has a greater impact on the setting than the taper. This is consistent with the single-factor experiments reported in [13,14]. But the analysis of Fig. 10 also shows that the simultaneous increase in taper and ovality can lead to an increase in carrying capacity to a level corresponding to the defect of a free shell. This is essentially a nonlinear effect, which could not be found by single-factor experiments.

Further study of the nonlinear interaction of defects (Fig. 11) showed that in the developed cone-of-roundness of the lower shell end has a more significant impact on the setting of critical effort than the roundness of the upper end. The joint increase in roundness of ends leads to an increase in carrying capacity, which is also an essentially nonlinear effect and is in good agreement with the results shown in Fig. 10.

Subcritical deformation has been studied in thin-walled shells with an oval on the lower and upper end being equal to 0.84 and 0.96, respectively, and taper equal to $0^\circ 56'$ and $2^\circ 16'$, respectively. The selected values α, $\left(a/b\right)$, $\left(a/b\right)_B$ and $\left(a/b\right)_H$ correspond to characteristic points of the models [13,14].

A qualitative analysis of the effect of displacement fields on the results of the holographic experiment suggests an important role played by the strain state of shells under non-uniform roundness in the district and in the longitudinal direction (Fig. 12).

With the increase of α up to $2^\circ 16'$ heterogeneity of the radial deflection is shifted to the lower end. A comparison of interferograms obtained at different load levels shows that an increase in the last number of fringes decreases with equal values of the additional load, and this indicates the hardening of structures, possibly caused by high deformability of the shell at the beginning of loading. The deformation of the shell in the experiment depends on the character of ends, and imperfections differ significantly on the panels of varying curvature.

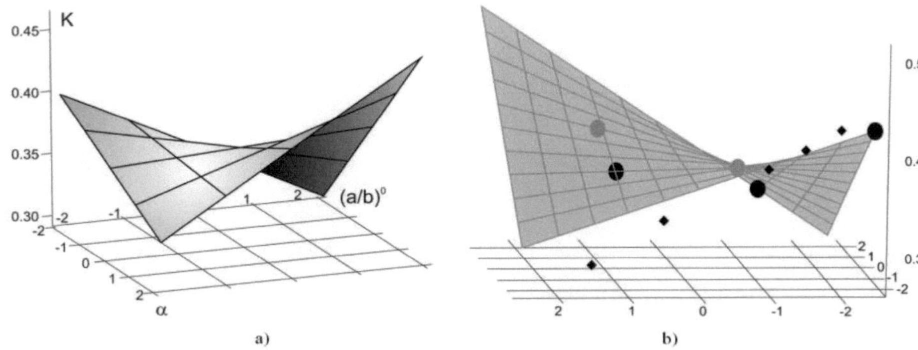

Figure 10. The surfaces of the pair interactions of imperfections α and a/b: (two-factor model (a), comparison between model, single-factor experiment [13] (rhombus) and MMPC calculations (circles) (b).

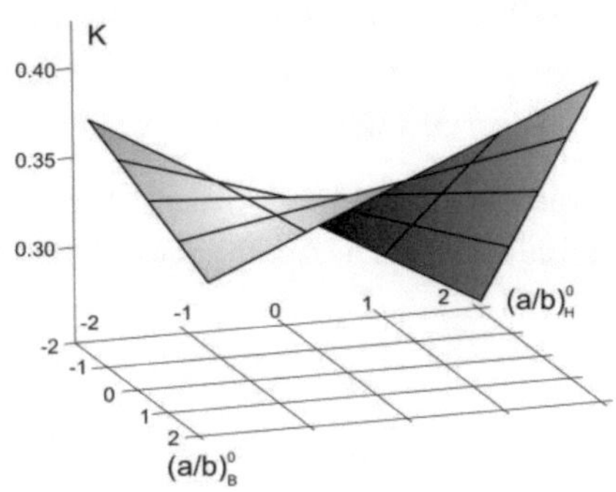

Figure 11. Surfaces of pair interactions with imperfections $\left(a/b\right)^{\circ}_{H}, \left(a/b\right)^{\circ}_{B}$.

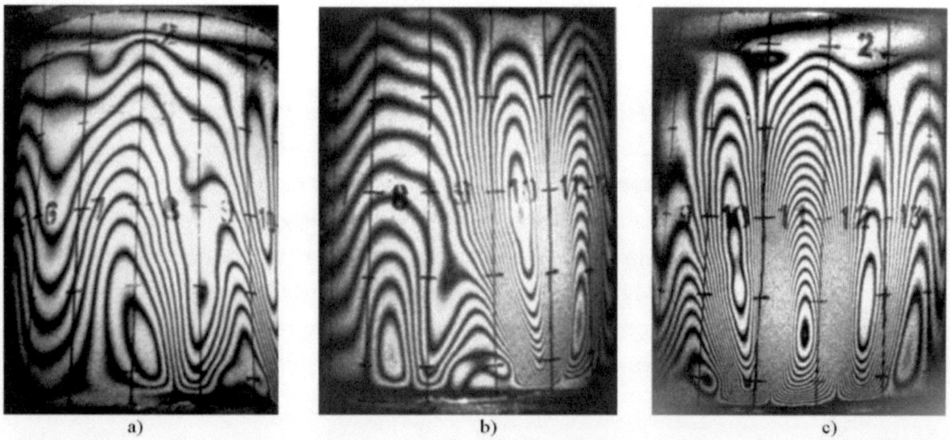

Figure 12. Interferogram envelope with the taper and ovality of the small curvature of the panel (a), the joint panel zone (b), and the larger curvature of the panel (c).

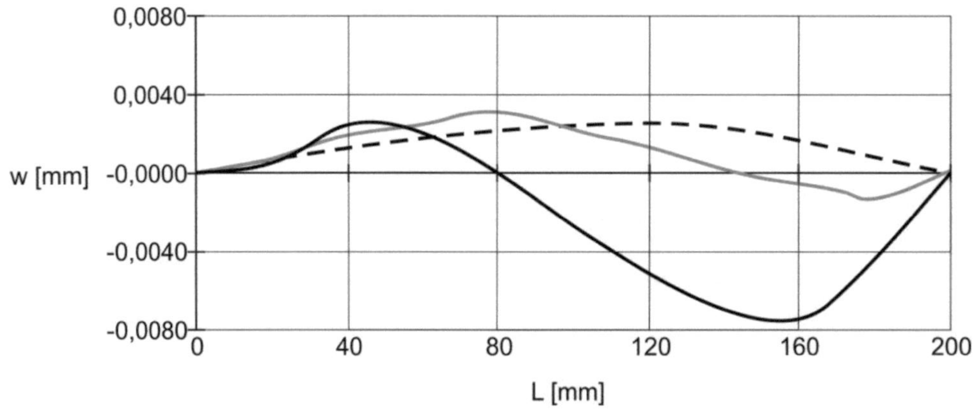

Figure 13. The deformation of the shell with ovality and taper. Solid curves correspond to the middle of panels: black – large curvature, gray – small curvature, dash – forming at the junction of the panels; positive direction goes toward the center of curvature

The change of $\left(a/b\right)_H$ from 0.96 to 0.84 significantly (1.2-1.4 fold) increases compliance of the membranes, while maintaining the overall picture of the distribution of displacements in the circumferential direction and increasing heterogeneity in the longitudinal direction.

The field of displacements was explained semi-automatically on the basis of the above algorithm. The forms of the radial deflection of some shell generatrixes are shown in Fig. 13.

8. Conclusions

A modified method of the parameter continuation (MMPC) is proposed. This method enables simplification of the calculations both at the stage of constructing the model, and also within its continued use due to precise values of the Taylor coefficients for the solution of the degree not exceeding the number of approximation.

The expression to calculate approximations by the MMPC in the general case and with the nonlinearity type of products and squares of the desired functions is presented.

The application of fractional-rational transformation for the polynomial approximation in the form of the 1-D and 2-D PAs used for increasing the degree of convergence and for the analytical continuation of the approximation in the region of its meromorphy was analyzed. It was concluded that such a transformation is justified if it is applied to polynomials which depend on the variable of integration. We used 2-D PAs for the independent variable and for the artificial parameter applying the scheme proposed by V. Vavilov. In this paper it is shown that this transformation provides a satisfactory quality for the approximation behavior and minimizes its error, in spite of the fact that the use of 2-D PAs requires a further theoretical justification.

The estimation of stability using MMPC approximation is also proposed. A study of numerical results was conducted by applying the methods for three model examples which were perturbed with a natural small parameter. It is shown that the application of PAs provides them with sufficient accuracy in the studied area. This paper shows the advantage of approximations which were obtained based on the MMPC.

Calculations of nonlinear deformation and stability of elastic flexible circular cylindrical shell under uniform external pressures and of free oscillations of simply supported stringer shell demonstrated the efficiency and accuracy of the proposed method.

The methodology and results of a holographic experiment with thin low-conical shells having oval ends are presented. They show good agreement with calculation results.

Author details

Igor Andrianov
Institute of General Mechanics, RWTH Aachen University, Templergraben, Aachen, Germany

Jan Awrejcewicz
Lodz University of Technology, Department of Automation and Biomechanics, Stefanowski Str., Lodz, Poland

Victor Olevs'kyy
Ukrainian State Chemistry and Technology University, Gagarina av., 8, UA-49070, Dnipropetrovs'k, Ukraine

Acknowledgement

J. Awrejcewicz input to this chapter was supported by the Alexander von Humboldt Award.

9. References

[1] Baker GA Jr, Graves-Morris P (1996) Padé Approximants. Encyclopedia of Mathematics and Its Applications. Cambridge: Cambridge University Press, 2nd ed., v. 59.
[2] Vavilov VV, Tchobanou MK, Tchobanou PM (2002) Design of multidimensional recursive systems through Padé type rational approximation, Nonlinear Analysis: Modelling and Control, 7(1): 105-125.
[3] Wasov W (1965) Asymptotic Expansions for Ordinary Differential Equations. New York: John Wiley & Sons.
[4] Obraztsov IF, Nerubaylo BV, Andrianov IV (1991) Asymptotic Methods in Structural Mechanics of Thin-Walled Structures. Moscow: Mashinostroyenie.
[5] Adomian G (1989) A review of the decomposition method and some recent results for nonlinear equations. Comp. Math. Appl., 21: 101-127.
[6] Abassy TA, El-Tawil MA, Saleh HK (2007) The solution of Burgers' and good Boussinesq equations using ADM–Padé technique. Chaos, Solitons and Fractals, 32: 1008-1026.

[7] He JH (2008) Recent developments of the homotopy perturbation method. Top. Meth. Nonlin. Anal., 31: 205–209.

[8] Grigolyuk EE, Shalashilin VI (1991) Problems of Nonlinear Deformation: The Continuation Method Applied to Nonlinear Problems in Solid Mechanics. Dordrecht: Kluwer.

[9] Andrianov IV, Kholod EG, Olevsky VI (1996) Approximate non-linear boundary value problems of reinforced shell dynamics. J. Sound Vibr., 194(3): 369-387.

[10] Vest Ch (1979) Holographic interferometry. New York: John Wiley & Sons.

[11] Mossakovskii VI, Mil'tsyn AM and Olevskii VI (1990) Deformation and stability of technologically imperfect cylindrical shells in a nonuniform stress state. Strength of Materials, 22(12): 1745-1750.

[12] Mossakovskii, V.I. Mil'tsyn AM, Selivanov YuM and Olevskii VI (1994) Automating the analysis of results of a holographic experiment. Strength of Materials, 26(5): 385-391.

[13] Krasovsky VL (1997) On buckling mechanism of real thin-walled cylinders at axial compression. Proc. of the VIII Symposium on Stability of Structures. Zakopane (Poland): 145-150.

[14] Preobrazhenskiy IN, Grishchak VZ (1986) Stability and Vibration of Conical Shells. Moscow: Mashinostroyenie.

[15] Mil'tsyn AM (1992) The influence of technological imperfections on the stability of thin shells (multivariate approach) Ch. I. Mechanics of Solids, 6: 181-188.

[16] Mil'tsyn AM (1993) Nonlinear interaction of technological imperfections and their influence on the stability of thin shells (multivariate approach), Ch. II. Mechanics of Solids, 1: 178-184.

[17] Andreev LV, Obodan NI, Lebedev AG (1988) Stability of Shells under Nonaxisymmetric Deformation. Moscow: Nauka. 208 p.

Chapter 2

Floquet Exponents and Bifurcations in Switched Converters

John Alexander Taborda, Fabiola Angulo and Gerard Olivar

Additional information is available at the end of the chapter

http://dx.doi.org/10.5772/51330

1. Introduction

Switched power converters are finding wide applications in the area of electrical energy conditioning. Many electronic devices have power converters to achieve high conversion efficiency and therefore low heat waste. Some of them are: drivers for industrial motion control, battery chargers, uninterruptible power supplies (UPS), electric vehicles, laptops, gadgets and mobile phones. Therefore control of power converters in order to optimize conversion efficiency is a current and challenging research topic. Pulsewidth modulation (*PWM*) is the most used method to control power converters [11]-[12].

Digital-PWM controllers are a novel alternative to control power converters. These controllers have many advantages as programmability, high flexibility, reliability and easy implementation of advanced control algorithms. They can be designed with *delays* in the measured variables in order to guarantee the necessary computing time of the signal control. However, performance of PWM controllers is affected by delays.

In this chapter, we investigate the incidence of *delays* in a digital-PWM controller based on two novel techniques: *Zero Average Dynamics* (ZAD) and *Fixed-Point Inducting Control* (FPIC). Both control strategies have been developed, applied and widely analyzed in the last decade [5].

Floquet theory and *smooth bifurcation theory* can be used to define stability regions and to find optimum parameter sets (see for example [6]-[9]). In our case, three parameters should be tuned in the digital-PWM controller. Each parameter is denoted as: k_s in ZAD strategy, N in FPIC technique and τ is the number of delay periods in the measured variables [1].

The 3D-parameter space (k_s,N,τ) of the delayed PWM controller is analyzed and stability regions are bounded by *Flip*, *Fold* and *Neimark-Sacker* transitions. The presence of the three smooth bifurcations in the same nonlinear circuit is not common and this fact has not been reported widely in Digital-PWM switched converters.

INTECH
open science | open minds

© 2012 The Author(s). Licensee InTech. This chapter is distributed under the terms of the Creative Commons Attribution License http://creativecommons.org/licenses/by/3.0), which permits unrestricted use, distribution, and reproduction in any medium, provided the original work is properly cited.

Stability, bifurcations and transient response of switched converters with *Delayed PWM controllers* can be studied more efficiently using an analysis of disturbances based on *Floquet theory*. We show that this procedure can be generalized to compute Floquet exponents for any number of delays (τ) in the control law (d_k, so-called *duty cycle*). We compare this approach with other methods which determine stability in switched converters. One of them is the computation of characteristic multipliers based on the jacobian matrix. Another one is the computation of Lyapunov exponents using a numeric routine. Each method gives equivalent information. However Floquet approach is the most appropriated when delays appear since this method does not require the evaluation of the jacobian matrix (its dimension increases when the number of delays is higher). The other two methods have this disadvantage [13].

The chapter is organized as follows. In *Section 2* we present the general procedure to compute Floquet exponents in PWM switched converters. The particular case of a buck converter controlled with digital-PWM controller based on ZAD, FPIC and DELAY schemes, is presented in *Section 3*. The stability of fixed points in delayed PWM switched converters is discussed in *Section 4*, while fold, flip and Neimark-Sacker bifurcations are presented in *Section 5*. Finally, the conclusions and future work are presented in *Section 6*.

2. Floquet-based procedure in PWM switched converters

We assume that a PWM switched converter can be modelled as a piecewise-linear dynamical system, as it is written in equation (1). $\mathbf{x}$ is the state $(nx1)$-vector, $\mathbf{A}$ is the state (nxn)-matrix, $\mathbf{B}$ is the input $(nx1)$-vector, $\mathbf{C}$ is the output $(1xn)$-vector, y is the scalar output and u_{PWM} is the control signal.

$$\begin{cases} \dot{\mathbf{x}} = \mathbf{A}\mathbf{x} + \mathbf{B}u_{PWM} \\ y = \mathbf{C}\mathbf{x} \end{cases} \tag{1}$$

The PWM signal is a function of d_k and the duty cycle is a function of the delayed state variables ($d_k = f(\mathbf{x}(k-\tau))$). We consider the centered scheme given by equation (2).

$$u_{PWM}(t) = \begin{cases} u_l & \text{if } k \le t \le k + d_k/2 \\ u_c & \text{if } k + d_k/2 < t < k + 1 - d_k/2 \\ u_r & \text{if } k + 1 - d_k/2 \le t \le k + 1 \end{cases} \tag{2}$$

First, we define the PWM Switched System including the discontinuity by using a unit step function (θ), as in equation (3) where $t \in \{0, (\tau+1)T\}$, $t_{s1} = (\tau + d_0/2)T$, $t_{s2} = (\tau + 1 - d_0/2)T$, $\Delta u_{cl} = u_c - u_l$ and $\Delta u_{rc} = u_r - u_c$.

$$\dot{\mathbf{x}} = \mathbf{A}\mathbf{x} + u_l\mathbf{B} + \Delta u_{cl}\theta(t - t_{s1})\mathbf{B} + \Delta u_{rc}\theta(t - t_{s2})\mathbf{B} \tag{3}$$

According to Floquet theory we define, for the perturbed solution,

$$\mathbf{x} = \mathbf{x}^* + e^{\mu T}\mathbf{p}(t),$$

where

$\mu \in C$ is the so-called Floquet exponent and $\mathbf{p}(t+T) = \mathbf{p}(t)$ is an associated T-periodic function.

Replacing the perturbed solution in equation (3) and neglecting the periodic solution, the variational equation of the system is obtained by equation (4) where $t^*_{s1} = \left(\tau + \frac{d^*}{2}\right)T$, $t^*_{s2} = \left(\tau + 1 - \frac{d^*}{2}\right)T$ and $\mathbf{H}$ is a (nxn)-matrix which depends on the time differential related to the perturbation of the PWM signal. Matrix $\mathbf{H}$ depends on the control strategy.

$$\dot{\mathbf{p}} = (\mathbf{A} - \mu\mathbf{I})\mathbf{p} + \left(\Delta u_{rc} e^{-\mu(t-t^*_{s2})}\delta\left(t - t^*_{s2}\right) - \Delta u_{cl} e^{-\mu(t-t^*_{s1})}\delta\left(t - t^*_{s1}\right)\right)\mathbf{H}\mathbf{p}(0) \quad (4)$$

The solution of the variational equation is stated in (5) where $z = e^{-\mu T}$.

$$\mathbf{p}\left(1\right) = \left(z e^{\mathbf{A}T} + z^{\tau+1}\left(\Delta u_{rc} e^{\mathbf{A}T\frac{d^*}{2}} - \Delta u_{lc} e^{\mathbf{A}T\left(1-\frac{d^*}{2}\right)}\right)\mathbf{H}\right)\mathbf{p}(0) \quad (5)$$

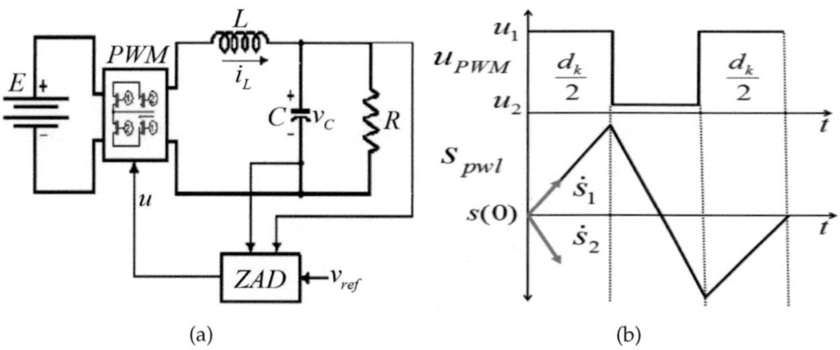

Figure 1. (a). Scheme of a PWM-controlled Buck converter with ZAD-strategy. (b). Piecewise-linear error dynamic (s_{pwl}) in a sampling period.

3. Floquet exponents in a synchronous buck converter

The relevant roles of *Floquet exponents* on analysis, design and control of PWM switched converters are analyzed in a synchronous buck converter. Its main feature is that the output value V_o is lower than the source E (step down converter). Figure 1 (a) shows a scheme of buck converter controlled with ZAD strategy.

The mathematical model for the synchronous buck converter can be written in compact form as:

$$\begin{pmatrix} \dot{x}_1 \\ \dot{x}_2 \end{pmatrix} = \begin{pmatrix} \frac{-1}{RC} & \frac{1}{C} \\ \frac{-1}{L} & \frac{-r_L}{L} \end{pmatrix} \begin{pmatrix} x_1 \\ x_2 \end{pmatrix} + \begin{pmatrix} 0 \\ \frac{E}{L} \end{pmatrix} u_{PWM} \quad (6)$$

where $x_1 = v_C$, $x_2 = i_L$ and u_{PWM} belongs to the discrete set $\{-1, 1\}$.

For the *ZAD condition,* a piecewise-linear function is defined as equation (7). Figure 1 (b) shows a scheme of s_{pwl} in a period sampling.

$$s_{pwl}(t) = \begin{cases} s_1 + (t - kT)\dot{s}_1 & \text{if } kT \le t \le kT + \frac{d_k}{2} \\ s_2 + (t - kT + \frac{d_k}{2})\dot{s}_2 & \text{if } kT + \frac{d_k}{2} < t < kT + (T - \frac{d_k}{2}) \\ s_3 + (t - kT + T + \frac{d_k}{2})\dot{s}_1 & \text{if } kT + (T - \frac{d_k}{2}) \le t \le (k+1)T \end{cases} \quad (7)$$

where

$$\dot{s}_1 = (\dot{x}_1 + k_s \ddot{x}_1)\Big|_{x=x(kT),u=1} \qquad \dot{s}_2 = (\dot{x}_1 + k_s \ddot{x}_1)\Big|_{x=x(kT),u=0}$$
$$s_1 = (x_1 - ref + k_s \dot{x}_1)\Big|_{x=x(kT),u=1} \quad s_2 = \tfrac{d}{2}\dot{s}_1 + s_1 \; s_3 = s_1 + (T-d)\dot{s}_2 \tag{8}$$

and k_s is a positive constant. Therefore, the zero average condition is

$$\int_{kT}^{(k+1)T} s_{pwl}(t)dt = 0 \tag{9}$$

Now, finding d_k means solving equation (9) and redefining the duty cycle as d_{zad}

$$d_{zad} = \frac{2s_1 + T\dot{s}_2}{\dot{s}_2 - \dot{s}_1} \tag{10}$$

The FPIC control law is given by equation (11), where N is the control parameter and d_{ss} is the duty cycle when the stationary state is reached.

$$d_k = \frac{d_{zad} + Nd_{ss}}{N+1} \tag{11}$$

This result can be expressed as a linear combination of the state variables, where c_1, c_2 and c_3 are constants.

$$d_k = c_1 x_1(k-\tau) + c_2 x_2(k-\tau) + c_3 \tag{12}$$

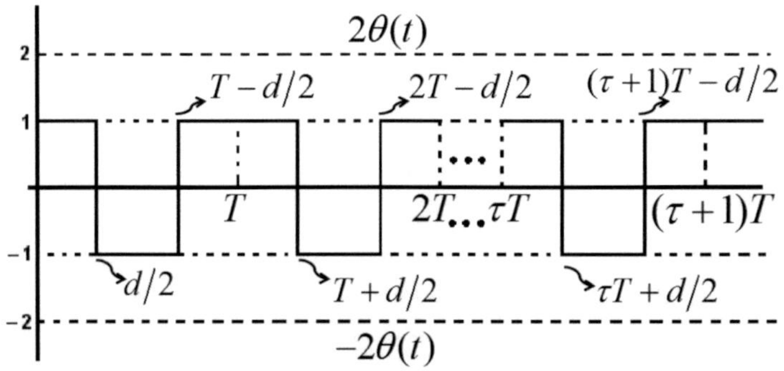

Figure 2. Scheme of centered PWM function depending on delays (τ).

Now, we should define the variational equation of the buck converter using the general procedure described in *Section 2*. Basically, we apply to the periodic solution ($\mathbf{x}^*$) an appropriate perturbation using exponential functions ($e^{\mu t}$). The stability of the digital-PWM power converter can be inferred studying the behavior of the perturbation. If the real part of the exponent μ is positive, the perturbation will tend to infinity and the solution will be unstable. If the real part of the exponent μ is negative, the perturbation will tend to zero and the solution will be stable.

Therefore, we can find the variational equation based on the dynamical equations of the system. We use a more compact expression applying the following change of variables: $x_1 = v_C/E$, $x_2 = \frac{1}{E}\sqrt{\frac{L}{C}}i_L$ and $t = \tau/\sqrt{LC}$, thus $\gamma = \frac{1}{R}\sqrt{\frac{L}{C}}$, $\beta = r_L\sqrt{\frac{C}{L}}$ and the sampling period is $T = T_c/\sqrt{LC}$ [10].

$$\begin{pmatrix} \dot{x_1} \\ \dot{x_2} \end{pmatrix} = \begin{pmatrix} -\gamma & 1 \\ -1 & -\beta \end{pmatrix} \begin{pmatrix} x_1 \\ x_2 \end{pmatrix} + \begin{pmatrix} 0 \\ 1 \end{pmatrix} u_{PWM} \tag{13}$$

We use the parameter values of an experimental prototype reported in [3]. We fix $R = 20\Omega$, $C = 40\mu$F, L=2mH, $r_L = 0\Omega$, $v_{ref} = 32$V, E=40V and the sampling period $T_c = 50\mu s$.. Therefore, the dimensionless parameters are $\gamma = 0.35$, $\beta = 0$ and $T = 0.1767$.

For simplicity, in the remainder of *Section* 3, we note d_k as d with $d \in [0, T]$, and we present the procedure for one period sampling $t \in [\tau T, (\tau+1)T]$ (general notation was used in *Section* 2). The signal control u_{PWM} is defined as equation (14), where $t_{s1} = \left(\tau T + \frac{d}{2}\right)$ and $t_{s2} = \left((\tau+1)T - \frac{d}{2}\right)$.

$$u_{PWM} = \begin{cases} 1 & \tau T < t \leqslant t_{s1} \\ -1 & t_{s1} < t \leqslant t_{s2} \\ 1 & t_{s2} < t \leqslant (\tau+1)T \end{cases} \tag{14}$$

Figure 2 illustrates the mechanism to model the centered pulse (u_{PWM}) using unit step

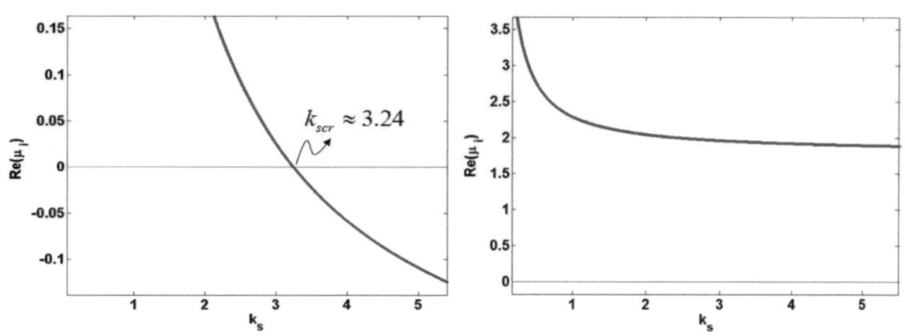

(a) $\tau = 0, N = 0$ (b) $\tau = 1, N = 0$

Figure 3. The evolution of the real part of the Floquet exponents when k_s is varied in the ZAD controller. (a). without delay and without FPIC, (b). with one delay and without FPIC

functions $\theta(t)$. The duty cycle depend on state variables in the instant $t = (k-\tau)T$. Equation (15) shows as two unit step functions can be used to model the centered pulse.

$$\begin{pmatrix} \dot{x_1} \\ \dot{x_2} \end{pmatrix} = \begin{pmatrix} -\gamma & 1 \\ -1 & -\beta \end{pmatrix} \begin{pmatrix} x_1 \\ x_2 \end{pmatrix} + \begin{pmatrix} 0 \\ 1 \end{pmatrix} - 2\theta\left(t - t_{s1}\right) \begin{pmatrix} 0 \\ 1 \end{pmatrix} + 2\theta\left(t - t_{s2}\right) \begin{pmatrix} 0 \\ 1 \end{pmatrix} \tag{15}$$

Equation (15) implies that duty cycle can be defined in function of state variables in an initial instant ($t = 0$) because the delay information was included in the unit step functions.

Therefore, we can define $(d/2)$ such as equation (16) where c_1, c_2 and c_3 are given by equations (17), (18) and (19), respectively.

$$\frac{d}{2} = c_1 x_1(0) + c_2 x_2(0) + c_3 \tag{16}$$

$$c_1 = \frac{2 - \gamma(2k_s + T(1 - \gamma k_s)) - k_s T}{-4k_s(N+1)} \tag{17}$$

$$c_2 = \frac{2k_s + T(1 - k_s(\gamma + \beta))}{-4k_s(N+1)} \tag{18}$$

$$c_3 = \frac{2ref + k_s T}{4k_s} + \frac{Nd_{ss}}{2(N+1)} \tag{19}$$

Let the perturbed periodic orbit be

$$x_1(t) = x_1^*(t) + e^{\mu t} p_1(t), \qquad x_2(t) = x_2^*(t) + e^{\mu t} p_2(t),$$

where the superstar labels are the periodic solutions and $e^{\mu t} p_1(t)$, $e^{\mu t} p_2(t)$ are the perturbations. Then, we replace the perturbed periodic orbit in equation (15).

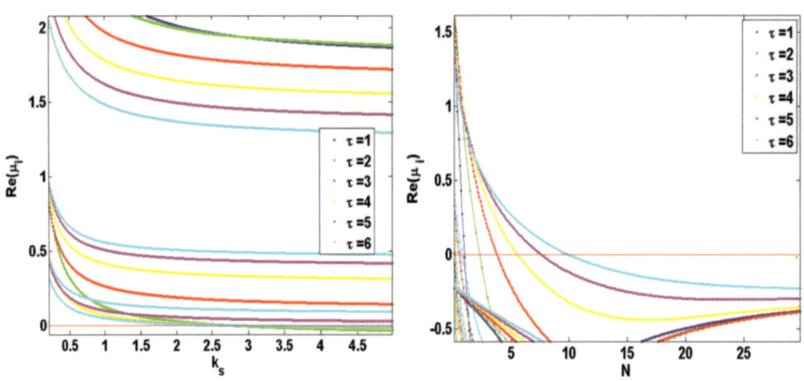

(a) ZAD scheme (without FPIC)   (b) ZAD-FPIC scheme

Figure 4. The evolution of the real part of the Floquet exponents for several delays in the Digital-PWM converter based on ZAD or ZAD-FPIC techniques. (a). k_s is varied between $[0;5]$ with $N = 0$, (b). N is varied between $[0;30]$ with $k_s = 4.5$.

$\dot{x}_1^* + \mu e^{\mu t} p_1 + e^{\mu t} \dot{p}_1 = -\gamma(x_1^* + e^{\mu t} p_1) + (x_2^* + e^{\mu t} p_2)$

$\dot{x}_2^* + \mu e^{\mu t} p_2 + e^{\mu t} \dot{p}_2 = -(x_1^* + e^{\mu t} p_1) - \beta(x_2^* + e^{\mu t} p_2) + 1 - 2\theta\left(t - \left(\tau T + \frac{d}{2}\right)\right) + 2\theta\left(t - \left((\tau+1)T - \frac{d}{2}\right)\right)$

Unit step functions are replaced as follows. The periodic solution of duty cycle is noted:

$\frac{d^*}{2} = c_1 x_1^*(0) + c_2 x_2^*(0) + c_3$

then in the first unit step function,

$\theta\left(t - t_{s1}^*\right) = \theta\left(t - \left(\tau T + \frac{d^*}{2}\right)\right) = \theta\left(t - \left(\tau T + c_1 x_1^*(0) + c_2 x_2^*(0) + c_3\right)\right)$

therefore, the step function in function of the perturbed periodic solution is

$$\theta\left(t-\left(\tau T+\frac{d}{2}\right)\right)=\theta\left(t-\left(\tau T+c_1(x_1^*(0)+p_1(0))+c_2(x_2^*(0)+p_2(0))+c_3\right)\right)$$

computing a first order Taylor expansion approximation of the unity step function, we obtain

$$\theta\left(t-\left(\tau T+\frac{d}{2}\right)\right)=\theta\left(t-\left(\tau T+\frac{d^*}{2}\right)\right)-\delta\left(t-\left(\tau T+\frac{d^*}{2}\right)\right)(c_1p_1(0)+c_2p_2(0))$$

where $\delta(t)$ is the Dirac delta function with $\delta(t)=\frac{d\theta(t)}{dt}$. The same considerations are applied in the second unity step function $\theta(t-t_{s2})$. The dynamic of the perturbed periodic solution is

$$\begin{aligned}\dot{x}_1^*+\mu e^{\mu t}p_1+e^{\mu t}\dot{p}_1&=-\gamma x_1^*-\gamma e^{\mu t}p_1+x_2^*+e^{\mu t}p_2\\ \dot{x}_2^*+\mu e^{\mu t}p_2+e^{\mu t}\dot{p}_2&=-x_1^*-e^{\mu t}p_1-\beta x_2^*-\beta e^{\mu t}p_2+1-2\theta\left(t-\left(\tau T+\frac{d^*}{2}\right)\right)+\\ &2\delta\left(t-\left(\tau T+\frac{d^*}{2}\right)\right)(c_1p_1(0)+c_2p_2(0))+2\theta\left(t-\left((\tau+1)T-\frac{d^*}{2}\right)\right)\\ &+2\delta\left(t-\left((\tau+1)T-\frac{d^*}{2}\right)\right)(c_1p_1(0)+c_2p_2(0))\end{aligned}$$

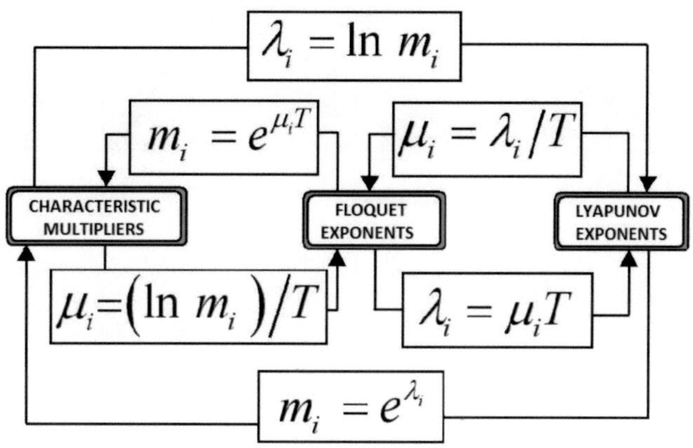

Figure 5. Scheme of equivalent transformations between Floquet exponents (μ_i), Lyapunov exponents (λ_i) and characteristics multipliers (m_i).

Neglecting the periodic solution $\mathbf{x}^*$, we obtain the dynamic of the perturbation.

$$\begin{aligned}\mu e^{\mu t}p_1+e^{\mu t}\dot{p}_1&=-\gamma e^{\mu t}p_1+e^{\mu t}p_2\\ \mu e^{\mu t}p_2+e^{\mu t}\dot{p}_2&=-e^{\mu t}p_1-\beta e^{\mu t}p_2+2\delta\left(t-\left(\tau T+\frac{d^*}{2}\right)\right)(c_1p_1(0)+c_2p_2(0))+\\ &2\delta\left(t-\left((\tau+1)T-\frac{d^*}{2}\right)\right)(c_1p_1(0)+c_2p_2(0))\end{aligned}$$

multiplying both sides of the equation by $e^{-\mu t}$, we obtain:

$$\dot{p}_1=-\mu p_1-\gamma p_1+p_2$$

$$\dot{p}_2 = -p_1 - \mu p_2 - \beta p_2 + 2e^{-\mu t}\delta\left(t - \left(\tau T + \tfrac{d^*}{2}\right)\right)\left(c_1 p_1(0) + c_2 p_2(0)\right) + 2e^{-\mu t}\delta\left(t - \left((\tau+1)T - \tfrac{d^*}{2}\right)\right)\left(c_1 p_1(0) + c_2 p_2(0)\right)$$

Simplifying the expressions and writing in matrix notation we obtain the variational equation of the buck converter. Note that the terms $e^{-\mu t}$ only have sense in $t = \vartheta$ when these are multiplied by Dirac delta functions $\delta(t - \vartheta)$.

$$\dot{\mathbf{p}} = \begin{pmatrix} -\gamma - \mu & 1 \\ -1 & -\beta - \mu \end{pmatrix} \mathbf{p} + e^{-\mu t^*_{s1}}\delta\left(t - t^*_{s1}\right)\mathbf{Hp}(0) + e^{-\mu t^*_{s2}}\delta\left(t - t^*_{s2}\right)\mathbf{Hp}(0) \tag{20}$$

Therefore, equation (20) is the variational equation where $t^*_{s1} = \tau T + \frac{d^*}{2}$, $t^*_{s2} = (\tau+1)T - \frac{d^*}{2}$ and,

$$\mathbf{H} = \begin{pmatrix} 0 & 0 \\ 2c_1 & 2c_2 \end{pmatrix}, \dot{\mathbf{p}} = \begin{pmatrix} \dot{p}_1 \\ \dot{p}_2 \end{pmatrix}, \mathbf{p} = \begin{pmatrix} p_1 \\ p_2 \end{pmatrix}, \mathbf{a}_e(0) = \begin{pmatrix} p_1(0) \\ p_2(0) \end{pmatrix}$$

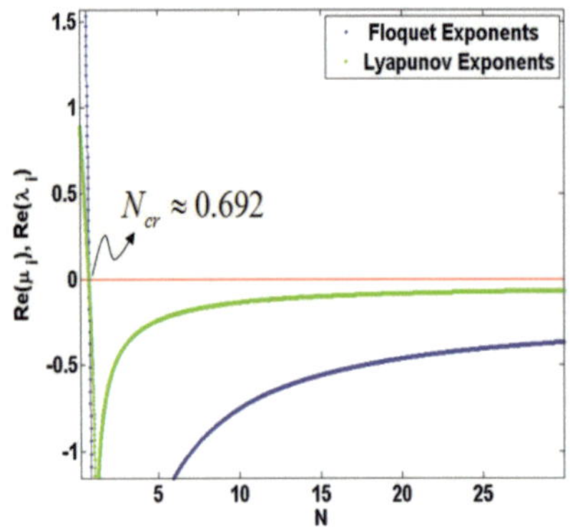

(a) $\tau = 0, k_s = 0.125$

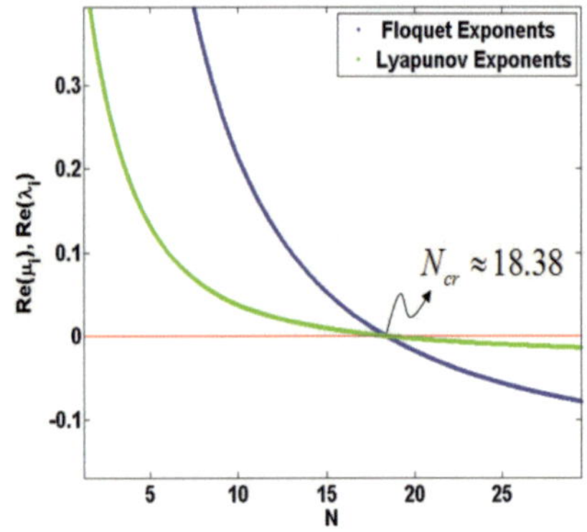

(b) $\tau = 1, k_s = 0.125$

Figure 6. The evolution of the real part of the Floquet and Lyapunov exponents when N is varied. (a). without delay, (b). with one delay.

Note that equation (20) can be written in a compact form as:

$$\dot{\mathbf{p}} = \mathbf{Mp} + e^{-\mu t}\left[\delta\left(t - (\tau T + \frac{d}{2})\right) + \delta\left(t - (\tau+1)T + \frac{d}{2}\right)\right]\mathbf{Hp}(0) \tag{21}$$

with $\mathbf{M}$ and $\mathbf{H}$ defined according to the equation (20), and again, for simplicity we note $d = d^*$. For solving this piecewise-smooth ordinary differential equation, we write $z = e^{-\mu T}$, $\mathbf{M} = \mathbf{M}_1 + \mathbf{M}_2$ with $e^{\mathbf{M}t} = e^{\mathbf{M}_1 t}e^{\mathbf{M}_2 t}$.

$$\mathbf{M}_1 = \begin{pmatrix} -\frac{\gamma}{2} - \frac{\beta}{2} - \mu & 0 \\ 0 & -\frac{\gamma}{2} - \frac{\beta}{2} - \mu \end{pmatrix} \tag{22}$$

$$\mathbf{M}_2 = \begin{pmatrix} -\frac{\gamma}{2} + \frac{\beta}{2} & 1 \\ -1 & \frac{\gamma}{2} - \frac{\beta}{2} \end{pmatrix} \tag{23}$$

The particular selection of $\mathbf{M}_1$ and $\mathbf{M}_2$ allows a easier solution of state-transition matrix $e^{\mathbf{M}t}$. The first exponential matrix $e^{\mathbf{M}_1 t}$ is computed using the identity matrix, while the second exponential matrix $e^{\mathbf{M}_2 t}$ is computed using sine and cosine functions.

$$e^{\mathbf{M}_1 t} = e^{-(\frac{\gamma}{2}+\frac{\beta}{2}+\mu)t}\mathbf{I}$$

$$e^{\mathbf{M}_2 t} = \begin{pmatrix} -\frac{\alpha_1}{\alpha_2} sen(\alpha_2 t) + \cos(\alpha_2 t) & \frac{1}{\alpha_2} sen(\alpha_2 t) \\ -\frac{1}{\alpha_2} sen(\alpha_2 t) & \frac{\alpha_1}{\alpha_2} sen(\alpha_2 t) + \cos(\alpha_2 t) \end{pmatrix}$$

where $\alpha_1 = \left(\frac{\gamma}{2} - \frac{\beta}{2}\right)$ and $\alpha_2 = \sqrt{1 - \alpha_1^2}$.

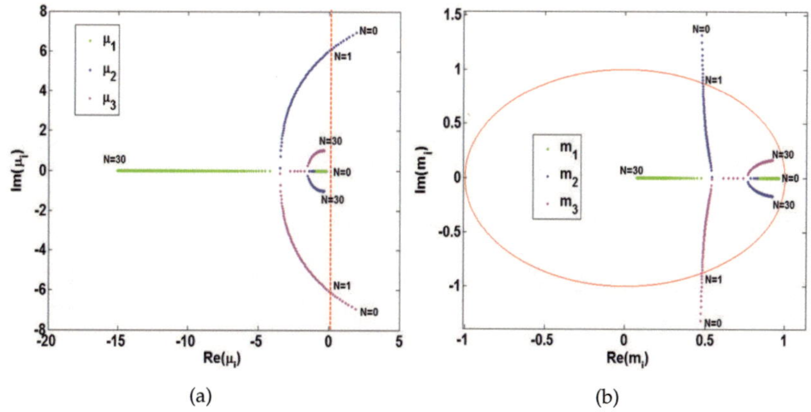

Figure 7. Locus of Floquet exponents and characteristic multipliers when N is varied between $[0;30]$ for $\tau = 1$. (a). Floquet locus, (b). Characteristic multipliers locus.

The piecewise-smooth ordinary differential equation can be solved in each interval with special attention in the discontinuities due to Dirac delta functions.

1). Initially, we compute the solution between $t = 0$ and $t = \frac{d}{2}$

$$\mathbf{p}\left(\frac{d}{2}\right)_- = e^{\mathbf{M}_1 \frac{d}{2}} e^{\mathbf{M}_2 \frac{d}{2}} \mathbf{p}(0)$$

then we compute $\mathbf{p}\left(\frac{d}{2}\right)_+$ integrating about the discontinuity as follows.

$$\mathbf{p}\left(\frac{d}{2}\right)_+ = \mathbf{p}\left(\frac{d}{2}\right)_- + e^{-\mu\left(\tau T + \frac{d}{2}\right)}\mathbf{H}\mathbf{p}$$

$$\mathbf{p}\left(\frac{d}{2}\right)_+ = \left(e^{\mathbf{M}_1 \frac{d}{2}} e^{\mathbf{M}_2 \frac{d}{2}} + e^{-\mu\left(\tau T + \frac{d}{2}\right)}\mathbf{H}\right)\mathbf{p}(0)$$

2). Now, we compute the solution in the interval between $t = \left(\frac{d}{2}\right)$ and $t = \left(T - \frac{d}{2}\right)$.

$$\mathbf{p}\left(T-\tfrac{d}{2}\right)_{-} = e^{\mathbf{M}_1(T-d)}e^{\mathbf{M}_2(T-d)}\mathbf{p}\left(\tfrac{d}{2}\right)_{+}$$

$$\mathbf{p}\left(T-\tfrac{d}{2}\right)_{-} = e^{\mathbf{M}_1(T-d)}e^{\mathbf{M}_2(T-d)}\left(e^{\mathbf{M}_1\frac{d}{2}}e^{\mathbf{M}_2\frac{d}{2}} + e^{-\mu\left(\tau T+\frac{d}{2}\right)}\mathbf{H}\right)\mathbf{p}\left(0\right)$$

in the discontinuity,

$$\mathbf{p}\left(T-\tfrac{d}{2}\right)_{+} = \mathbf{p}\left(T-\tfrac{d}{2}\right)_{-} + e^{-\mu\left((\tau+1)T-\frac{d}{2}\right)}\mathbf{H}\mathbf{p}\left(0\right)$$

$$\mathbf{p}\left(T-\tfrac{d}{2}\right)_{+} = \left(e^{\mathbf{M}_1(T-d)}e^{\mathbf{M}_2(T-d)}e^{\mathbf{M}_1\frac{d}{2}}e^{\mathbf{M}_2\frac{d}{2}} + e^{\mathbf{M}_1(T-d)}e^{\mathbf{M}_2(T-d)}e^{-\mu\left((\tau+1)T-\frac{d}{2}\right)}\mathbf{H}\right)\mathbf{p}\left(0\right)$$

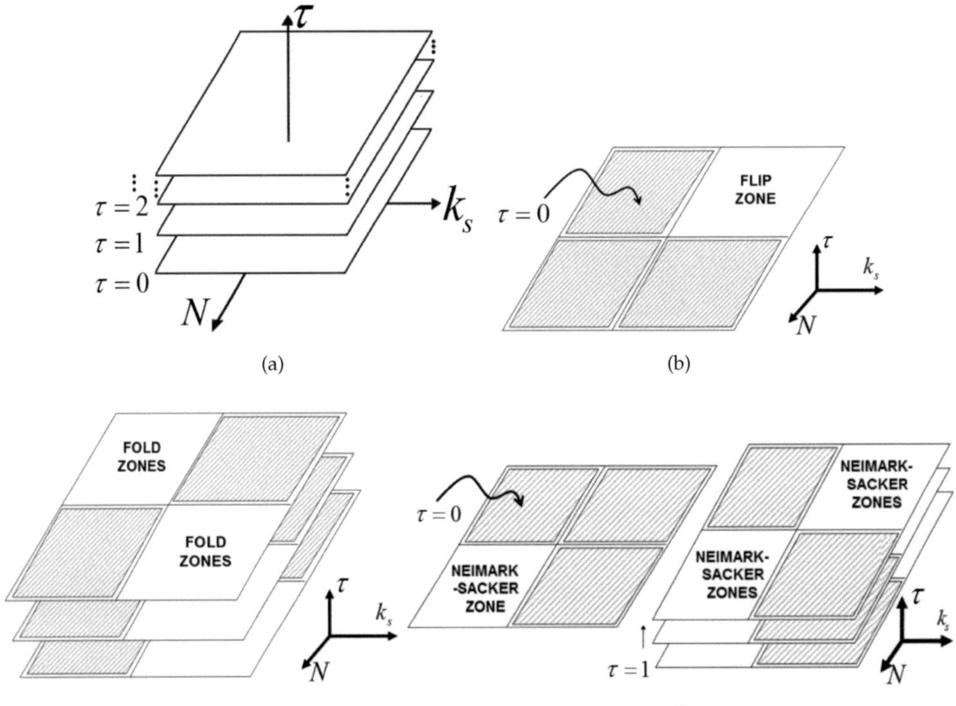

Figure 8. 3D parameter space of Digital-PWM switched converter and bifurcation zones. (a). (k_s, N, τ) space where the lines $k_s = 0$ and $N = -1$ separate each plane in four sub-spaces. (b). flip bifurcation zone. (c). fold bifurcation zones. (d). Neimark-Sacker bifurcation zones.

3). Finally, we compute the solution in the last interval between $t = (T - \frac{d}{2})$ and T.

$$\mathbf{p}\left(T\right) = e^{\mathbf{M}_1\frac{d}{2}}e^{\mathbf{M}_2\frac{d}{2}}\mathbf{p}\left(T-\tfrac{d}{2}\right)_{+}$$

$$\mathbf{p}\left(T-\tfrac{d}{2}\right)_{+} = \left(e^{\mathbf{M}_1(T-d)}e^{\mathbf{M}_2(T-d)}e^{\mathbf{M}_1\frac{d}{2}}e^{\mathbf{M}_2\frac{d}{2}} +\right.$$

$$e^{\mathbf{M}_1(T-d)}e^{\mathbf{M}_2(T-d)}e^{-\mu\left((\tau+1)T-\frac{d}{2}\right)}\mathbf{H}\Big)\,\mathbf{p}\,(0)$$

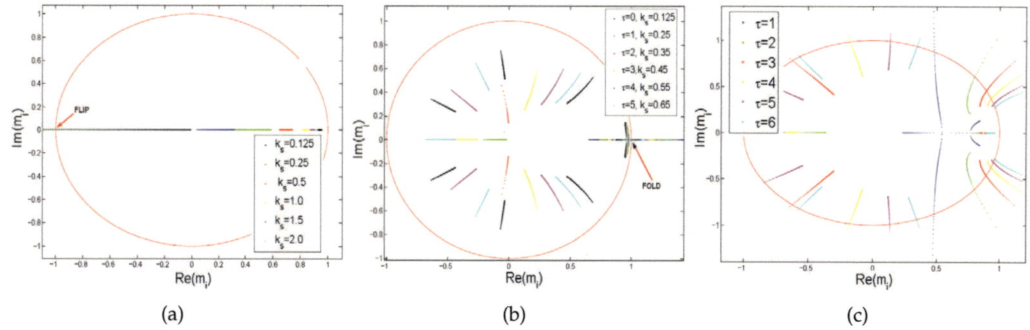

Figure 9. Evolution of characteristic multipliers in several bifurcation zones. (a). flip bifurcation zone $(\tau = 0, k_s > 0\ N > -1)$. (b). fold bifurcation zones $(\tau > 0, k_s > 0\ N < -1)$. (c). Neimark-Sacker bifurcation zones $(\tau > 1, k_s > 0\ N > -1)$.

Equation (24) is the solution of the variational system where $z = e^{-\mu T}$.

$$\mathbf{p}\,(T) = \mathbf{Q}_1\mathbf{p}_e\,(0) \tag{24}$$

The matrix $\mathbf{Q}_1$ is given by equation (25), where $\alpha_3 = \frac{\gamma}{2} + \frac{\beta}{2}$.

$$\mathbf{Q}_1 = \left(ze^{-\alpha_3 T}e^{\mathbf{M}_2 T} + z^{(\tau+1)}e^{-\alpha_3\left(T-\frac{d}{2}\right)}e^{\mathbf{M}_2\left(T-\frac{d}{2}\right)}\mathbf{H} + z^{(\tau+1)}e^{-\alpha_3\left(\frac{d}{2}\right)}e^{\mathbf{M}_2\left(\frac{d}{2}\right)}\mathbf{H}\right) \tag{25}$$

The existence of T-periodic solution, i.e. $\mathbf{p}\,(T) = \mathbf{p}\,(0)$, depends on equation (26) is satisfied.

$$(\mathbf{Q}_1 - I)\,\mathbf{p}\,(0) = \mathbf{0} \tag{26}$$

The stability of the periodic solution depends on whether the real part of each Floquet exponent is negative or not. If $\det(\mathbf{Q}_1 - \mathbf{I}) = 0$ implies μ negative.

We fix the parameter values in $\gamma = 0.35$; $\beta = 0$; $T = 0.1767$; $k_s = 4.5$; $N = 0$ and $ref = 0.8$. Equation (27) shows equation (26) in function of the number of delays (τ). The equation is a polynomial in z of degree $2(\tau + 1)$.

$$0.1e - 20z^{2(\tau+1)} - 1.867z^{(\tau+2)} + 1.911z^{(\tau+1)} + 0.93z^2 - 1.899z + 1 = 0 \tag{27}$$

Assuming real-time behavior, i.e. $\tau = 0$, the determinant is the second order polynomial of equation (28). The evolution of the real part of the Floquet exponents as parameter k_s varies is displayed in figure $3(a)$. The T-periodic solution is stable for $k_s > 3.24$.

$$\left(\frac{1e-10}{k_s^2} - \frac{0.00215387}{k_s} - 0.9466771\right)z^2 + \left(\frac{0.3443753}{k_s} - 0.0535075\right)z + 1 = 0 \tag{28}$$

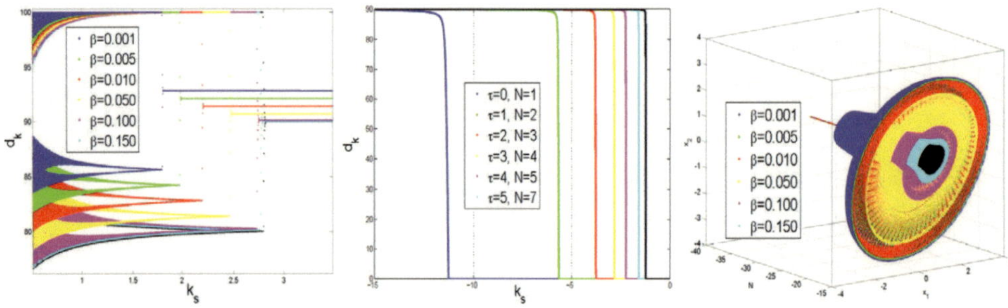

(a) (b) (c)

Figure 10. Examples of bifurcation diagrams in each zone. (a). flip bifurcation diagram ($\tau = 0, k_s > 0$ $N > -1$). (b). fold bifurcation diagram ($\tau > 0, k_s > 0$ $N < -1$). (c). Neimark-Sacker bifurcation diagram ($\tau > 1, k_s > 0$ $N > -1$).

Assuming one-delay period, i.e. $\tau = 1$, the determinant is the forth order polynomial of equation (29). The evolution of the real part of the Floquet exponents as parameter k_s varies is displayed in figure 3(b). In this case, one Floquet exponent has positive real part for any k_s. Therefore, ZAD strategy should be combined with FPIC ($N \neq 0$) to reach stable solutions.

$$\frac{1e-9}{k_s^2}z^4 - (1.8867 + \frac{0.002154}{k_s})z^3 + (\frac{0.344375}{k_s} + 2.79635)z^2 - 1.90983z + 1 = 0 \qquad (29)$$

For $\tau > 1$, ZAD strategy is not sufficient to stabilize 1T-periodic orbit and ZAD-FPIC scheme

Control Parameters	Stability Limit
$(\tau = 0), (k_s = 4.5)$	$N_{cr} \approx 0$
$(\tau = 1), (k_s = 4.5)$	$N_{cr} \approx 0.99$
$(\tau = 2), (k_s = 4.5)$	$N_{cr} \approx 2.32$
$(\tau = 3), (k_s = 4.5)$	$N_{cr} \approx 3.79$
$(\tau = 4), (k_s = 4.5)$	$N_{cr} \approx 5.53$
$(\tau = 5), (k_s = 4.5)$	$N_{cr} \approx 7.55$
$(\tau = 6), (k_s = 4.5)$	$N_{cr} \approx 9.89$

Table 1. Critical value of stability (N_{cr}) of buck converter controlled with ZAD-FPIC with several delay numbers. Figure 4 (b) shows the evolution of real part of Floquet exponents.

is necessary. Figure 4(a) shows the evolution of Floquet exponents when k_s is varied for several delay numbers and $N = 0$. The number of Floquet exponents with positive real part increases as the delay number grows.

Figure 4 (b) shows the results of ZAD-FPIC scheme when $k_s = 4.5$ and N is varied between $[0; 30]$. The critical value of stability (N_{cr}) increases as the delay number grows. Table 1 summarizes the behavior of critical value for different delays.

The behavior of the critical value is similar when N is fixed and k_s is varied for several delay numbers. The value k_{scr} increases as the delay number grows. Table 2 shows this condition.

Control Parameters	Stability Limit
$(\tau = 0, N = 0)$	$k_s \approx 3.24$
$(\tau = 1, N = 2)$	$k_s \approx 0.46$
$(\tau = 2, N = 3)$	$k_s \approx 1.19$
$(\tau = 3, N = 4)$	$k_s \approx 2.99$
$(\tau = 4, N = 6)$	$k_s \approx 2.72$
$(\tau = 5, N = 8)$	$k_s \approx 3.25$
$(\tau = 6, N = 10)$	$k_s \approx 4.21$

Table 2. Critical value of stability (k_{scr}) of buck converter controlled with ZAD-FPIC with several delay numbers.($\tau = 0$ to $\tau = 6$)

4. Stability of fixed points in delayed PWM switched converters

In previous section, we show that the procedure based on variational equation can be used to compute Floquet exponents for any number of delays (τ). In this section, we compare this approach with other methods which determine stability in switched converters. One of them is the computation of characteristic multipliers based on the jacobian matrix. Another one is the computation of Lyapunov exponents using a numeric routine. Each method gives equivalent information. However Floquet approach is the most appropriated when delays appear since this method does not require the evaluation of the jacobian matrix (its dimension increases when the number of delays is higher). The other two methods have this disadvantage [13].

4.1. Stability of 1-periodic orbit using Jacobian matrix

The evaluation of the jacobian matrix is necessary to compute characteristic multipliers and Lyapunov exponents in PWM switched converters. The dimension of the jacobian matrix depends on the delay number considered in the control law. The order of Jacobian matrix is $2(\tau+1)$.

Poincaré map of the PWM switched converter can be used to determine the stability of 1-periodic orbit. Equation (30) presents the Poincaré map of synchronous buck converter with centered PWM control.

$$\mathbf{x}((k+1)T) = e^{\mathbf{A}T}\mathbf{x}(kT) + (e^{\mathbf{A}(T-d_k/2)} + \mathbf{I})\mathbf{A}^{-1}(e^{\mathbf{A}d_k/2} - \mathbf{I})\mathbf{B} - e^{\mathbf{A}d_k/2}\mathbf{A}^{-1}(e^{\mathbf{A}(T-d_k)} - \mathbf{I})\mathbf{B} \quad (30)$$

Real-time control law implies that the duty cycle d_k depends on state variables in the instant kT, i.e., $d_k = c_1x_1(kT) + c_2x_2(kT) + c_3$. Therefore, Poincaré map (30) can be written as follows.

$$\begin{aligned} x_1((k+1)T) &= f_1(x_1(kT), x_2(kT)) \\ x_2((k+1)T) &= f_2(x_1(kT), x_2(kT)) \end{aligned} \quad (31)$$

Jacobian matrix of the system with $\tau = 0$ can be computed with equation (32).

$$\mathbf{A}_{n0} = \begin{bmatrix} \frac{\partial f_1}{\partial x_1(kT)} & \frac{\partial f_1}{\partial x_2(kT)} \\ \frac{\partial f_2}{\partial x_1(kT)} & \frac{\partial f_2}{\partial x_2(kT)} \end{bmatrix} \quad (32)$$

The matrix $\mathbf{A}_{n0}$ should be evaluated in the fixed point ($\mathbf{A}_{n0} = \left(\frac{\partial \mathbf{f}}{\partial \mathbf{x}_i}\right)_{F.P.}$). In this case, we define $F.P.$ as $(ref, \gamma ref)$. Its eigenvalues (or characteristic multipliers) determine stability properties of the fixed point. The 1-periodic orbit is asymptotically stable if all characteristic multipliers have magnitude less than one ($|m_i| < 1$); it is unstable if at least one eigenvalue has magnitude greater than one ($|m_i| > 1$).

One-delay control law implies that the duty cycle d_k depends on state variables in the instant $(k-1)T$, i.e., $d_k = c_1 x_1((k-1)T) + c_2 x_2((k-1)T) + c_3$. Two additional state variables can be defined $x_3(kT) = x_1((k-1)T)$ and $x_4(kT) = x_2((k-1)T)$. Therefore, $d_k = c_1 x_3(kT) + c_2 x_4(kT) + c_3$.

In this case, Poincaré map (30) can be written as equation (33).

$$\begin{aligned} x_1((k+1)T) &= f_1(x_1(kT), x_2(kT), x_3(kT), x_4(kT)) \\ x_2((k+1)T) &= f_2(x_1(kT), x_2(kT), x_3(kT), x_4(kT)) \\ x_3((k+1)T) &= x_1(kT) \\ x_4((k+1)T) &= x_2(kT) \end{aligned} \tag{33}$$

Jacobian matrix of the system with $\tau = 1$ can be computed with equation (34).

$$\mathbf{A}_{n1} = \begin{bmatrix} \frac{\partial f_1}{\partial x_1(kT)} & \frac{\partial f_1}{\partial x_2(kT)} & \frac{\partial f_1}{\partial x_3(kT)} & \frac{\partial f_1}{\partial x_4(kT)} \\ \frac{\partial f_2}{\partial x_1(kT)} & \frac{\partial f_2}{\partial x_2(kT)} & \frac{\partial f_2}{\partial x_3(kT)} & \frac{\partial f_2}{\partial x_4(kT)} \\ 1 & 0 & 0 & 0 \\ 0 & 1 & 0 & 0 \end{bmatrix} \tag{34}$$

The matrix $\mathbf{A}_{n1}$ should be evaluated in the fixed point ($\mathbf{A}_{n1} = \left(\frac{\partial \mathbf{f}}{\partial \mathbf{x}_i}\right)_{F.P.}$). In this case, we define $F.P.$ as $(ref, \gamma ref, ref, \gamma ref)$.

Four characteristic multipliers are computed. The 1-periodic orbit is asymptotically stable if the four characteristic multipliers have magnitude less than one.

Characteristic multipliers for delayed PWM control law with $\tau > 1$ can be computed following the same procedure. However, The order of Jacobian matrix increases as the delay number grows.

Now, we compute Lyapunov exponents using a numeric routine. This algorithm is based on the definition of Lyapunov exponents. Equation (35) synthesizes this procedure. Poincaré map is used to compute the values of state variables. Jacobian matrix should be known to compute the eigenvalues q_i in each iteration k.

$$\lambda_i = \underset{M\to\infty}{Lim} \left\{ \frac{1}{M} \sum_{k=0}^{M} \log \left| q_i \left(A_{nj} \left(x(k) \right) \right) \right| \right\} \tag{35}$$

4.2. Equivalence between stability methods

Floquet exponents, Lyapunov exponents and characteristic multipliers are interconnected to each other. Mathematical expressions to relate each approach are synthesized in figure 5. For example, we can compute the Floquet exponents (μ_i) for any delay and later we can apply the relations $m_i = e^{\mu_i T}$ and $\lambda_i = \mu_i T$ to find characteristic multipliers and Lyapunov exponents, respectively.

Figure 6 shows the evolution of Floquet and Lyapunov exponents when the duty cycle is computed without delay and with one delay. The critic values are the equals using any method. Therefore, both methods give the same information.

Figure 7 shows the evolution of Floquet exponents and characteristic multipliers in the complex plane of each representation. In both cases, the parameter N is varied in the range $[0; 30]$ with $k_s = 4.5$ and $\tau = 1$. Imaginary axis is the stability limit of the Floquet exponents locus, while unity circle is the stability limit of the characteristic multipliers locus.

5. Bifurcations in Buck converter with delayed ZAD-FPIC

In this section, we analyze types of bifurcations in the buck converter controlled with Delayed ZAD-FPIC scheme using the procedure based on Floquet exponents described in previous sections. We transform Floquet exponents in characteristic multipliers using the equivalences shown in figure 5.

If at least one characteristic multiplier is outside of the unit circle then the system has an unstable fixed point and nonlinear phenomena as quasi-periodicity and chaos could be present. In the boundary, the smooth bifurcations (flip, fold and Neimark-Sacker) are present. The presence of the three smooth bifurcations in the same converter is not common and this fact has not been reported widely in Digital-PWM switched converters [14].

Control parameters k_s and N can be varied in $\mathbb{R}$ with the exception of $k_s = 0$ and $N+1 = 0$ (because the control law is not defined there). Parameter τ can be varied in $\mathbb{Z}$. The 3D-parameter space (k_s, N, τ) is discontinuous due to the discrete delays ($\tau = 0, 1, 2, 3,$) and the undefined planes ($k_s = 0$ and $N + 1 = 0$). Figure 8 (a) shows a representation of the control parameter space.

The two-dimensional plane (k_s, N) can be divided into four regions: region I: $k_s > 0$ and $N > -1$; region II: $k_s < 0$ and $N > -1$; region III: $k_s < 0$ and $N < -1$; and region IV: $k_s > 0$ and $N < -1$. Fold zones, flip zones and Neimark-Sacker zones can be identified in the control space. The fold bifurcation is an alarm for duty cycle saturation in $d = 0\%$ or $d = 100\%$; the flip bifurcation signals a doubling period and the Neimark-Sacker bifurcation is related to 2D-torus birth.

Computer simulations are given for the purpose of illustration and verification. Next, we present the three bifurcations types in the 3D-parameter space.

5.1. Flip bifurcations in (k_s, N, τ) space

This bifurcation is associated with the appearance of a negative real characteristic multiplier in the unit cycle boundary ($m_i = -1$). Figure 9 (a) shows the evolution of characteristic multipliers when N is varied in a positive range for several k_s values.

5.1.1. Control subspaces

The flip bifurcations have been detected in the following $(k_s - N)$ plane: Subspace I when $\tau = 0$. The flip zone in the plane $\tau = 0$ is presented in figure 8 (b).

5.1.2. Characteristics near to flip bifurcation

Before the flip bifurcation the converter has a stable fixed point or T-periodic orbit. After the flip bifurcation the T-periodic orbit is unstable and the converter has a stable 2T-periodic orbit. Successive flip bifurcations and border-collision bifurcations are presented until the chaos formation. More details can be found in [2], [4], [15]. An illustrative example is shown in figure 10 (a).

5.2. Fold bifurcations in (k_s, N, τ) space

This bifurcation is associated with the appearance of a positive real characteristic multiplier in the unit cycle boundary $(m_i = 1)$. Figure 9 (b) shows the evolution of characteristic multipliers when N is varied in a negative range for several k_s and τ values.

5.2.1. Control subspaces

The fold bifurcations have been detected in the following $(k_s - N)$ planes: Subspaces II and IV for (see figure 8 (c)).

5.2.2. Characteristics near to fold bifurcation

Before the fold bifurcation the converter has two fixed points: one stable and other unstable. The stable fixed point is near to reference value. After the fold bifurcation the converter has not fixed points and the output is saturated. An illustrative case is presented in figure 10 (b).

5.3. Neimark-Sacker bifurcations in (k_s, N, τ) space

This bifurcation is associated with the appearance of two conjugate complex characteristic multipliers in the unit cycle boundary. Figure 9 (c) shows the evolution of characteristic multipliers when N and k_s are varied in positive ranges for $\tau > 1$.

5.3.1. Control subspaces

The Neimark-Sacker bifurcations have been detected in the following $(k_s - N)$ planes: Subspace III for $\tau = 0$ and Subspace I and III for $\tau > 0$. The control subspaces with Neimark-Sacker bifurcations are presented in figure 8 (d).

5.3.2. Characteristics near to Neimark-Sacker bifurcation

Before the Neimark-Sacker bifurcation the converter has a stable fixed point or T-periodic orbit. After the Neimark-Sacker bifurcation the converter has quasi-periodic behavior and 2D-torus birth. The bifurcation diagram and the characteristic multipliers in the Neimark-Sacker transition are shown in figure 10 (c).

6. Conclusions and future work

In this chapter, we have presented a generalized procedure to compute Floquet exponents for any number of delays (τ) in the control law of PWM switched converters. We have investigated the incidence of *delays* in a digital-PWM controller based on two novel techniques: *Zero Average Dynamics* (ZAD) and *Fixed-Point Inducting Control* (FPIC). Principles of *Floquet theory* and *smooth bifurcation theory* were used to define stability regions. The incidence of control parameters in transient response of PWM switched converters will be analyzed in a future work using a similar Floquet-based procedure.

Author details

John Alexander Taborda
Universidad del Magdalena - Facultad de Ingeniería - Programa de Ingeniería Electrónica - Magma Ingeniería - Santa Marta D.T.C.H., 2121630, Colombia

Fabiola Angulo and Gerard Olivar
Universidad Nacional de Colombia - Sede Manizales - Facultad de Ingeniería y Arquitectura - Departamento de Ingeniería Eléctrica, Electrónica y Computación - Percepción y Control Inteligente - Bloque Q, Campus La Nubia, Manizales, 170003 - Colombia

7. References

[1] Angulo F., "Dynamical analysis of PWM-controlled power electronic converters based on the zero average dynamics (ZAD) strategy", *Ph. D. Thesis. Technical University of Catalonia (in Spanish)*, 2004.
[2] Angulo F., Fossas E. & Olivar G. "Transition From Periodicity to Chaos in a PWM-Controlled Buck Converter with ZAD strategy", *International Journal of Bifurcations and Chaos*, 15, 10, 3245–3264, 2005.
[3] Angulo F., Ocampo C., Olivar G. & Ramos R., *Nonlinear and Nonsmooth Dynamics in a DC-DC Buck Convreter: Two experimental set-ups* Nonlinear Dynamics, Vol. 46, No. 3, pp. 239-257, 2006.
[4] Angulo F., Olivar G. & Taborda J.A., "Continuation of periodic orbits in a ZAD-strategy controlled buck converter", *Chaos, Solitons and Fractals*, 38, 348–363, 2008.
[5] Angulo F., Hoyos F.E. & Taborda J.A., "Principios de la Estrategia de Control Zero Average Dynamics (ZAD)" (in Spanish). Lambert Academic Publishing, Madrid, España, 2011.
[6] Awrejcewicz J., "Bifurcation and Chaos in Simple Dynamical Systems". World Scientific, Singapore, 1989.
[7] Awrejcewicz J., "Three routes to chaos in simple sinusoidally driven oscillators", Journal of Applied Mathematics and Mechanics ZAMM, 71 (2), 1991, 71-79.
[8] Awrejcewicz J., "Numerical analysis of the oscillations of human vocal cords", Nonlinear Dynamics, 2, 1991, 35-52.
[9] Awrejcewicz J., Kudra G. & Lamarque C-H., "Investigation of triple pendulum with impacts using fundamental solution matrices", International Journal of Bifurcation and Chaos, 14 (12), 2004, 4191-4213.

[10] Fossas E. & Zinober A., *Adaptive tracking control of nonlinear power converters* In Proceedings IFAC Workshop on Adaptation in Control and Signal Processing. Connobio. Italy. pp 264-266, 2001.
[11] Hart D.,"Power Electronics". Prentice Hall, Madrid, España, 2001.
[12] Mohan N., Undeland T. & Robbins W., "Power Electronics: Converters, applications and design", J. Wiley, 1995.
[13] Taborda J.A., "Bifurcation analysis in second-order systems with PWM based on zero average dynamics (ZAD) strategy.", *Master Thesis. National University of Colombia (in Spanish)*, Manizales, Colombia, 2006.
[14] Taborda J.A., Angulo F. & Olivar G., "Smooth Bifurcations in 3D-parameter Space of Digital-PWM Switched Converter", In Proceedings of 2nd IEEE Latin American symposium on circuits and Systems - LASCAS 2011. Bogotá. Colombia, 2011.
[15] Taborda J.A., Angulo F. & Olivar G., "Characterization of chaotic attractors inside band-merging scenario in ZAD-controlled buck converter", Special Issue of 2010 DDays, *International Journal of Bifurcations and Chaos*, (to appear) .

Chapter 3

On an Overview of Nonlinear and Chaotic Behavior and Their Controls of an Atomic Force Microscopy (AFM) Vibrating Problem

José Manoel Balthazar, Angelo Marcelo Tusset, Atila Madureira Bueno and Bento Rodrigues de Pontes Junior

Additional information is available at the end of the chapter

http://dx.doi.org/10.5772/51834

1. Introduction

It is known that in Brazil and in the whole world, the development of mathematical modeling techniques for the study of problems involving new technologies and the design of nonlinear structures is an emerging area in Engineering Science. To background to the general problems of non-linear dynamics, see the comprehensive monographs: (Awrejcewicz, 1991) and (Awrejcewicz and Lamarque, 2003).

This chapter addresses this issue, involving the dynamical behavior of Electrical Mechanical Systems (EMS), that are systems responsible for electromechanical energy conversion. It is common knowledge that an electromechanical system can be classified in three different groups, according to scales dimensions: *i*) Macro-Systems >1mm, *ii*) Micro-Systems <1mm and >0.1μm, and *iii*) Nano-Systems <0.1 μm and >0.1nm.

The Macro-Systems are characterized by being visible without using microscope and they are present in our Day to Day activities. The Micro-Systems are generally constructed on silicon chips, and are employed in manufacturing integrated electronic circuits. The Nano-Systems can only be seen and manipulated with the use of the Atomic Force Microscope.

In this research field - MEMS and NEMS -, the researches are based on the fact that, in opposition to traditional macro-scale problems, the atomic forces effects and surface phenomena are preponderant over the forces of inertia and gravity. It is also known that the manufacture of these new products have been based on the binomial consuming time and an enormous amount of financial resources. Thus, it becomes necessary some investment in an interface, which addresses the design and manufacturing that, gives designers of

INTECH
open science | open minds

© 2012 The Author(s). Licensee InTech. This chapter is distributed under the terms of the Creative Commons Attribution License http://creativecommons.org/licenses/by/3.0), which permits unrestricted use, distribution, and reproduction in any medium, provided the original work is properly cited.

"MEMS" and "NEMS" the proper tools of mathematical modeling and simulations as well as qualitative dynamic analysis. It should be considering a macro-model, which agrees with the simulation results on a physical level, and with the experimental results, obtained from test structures in a lab. Many surveys have been developed in major research centers in Brazil and throughout the world in this direction. It is announced that several research groups in Brazil are involved with this theme, but most of the research carried out or in development focuses on experimentation or in the use of the numerical method - called finite element method. It is known that the nonlinear phenomena are prevalent in MEMS and NEMS, which is a strong motivation for the work of our research group.

The main idea of this chapter lies in the study of "State of the Art" of non-linear dynamic models and the design of their controls, since this new area of research is of paramount importance. This chapter deals with two boundary problems belonging to this line of research, in development at the university of Rio Claro and Bauru UNESP, for us, through research groups and development projects for Graduate (Masters) and Post–Doctor. Mainly for the Atomic Force Microscope.

2. General aspects of the Atomic Force Microscope (AFM)

In Binnig et al. (1986) the atomic force microscope ("AFM") was presented. The AFM has been described as one of the most efficient tools for obtaining high-resolution images of the samples, by exploiting its surface, both in air and in liquid media. At the core of the "AFM", there is a tip, mounted on the end of a micro cantilever, while the tip vibrates to scan the sample, its vibration is detected by a laser system, which emit signals to a photo detector that generate images of the object under examination at high resolution. This movement may vary depending on the need and on the type of material being analyzed. These variations include techniques such as static contact mode and dynamic techniques, such as non-contact mode and intermittent or tapping mode. The principle of operation of the "AFM" is to the measure the deflection of the microcantilever on whose free end the probe is mounted. The deflections (analyzed during the scan) are caused by forces acting between the tip and sample. These forces act on medium to large distances - typically ≥ 100 Å – such as attractive Van der Waals forces, magnetic forces and Coulomb forces (Eisenschitz, 1930). It can be said that when the tip approaches to the sample, it is first attracted by the surface, due to a wide range of attractive forces in the region. This attraction increases when the probe is very close to the sample. However, when the atoms of the tip and sample become very close, the orbital electrons begin to repel. The forces are canceled when the distance between the atoms is of the order of several angstroms (the distance of the order characteristic of a chemical bond). When the forces become positive, it can be said that the atoms of the tip and the sample are in contact and ultimately the repulsive forces dominate.

In Figures: 1,2 and 3, the main characteristics and the operation of an AFM can be observed. Figure 3; display the basic configuration of an AFM. The micro cantilever is V-shaped or rectangular with a sharp tip. This tip is usually of pyramidal or conical shape (Figure 2).

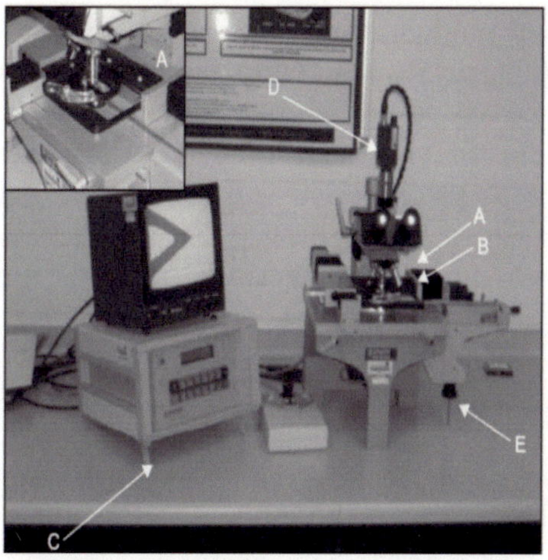

Figure 1. AFM photo (source: Bowen and Hilal 2009)

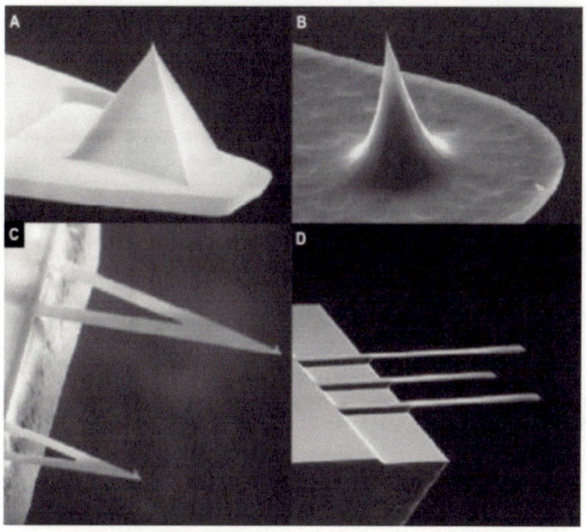

Figure 2. (source: Bowen and Hilal 2009)

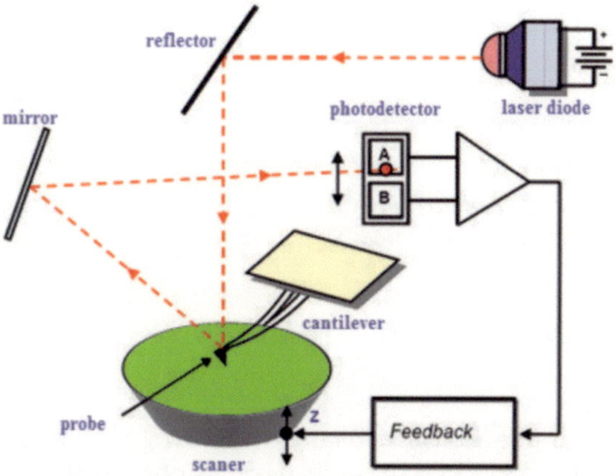

Figure 3. Control of the positioning of the scanner. Adapted from : (Cidade et al. 2003)

Typically, probes are made predominantly of silicon nitride (Si3N4); its upper surface is coated with a thin reflective surface, generally gold (Au) or aluminum (Al). The probe is brought to inside and out of contact with the sample surface, by using a piezo-crystal (Hilal and Bowen, 2009). In the illustration of a typical micro-manipulator, shown in Figure 3, it is possible to observe that it consists of a movable stage mounted under an optical microscope. The movement of the base can be controlled by an electronic control console. Large deflections are required to achieve high sensitivity. Therefore, the spring should be very "soft" (slightly stiff).

2.1. Operation modes in AFM

The probe mounted on the AFM performs the scan on the sample in a raster fashion. The movement of the microcantilever over the sample is carried out by the piezoelectric scanner, which comprises piezoelectric material that expands and contracts according to the applied voltage. There are several modes of operation for scanning and mapping surface. These modes include non-contact, contact and intermittent contact modes. These three modes of operation differ from each other, basically, by the tip and sample distance.

The Lennard-Jones potential describes the relationship of the tip and sample interaction forces as depending on the tip and sample surface distance, considering the potential energy of a pair of particles, and is given by:

$$U(r) = 4\varepsilon\left[\left(\frac{\sigma}{r}\right)^{12} - \left(\frac{\sigma}{r}\right)^{6}\right] \tag{1}$$

where ε and σ are constants depending on the sample properties, σ is approximately equal to the diameter of the particles involved. Deriving potential function (U) in relation to the distance (r) gives an expression for the force (F) versus distance (r) (Equation (1a)). This force is represented in Figure 4.

$$F(r) = -\frac{\partial U}{\partial r} = 24\varepsilon\left[\frac{2\sigma^{12}}{r^{13}} - \frac{\sigma^{6}}{r^{7}}\right] \tag{1a}$$

The region above the r-axis corresponds to the region where the repulsive forces dominate (contact region). The region below the r-axis corresponds to the region where attractive forces dominate (non-contact region). Also in red, it can be seen the distance region that the tapping-mode technique is applied.

3. Mathematical modeling of the Atomic Force Microscope (AFM)

The mathematical models governing the dynamics of the AFM micro-beams, usually result from the discretization of the classical beam equations, based on its mode of vibration, leading to one or more degrees of freedom. Several models of this kind are described in the literature, e.g., Wang et al. (2009), Garcia and San Paulo, (2000); (Laxminarayana and Jalili,

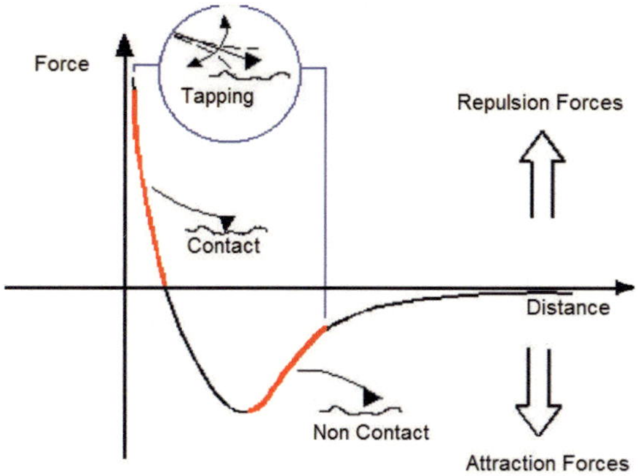

Figure 4. Force (F) versus Distance (r), (Source: Cidade et al., 2003).

2004), (Hu and Raman, 2007) (Raman et al. 2008), (Ashab et al. 1999), (Farrokh et al. 2009), (Lozano and Garcia, 2008) among others. Most of the mathematical models are linear mass-spring-damper systems incorporating nonlinear interaction between the tip and sample (Paulo & Garcia, 2002). Different AFM techniques provide a number of possibilities for topographical images of the samples, generating a wide range of information. In this chapter models for intermittent contact mode are discussed.

3.1. Mathematical modeling of AFM: with inclusion of the cubic (spring) term

The physical model of the AFM tip-sample interaction can be considered as shown in Figure 5 (Wang, Father and Yau, 2009). The microcantilever-tip-sample system is regarded as a sphere of radius R_s and mass m_s, suspended by a spring of stiffness $k = k_{l_s} + k_{nl_s}$, where k_{l_s} and k_{nl_s} are the linear and nonlinear stiffness. The van der Waals potential for the sample-sphere system is given by:

$$P = -\frac{A_c R_c}{6\left(Z_b + X\right)} + \frac{1}{2} k_{l_s} X^2 + \frac{1}{4} k_{nl_s} X^4 \tag{2}$$

The energy of the system scaled by the mass of the cantilever is given by $E(X, X', Z)$:

$$E = \frac{1}{2}\dot{X}^2 + \frac{1}{2}\omega_1^2 X^2 + \frac{1}{4}\omega_2^2 X^4 - \frac{D\omega_1^2}{(Z_b - X)} \tag{3}$$

Replacing $X_1 = X$ and $X_2 = \dot{X}$, then, from equation (3) results:

$$\begin{aligned} X'_1 &= \frac{\partial E}{\partial X_2} \\ X'_2 &= -\frac{\partial E}{\partial X_1} \end{aligned} \tag{4}$$

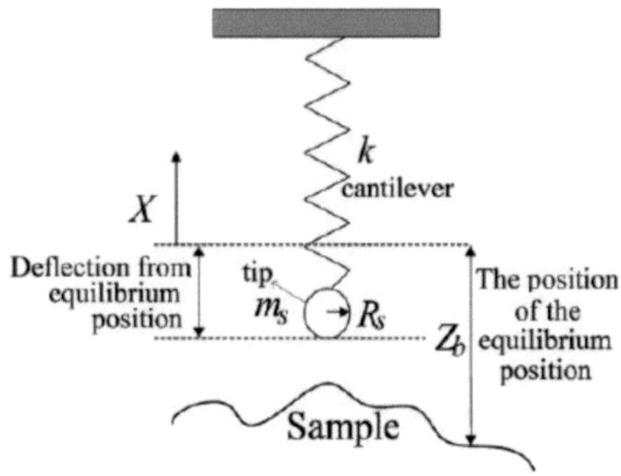

Figure 5. Model of an AFM (Source: Wang, Father and Yau (2009))

The dynamic AFM system in Figure 5 is obtained replacing (2) and (3) into (4):

$$\begin{cases} \dot{X}_1 = X_2 \\ \dot{X}_2 = -\omega_1^2 X_1 - \omega_2^2 X_1^3 - \dfrac{D\omega_1^2}{(Z_b + X_1)^2} \end{cases} \tag{5}$$

where Z_b the distance from the equilibrium position. The molecular diameter is $D = \dfrac{A_H R}{6k}$, where A_h is the Hamaker constant and R is the sphere radius. Considering only attractive Van der Waals force, and that the cantilever is being excited by $mf\cos(wt)$, where w is the natural frequency, the system equations can be written as:

$$\begin{aligned} \dot{X}_1 &= X_2 \\ \dot{X}_2 &= -\omega_1^2 X_1 - \omega_2^2 X_1^3 - \frac{D\omega_1^2}{\left(Z_b - X_1\right)^2} - f\cos wt - \phi X_2 \end{aligned} \tag{6}$$

Where $\phi X_2 = \left(a_4 \cos b - a_5 X_2\right)'$ is the damping force. Considering the relations: $x_1 = \dfrac{X_1}{Z_s}$, $x_2 = \dfrac{X_2}{\omega_s Z_s}$, $z = \dfrac{Z_b}{Z_s}$, $Z_s = \dfrac{3}{2}(2D)^{\frac{1}{3}}$ and $\tau = wt$, the system (6) may be rewritten in the following dimensionless form:

$$\begin{aligned} \dot{x}_1 &= x_2 \\ \dot{x}_2 &= -a_1 x_1 - a_2 x_1^3 - \frac{b}{\left(z + x_1\right)^2} + c\sin\tau \end{aligned} \tag{7}$$

3.2. AFM Mathematical modeling: intermittent mode and hydrodynamic damping

The microcantilever schematic diagram of the AFM operating in intermittent mode can be seen in Figure 6. The base of the microcantilever is excited by a piezoelectric actuator generating a displacement $f\cos(wt)$. According to (Zhang et al., 2009), considering only the first vibration mode, the (AFM) can be modeled as a spring-mass-damper, as shown in Figure 7. The "tip" is considered as being a of radius R and Z_0 is the distance from the equilibrium position of the cantilever to the sample. The position of the cantilever is given by x, measured from the equilibrium position. According to (Rutzel et al., 2003) the tip-sample interaction can be modelled as a sphere- flat surface interaction, given by:

$$U(x,z_0)=\frac{A_1R}{1260\left(z_0+x\right)^7}-\frac{A_2R}{6\left(z_0+x\right)} \tag{8}$$

where $U(x,z_0)$ is the Lennard-Jones potential(LJ), $A_1=\pi^2\rho_1\rho_2c_1$ and $A_2=\pi^2\rho_1\rho_2c_2$ are the Hamaker constant for the attractive and repulsive potential, respectively, with ρ_1 and ρ_2 the densities of the interacting components, and c_1 and c_2 are constants from interaction. It should be noted that, when the "cantilever" is close to the sample, attractive van der Waals

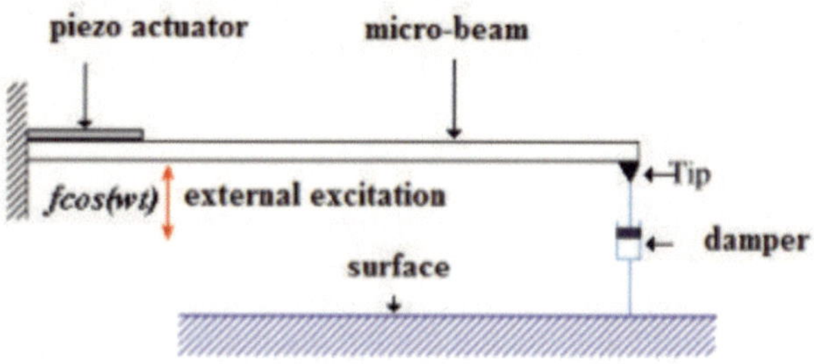

Figure 6. Microcantilver-tip-sample system

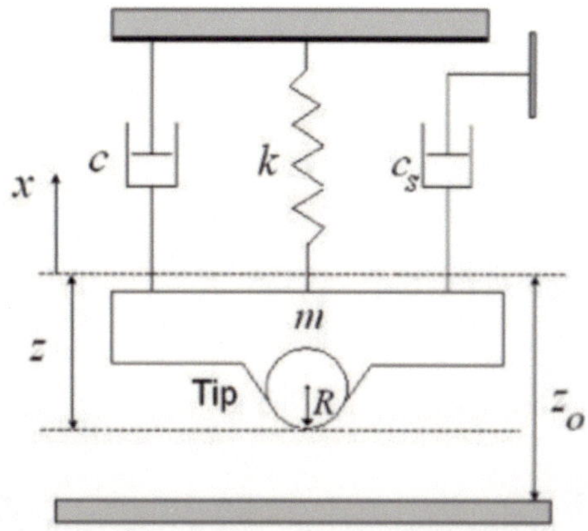

Figure 7. Physical model (Source: Zhang et al., 2009)

force must be considered. These forces can be represented as the sum of the attractive and repulsive forces (Rutzel et al., 2003), expressed by:

$$F = -\frac{\partial U}{\partial(x+z_0)} = \frac{A_1 R}{180\left(z_0+x\right)^8} - \frac{A_2 R}{6\left(z_0+x\right)^2} \tag{9}$$

In the intermittent mode (TM-AFM) the probe touches the surface of the sample at the point of maximum amplitude of oscillation. During the scanning the microcantilever is driven to oscillate according to the force $f\cos wt$, resulting that the tip-sample contact generates the force F. The contact between the tip and sample is delicate, and this mode of operation is suitable for fragile samples. Then, considering the Lagrangian $L = T - V$ and the Euler-Lagrange equation $\frac{d}{dt}\left(\frac{\partial L}{\partial \dot{q}_i}\right) - \frac{\partial L}{\partial q_i} = Q_i$, where

$$T = \frac{1}{2} m\dot{x}^2,\; V = \frac{1}{2}k_l x^2 + \frac{1}{4}k_{nl}x^4 \text{ and } Q_i = F + c\dot{x} + c_s\dot{x} + f\cos wt \tag{10}$$

are the kinetic energy, the gravitational potential energy, and nonconservative forces, respectively, the equation of motion for the microcantilever tip displacement x is given by:

$$m\ddot{x} + c\dot{x} + k_l x + k_{nl}x^3 = \frac{A_1 R}{180\left(z_0+x\right)^8} - \frac{A_2 R}{6\left(z_0+x\right)^2} + \frac{\mu_{eff}B^3 L}{\left(x+z_0\right)^3}\dot{x} + f\cos wt \tag{11}$$

with:

$$c_s\dot{x} = \frac{\mu_{eff}B^3 L}{\left(x+z_0\right)^3}\dot{x} \tag{12}$$

where μ_{eff} is an effective coefficient of viscosity, B the width of the cantilever and L is the length of the cantilever (Zhang et al. (2009)).

Defining:

$$Z_s = \left(\frac{2}{3}\right)(2D)^{1/3} \tag{13}$$

where $D = \frac{A_2 R}{6k}$ and considering the following relations in (11)

$\tau = w_1 t$, $y = \frac{x}{z_s}$, $\dot{y} = \frac{\dot{x}}{w_1 z_s}$, $a = \frac{z_0}{z_s}$, $b = \frac{k_l}{k_{at}}$, $c = \frac{k_{nl}}{k_{at}} z_s^2$, $d = \frac{4}{27}$, $e = \frac{2}{405}\left(\frac{a}{z_s}\right)^6$, $g = \frac{f_0}{k_{at} z_s}$, $p = \frac{\mu_{eff}B^3 L}{m w_1 z_s^3}$, $\Omega = \frac{w}{w_1}$, $r = \frac{1}{Q}$. Equation (11) can be rewritten in the dimensionless form:

$$\ddot{y}+r\dot{y}+by+cy^3=-\frac{d}{(a+y)^2}+\frac{e}{(a+y)^8}+g\cos\Omega\tau-\frac{p}{(a+y)^3}\dot{y} \tag{14}$$

Writing equation (14) into state space form results:

$$\begin{aligned}\dot{x}_1&=x_2\\ \dot{x}_2&=-rx_2-bx_1-cx_1^{\,3}-\frac{d}{(a+x_1)^2}+\frac{e}{(a+x_1)^8}+g\cos\Omega x_3-\frac{p}{(a+x_1)^3}x_2\end{aligned} \tag{15}$$

where: $x_1=y$, $x_2=\dot{y}$ and $x_3=\tau$.

Considering the values of parameters: $\Omega=1$; $\mathrm{r}=0.1$; $\mathrm{b}=1$; $\mathrm{c}=0.35$; $\mathrm{d}=4/27$; $\mathrm{e}=0.0001$; $\mathrm{g}=0.2$; $\mathrm{p}=0.005$ e $\mathrm{a}=1.6$ (obtained by Zhang et al. (2009)). The displacement can be seen in Figure 8a and the phase portrait can be seen in Figure 8b.

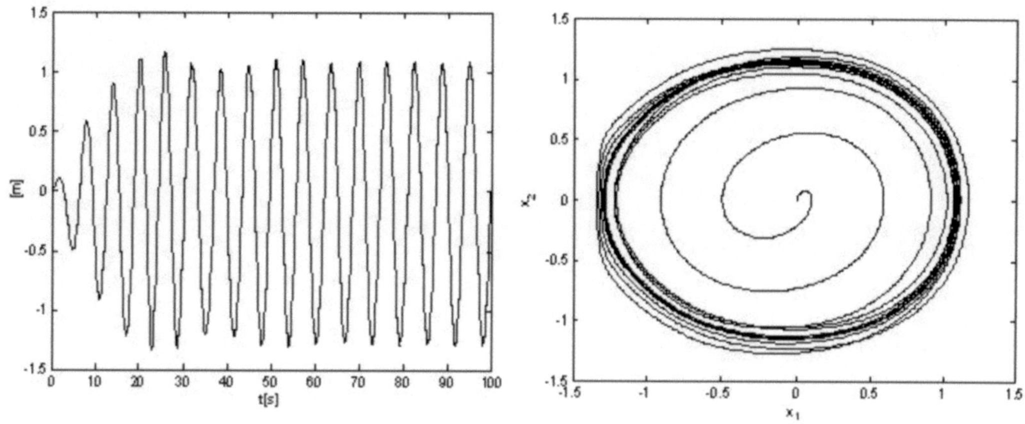

Figure 8. (a): Tip displacement (b): Phase Portrait

4. Chaos in mathematical model with cubic spring

According to Ashhab (1999) the chaotic behavior of AFM depends on the damping of the excitation and on the distance between the tip and the sample, suggesting that a feedback control of the states can be used to eliminate the possibility of chaotic behavior. According to that, considering the system (7) in nondimensional form:

$$\begin{aligned}\dot{x}_1&=x_2\\ \dot{x}_2&=-\alpha a_1x_1-\alpha a_2x_1^3-\frac{b}{(z+x_1)^2}+\delta\sin\tau\end{aligned} \tag{16}$$

with the parameters: $\alpha=0.14668$; $b=0.17602$, $\delta=2.6364$ e $z=2.5$, $a_1=1$, $a_2=14.5$, numerically simulation results can be seen Figure 9. Additionally, The FFT and the Lyapunov exponents ($\lambda_1=0.336$, $\lambda_2=-0.336$) are shown in Figure 10.

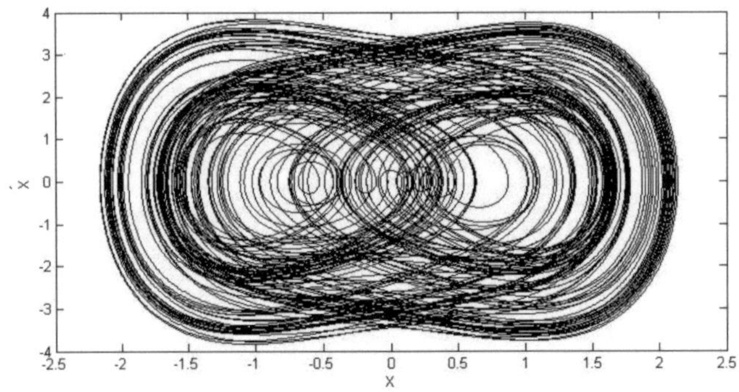

Figure 9. Phase diagram

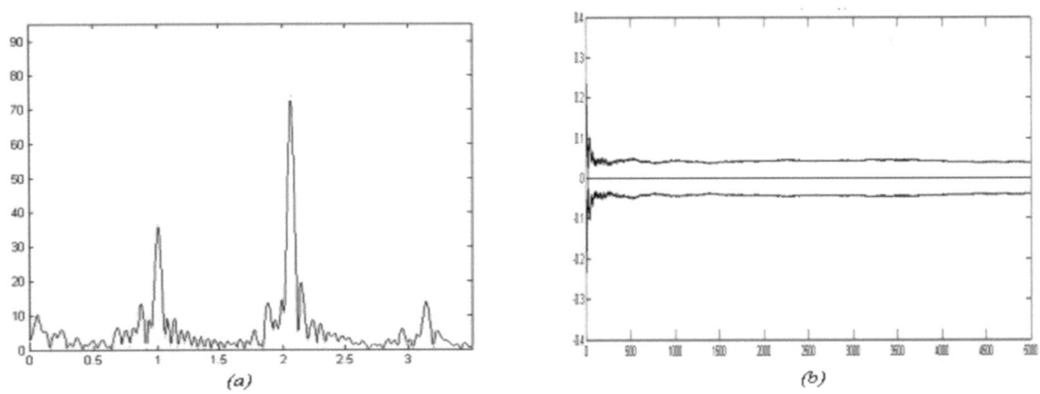

Figure 10. (a): FFT, (b): Lyapunov exponent

5. Chaos with hydrodynamic damping in TM-AFM

The elastic constant of the cantilever k_c must be less than the effective elastic constant of the interatomic coupling k_{at} of the sample. Thus the elastic constant of the spring must be $K < K_{at}$, with $K_{at} = w_{at}^2 m_{at}$. Typical atomic vibration frequencies are $\omega_{at} = 10^{13} Hz$ and atomic masses are of order 10^{-25} kg and $K < 10$ [N / m]. Considering the case of $K < K_{at}$ and rewriting the equation into state space results:

$$\begin{aligned} \dot{x}_1 &= x_2 \\ \dot{x}_2 &= -rx_2 - bx_1 - cx_1^{\,3} - \frac{d}{\left(a+x_1\right)^2} + \frac{e}{\left(a+x_1\right)^8} + g\cos\Omega x_3 - \frac{p}{\left(a+x_1\right)^3} x_2 \end{aligned} \tag{17}$$

where $x_1 = y$, $x_2 = \dot{y}$. The phase diagram can be observed in Figure 11a. For the parameters values: $\Omega = 1$; $\mathrm{r} = 0.1$; $\mathrm{b} = 0.05$; $\mathrm{c} = 0.35$; $\mathrm{d} = 4/27$; $\mathrm{e} = 0.0001$; $\mathrm{g} = 0.2$; $\mathrm{p} = 0.005$ and $\mathrm{a} = 1.6$. The Lyapunov exponents ($\lambda_1 = -0.23$, $\lambda_2 = 0$; $\lambda_3 = 0.1$), can be seen in Figure 11b, indicating that the system has a chaotic attractor.

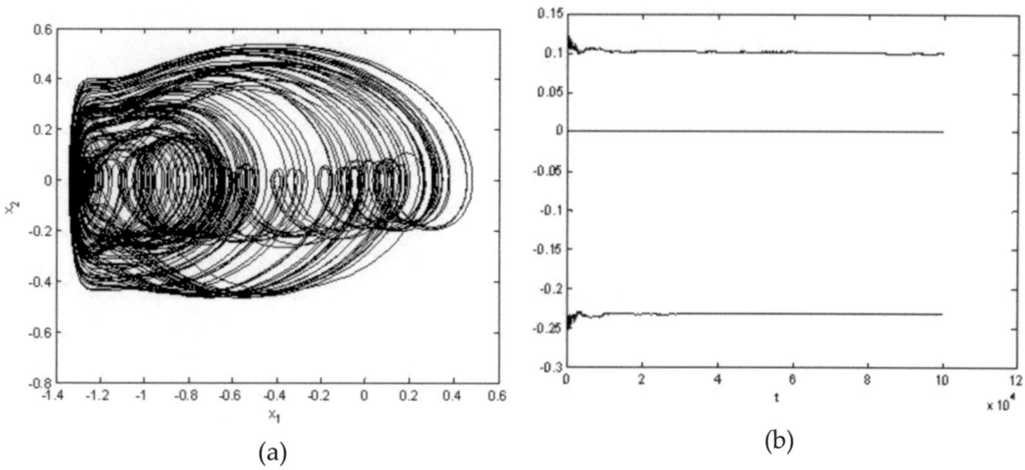

Figure 11. (a): Phase diagram (b): Lyapunov exponents

6. Scanner position control

A laser beam focus on the top of the microcantilever and the reflection is detected by a photodiode. The light is converted into an electrical signal, and stored in the computer as a reference. An oscillation of the microcantilever deflects the laser beam on the photodiode, allowing the system to compute the microcantilever motion. The error signals are then forwarded, and the piezoelectric scanner moves vertically to scan the sample, as shown in Figure 3. The control techniques are diverse: PID or PD, sliding mode, LQR, or other.

6.1. Feedback control for the model with hydrodynamic damping

Considering the model (17) with the inclusion of control: F_u .

$$\begin{aligned} \dot{x}_1 &= x_2 \\ \dot{x}_2 &= -rx_2 - bx_1 - cx_1{}^3 - \frac{d}{\left(a+x_1\right)^2} + \frac{e}{\left(a+x_1\right)^8} + g\cos\Omega x_3 - \frac{p}{\left(a+x_1\right)^3} x_2 + F_u \end{aligned} \tag{18}$$

and defining a periodic orbit as a function of $\tilde{x}(t)$. The desired regime is given by:

$$\ddot{\tilde{y}} = -r\dot{\tilde{y}} - b\tilde{y} - c\tilde{y}^3 - \frac{d}{\left(a+\tilde{y}\right)^2} + \frac{e}{\left(a+\tilde{y}\right)^8} + g\cos\Omega\tau - \frac{p}{\left(a+\tilde{y}\right)^3}\dot{\tilde{y}} + \tilde{u} \tag{19}$$

Since $\tilde{u}$ control the system in the desired trajectory, and $\tilde{x}(t)$ is a solution of (19), without the term control F_u , then $\tilde{u}=0$, resulting:

$$\tilde{y} = a_0 + a_1\cos(\Omega\tau) + b_1 sen(\Omega\tau) + a_2\cos(2\Omega\tau) + b_2 sen(2\Omega\tau) + \ldots \tag{20}$$

The feedforward control $\tilde{u}$ is given by:

$$\tilde{u} = \dddot{\tilde{y}} + r\ddot{\tilde{y}} + b\tilde{y} + c\tilde{y}^3 + \frac{d}{(a+\tilde{y})^2} - \frac{e}{(a+\tilde{y})^8} - g\cos\Omega\tau + \frac{p}{(a+\tilde{y})^3}\dot{\tilde{y}} \tag{21}$$

Replacing (19) in (18) and defining the deviation from the desired trajectory as:

$$e = \begin{bmatrix} x_1 - \tilde{x}_1 \\ x_2 - \tilde{x}_2 \end{bmatrix} \tag{22}$$

The system (18) can be represented as follows:

$$\begin{aligned} \dot{e}_1 &= e_2 \\ \dot{e}_2 &= -re_2 - be_1 - c(e_1+\tilde{x}_1)^3 + c\tilde{x}_1{}^3 - \frac{d}{(a+e_1+\tilde{x}_1)^2} + \frac{d}{(a+\tilde{x}_1)^2} + \frac{e}{(a+e_1+\tilde{x}_1)^8} - \frac{e}{(a+\tilde{x}_1)^8} - \\ &\frac{p(e_2+\tilde{x}_2)}{(a+e_1+\tilde{x}_1)^3} + \frac{p\tilde{x}_2}{(a+\tilde{x}_1)^3} + u \end{aligned} \tag{23}$$

with $u = F_u - \tilde{u}$, and the feedback control: $u = -Ke$. The system (23) can be represented in deviations as:

$$\dot{e} = Ae + g(x) - g(\tilde{x}) + Bu \tag{24}$$

Representing the system (24) in the form:

$$\begin{bmatrix} \dot{e}_1 \\ \dot{e}_2 \end{bmatrix} = \begin{bmatrix} 0 & 1 \\ -b & -r \end{bmatrix} \begin{bmatrix} e_1 \\ e_2 \end{bmatrix} + g(x) - g(\tilde{x}) + \begin{bmatrix} 0 \\ 1 \end{bmatrix} u \tag{25}$$

where

$$g(x) - g(\tilde{x}) = \begin{bmatrix} 0 \\ -c\left[(e_1+\tilde{x}_1)^3 - \tilde{x}_1{}^3\right] - \frac{d}{(a+e_1+\tilde{x}_1)^2} + \frac{d}{(a+\tilde{x}_1)^2} \end{bmatrix} + \begin{bmatrix} 0 \\ \frac{e}{(a+e_1+\tilde{x}_1)^8} - \frac{e}{(a+\tilde{x}_1)^8} - \frac{p(e_2+\tilde{x}_2)}{(a+e_1+\tilde{x}_1)^3} + \frac{p\tilde{x}_2}{(a+\tilde{x}_1)^3} \end{bmatrix}$$

Defining the desired trajectory as the periodic orbit, with amplitude less than (a) and frequency equal to (Ω), then:

$$\tilde{x} = 1.3\sin(t) \tag{26}$$

Considering the parameters values: $\Omega = 1$; $\mathrm{r} = 0.1$; $\mathrm{b} = 0.05$; $\mathrm{c} = 0.35$; $\mathrm{d} = 4/27$; $\mathrm{e} = 0.0001$; $\mathrm{g} = 0.2$; $\mathrm{p} = 0.005$ e $\mathrm{a} = 1.6$ the matrices A and B, are given by:

$A=\begin{bmatrix}0 & 1\\ -0.05 & -0.1\end{bmatrix}$, $B=\begin{bmatrix}0\\1\end{bmatrix}$, and defining $Q=\begin{bmatrix}10 & 0\\ 0 & 10\end{bmatrix}$, $R=\begin{bmatrix}1\end{bmatrix}$, and using the matlab(R) the control u is obtained. In Figure 12 the tip displacement without and with control are shown.

$$u=-k_{11}e_1-k_{12}e=-3.1127e_1-3.9293e_2 \tag{27}$$

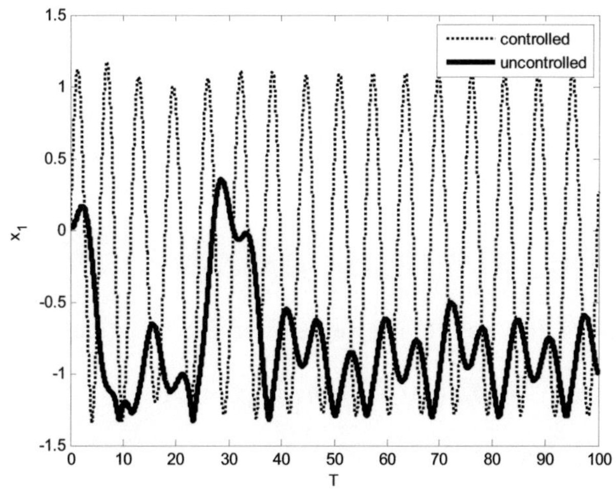

Figure 12. Displacement of the tip to the system without control and with control

6.2. Feedback control for a model with cubic spring

Considering the following parameters: $\alpha=0.14668$, $b=0.17602$, $\delta=2.6364$, $z=2.5$, $a_1=1$, $a_2=14.5$, and the control U in (16) results:

$$\begin{aligned}\dot{x}_1&=x_2\\ \dot{x}_2&=-\alpha a_1x_1-\beta a_2x_1^3-\frac{b}{\left(z+x_1\right)^2}+\delta\sin b+U\end{aligned} \tag{28}$$

where: $U=\tilde{u}+u$, u is the feedback control, $\tilde{u}$ is the feedforward control, given by :

$$\tilde{u}=\dot{\tilde{x}}_2+\alpha a_1\tilde{x}_1+\alpha a_2\tilde{x}_1^3+\frac{b}{\left(z+\tilde{x}_1\right)^2}-\delta\sin b \tag{29}$$

Replacing (29) in (28) and defining the deviation from the desired trajectory as:

$$y=(x-\tilde{x}) \tag{30}$$

where $\tilde{x}$ is the desired orbit, and rewriting the system in deviations, results:

$$\dot{y}_1 = y_2$$
$$\dot{y}_2 = -\alpha a_1 y_1 - \alpha a_2 \left(y_1 + \tilde{x}_1\right)^3 + \alpha a_2 \tilde{x}_1^{\ 3} - \frac{b}{\left(z + y_1 + \tilde{x}_1\right)^2} + \frac{b}{\left(z + \tilde{x}_1\right)^2} + u \tag{31}$$

Considering the system (31) in the following way:

$$\dot{y} = Ay + g(x) - g(\tilde{x}) + Bu \tag{32}$$

where: $y = \begin{bmatrix} y_1 \\ y_2 \end{bmatrix}; A = \begin{bmatrix} 0 & 1 \\ -\alpha a_1 & 0 \end{bmatrix}; B = \begin{bmatrix} 0 \\ 1 \end{bmatrix}$ and

$g(x) - g(\tilde{x}) = \begin{bmatrix} 0 \\ -\alpha a_2 \left(y_1 + \tilde{x}_1\right)^3 + \alpha a_2 \tilde{x}_1^{\ 3} - \dfrac{b}{\left(z + y_1 + \tilde{x}_1\right)^2} + \dfrac{b}{\left(z + \tilde{x}_1\right)^2} \end{bmatrix}$. The control u is obtained by solving the following equation:

$$u = -R^{-1} B^T P y \tag{33}$$

where P is a symmetric matrix, solution of the reduced Riccati equation:

$$PA + A^T P - PBR^{-1} B^T P + Q = 0 \tag{34}$$

Defining the desired trajectory as:

$$\tilde{x} = 2\cos(t) \tag{35}$$

The matrices $A = \begin{bmatrix} 0 & 1 \\ -0.14668 & 0 \end{bmatrix}$, $B = \begin{bmatrix} 0 \\ 1 \end{bmatrix}$, and defining Q and R as $Q = \begin{bmatrix} 250 & 0 \\ 0 & 20 \end{bmatrix}$, $R = \begin{bmatrix} 0.1 \end{bmatrix}$, results after using the matlab(R) to obtain u:

$$u = -49.8535 y_1 - 17.3120 y_2 \tag{36}$$

In Figure 13 it can observed the tip displacement with and without control.

7. AFM mathematical modeling with Phase-Locked Loops (PLLS)

As mentioned above, the Atomic Force Microscopy started in 1986 when the Atomic Force Microscope (AFM) was invented by Binnig in 1986. Since then many results have been obtained by simple contact measurements. However, the AFM cannot generate truly atomic resolution images, by simple contact measurement, in a stable operation. Besides, since 1995, using noncontact techniques, it was possible to obtain atomic resolution images, with stable operation, under attractive regime at room temperature (Giessible, 1995; Morita et. al., 2009). Noncontact AFM operates in static and dynamic modes. In the static mode the tip-sample interaction forces are translated into measured microcantilever deflections, and the image is a map $z(x,y,F_{ts})$ with F_{ts} constant.

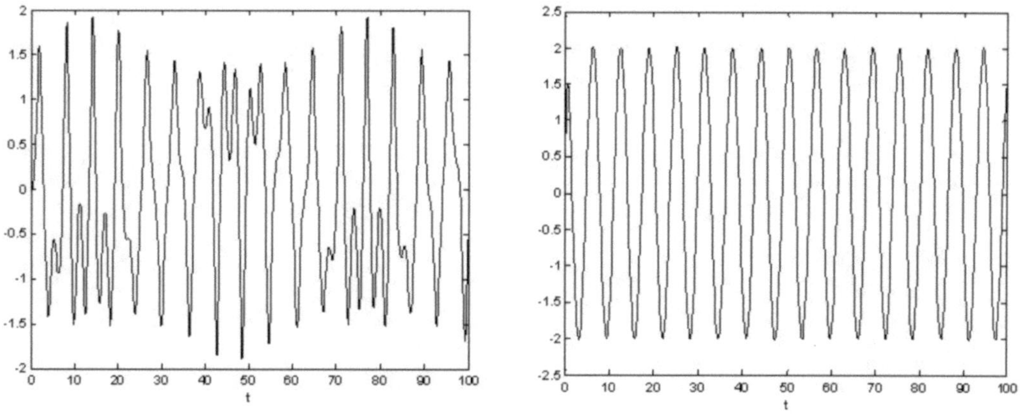

(a) (b)

Figure 13. (a): Tip displacement without control (b): Tip Displacement with control

In the dynamic mode the microcantilever is deliberately vibrated. The Amplitude Modulated AFM and the Frequency Modulated AFM are the most important techniques. In both AM-AFM and FM-AFM the amplitude and frequency of the microcantilever are kept constant by two control loops. The AGC (Automatic Gain Control) and the ADC (Automatic Distance Control). The AGC controls the amplitude of oscillation and the ADC controls the frequency by adjusting the distance between tip and sample. The oscillatory behavior of the microcantilever is illustrated in Figure 14. The FM-AFM block is shown in Figure 15.

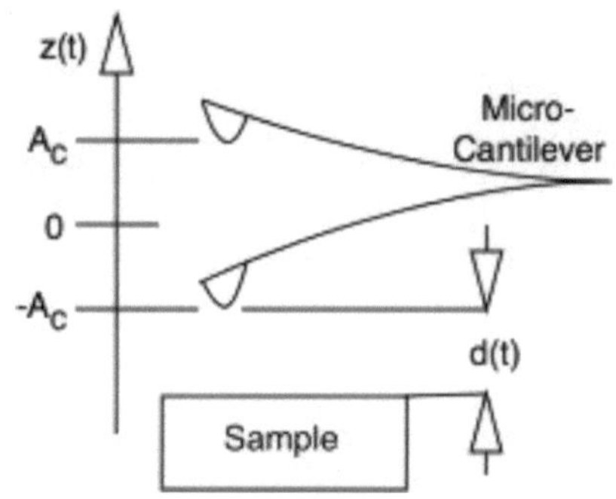

Figure 14. Microcantilever oscillatory behavior. Source: (Bueno et al., 2011).

In the FM-AFM the control signal of the AGC loop is used to generate the dissipation images and the ADC control signal is used to generate the topographic images. The FM-AFM improved image resolution and for surface studies in vacuum is the preferred AFM technique (Morita et. al., 2009; Bhushan, 2004). From Figure 15 it can be seen that the PLL generates the feedback signal for both control loops, therefore the PLL performance is vital to the FM-AFM. The PLL is a closed loop control system that synchronizes a local oscillator to a sinusoidal input. The PLLs are composed of a phase detector (usually a multiplier circuit), of a low-pass filter and of a VCO (Bueno et. al., 2010; Bueno et al., 2011), as it can be seen in Figure 16, and additionally, shows the PM and AM outputs used in the AFM system.

The AGC loop also depends on the amplitude detector output, shown in Figure 17. The amplitude detector is composed of diode followed by a first-order low-pass filter. The circuit holds the output A(t) for a while, allowing the AGC to determine the control signal.

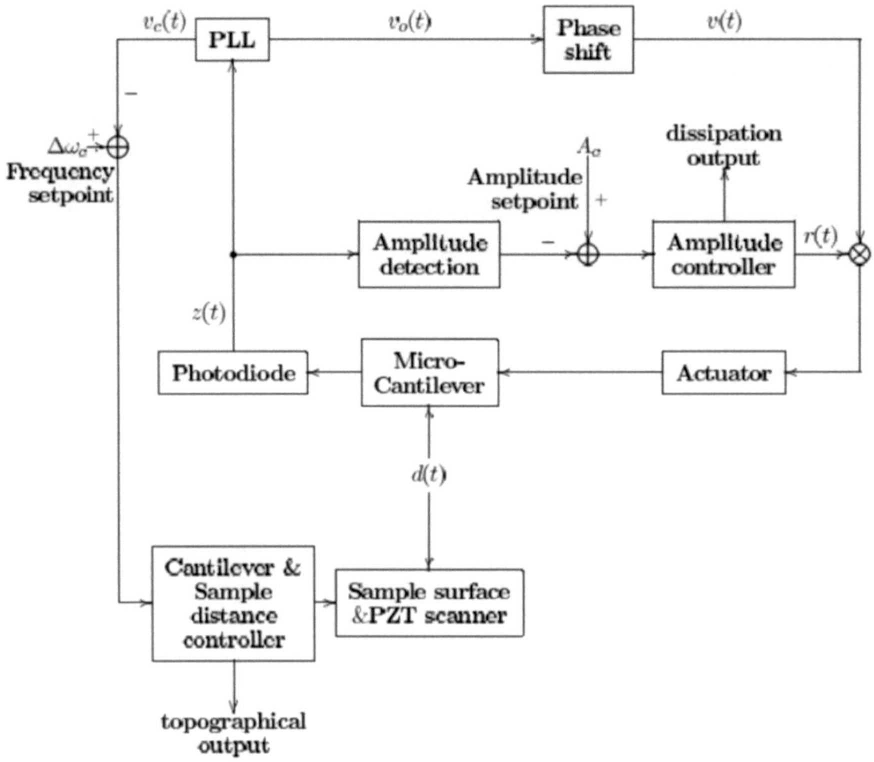

Figure 15. Block Diagram of the FM-AFM control system. Source: (Bueno et al., 2011).

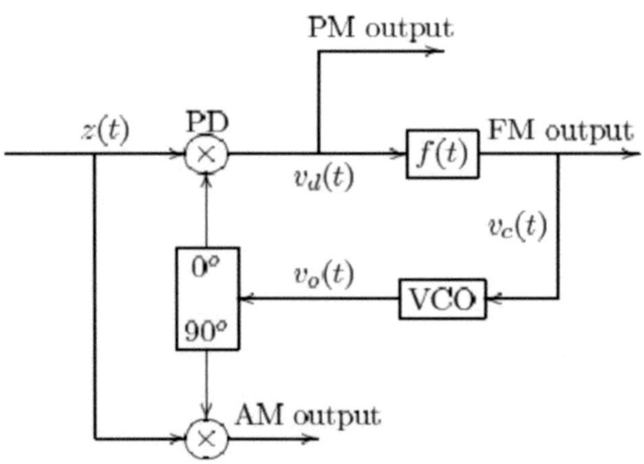

Figure 16. PLL block diagram

7.1. Mathematical model of the FM-AFM

The mathematical model of the FM-AFM considers the microcantiler dynamics, the tip-sample interaction, the amplitude detector circuit and the PLL. The microcantilever is assumed to be a second order system with natural frequency ω_c and damping coefficient γ. Concerning to tip-sample interaction, there are short, medium and long range forces. Since the FM-AFM operates in long-range distance, the predominant force is the Van Der Waals $\frac{A_H}{6(d(t)+z(t))^2}$, where AH is the Hamaker constant and d(t) is the tip-sample distance, Figure 7.1. Besides, the microcatilever is excited by an external forcing signal with a previously determined amplitude. The signal is a sinusoid with phase $\omega_c t+\varphi(t)$ and amplitude $r(t)v_o$, where $r(t)$ is the AGC signal, and v_0 is constant. The tip-sample interaction forces cause modulations both in the amplitude and in the frequency of oscillation of the AFM microcantilever. The modulations are detected by the PLL and used by the AGC and by the ADC, in order to control the microcantilever, drivin it to oscillate according to $z(t)=A(t)sen(\omega_c t+\varphi_c(t))$. The microcantilever mathetamical model is given by equation 37.

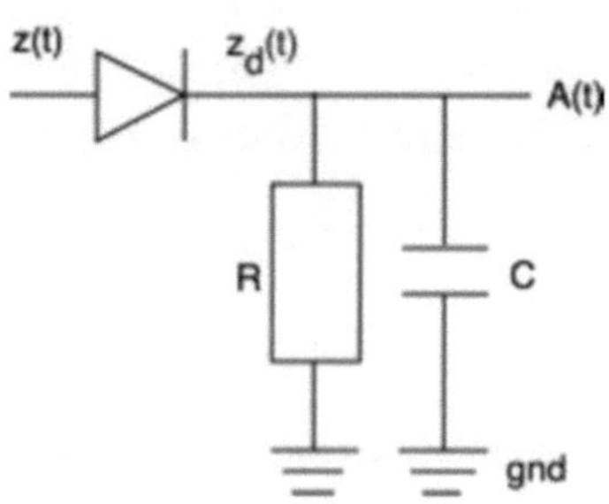

Figure 17. Amplitude detector. Source (Bueno et al., 2011)

The mathematical model of the amplitude detector is given by equation 38, where $\tau_d=\frac{1}{RC}$ and $z_d=\begin{cases} z(t), & z(t)>0 \\ 0, & z(t)\le 0 \end{cases}$. The PLL model can be seen in many works in the literature. The mathematical model in equation 39 follows Bueno et al., 2010 and 2011. Considering the filter transfer function $f(t)=\frac{\alpha_0}{s^2+\beta_1 s+\beta_0}$ and the gain $G=\frac{1}{2}k_m k_o v_o A_c$, where k_m is the phase detector gain, k_o is the VCO gain, v_o is the VCO output amplitude and A_c is the nominal microcantilever amplitude of oscillation, Figure 14. Equations 40 and 41 represent the AGC and ADC, respectively. Equations 37 to 41 are the model of the FM-AFM.

$$\ddot{z}+\gamma\dot{z}(t)+\omega_c^2 z(t)=r(t)v_o sen(\omega_c t+\varphi_o(t))+\frac{A_H}{6(d(t)+z(t))^2} \tag{37}$$

$$\dot{A}(t) + \tau_d A(t) = \tau_d z_d(t) \tag{38}$$

$$\dddot{\varphi}_o(t) + \beta_1 \ddot{\varphi}_o(t) + \beta_0 \dot{\varphi}_o(t) + \alpha_0 G sen\left(\varphi_o(t) - \varphi_c(t)\right) = 0 \tag{39}$$

$$r(t) = \Phi_{AGC}\left(A_c - A(t)\right) \tag{40}$$

$$d(t) = \Phi_{ADC}\left(\Delta\omega_c - \dot{\varphi}_o(t)\right) \tag{41}$$

7.2. Local stability and PLL design for FM-AFM

From equation 39, and considering the phase error $\vartheta = \varphi_o - \varphi_c$, results that:

$$\dddot{\vartheta} + \beta_1 \ddot{\vartheta} + \beta_0 \dot{\vartheta} + \alpha_0 G sen(\vartheta) = \dddot{\varphi}_c + \beta_1 \ddot{\varphi}_c + \beta_0 \dot{\varphi}_c \tag{42}$$

that represents the phase erros between the microcantilever oscillation and the PLL. The PLL behavior analysis is conveniently perfomed considering the cylindric state space, considering $\vartheta \in (-\pi, \pi]$. In that case, the synchronous state, corresponding to an asymptocally stable equilibrium point of equation 42 (See Bueno et al., 2010 and Bueno et al., 2011), corresponds to a Constant phase error ϑ and to null frequency and acceleration errors, i.e., $\dddot{\vartheta} = \ddot{\vartheta} = \dot{\vartheta} = 0$. For small phase erros it can be considered that $sen(\vartheta) \approx \vartheta$ in (42). In addition, considering $\varphi_c = \Omega t$, (42) can be rewritten as:

$$\dddot{\vartheta} + \beta_1 \ddot{\vartheta} + \beta_0 \dot{\vartheta} + \alpha_0 G \vartheta = \beta_0 \Omega \tag{43}$$

that represents the PLL linear response to a frequency shift (step) of amplitude Ω. The local stability of equation 42 can be determined by the position of the poles of equation 43, or by the Routh-Hurwitz criterion (See Bueno et al., 2010 and Bueno et al., 2011; Ogata, 1993). Therefore, considering that the coefficients of the filter are all positive and real, the poles of equation 43 have negative real parts if:

$$G < \frac{\beta_0 \beta_1}{\alpha_0}. \tag{44}$$

Considering the filter coefficients $\alpha_0 = \beta_0 = \omega_n^2$ and $\beta_1 = 2\xi\omega_n$, where ξ is the damping factor and ω_n the natural frequency, then, from equation 44, results:

$$G < 2\xi\omega_n. \tag{45}$$

Equation 45 establishes a design criterion that assure the local stability of the PLL, i.e., for small phase and frequency steps the PLL synchronizes to the microcantilever oscillation. Additionally, from the design parameters ξ and ω_n the loop gain G has a superior bound, and can be determined in order to satisfy the requirements of performance and stability.

Despite the good transient response and high frequency noise rejection - such as the double frequency jiter -, provided by the all pole filter the steady state response may need improvement. The PLL must demodulate an FSK (Frequency Shift Keying) signal, that actually is a frequency step. In order to track a frequency shift the loop filter must have at least a pure integration, i.e., the PLL must be at least a type 2 system (Bueno et al., 2010 and Bueno et al., 2011; Bueno, 2009, Ogata, 1993). According to that, and considering the loop filter $F(s) = \dfrac{as+1}{s(bs+1)}$ the stability of the PLL is assured only if $a > b$.

Figure 18 illustrates the PLL response to an FSK signal, showing the PLL FM output (figure 16). After the transient the mean value of the FM output is the same value of the FSK signal. The oscillation is due to the Double frequency jitter (Bueno et al., 2010 and Bueno, 2009). This shows that the PLL design must provide strong damping to noise and to the double frequency jitter. Besides, if the PLL is not at least of type 2 the PLL presents steady state error in the FSK demodulation, impairing the AFM imaging process. The PLL perfomance was analysed and presented under the FM-AFM perspective. Besides, a PLL design method was shown and illustrated by simulation, making clear the PLL performance importantance in the AFM control system.

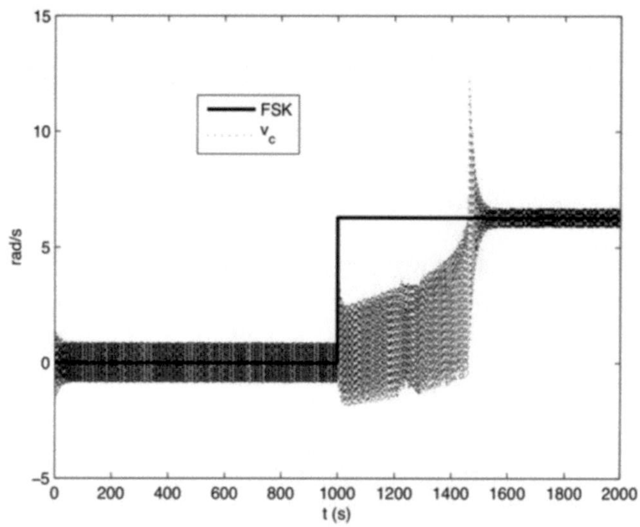

Figure 18. PLL response to a FSK signal.

8. Conclusions

This chapter deals with emergent problems in the Engineering Science research, presenting study and research related to NEMS systems, specially microcantilevers with many modes of vibration, for which the tip-sample interaction forces are highly nonlinear, impairing the stability of the latent image, while the others modes of vibration can be explored in order to improve the AFM performance.

In the context of this work, the following specific problem have been approached: The understanding of the relations of the properties and the structure of the nanoscopic and molecular materials, through the atomic force microscopy, using microcantilevers, giving subsidiary information to next generation of microscopy instrumentation.

Author details

José Manoel Balthazar
UNESP - Univ Estadual Paulista, Rio Claro, SP, Brasil
UNESP - Univ Estadual Paulista Paulista, Bauru, SP, Brasil

Angelo Marcelo Tusset and Atila Madureira Bueno
UTFPR - Universidade Técnica Federal do Paraná, Ponta Grossa, PR, Brasil

Bento Rodrigues de Pontes Junior
UNESP - Univ Estadual Paulista Paulista, Bauru, SP, Brasil

9. References

[1] Awrejcewicz, J. Bifurcation and Chaos in Coupled Oscillators. World Scientific, Singapore, 1991.

[2] Awrejcewicz, J., Lamarque, C.H. Bifurcation and Chaos in Nonsmooth Mechanical Systems. World Scientific Publishing, Singapore 2003.

[3] Alves, J.R.; Pontes Junior, B.R.; Balthazar, J.M., Modeling and dynamics of an electromechanical absorber applied to vibration control: performance and limitations. 21st Brazilian Congress of Mechanical Engineering, October 24-28, 2011, Natal, RN, Brazil, 2011.

[4] Alves, J.R. & Pontes Junior, B.R., "Modelagem e Dinâmica de um Absorvedor Eletromecânico Aplicado no Controle de Vibrações." (Relatório de Pesquisa - FAPESP), Departamento de Engenharia Mecânica, Faculdade de Engenharia, UNESP, Bauru. 2011. 70p.

[5] Alves, J.R., Pontes Junior, B.R., Modelagem e dinâmica de um absorvedor eletromecânico aplicado no controle de vibrações. ANAIS do CIC2010 - Congresso de Iniciação Científica da UNESP, Bauru, p. 1-4, 2010

[6] Anton, S.R. and Sodano, H.A. 2007. ''A Review of Power Harvesting Using Piezoelectric Materials (2003—2006),'' Smart Mater. Struct., 16:R1 R21.

[7] Ashhab M, Salapaka MV, Dahleh M, Mezic I. Dynamical analysis and control of microcantilevers. Automatica 1999; 35:1663–70.

[8] Ashhab M, M. V. Salapaka, M. Dahleh, I. Mezic, "Melnikov-based dynamical analysis of microcantilevers in scanning probe microscopy", Nonlin. Dyn, 20, 197-220. 1999.

[9] Binnig, G.: Atomic Force Microscope and Method for Imaging Surfaces with Atomic Resolution. US Patent. n: 4, 724, 318, (1986).

[10] Binnig, G., Gerber, Ch., Quate, C. F.: Atomic Force Microscope. Phys. Rev. Lett. 56, pp: 930-933, (1986).

[11] Beeby, S.P., Tudor, M.J., White, N.M., "Energy harvesting vibration sources for microsystems applications", Measurement Science and Technology 17, 2006, pp 175-195
[12] Bishop, R. H. The Mechatronics Handbook, CRC Press, 2002, ISBN: 0849300665.
[13] Bowen, W.R.; Hilal, N., Atomic Force Microscopy in Process Engineering - An Introduction to AFM for Improved Processes and Products, Elsevier, 2009.
[14] Bhushan, B; Handbook of Nanotechnology, Springer, Berlin, 2004.
[15] Bueno, A.M.; Ferreira, A. A.; Piqueira, J.R.C.; Modeling and Filtering Double-Frequency Jitter in One-Way Master Slave Chain Networks. Circuits and Systems I: Regular Papers, IEEE Transactions on 57 3104 -3111, 2010.
[16] Bueno, A. M.; Balthazar, J. M.; Piqueira, J. R. C. Phase-Locked Loop design applied to frequency-modulated atomic force microscope. Communications in Nonlinear Science and Numerical Simulation, (16)3835 – 3843, 2011.
[17] Bueno, A.M.; Ferreira, A. A.; Piqueira, J.R.C.; Modeling and measuring. Communications In Nonlinear Science and Numerical Simulation 14 1854-1860, 2009.
[18] Cidade, G. A. G.; Silva Neto, A. e Roberty, N. C Restauração de Imagens com Aplicações em Biologia e Engenharia: Problemas Inversos em Nanociência e Nanotecnologia Notas em Matemática Aplicada; 3 - Sao Carlos, SP : SBMAC, 2003, xiv, 88 p.
[19] Chtiba, M. O., Choura, S., Nayfeh, A.H., El-Borgia, S. 2010.Vibration confinement and energy harvesting in flexible structures using collocated absorbers and piezoelectric devices, Journal of Sound and Vibration 329 (2010) 261–276.
[20] Cottone, F., 2007, "Nonlinear Piezoelectric Generators for Vibration Energy Harvesting", Universita' Degli Studi Di Perugia, Dottorato Di Ricerca In Fisica, XX Ciclo.
[21] De Marqui Jr, C., Erturk, A., Inman, D. J. 2009. An electromechanical finite element model for piezoelectric energy harvester plates, Journal of Sound and Vibration 327 (2009) 9–25.
[22] Du Toit, N.E. and Wardle, B.L. 2007. ''Experimental Verification of Models for Microfabricated Piezoelectric Vibration Energy Harvesters,'' AIAA Journal, 45:1126—1137.
[23] De Martini, B.E., Rhoads, J.F., Turner, K.L., Shaw, S.W., Moehlis, J., "Linear and Nonlinear Tuning of Parametrically Excited MEMS Oscillators" in Journal of Microelectromechanical Systems, vol. 16, no. 2, 2007, pp 310-318
[24] Erturk, A., Inman, D.J., "On mechanical modeling of cantilevered piezoelectric vibration energy harvesters", Journal of Intelligent Material Systems and Structures, 2008
[25] Halliday, D., Resnick, R., Walker, J., Fundamentals of Physics, v. 3, 7th ed, Jonh Wiley and Sons, 2005.
[26] Felix, Jl.P. and Balthazar, J.M, Comments on a nonlinear and nonideal electromechanical damping vibration absorber, Sommerfeld effect and energy transfer. Nonlinear Dynamics, (2009) Volume 55, Numbers 1-2, 1-11, DOI: 10.1007/s11071-008-9340-8
[27] Farrokh, A., Fathipour, M., Yazdanpanah, M.J., High precision imaging for noncontact mode atomic force microscope using an adaptive nonlinear observer and output state

feedback controller, Digest Journal of Nanomaterials and Biostructures Vol. 4, No.3, September 2009, p. 429-442.

[28] Garcia, R., Pérez, R.: Dynamic atomic force microscopy method. Surface Science Report, 47, pp: 197-301, (2002).

[29] Garcia, R., San Paulo, A.: Dynamics of a vibrating tip near or in intermittent contact with a surface. Physical Review B, 61, pp: R13381-R13384, (2000).

[30] Giessibl, F. Atomic Resolution of the Silicon (111)-(7x7) Surface by Atomic Force Microscopy. Science 267 68--71 1995.

[31] Hu, S., Raman, A.: Analytical formulas and scaling laws for peak interaction forces in dynamic atomic force microscopy. Applied Physics Letters, 91, pp: 123106(1-3), (2007).

[32] Iossaqui, J. G., Uso de Absorvedores de Vibrações Eletromecânicos Lineares e Não-Lineares em Sistemas Não-Lineares e Não-Ideais, Bauru: Faculdade de Engenharia, UNESP-Universidade Estadual Paulista, 2009, 106 p., Dissertação (Mestrado).

[33] Iossaqui, J. G.; Balthazar, J. M.; Pontes, B. R. J; Felix, J. L. P.. Atuação de Absorvedores Eletromecânicos de Vibrações Não-Lineares e Não-Ideais. In: V Congresso Nacional de Engenharia Mecânica (CONEM 2008), 2008, Salvador. Anais do V Congresso Nacional de Engenharia Mecânica (CONEM 2008). Rio de Janeiro : ABCM, 2008-a. v. 1. p. 1-10.

[34] Iossaqui, J. G.; Balthazar, J. M.; Pontes, B. R. J; Felix, J. L. P.. On a Passive Control in a Nonlinear and a Nonideal System. In: 7º Congresso Temático de Dinâmica, Controle e Aplicações - DINCON 2008, 2008, Presidente Prudente. Anais do 7º Congresso Temático de Dinâmica, Controle e Aplicações - DINCON 2008. São Carlos : SBMAC, 2008-b. v. 1. p. 1-6.

[35] Iossaqui, J. G., Balthazar, J. M.;; Pontes, B. R. J; Felix, J. L. P.. Suppresing Chaotic Behaviour in A Double-Well Oscillator with Limited Power Supply Using Eleteromechanical Damper Device. In: Joint Conference on Mechanics and Materials, 2009, Blacksburg, VA. Proceedings of 2009 Joint Conference on Mechanics and Materials. Blacksburg : Virginia TECH, 2009.

[36] Jalili, N., Laxminarayana,K. A review of atomic force microscopy imaging systems: application to molecular metrology and biological sciences. Mechatronics 14:8, 907–945. 2004.

[37] Jalili, N.; Dadfarnia, M.; Dawson, D. M.; Distributed parameters base modeling and vibration analysis of microcantilevers used in atomic force microscopy, Proceedings of the ASME 2009 International Design Engineering Technical Conferences & Computers and Information in Engineering Conference, IDETC/CIE 2009, August 30 - September 2, 2009, San Diego, California, USA

[38] Krylov, S., Harari, I., Cohen, Y., "Stabilization of electrostatically actuated microstructures using parametric excitation", Journal of Micromechanics and Microengineering 15, 2005, pp 1188–1204

[39] Kuroda,M, H. Yabuno, K. Hayshi, K. Ashida, "Amplitude Control in a van der Pol-Type Self-Excited AFM Microcantililever", Journal of System Design and Dynamics. Vol. 2, n. 3, 2008.

[40] Liu, S., Davidson, A., Lin, Q., "Simulation studies on nonlinear dynamics and chaos in a MEMS cantilever control system" in Journal of Micromechanics and Microengineering 14, 2004, pp 1064–1073

[41] Moon, F. C. Chaotic and Fractal Dynamics, New Jersey: Wiley, 1992

[42] Moon, F.C., 1998, "Applied Dynamics: With Applications to Multibody and Mechatronic Systems", Ed. Wiley-Interscience, Canada, 492 p

[43] Lennard-Jones, J. E. (1924), "On the Determination of Molecular Fields", Proc. R. Soc. Lond. A 106 (738): 463–477, doi:10.1098/rspa.1924.0082.

[44] Lozano, J. R., Garcia, R.: Theory of Multifrequency Atomic Force Microscopy. Physical Review Letters. PRL 100, pp: 076102(1-4), (2008).

[45] Mauricio, M. H. P.: Microscopia de Ponta de Prova (Scanning Probe Microscopy): AFM - Microscopia de Força Atômica (2011) http://www.dema.pucrio.br/cursos/micquant/SPM.pdf, acesso em 10 de Fev., 2011.

[46] Morita, S.; Wiesendanger, R.; Meyer, E.; Giessibl, F. J.; Noncontact atomic force microscopy, Berlin:Springer, 2009.

[47] Ogata, K; Engenharia De Controle Moderno.1993.

[48] Preumont, A. Mechatronics Dynamics of Electromechanical and Piezoelectric Systems. Netherlands: Springer, 1999.

[49] Priya, S., Inman, D.J. 2009 Energy Harvesting Technologies, Springer Science Business Media, LLC 2009.

[50] Paulo, A.S., R. Garcia, "Unifying theory of tapping-mode atomic-force microscopy", Phys. Rev. B 66, 041406 (R), 2002.

[51] Quinn, D.D., Vakakis, A.F., Bergman, L.A., "Vibration-based energy harvesting with essential nonlinearities" in Proceedings of the ASME International Design Engineering Technical Conferences & Computers and Information in Engineering Conference, 2007

[52] Raman, A., Melcher, J., Tung, R.: Cantilever Dynamics in Atomic Force Microscopy. Nanotoday, vol: 3, pp: 20-27, (2008).

[53] Rafikov, M., Balthazar J. M., 2008. "On control and synchronization in chaotic and hyperchaotic systems", Communications in Nonlinear Science & Numerical Simulation, vol: 13, pp. 1246-1255.

[54] Roundy, S., Wright, P.K., Rabaey, J., "A study of low level vibrations as a power source for wireless sensor nodes", Comput. Commun. 26 (2003) 1131–1144.

[55] Richard W., B. And Nidal H. Microscopy in process engineering, An introdution to afm for improved processes and products, Elsevier; 2009 USA.

[56] Rutzel, S.; Lee, S. I.; Raman, A. Nonlinear dynamics of atomic-force-microscope probes driven is Lennard-Jones potentials. Proc. R. Soc. Lond, 459, 1925-1948, 2003

[57] Salvadori, M. C. B. S.: Microscopia de Força Atômica e Tunelamento. (2010) http://fluidos.if.usp.br/v1/iiiev/msalvadori_imfcx08.pdf, last accessed March 10, 2010

[58] Sebastian A., A. Gannepalii, M. V. Gannepali. M. V. Salapka. A review of the systems approach to the analysis of dynamic-mode atomic force microscopy. IEEE Transactions on Control Systems Technology, 15(15), 952- 959 (2007).

[59] Sinclair, I. R., Sensors and Transducers 3º ed, New York: Butterworth-Heinemann, 2001.

[60] Sodano, H.A., Inman, D.J., Park, G., "Comparison of piezoelectric energy harvesting devices for recharging batteries", Journal of Intelligent Material Systems And Structures, Vol. 16, 2005

[61] Sodano, H.A., Inman, D.J. and Park, G. 2004. ''A Review of Power Harvesting from Vibration Using Piezoelectric Materials,'' Shock Vib. Dig., 36:197—205.

[62] Sodano, H.A., Park, G. and Inman, D.J. 2004. ''Estimation of Electric Charge Output for Piezoelectric Energy Harvesting,'' Strain, 40:49 58.

[63] Triplett, A., Quinn, D. D. 2008. The Role of Non-Linear Piezoelectricity in Vibration-based Energy Harvesting, Twelfth Conference on Nonlinear Vibrations, Stability, and Dynamics of Structures—June 1–5, 2008.

[64] Triplett, A., Quinn, D. D. 2009. The Effect of Non-linear Piezoelectric Coupling on Vibration-based Energy Harvesting Journal of Intelligent Material Systems and Structures, Vol. 20 p. 1959-1967—November 2009.

[65] Twiefel, J., Richter, B., Sattel, T. And Wallaschek, J. 2008. ''Power Output Estimation and Experimental Validation for Piezoelectric Energy Harvesting Systems,'' J. Electroceram., 20:203—208.

[66] Zanette, S. I.; Funcionamento de um microscópio de força atômica.Notas do curso ministrado no CBPF,1997. DCP/Centro Brasileiro de Pesquisas Físicas/MCT,1997.

[67] Zhang, W.M.; Meng, G., Zhou, J. B.; Chen, J. Y. Nonlinear Dynamics and Chaos of Microcantilever-Based TM-AFMs with Squeeze Film Damping Effects. Sensors, 9, 3854-3874, 2009.

[68] Wang,C.C., N. S. Pai and H. T. Yau. "Chaos control in AFM system using sliding mode control backstepping desing", Comun Nonlinear Sci Numer Simulat, pp 1-11, 2009

[69] Wiesenganger, R. "Scanning probe microscopy and spectroscopy", Great Britain: Canbridge University Press, 1994.

[70] Yamapi, R. Dynamics and sychronization of electromechanical devices with a Duffing nonlinearity. Ph.D. Thesis, University of Abomey-Calavi, Bénin, 2003.

Chapter 4

Dynamics of a Pendulum of Variable Length and Similar Problems

A. O. Belyakov and A. P. Seyranian

Additional information is available at the end of the chapter

http://dx.doi.org/10.5772/48168

1. Introduction

In this chapter we study three mechanical problems: dynamics of a pendulum of variable length, rotations of a pendulum with elliptically moving pivot and twirling of a hula-hoop presented in three subsequent sections. The dynamics of these mechanical systems is described by similar equations and is studied with the use of common methods. The material of the chapter is based on publications of the authors [1-7] with the renewed analytical and numerical results. The methodological peculiarity of this work is in the assumption of quasi-linearity of the systems which allows us to derive higher order approximations by the averaging method. All the approximate solutions are compared with the results of numerical simulation demonstrating good agreement. Supplementary, in Appendix (section 5) we briefly presented the method of averaging with higher order approximations which is used in sections 2, 3, and 4.

2. Pendulum with periodically variable length

Oscillations of a pendulum with periodically variable length (PPVL) is the classical problem of mechanics. Usually, the PPVL is associated with a child's swing, see Fig. 1. Everyone can remember that to swing a swing one must crouch when passing through the middle vertical position and straighten up at the extreme positions, i.e. perform oscillations with a frequency which is approximately twice the natural frequency of the swing. Among previous works we cite [8–15] in which analytical and numerical results on dynamic behavior of the PPVL were presented.

The present section is devoted to the study of regular and chaotic motions of the PPVL. Asymptotic expressions for boundaries of instability domains near resonance frequencies are derived. Domains for oscillation, rotation, and oscillation-rotation motions in parameter space are found analytically and compared with numerical study. Chaotic motions of the pendulum depending on problem parameters are investigated numerically. Here we extend our results published in [1–4] in investigating dynamics of this rather simple but interesting mechanical system.

INTECH
open science | open minds

© 2012 The Author(s). Licensee InTech. This chapter is distributed under the terms of the Creative Commons Attribution License http://creativecommons.org/licenses/by/3.0), which permits unrestricted use, distribution, and reproduction in any medium, provided the original work is properly cited.

2.1. Main relations

Equation for motion of the swing can be derived with the use of angular momentum alteration theorem, see [8–11]. Taking into account also linear damping forces we obtain

$$\frac{d}{dt}\left(ml^2\frac{d\theta}{dt}\right) + \gamma l^2\frac{d\theta}{dt} + mgl\sin\theta = 0, \tag{1}$$

where m is the mass, l is the length, θ is the angle of the pendulum deviation from the vertical position, g is the acceleration due to gravity, and t is the time, Fig. 1.

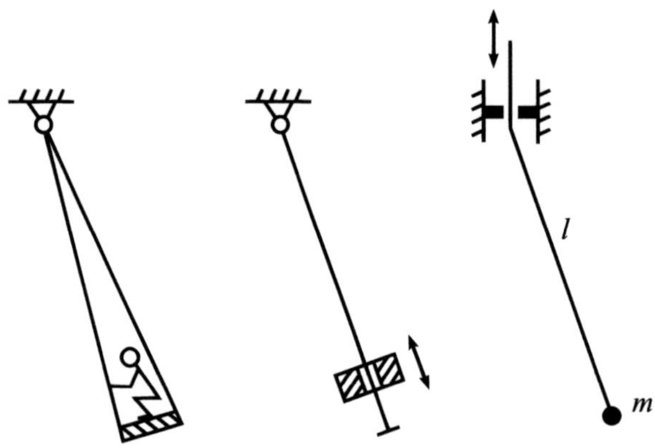

Figure 1. Schemes of the pendulum with periodically varying length.

It is assumed that the length of the pendulum changes according to the periodic law

$$l = l_0 + a\varphi(\Omega t) > 0, \tag{2}$$

where l_0 is the mean pendulum length, a and Ω are the amplitude and frequency of the excitation, $\varphi(\tau)$ is the smooth periodic function with period 2π and zero mean value.

We introduce new time $\tau = \Omega t$ and three dimensionless parameters

$$\varepsilon = \frac{a}{l_0}, \quad \omega = \frac{\Omega_0}{\Omega}, \quad \beta = \frac{\gamma}{m\Omega_0}, \tag{3}$$

where $\Omega_0 = \sqrt{\frac{g}{l_0}}$ is the eigenfrequency of the pendulum with constant length $l = l_0$ and zero damping. In this notations equation (1) takes the form

$$\ddot{\theta} + \left(\frac{2\varepsilon\dot{\varphi}(\tau)}{1+\varepsilon\varphi(\tau)} + \beta\omega\right)\dot{\theta} + \frac{\omega^2\sin(\theta)}{1+\varepsilon\varphi(\tau)} = 0, \tag{4}$$

where the upper dot denotes differentiation with respect to new time τ.

Stability and oscillations of the system governed by equation (4) will be studied in the following subsections via analytically under the assumption that the excitation amplitude ε and the damping coefficient ε are small. For rotational orbits we will also assume the smallness of the frequency ω which means high excitation frequency compared with the eigenfrequency Ω_0.

2.2. Instability of the vertical position

It is convenient to change the variable by the substitution

$$\eta = \theta\,(1 + \varepsilon\varphi(\tau)). \tag{5}$$

Using this substitution in equation (4) and multiplying it by $1 + \varepsilon\varphi(\tau)$ we obtain the equation for η as

$$\ddot{\eta} + \beta\omega\dot{\eta} - \frac{\varepsilon\,(\ddot{\varphi}(\tau) + \beta\omega\dot{\varphi}(\tau))}{1 + \varepsilon\varphi(\tau)}\eta + \omega^2 \sin\left(\frac{\eta}{1 + \varepsilon\varphi(\tau)}\right) = 0. \tag{6}$$

This equation is useful for stability study of the vertical position of the pendulum as well as analysis of small oscillations.

Let us analyze the stability of the trivial solution $\eta = 0$ of the nonlinear equation (6). Its stability with respect to the variable η is equivalent to that of the equation (4) with respect to θ due to relation (5). According to Lyapunov's theorem on stability based on a linear approximation for a system with periodic coefficients the stability (instability) of the solution $\eta = 0$ of equation (6) is determined by the stability (instability) of the linearized equation

$$\ddot{\eta} + \beta\omega\dot{\eta} + \frac{\omega^2 - \varepsilon\,(\ddot{\varphi}(\tau) + \beta\omega\dot{\varphi}(\tau))}{1 + \varepsilon\varphi(\tau)}\,\eta = 0. \tag{7}$$

This equation explicitly depends on three parameters: ε, β and ω. Expanding the ratio in (7) into Taylor series and keeping only first order terms with respect to ε and β we obtain

$$\ddot{\eta} + \beta\omega\dot{\eta} + \left[\omega^2 - \varepsilon(\ddot{\varphi}(\tau) + \omega^2\varphi(\tau))\right]\eta = 0. \tag{8}$$

This is a Hill's equation with damping with the periodic function $-(\ddot{\varphi}(\tau) + \omega^2\varphi(\tau))$. It is known that instability (i.e. parametric resonance) occurs near the frequencies $\omega = k/2$, where $k = 1, 2, \ldots$. Instability domains in the vicinity of these frequencies were obtained in [16, 17] analytically. In three-dimensional space of the parameters ε, β and ω, these domains are described by half-cones

$$(\beta/2)^2 + (2\omega/k - 1)^2 < r_k^2\varepsilon^2, \quad \beta \geq 0, \quad k = 1, 2, \ldots, \tag{9}$$

where $r_k = \frac{3}{4}\sqrt{a_k^2 + b_k^2}$ is expressed through the Fourier coefficients of the periodic function $\varphi(\tau)$

$$a_k = \frac{1}{\pi}\int_0^{2\pi} \varphi(\tau)\cos(k\tau)d\tau, \quad b_k = \frac{1}{\pi}\int_0^{2\pi} \varphi(\tau)\sin(k\tau)d\tau. \tag{10}$$

Inequalities (9) give us the first approximation of the instability domains of the vertical position of the swing. These inequalities were obtained in [1] using different variables.

Note that each k-th resonant domain in relations (9) depends only on k-th Fourier coefficients of the periodic excitation function. Particularly, for $\varphi(\tau) = \cos(\tau)$, $k = 1$ we obtain $a_1 = 1$, $b_1 = 0$, and $r_1 = 3/4$. Thus, the first instability domain takes the form

$$\beta^2/4 + (2\omega - 1)^2 < 9\varepsilon^2/16, \quad \beta \geq 0. \tag{11}$$

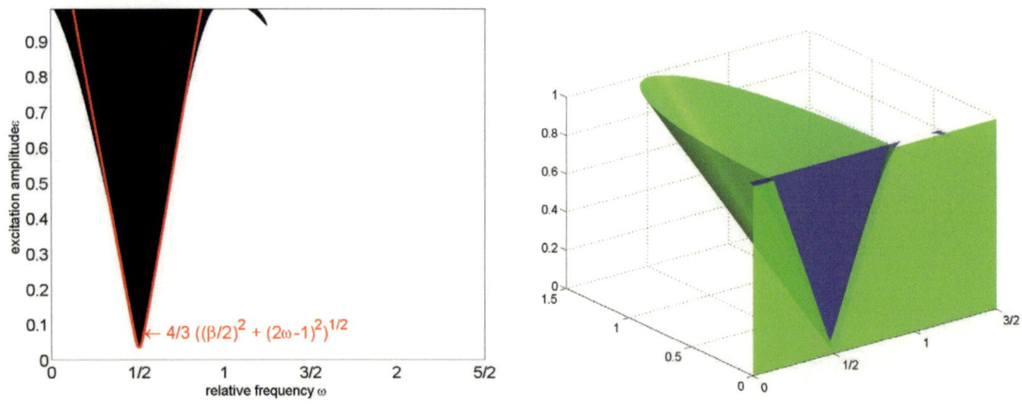

Figure 2. Instability domain (9) of PPVL (red line, left) in comparison with numerical results (black area) on parameter plane (ω, ε) at $\beta = 0.05$. Half cone instability domain (9) of PPVL (green surface, right) compared with the same numerical results (blue plane) in the parameter space $(\omega, \beta, \varepsilon)$.

The boundary of the first instability domain ($k = 1$) is presented in Fig. 2 by the solid red line demonstrating a good agreement with the numerically obtained instability domain which is marked black. These boundaries are also drawn in Fig. 5 and 6 by solid white lines. It is easy to see from (8) that for the second resonance domain ($k = 2$, $\omega = 1$) the excitation function $-(\ddot{\varphi}(\tau) + \omega^2\varphi(\tau))$ is zero for $\varphi(\tau) = \cos(\tau)$. This explains why the second resonance domain is empty, and the numerical results confirm this conclusion, see Fig. 2. Inside the instability domains (9) the vertical position $\eta = 0$ becomes unstable and motion of the system can be either regular (limit cycle, regular rotation) or chaotic.

2.3. Limit cycle

When the excitation amplitude ε is small, we can expect that the oscillation amplitude θ in equation (4) will be also small. We suppose that ε and β are small parameters of the same order, and $\theta = O(\sqrt{\varepsilon})$. Then, we introduce notation $\tilde{\beta} = \beta/\varepsilon$, $\vartheta = \theta/\sqrt{\varepsilon}$ and expand the sine into Taylor's series around zero in equation (4) keeping only three terms. Thus, equation (4) takes the following form

$$\ddot{\vartheta} + \omega^2\vartheta = \varepsilon f_1(\vartheta, \dot{\vartheta}, \tau) + \varepsilon^2 f_2(\vartheta, \dot{\vartheta}, \tau) + \ldots, \tag{12}$$

where

$$f_1(\vartheta, \dot{\vartheta}, \tau) = \omega^2 \left(\varphi(\tau)\,\vartheta + \frac{\vartheta^3}{6}\right) - \left(\tilde{\beta}\omega + 2\dot{\varphi}(\tau)\right)\dot{\vartheta}, \tag{13}$$

$$f_2(\vartheta, \dot{\vartheta}, \tau) = 2\dot{\varphi}(\tau)\varphi(\tau)\,\dot{\vartheta} - \omega^2 \left(\varphi(\tau)\left(\varphi(\tau)\,\vartheta + \frac{\vartheta^3}{6}\right) + \frac{\vartheta^5}{120}\right). \tag{14}$$

We study the parametric excitation of nonlinear system (12) with the periodic function $\varphi(\tau) = \cos\tau$ at the first resonance frequency, $\omega - 1/2 = O(\varepsilon)$. To solve equation (12) we use the method of averaging [9, 18–20]. For that purpose we write (12) in the Bogolubov's *standard form* of first order differential equations with small right-hand sides. First, we use *Poincaré variables* $q(\tau)$ and $\psi(\tau)$ defined via the following solution of the *generating equation* $\ddot{\vartheta} + \omega^2\vartheta = 0$ which is equation (12) with $\varepsilon = 0$, when $\dot{q} = 0$ and $\dot{\psi} = \omega$

$$\vartheta = q\cos(\psi), \quad \dot{\vartheta} = -\omega q \sin(\psi). \tag{15}$$

We can express Poincaré variables $q(\tau)$ and $\psi(\tau)$ via ϑ and $\dot{\vartheta}$ from (15) as $q^2 = \vartheta^2 + \dot{\vartheta}^2/\omega^2$ and $\psi = \arctan\left(-\frac{\dot{\vartheta}}{\vartheta\omega}\right)$. We differentiate these equations with respect to time and substitute expressions for $\ddot{\vartheta}$, $\dot{\vartheta}$, and ϑ in terms of q and ψ obtained from (12), and (15). Then, in the resonant case we have equations for the *slow amplitude* $q(\tau)$ and *phase shift* $\zeta(\tau) = \psi(\tau) - \frac{1}{2}\tau$

$$\dot{q} = -\frac{\sin(\frac{\tau}{2}+\zeta)}{\omega} f\left(q\cos\left(\frac{\tau}{2}+\zeta\right), -q\omega\sin\left(\frac{\tau}{2}+\zeta\right), \tau\right), \tag{16}$$

$$\dot{\zeta} = \omega - \frac{1}{2} - \frac{\cos(\frac{\tau}{2}+\zeta)}{\omega q} f\left(q\cos\left(\frac{\tau}{2}+\zeta\right), -q\omega\sin\left(\frac{\tau}{2}+\zeta\right), \tau\right), \tag{17}$$

where $f(\vartheta, \dot{\vartheta}, \tau) = \varepsilon f_1(\vartheta, \dot{\vartheta}, \tau) + \varepsilon^2 f_2(\vartheta, \dot{\vartheta}, \tau) + o(\varepsilon^2)$. System (16)-(17) has small right hand sides because we assumed that $\omega - 1/2 = O(\varepsilon)$. As a result of averaging in the second approximation, see (121) in the Appendix, we get the system of *averaged differential equations*

$$\dot{Q} = \varepsilon Q\left(\frac{2-\omega}{4}\sin(2Z) - \frac{\tilde{\beta}\omega}{2}\right) + \varepsilon^2 Q\omega\left(5Q^2\frac{2-\omega}{192}\sin(2\,Z) + \frac{\tilde{\beta}\omega}{4}\cos(2Z)\right) + o(\varepsilon^2), \tag{18}$$

$$\begin{aligned}\dot{Z} &= \omega - \frac{1}{2} + \varepsilon\left(\frac{2-\omega}{4}\cos(2Z) - \frac{Q^2\omega}{16}\right) \\ &+ \varepsilon^2\left(5Q^2\omega\frac{2-\omega}{96}\cos(2Z) - \frac{\tilde{\beta}\omega^2}{4}\sin(2Z) - Q^4\omega\frac{17\omega-4}{1536} - \frac{(2-\omega)^2}{32} - \frac{\omega^2\tilde{\beta}^2}{4}\right) + o(\varepsilon^2),\end{aligned} \tag{19}$$

where Q and Z are the *averaged variables* corresponding to q and ζ. This system gives steady solutions for $\dot{Q} = 0$, $\dot{Z} = 0$. Thus, besides the trivial one $Q = 0$, in the first approximation we obtain from system (18)-(19) expressions for the averaged amplitude and phase shift as

$$\varepsilon Q_{\{1\}}^2 = \frac{4}{\omega}\left(4\omega - 2 \mp \sqrt{\varepsilon^2(2-\omega)^2 - 4\beta^2\omega^2}\right), \tag{20}$$

$$Z_{\{1\}} = \frac{1}{2}\arctan\left(\frac{\mp 2\beta\omega}{\sqrt{\varepsilon^2(2-\omega)^2 - 4\beta^2\omega^2}}\right) + \pi j, \tag{21}$$

where $j = \ldots, -1, 0, 1, 2, \ldots$ and "arctan" gives the major function value lying between zero and π; subindex "{1}" denotes the order of approximation with which the corresponding variable is obtained. Solution of system (16)-(17) in the first approximation is $q = Q_{\{1\}} + o(1)$, $\zeta = Z_{\{1\}} + o(1)$ so the solution of (4) is $\theta = \sqrt{\varepsilon}Q_{\{1\}}\cos(\tau/2 + Z_{\{1\}}) + o(\sqrt{\varepsilon})$.

Solution of system (16)-(17) in the second approximation is the following, see (119) and (121),

$$\begin{aligned}q &= Q_{\{2\}} + \varepsilon Q_{\{2\}}\left(\frac{\tilde{\beta}\omega}{2}\sin\left(\tau + 2Z_{\{2\}}\right) - \cos(\tau) + \frac{\omega+2}{8}\cos(2\tau + 2Z_2)\right) \\ &+ \varepsilon Q_{\{2\}}^3\left(\frac{\omega}{24}\cos\left(\tau + 2Z_{\{2\}}\right) + \frac{\omega}{96}\cos\left(2\tau + 4Z_{\{2\}}\right)\right) + o(\varepsilon),\end{aligned} \tag{22}$$

$$\begin{aligned}\zeta &= Z_{\{2\}} + \varepsilon\left(\frac{\tilde{\beta}\omega}{2}\cos\left(\tau + 2Z_{\{2\}}\right) - \frac{\omega}{2}\sin(\tau) - \frac{\omega+2}{8}\sin\left(2\tau + 2Z_{\{2\}}\right)\right) \\ &- \varepsilon Q_{\{2\}}^2\left(\frac{\omega}{12}\sin\left(\tau + 2Z_{\{2\}}\right) + \frac{\omega}{96}\sin\left(2\tau + 4Z_{\{2\}}\right)\right) + o(\varepsilon),\end{aligned} \tag{23}$$

where $Q_{\{2\}}$ and $Z_{\{2\}}$ are the steady state variables of system (18)-(19) in the second order approximation. Substitution of these expressions into (15) yields the second order approximate solution of (4) in the following form

$$\theta = \sqrt{\varepsilon} Q_{\{2\}} \cos\left(\frac{\tau}{2} + Z_{\{2\}}\right) + \beta\sqrt{\varepsilon} Q_{\{2\}} \frac{\omega}{2} \sin\left(\frac{\tau}{2} + Z_{\{2\}}\right)$$
$$- \varepsilon^{\frac{3}{2}} Q_{\{2\}} \left(\frac{2-\omega}{4} \cos\left(\frac{\tau}{2} - Z_{\{2\}}\right) + \frac{2+\omega}{8} \cos\left(\frac{3\tau}{2} + Z_{\{2\}}\right)\right)$$
$$+ \varepsilon^{\frac{3}{2}} Q^3_{\{2\}} \left(\frac{\omega}{16} \cos\left(\frac{\tau}{2} + Z_{\{2\}}\right) - \frac{\omega}{96} \cos\left(\frac{3\tau}{2} + 3Z_{\{2\}}\right)\right) + o(\varepsilon^2). \qquad (24)$$

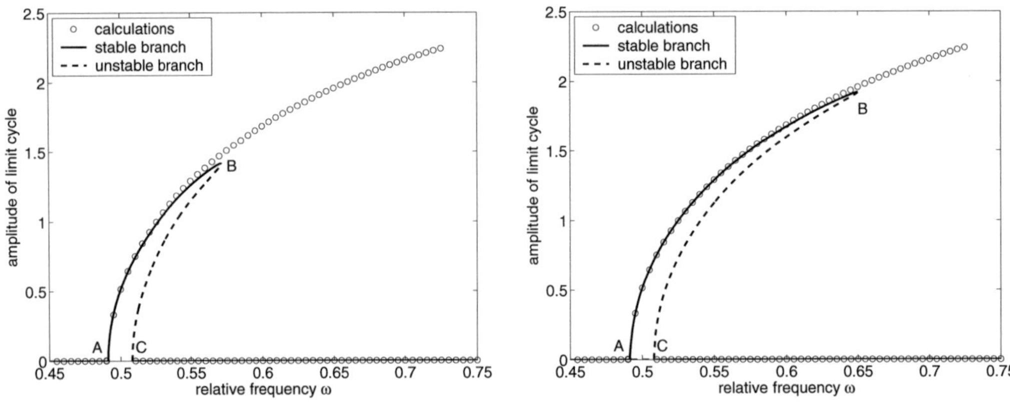

Figure 3. Frequency-response curve for the parameters $\varepsilon = 0.04$ and $\beta = 0.05$. Amplitude of the limit cycle in the first (left) and second (right) approximation compared with the results of numerical simulations (circles) depending on the relative excitation frequency ω. In the first approximation (left) the amplitude $\sqrt{\varepsilon} Q_{\{1\}}$ is described by (20). In the second approximation (right) the amplitude of solution (24) is calculated with the use of numerically obtained steady state $Q_{\{2\}}$ and $Z_{\{2\}}$ of system (18)-(19).

Fig. 3 shows better coincidence with the numerical simulations of the second order approximate solution (24) up to the amplitude equals $2 \approx 2\pi/3$ and the frequency mismatch is $\omega - \frac{1}{2} \approx 0.15$.

2.4. Regular rotations

We say that the system performs regular rotations if a nonzero average rotational velocity exists

$$b = \lim_{T\to\infty} \frac{1}{T} \int_0^T \dot{\theta} d\tau.$$

Velocity b is a rational number because regular motions can be observed only in resonance with excitation. Motion with fractional average velocity such as $|b| = 1/2$ in Fig. 4 a) is usually called *oscillation-rotation*. Let us first study monotone rotations, where velocity $\dot{\theta}$ has constant sign and integer average value b, see Fig. 4 b) and c).

In order to describe resonance rotations of the PPVL we use the method of averaging [9, 18, 19] which requires rewriting (4) in the Bogolubov's standard form. For that reason we assume

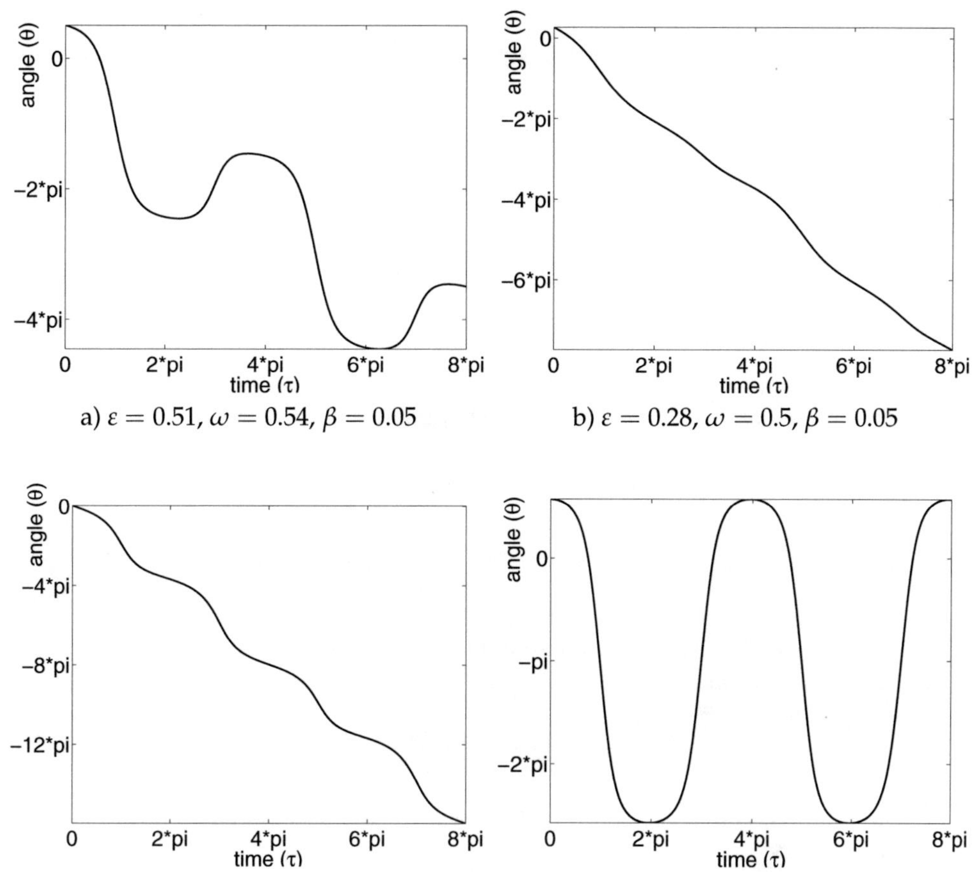

a) $\varepsilon = 0.51, \omega = 0.54, \beta = 0.05$

b) $\varepsilon = 0.28, \omega = 0.5, \beta = 0.05$

c) $\varepsilon = 0.43, \omega = 0.5, \beta = 0.05$

d) $\varepsilon = 0.59, \omega = 0.6, \beta = 0.05$

Figure 4. a) Regular rotation-oscillation with the mean angular velocity equal to one half of the excitation frequency, $b = -1/2$. b) Regular rotation with $b = -1$. c) Regular rotation with $b = -2$. d) Regular rotation-oscillation with $b = 0$.

that ε, β and ω are small parameters, ε being of order ω^2, and β of order ω^3, which makes the system quasi-linear.

We introduce the vector of *slow* variables x and the fast time $s = |b|\,\tau$, where $x_1 = \theta - b\,\tau$ is the *phase mismatch*, $x_2 = \frac{d\theta}{ds}$ is the velocity, $x_3 = 1 + \varepsilon\cos\left(\frac{s}{|b|}\right)$ is the excitation. From here the dot denotes derivative with respect to the new time s. Thus, equations (4) takes the standard form

$$\begin{aligned}
\dot{x}_1 &= x_2 - \operatorname{sign}(b),\\
\dot{x}_2 &= \left(\frac{2\varepsilon}{x_3}\sin\left(\frac{s}{|b|}\right) - \beta\omega\right)\frac{x_2}{|b|} - \frac{\omega^2}{b^2}\frac{\sin(x_1 + s)}{x_3},\\
\dot{x}_3 &= -\frac{\varepsilon}{|b|}\sin\left(\frac{s}{|b|}\right),
\end{aligned} \tag{25}$$

where it is assumed that $x_2 - \text{sign}(b)$ is of order ε, $\text{sign}(b) = 1$ if $b > 0$ and $\text{sign}(b) = -1$ if $b < 0$. With the method of averaging we can find the first, second and the following order approximations of equations (25).

Resonance rotation domains of PPVL for various $|b|$ are presented in Fig. 5. We see that greater values of relative rotational velocities $|b|$ are possible for higher excitation amplitudes ε. Numerically obtained rotational regimes are depicted in Fig. 5 by color points in parameter space (ω, ε) with $\beta = 0.05$ and initial conditions $\theta(0) = \pi$, $\dot{\theta}(0) = 0.05$. Domains of these points are well bounded below by analytically obtained curves for corresponding $|b|$.

2.4.1. *Rotations with relative velocity* $|b| = 1$

It is the third order approximation of averaged equation where regular rotations with $|b| = 1$ can be observed, see Fig. 4 b). In the third order approximation averaged equations take the following form, see (122),

$$\begin{aligned} \dot{X}_1 &= X_2 - b, \\ \dot{X}_2 &= -\frac{3\varepsilon\omega^2}{2}\sin(X_1) - \beta\omega X_2, \end{aligned} \tag{26}$$

where X_1 and X_2 are the averaged slow variables x_1 and x_2. Auxiliary variable $x_3 = 1 + \varepsilon\cos(s/b)$ has unit average $X_3 = 1$ and is excluded from the consideration. Excluding variable X_2 from the steady state conditions $\dot{X}_1 = 0$ and $\dot{X}_2 = 0$ in (26) we obtain the equation for the averaged phase mismatch X_1

$$\sin(X_1) = -b\frac{2\beta}{3\varepsilon\omega}. \tag{27}$$

Thus, it is clear from (27) that equation (26) has a steady state solution only if

$$\omega \geq \frac{2\beta}{3\varepsilon}. \tag{28}$$

Inequality (28) determines the domain in parameter space, where rotations with $|b| = 1$ can exist. The boundary of this domain is depicted with a bold dashed line in Fig. 5 on the parameter plane (ω, ε) for $\beta = 0.05$.

Stability of the solutions obtained from (27) was studied in [2]. There was found the condition for asymptotic stability $\cos(X_1) > 0$. Hence, if inequality (28) is strict, then there are asymptotically stable steady solutions

$$X_{1(1)} = -b\arcsin\left(\frac{2\beta}{3\omega\varepsilon}\right) + 2\pi k, \quad k = \dots, -1, 0, 1, 2, \dots \tag{29}$$

and unstable solutions

$$X_{1(2)} = \pi + b\arcsin\left(\frac{2\beta}{3\omega\varepsilon}\right) + 2\pi k, \quad k = \dots, -1, 0, 1, 2, \dots \tag{30}$$

Thus, we conclude that if the parameters satisfy strict inequality (28) there are two stable regular rotations $\theta = b\tau + X_{1(1)} + o(1)$ in opposite directions ($b = \pm 1$) and two unstable rotations $\theta = b\tau + X_{1(2)} + o(1)$ in opposite directions.

2.4.2. Rotations with relative velocity $|b| = 2$

Rotations with higher averaged velocities $|b| = 2, \ldots$ correspond to higher excitation amplitudes ε. That is why we consider the coefficient ω being of order ε, and β being of order ε^3. With this new ordering we obtain the sixth order approximation of the averaged equations for $|b| = 2$, see (128),

$$\begin{aligned} \dot{X}_1 &= X_2 - \frac{b}{2}, \\ \dot{X}_2 &= -\frac{9\varepsilon^2\omega^2}{16}\left(1 - \left(X_2 - \frac{b}{2}\right)^2 + \frac{\varepsilon^2}{27}\right)\sin(X_1) - \frac{\beta\omega}{2}X_2, \end{aligned} \tag{31}$$

which have steady state solutions determined by the following equation

$$\sin(X_1) = -b\frac{8\beta}{9\varepsilon^2\omega}\left(\frac{1}{1+\varepsilon^2/27}\right). \tag{32}$$

From equation (32) we get that the domain of rotations with $|b| = 2$ in the parameter space has the following boundary condition depicted in Fig. 5 with a bold solid line

$$\omega \geq \frac{8\beta}{9\varepsilon^2}\left(\frac{1}{1+\varepsilon^2/27}\right). \tag{33}$$

System (31) has similar structure to system (26). That is why stability condition for its steady state solutions appears to be the same: $\cos(X_1) > 0$. Hence, if inequality (33) is strict, we find from (32) that there are asymptotically stable steady solutions

$$X_{1(1)} = -\arcsin\left(\frac{8b\beta}{9\varepsilon^2\omega}\left(\frac{1}{1+\varepsilon^2/27}\right)\right) + 2\pi k, \quad k = \ldots, -1, 0, 1, 2, \ldots \tag{34}$$

and unstable steady solutions

$$X_{1(2)} = \pi + \arcsin\left(\frac{8b\beta}{9\varepsilon^2\omega}\left(\frac{1}{1+\varepsilon^2/27}\right)\right) + 2\pi k, \quad k = \ldots, -1, 0, 1, 2, \ldots \tag{35}$$

Thus, as in the previous case, if the parameters satisfy strict inequality (33) there are two stable regular rotations $\theta = b\tau + X_{1(1)} + o(1)$ in opposite directions ($b = \pm 2$) and two unstable rotations $\theta = b\tau + X_{1(2)} + o(1)$ in opposite directions.

2.5. Basins of attractions and transitions to chaos

In order to determine domains of chaos we calculate maximal Lyapunov exponents presented in Fig. 6. We recall that positive Lyapunov exponents correspond to chaotic motions. Note that chaotic motion includes passing through the upper vertical position, i.e. irregular oscillations-rotations. This is usually called tumbling chaos. We have observed two types of transition to chaos. The first type is when the system goes through the cascade of period doubling (PD) bifurcations occurring within the instability domain of the vertical position when the excitation amplitude ε increases, for example at $\omega = 0.5$ in Fig. 7(a). The second type is when chaos immediately appears after subcritical Andronov-Hopf (AH) bifurcation when the system enters the instability domain of the lower vertical position of PPVL, for

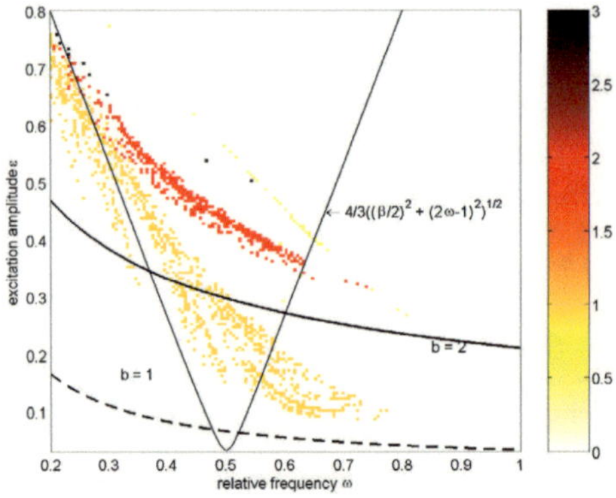

Figure 5. Absolute values $|b|$ of relative rotational velocities are shown with different colors on the plane of parameters ε and ω at the damping $\beta = 0.05$. The correspondence between the colors and values is shown by the color bar on the right. Approximate boundaries for rotations are drawn with bold dashed line (for $|b| = 1$) and bold solid line (for $|b| = 2$).

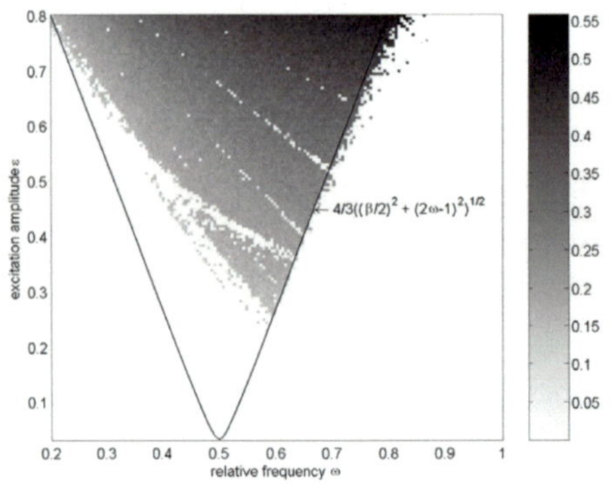

Figure 6. Maximal Lyapunov's exponents are shown on the plane of parameters ε and ω at the damping $\beta = 0.05$. The correspondence between the colors and values is shown by the color bar on the right, where white color distinguishes zero maximal Lyapunov's exponent which corresponds to regular regime. Positive Lyapunovs' exponents characterize chaotic motions.

example at $\omega = 0.67$; see Fig. 7(b). We can see the change of the system dynamics in its route to chaos along $\omega = 0.5$ in the bifurcation diagram shown in Fig. 7(a), where red points denote rotations with mean angular velocity equal to one excitation frequency ($|b| = 1$) and green points denote those equal to two excitation frequencies ($|b| = 2$). The domain with the most complex regular dynamics is surrounded by the red rectangle, where the system can have coexisting oscillations, rotations and rotations-oscillations.

Basins of attractions in Fig. 8 have been plotted using program Dynamics [21]. These basins track the changes of the system dynamics in its route to chaos along $\omega = 0.67$. In Fig. 8(a) the oscillatory attractor (limit cycle) coexists with stationary attractor (lower vertical position of PPVL). In Fig. 8(b) we can see the first emergence of two rotational attractors with counterrotations. This picture is in a good agreement with condition (18) for existence of rotational solutions $|b| = 1$, see Fig. 5. Closer to the boundary of chaotic region in Fig. 8(c) only stationary and rotational attractors remain. Note that the basins of rotational attractors are

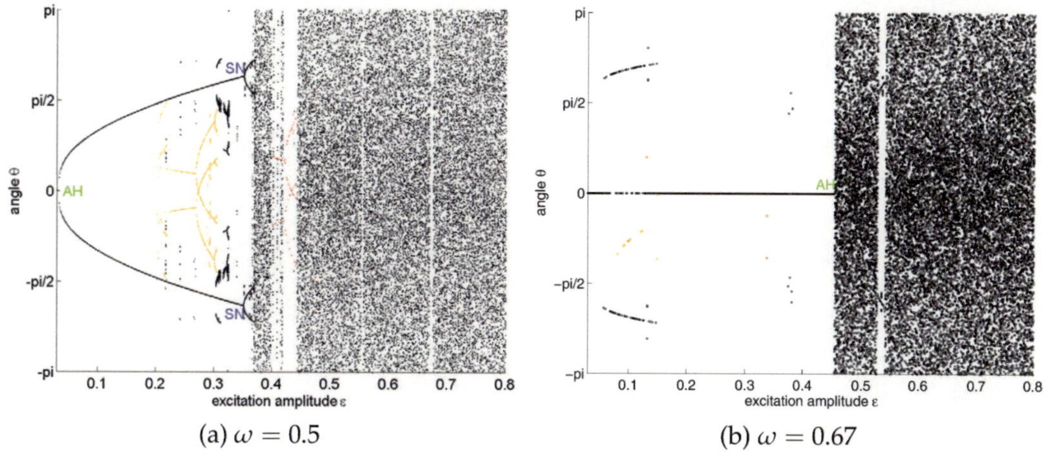

Figure 7. The bifurcation diagrams for different frequencies ω and the same damping $\beta = 0.05$ show two different types of transition to chaos. (a) After Andronov-Hopf bifurcation (AH) a limit cycle appears which experiences the saddle-node bifurcation (SN) and then the cascade of period-doubling bifurcations (PD). Regular rotations with relative mean angular velocity $|b| = 1$ are denoted by orange points and $|b| = 2$ by red. (b) After subcritical AH bifurcation of the vertical equilibrium the chaotic motion occurs immediately.

small which means that at $\omega = 0.67$ the transition to chaos through subcritical AH bifurcation is the most typical. In the middle of Fig. 8(d) the manifold of dark blue points reveals a typical strange attractor structure. The strange attractor inherits the basin of attraction from disappeared stationary attractor.

3. Elliptically excited pendulum

Elliptically excited pendulum (EEP) is a mathematical pendulum in the vertical plane whose pivot oscillates not only vertically but also horizontally with $\pi/2$ phase shift, so that the pivot has elliptical trajectory, see Fig. 9. EEP is a natural generalization of pendulum with vertically vibrating pivot that is one of the most studied classical systems with parametric excitation. It is often referred to simply as *parametric pendulum*, see e.g. [9, 22–27] and references therein. Stability and dynamics of EEP have been studied analytically and numerically in [28–30]. Approximate oscillatory and rotational solutions for EEP are the common examples in literature [31–34] on asymptotic methods. Sometimes EEP is presented in a slightly more general model of unbalanced rotor [31–33], where the phase shift between vertical and horizontal oscillations of the pivot can differ from $\pi/2$. EEP is also a special case of generally excited pendulum in [35]. The usual assumption for approximate solution in the literature is the smallness of dimensionless damping and pivot oscillation amplitudes in the EEP's equation of motion. We could find only one paper [36], where oscillations of EEP with high damping and yet small relative excitation were studied.

In this section we study rotations of EEP with not small excitation amplitudes and with both small and not small linear damping. Our analysis uses the exact solutions for EEP with the absence of gravity and with equal excitation amplitudes, when elliptical trajectory of the pivot

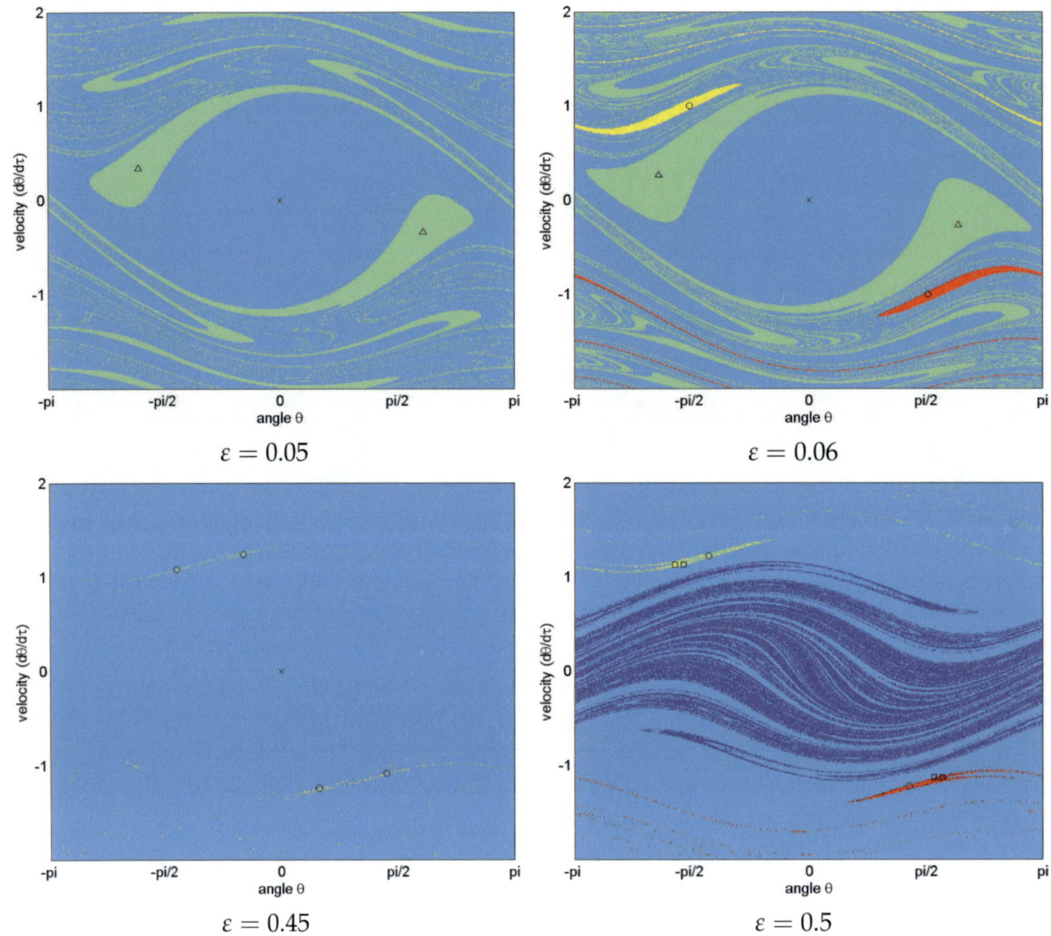

Figure 8. Basins of attractions in Poincare section for different excitation amplitude ε at the same frequency $\omega = 0.67$ and damping $\beta = 0.05$. $\triangle$ marks period-two oscillational attractors, $\bigcirc$ marks period-one rotational attractors, $\square$ marks period-two rotational attractors, $\times$ marks fixed points.

becomes circular. When there is no gravity the model of EEP coincides with that of hula-hoop, see section 4. The material of the section is based on the paper [5].

3.1. Main relations

Equation of EEP's motion can be derived with the use of angular momentum alteration theorem, see [30]

$$m\,l^2\frac{d^2\theta}{dt^2} + c\frac{d\theta}{dt} + m\,l\left(g\cos(\delta) - \frac{d^2y(t)}{dt^2}\right)\sin(\theta)$$

$$+\,m\,l\left(g\sin(\delta) - \frac{d^2x(t)}{dt^2}\right)\cos(\theta) = 0, \qquad (36)$$

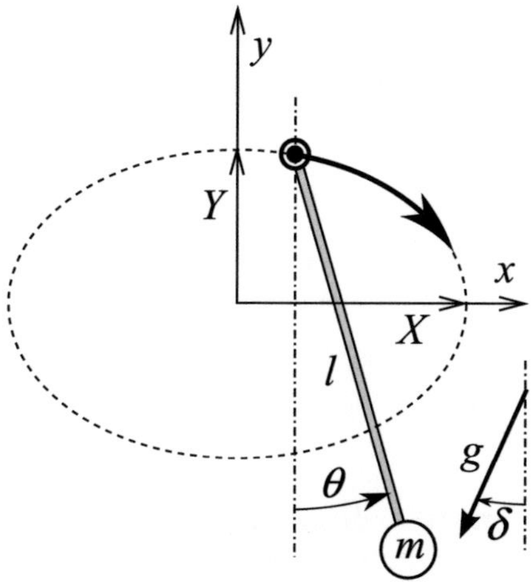

Figure 9. Scheme of the elliptically excited mathematical pendulum of length l. The pivot of the pendulum moves along the elliptic trajectory (dashed line) with semiaxes X and Y in the uniform gravitational field g.

where l is the distance between the pivot and the concentrated mass m; c is the viscous damping coefficient; θ is the angle of the pendulum deviation from the vertical position; t is time; g is gravitational acceleration at the angle δ with respect to the negative direction of the axis y.

It is assumed that the pivot of the pendulum moves according to the periodic law

$$x = X\sin(\Omega t), \quad y = Y\cos(\Omega t), \tag{37}$$

where X, Y, and Ω are the amplitudes and frequency of the excitation.

We introduce new time $\tau = \Omega t$ and the following dimensionless parameters

$$\varepsilon = \frac{Y-X}{2l}, \quad \mu = \frac{Y+X}{2l} > 0, \quad \omega = \frac{1}{\Omega}\sqrt{\frac{g}{l}}, \quad \beta = \frac{c}{m\,l^2\,\Omega}. \tag{38}$$

With this notation equation (36) with substituted (37) in it takes the following form

$$\ddot{\theta} + \beta\dot{\theta} + \mu\sin(\tau+\theta) = \varepsilon\sin(\tau-\theta) - \omega^2\sin(\theta+\delta), \tag{39}$$

where we use the formula $Y\cos(\Omega\,t)\sin(\theta) + X\sin(\Omega\,t)\cos(\theta) = \frac{Y+X}{2}\sin(\Omega\,t+\theta) - \frac{Y-X}{2}\sin(\Omega\,t-\theta)$.[1] Here the upper dot denotes differentiation with respect to the new time τ.

3.2. Exact rotational solution when $\varepsilon = 0$ and $\omega = 0$

Conditions $\varepsilon = \omega = 0$ mean that we find the mode of rotation for the circular excitation $X = Y$ with absence of gravity $g = 0$. In this case, we call equation (39) the *unperturbed equation*

[1] Note that this formula excludes the generalization $y = Y\cos(\Omega t + \Phi)$ which is considered e.g. in the model of unbalanced rotor [31–33]. Instead of Φ we introduce the angle δ of deviation of gravitational acceleration g from the vertical direction.

$$\ddot{\theta} + \beta\dot{\theta} + \mu \sin(\tau + \theta) = 0 \tag{40}$$

which has exact solutions

$$\theta = \theta_0 - \tau, \tag{41}$$

where constants θ_0 are defined by the following equality

$$\sin(\theta_0) = \frac{\beta}{\mu}, \tag{42}$$

provided that $|\beta| \leq \mu$.

To investigate the stability of these solutions we present the angle θ as $\theta = \theta_0 - \tau + \eta$, where $\eta = \eta(\tau)$ is a small addition, and substitute it in equation (40). Then linearizing (40) and using equality (42), we obtain the linear equation

$$\ddot{\eta} + \beta\dot{\eta} + \mu \cos(\theta_0)\eta = 0\,. \tag{43}$$

According to the Lyapunov stability theorem based on the linear approximation, solution (41) is asymptotically stable if all eigenvalues of linearized equation (43) have negative real parts. Which happens when all coefficients in (43) are positive

$$\beta > 0, \quad \mu \cos(\theta_0) > 0 \tag{44}$$

due to the Routh–Hurwitz conditions. From conditions (44), assumption $\mu > 0$ in (38), and equality (42), it follows for $\beta > 0$ that the solutions

$$\theta = \theta_0 - \tau, \quad \theta_0 = \arcsin\left(\frac{\beta}{\mu}\right) + 2\pi k \tag{45}$$

are asymptotically stable, while the solutions

$$\theta = \theta_0 - \tau, \quad \theta_0 = \pi - \arcsin\left(\frac{\beta}{\mu}\right) + 2\pi k \tag{46}$$

are unstable, where k is any integer number. For negative damping, $\beta < 0$, both these solutions are unstable. From now on we will assume that the following conditions are satisfied

$$0 < \beta < \mu, \tag{47}$$

which ensure the existence of stable rotational solution (45) as it is seen from (42) and (44). Indeed, in order to guarantee asymptotic stability β should be not only positive, but also strictly less than μ because of the second condition in (44), which can be transformed to inequality $\mu \cos(\theta_0) = \sqrt{\mu^2 - \beta^2} > 0$ with the use of the positive root for $\mu \cos(\theta_0)$ from (42).

3.3. Approximate rotational solutions when $\varepsilon \approx 0$ and $\omega \sim \sqrt{\varepsilon}$

We assume that values of ε and ω^2 are small of the same order of smallness, i.e. $\varepsilon \sim \omega^2 \ll 1$, so we can introduce new parameter $w = \omega^2/\varepsilon$. One can deduce from (38) and current assumptions that either gravity g is small or the frequency of excitation Ω is high with such damping c and mass m so that damping coefficient $\beta \sim 1$. All small terms are in the right-hand side of equation (39). To solve equation (39) we assume that general solution of equation (39) has the form

$$\theta = -\tau + \theta_0 + \varepsilon\theta_1 + \varepsilon^2\theta_2 + \dots \tag{48}$$

Next the general solution is substituted into equation (39), where sines are expanded into the Taylor series with respect to ε. By grouping together the terms with the same powers of ε and equating to zero, the set of differential equations is obtained

$$\ddot{\theta}_0 + \beta\dot{\theta}_0 + \mu\sin(\theta_0) = \beta, \tag{49}$$

$$\ddot{\theta}_1 + \beta\dot{\theta}_1 + \mu\cos(\theta_0)\,\theta_1 = \sin(2\tau - \theta_0) + w\sin(\tau - \theta_0 - \delta), \tag{50}$$

$$\ddot{\theta}_2 + \beta\dot{\theta}_2 + \mu\cos(\theta_0)\,\theta_2 = \mu\sin(\theta_0)\,\theta_1^2/2 - (\cos(2\tau - \theta_0) + w\cos(\tau - \theta_0 - \delta))\,\theta_1, \tag{51}$$

$$\dots$$

We have already found solution (41) for equation (49) in the previous section. Here we consider the same stable regular rotations 1:1 (with the period equal to the period of excitation) whose zero approximation is given by (45). Hence, θ_0 is a constant. Thus, equations (50) and (51) can be written in the following way

$$\ddot{\theta}_1 + \beta\dot{\theta}_1 + \sqrt{\mu^2 - \beta^2}\,\theta_1 = \sin(2\tau - \theta_0) + w\sin(\tau - \theta_0 - \delta) \tag{52}$$

$$\ddot{\theta}_2 + \beta\dot{\theta}_2 + \sqrt{\mu^2 - \beta^2}\,\theta_2 = \beta\theta_1^2/2 - (\cos(2\tau - \theta_0) + w\cos(\tau - \theta_0 - \delta))\,\theta_1, \tag{53}$$

where we denote $\mu\sin(\theta_0) = \beta$ and $\mu\cos(\theta_0) = \sqrt{\mu^2 - \beta^2}$ with the use of relation (42) and the second condition in (44).

3.3.1. *First order approximation*

In consequence of conditions (47) non-homogeneous linear differential equation (52) can be presented in the following form

$$\ddot{\theta}_1 + \beta\dot{\theta}_1 + \sqrt{\mu^2 - \beta^2}\,\theta_1 = A_1\cos(\tau) + B_1\sin(\tau) + A_2\cos(2\tau) + B_2\sin(2\tau) \tag{54}$$

where $A_1 = -w\cos(\delta)\beta/\mu - w\sin(\delta)\sqrt{1 - \beta^2/\mu^2}$, $B_1 = w\cos(\delta)\sqrt{1 - \beta^2/\mu^2} - w\sin(\delta)\beta/\mu$, $A_2 = -\beta/\mu$, $B_2 = \sqrt{1 - \beta^2/\mu^2}$, lower index denotes harmonics number. Equation (54) has a unique periodic solution

$$\theta_1(\tau) = a_1\cos(\tau) + b_1\sin(\tau) + a_2\cos(2\tau) + b_2\sin(2\tau)\,, \tag{55}$$

where $a_1 = -\frac{(1-\sqrt{\mu^2-\beta^2})A_1+\beta B_1}{\mu^2+1-2\sqrt{\mu^2-\beta^2}}$, $b_1 = -\frac{-\beta A_1+(1-\sqrt{\mu^2-\beta^2})B_1}{\mu^2+1-2\sqrt{\mu^2-\beta^2}}$, $a_2 = -\frac{(4-\sqrt{\mu^2-\beta^2})A_2+2\beta B_2}{3\beta^2+\mu^2+4(4-2\sqrt{\mu^2-\beta^2})}$, and $b_2 = \frac{-2\beta A_2+(4-\sqrt{\mu^2-\beta^2})B_2}{3\beta^2+\mu^2+4(4-2\sqrt{\mu^2-\beta^2})}$. Thus, the solution for (39) in the first approximation can be written as follows

$$\theta = -\tau + \theta_0 - \varepsilon \frac{2\beta\cos(2\tau - \theta_0) + \left(4 - \sqrt{\mu^2 - \beta^2}\right)\sin(2\tau - \theta_0)}{3\beta^2 + \mu^2 + 8(2 - \sqrt{\mu^2 - \beta^2})}$$
$$-\omega^2 \frac{\beta\cos(\tau - \theta_0 - \delta) + \left(1 - \sqrt{\mu^2 - \beta^2}\right)\sin(\tau - \theta_0 - \delta)}{\mu^2 + 1 - 2\sqrt{\mu^2 - \beta^2}} + o(\varepsilon), \tag{56}$$

where constant θ_0 is defined in (45).

3.3.2. Second order approximation

Equation (54) takes the following form

$$\ddot{\theta}_2 + \dot{\theta}_2 + \sqrt{\mu^2 - \beta^2}\theta_2 = \frac{A'_0}{2} + \sum_{n=1}^{4} \left(A'_n \cos(n\tau) + B'_n \sin(n\tau)\right), \tag{57}$$

where coefficients in the right-hand side are the following

$$\begin{aligned}
A'_0 &= (b_2^2 + b_1^2 + a_2^2 + a_1^2)\,\beta + (A_1 b_1 + A_2 b_2 - B_2 a_2 - B_1 a_1) \\
A'_1 &= (a_1 a_2 + b_1 b_2)\,\beta + (A_1 b_2 + A_2 b_1 - B_1 a_2 - B_2 a_1)/2 \\
A'_2 &= (a_1^2 - b_1^2)\,\beta/2 - (A_1 b_1 + B_1 a_1)/2 \\
A'_3 &= (a_1 a_2 - b_1 b_2)\,\beta - (A_2 b_1 + A_1 b_2 + B_1 a_2 + B_2 a_1)/2 \\
A'_4 &= (a_2^2 - b_2^2)\,\beta/2 - (A_2 b_2 + B_2 a_2)/2 \\
B'_1 &= (a_1 b_2 - b_1 a_2)\,\beta - (A_1 a_2 - A_2 a_1 + B_1 b_2 - B_2 b_1)/2 \\
B'_2 &= \beta a_1 b_1 + (A_1 a_1 - B_1 b_1)/2 \\
B'_3 &= (a_1 b_2 + b_1 a_2)\,\beta + (A_1 a_2 + A_2 a_1 - B_1 b_2 - B_2 b_1)/2 \\
B'_4 &= \beta a_2 b_2 + (A_2 a_2 - B_2 b_2)/2.
\end{aligned} \tag{58}$$

Periodic solution for equation (57) has the form

$$\theta_2(\tau) = \frac{A'_0}{2\sqrt{\mu^2 - \beta^2}} - \sum_{n=1}^{4} \frac{(n^2 - \sqrt{\mu^2 - \beta^2})A'_n + n\beta B'_n}{(n^2 - 1)\beta^2 + \mu^2 + n^2(n^2 - 2\sqrt{\mu^2 - \beta^2})}\cos(n\tau)$$
$$-\sum_{n=1}^{4} \frac{-n\beta A'_n + (n^2 - \sqrt{\mu^2 - \beta^2})B'_n}{(n^2 - 1)\beta^2 + \mu^2 + n^2(n^2 - 2\sqrt{\mu^2 - \beta^2})}\sin(n\tau). \tag{59}$$

Thus, second order approximate solution can be shortly written in the following form

$$\theta = -\tau + \theta_0 + \varepsilon\theta_1(\tau) + \varepsilon^2\theta_2(\tau) + o(\varepsilon^2), \tag{60}$$

where constant θ_0 is defined in (45), function θ_1 in (55), and function θ_2 in (59). In Fig. 10 it is shown how first and second order approximate solutions approach the numerical solution.

In this section the straightforward asymptotic method works since all solutions in each approximation converge to corresponding unique periodic solutions because damping β is not small. If damping β is small the analysis becomes more complicated since equation (49) takes the form $\ddot{\theta}_0 + \mu\sin(\theta_0) = 0$ and has a solution expressed in elliptic functions. One can simplify the analysis assuming that $\theta = -\tau + \varepsilon^{1/2}\theta_0 + \varepsilon\theta_1(\tau) + \varepsilon^{3/2}\theta_2(\tau) + \ldots$ instead of (48). We will use this assumption in the next section to apply the classical averaging technique to the problem of small damping $\beta \sim \sqrt{\varepsilon}$.

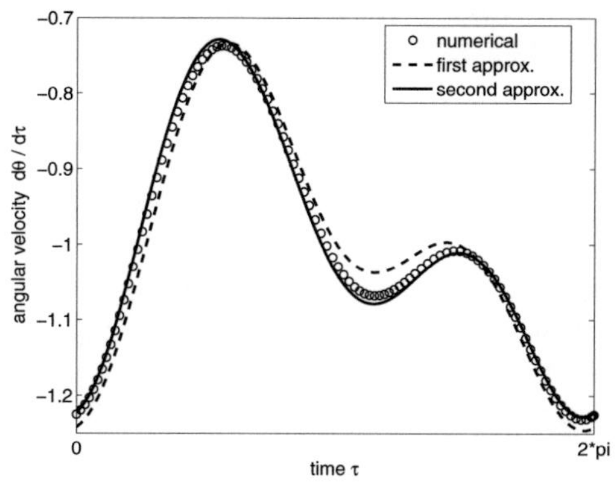

Figure 10. Angular velocities $\dot{\theta}$ calculated from the first order approximate solution (56), second order approximate solution (60), and results of numerical simulation, when damping coefficient β is not small. Parameters: $\delta = 0$, $\mu = 1$, $\omega = 0.3$, $\varepsilon = 0.2$, $\beta = 0.5$.

3.4. Approximate rotational solutions when $\varepsilon \approx 0$, $\omega \sim \sqrt{\varepsilon}$, and $\beta \sim \sqrt{\varepsilon}$

One can see in (38) that assumptions $\omega \sim \beta \sim \sqrt{\varepsilon}$ are valid for the high frequency of excitation $\Omega \sim 1/\sqrt{\varepsilon}$ with other parameters being of order 1. Another option is small gravity $g \sim \varepsilon$ along with small ratio $c/m \sim \sqrt{\varepsilon}$.

After change of variable $\theta = -\tau + \sqrt{\varepsilon}\vartheta$ equation (39) takes the following form

$$\ddot{\vartheta} + \mu\vartheta - \tilde{\beta} = \mu\left(\vartheta - \frac{\sin(\sqrt{\varepsilon}\vartheta)}{\sqrt{\varepsilon}}\right) - \sqrt{\varepsilon}\tilde{\beta}\dot{\vartheta} + \sqrt{\varepsilon}\sin(2\tau - \sqrt{\varepsilon}\vartheta) + \sqrt{\varepsilon}w\sin(\tau - \sqrt{\varepsilon}\vartheta - \delta) \quad (61)$$

with small right-hand side, where we denote $\tilde{\beta} = \beta/\sqrt{\varepsilon}$ and as in the previous section $w = \omega^2/\varepsilon$. With zero right-hand side equation (61) $\ddot{\vartheta} + \mu\vartheta - \tilde{\beta} = 0$ would describe harmonic oscillations about $\tilde{\beta}/\mu$ value with frequency $\sqrt{\mu}$. After Taylor's expansion of sines in the right-hand side of (61) about $\vartheta = 0$ we obtain the following equation

$$\begin{aligned} &\ddot{\vartheta} + (\mu + \varepsilon\cos(2\tau) + \varepsilon w\cos(\tau - \delta))\,\vartheta - \tilde{\beta} \\ &= \sqrt{\varepsilon}\left(\sin(2\tau) + w\sin(\tau - \delta) - \tilde{\beta}\dot{\vartheta}\right) + \varepsilon\mu\frac{\vartheta^3}{6} + o(\varepsilon), \end{aligned} \quad (62)$$

which describes oscillator with both basic and parametric excitations. To solve equation (62) we use the method of averaging [9, 18, 19]. For that purpose we write (62) in the *standard form* of first order differential equations with small right-hand sides. First, we use *Poincaré variables* q and ψ defined via the following solution of *generating system* $\ddot{\vartheta} + \mu\vartheta - \tilde{\beta} = 0$ which is (62) with $\varepsilon = 0$

$$\vartheta = \frac{\tilde{\beta}}{\mu} + q\cos(\psi), \quad \dot{\vartheta} = -\sqrt{\mu}q\sin(\psi). \quad (63)$$

In Poincaré variables equation (62) becomes a system of first order differential equations

$$\dot{q} = -\frac{\sin\psi}{\sqrt{\mu}}f(\tau, q, \psi), \quad \dot{\psi} = \sqrt{\mu} - \frac{\cos\psi}{q\sqrt{\mu}}f(\tau, q, \psi), \quad (64)$$

where small function $f(\tau, q, \psi) = \sqrt{\varepsilon} f_1(\tau, q, \psi) + \varepsilon f_2(\tau, q, \psi) + o(\varepsilon)$ is the right hand side of (61), where

$$f_1(\tau, q, \psi) = \sin(2\tau) + w\sin(\tau - \delta) + \tilde{\beta} q\sqrt{\mu}\sin(\psi), \tag{65}$$

$$f_2(\tau, q, \psi) = -\left(\cos(2\tau) + w\cos(\tau - \delta)\right)\left(\frac{\tilde{\beta}}{\mu} + q\cos(\psi)\right) + \frac{\mu}{6}\left(\frac{\tilde{\beta}}{\mu} + q\cos(\psi)\right)^3, \tag{66}$$

meaning that $f(\tau, q, \psi) = O(\sqrt{\varepsilon})$. Our next assumption is that $\sqrt{\mu} - 1 = O(\sqrt{\varepsilon})$ which means that excitation frequency is close to the first resonant frequency of basic excitation component $\sin(\tau - \delta)$ and to the first resonant frequency of parametric excitation component $\cos(2\tau)$ in equation (62). Thus, system (64) is transformed by $\psi = \zeta + \tau$ to the standard form

$$\dot{q} = -\frac{1}{\sqrt{\mu}}\sin(\zeta + \tau)\, f(\tau, q, \zeta + \tau), \tag{67}$$

$$\dot{\zeta} = \sqrt{\mu} - 1 - \frac{1}{q\sqrt{\mu}}\cos(\zeta + \tau)\, f(\tau, q, \zeta + \tau), \tag{68}$$

with small right-hand side, where new slow variable ζ is often referred to as *phase mismatch.*

In the second approximation so called *averaged equations* can be obtained from the system of equations (67) and (68) as follows, see (121) in the Appendix,

$$\begin{aligned}\dot{Q} = &-\sqrt{\varepsilon}\left(\frac{w}{2\sqrt{\mu}}\cos(Z + \delta) + \frac{\tilde{\beta}}{2}Q\right)\\ &+ \varepsilon\left(\frac{w\tilde{\beta}}{8\sqrt{\mu}}\left(\frac{4}{\mu} - 1\right)\sin(Z + \delta) + \frac{\sin(2Z)}{4\sqrt{\mu}}Q\right) + o(\varepsilon),\end{aligned} \tag{69}$$

$$\begin{aligned}\dot{Z} = &\sqrt{\mu} - 1 + \frac{\sqrt{\varepsilon}w}{2\sqrt{\mu}Q}\sin(Z + \delta)\\ &+ \varepsilon\left(\frac{w\tilde{\beta}}{8\sqrt{\mu}Q}\left(\frac{4}{\mu} - 1\right)\cos(Z + \delta) - \frac{\tilde{\beta}^2}{8}\left(\frac{2}{\mu\sqrt{\mu}} + 1\right) + \frac{\cos(2Z)}{4\sqrt{\mu}} - \frac{\sqrt{\mu}}{16}Q^2\right) + o(\varepsilon),\end{aligned} \tag{70}$$

where Q and Z are the *averaged variables* corresponding to q and ζ.

3.4.1. First order approximation

Stationary solutions ($\dot{Q} = 0$, $\dot{Z} = 0$) of (69)-(70) in the first approximation are the following

$$Q_{\{1\}}^2 = \frac{\omega^2/\varepsilon}{\mu\left(4(\sqrt{\mu} - 1)^2 + \beta^2\right)}, \quad Z_{\{1\}} = \arctan\left(\frac{2(\mu - 1)}{\beta}\right) - \delta + 2\pi k, \tag{71}$$

where we have substituted back $w = \omega^2/\varepsilon$ and $\tilde{\beta} = \beta/\sqrt{\varepsilon}$. Symbol arctan stands for the principal value of the function on the interval from 0 to π. Note that the phase $Z_{\{1\}}$ is determined to within 2π rather than π, since the functions $\sin(Z_{\{1\}})$ and $\cos(Z_{\{1\}})$ obtained from equations (69) and (70) determine $Z_{\{1\}}$ up to an additive term $2\pi k$. Solution of system (67)-(68) in the first approximation is $q = Q_{\{1\}} + o(1)$, $\zeta = Z_{\{1\}} + o(1)$ so the solution of (39) is the following

$$\theta = -\tau + \frac{\beta}{\mu} + \sqrt{\varepsilon}Q_{\{1\}}\cos(Z_{\{1\}} + \tau) + o(\sqrt{\varepsilon}), \tag{72}$$

which does not contain higher harmonics observed numerically. That is why we need to proceed to the second order approximation.

3.4.2. Second order approximation

In the second approximation averaged equations can be obtained as described in Appendix. Stationary solutions ($\dot{Q} = 0$, $\dot{Z} = 0$) of (69), (70) in the second approximation can be found numerically or with the absence of gravity ($\omega = 0$) analytically. Solution of system (67), (68) in the second approximation is the following, see (119),

$$q = Q_{\{2\}} + \frac{\sqrt{\varepsilon}}{2\sqrt{\mu}}\left(-\sin\left(\tau - Z_{\{2\}}\right) + \frac{w}{2}\sin\left(2\tau + Z_{\{2\}} - \delta\right) + \frac{1}{3}\sin\left(3\tau + Z_{\{2\}}\right)\right)$$
$$+\sqrt{\varepsilon}\frac{\tilde{\beta}\,Q_{\{2\}}}{4}\sin\left(2\tau + 2Z_{\{2\}}\right) + o(\sqrt{\varepsilon}), \tag{73}$$

$$\zeta = Z_{\{2\}} + \frac{\sqrt{\varepsilon}}{2\sqrt{\mu}Q_{\{2\}}}\left(\cos\left(\tau - Z_{\{2\}}\right) + \frac{w}{2}\cos\left(2\tau + Z_{\{2\}} - \delta\right) + \frac{1}{3}\cos\left(3\tau + Z_{\{2\}}\right)\right)$$
$$+\sqrt{\varepsilon}\frac{\tilde{\beta}}{4}\cos\left(2\tau + 2Z_{\{2\}}\right) + o(\sqrt{\varepsilon})\,. \tag{74}$$

Substitution of these expressions into (63) yields the second order approximate solution of (61), which after changes of variable $\theta = -\tau + \sqrt{\varepsilon}\vartheta$ and parameters $w = \omega^2/\varepsilon$, $\tilde{\beta} = \beta/\sqrt{\varepsilon}$ results in the approximate solution of the original equation (39)

$$\theta = -\tau + \frac{\beta}{\mu} + \sqrt{\varepsilon}Q_{\{2\}}\cos(Z_{\{2\}} + \tau) + \sqrt{\varepsilon}\frac{\beta\,Q_{\{2\}}}{4}\sin\left(Z_{\{2\}} + \tau\right)$$
$$+\frac{\omega^2\sin(\tau - \delta)}{4\sqrt{\mu}} - \frac{\varepsilon\sin(2\tau)}{3\sqrt{\mu}} + o(\varepsilon). \tag{75}$$

Agreement of solution (75) with the numerical experiment is shown in Fig. 11. We see that the amplitude of angular velocity oscillations is much higher than that for not small β in Fig. 10.

4. Twirling of a hula-hoop

A hula-hoop is a popular toy – a thin hoop that is twirled around the waist, limbs or neck. In recent decades it is widely used as an implement for fitness and gymnastic performances.[2] To twirl a hula-hoop the waist of a gymnast carries out a periodic motion in the horizontal plane. For the sake of simplicity we consider the two-dimensional problem disregarding the vertical motion of the hula-hoop. We assume that the waist is a circle and its center moves along an elliptic trajectory close to a circle.

Previously considered was the simple case in which a hula-hoop is treated as a pendulum with the pivot oscillating along a line, see [37, 38]. The stationary rotations of a hula-hoop

[2] The same model lies in the basis of some industrial machinery such as vibrating cone crushers designed for crushing hard brittle materials, see [33].

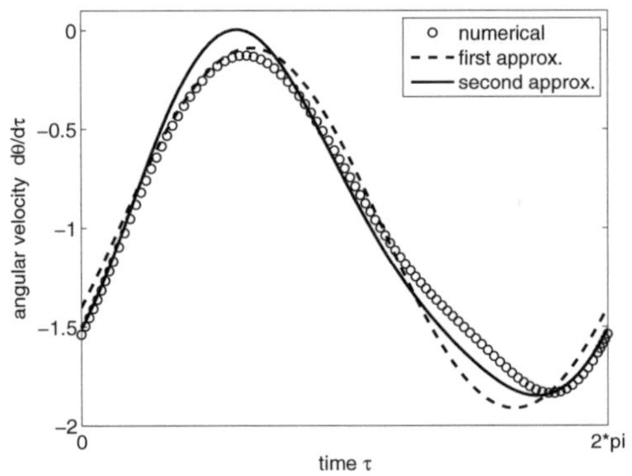

Figure 11. Angular velocity $\dot{\theta}$ calculated from the first order approximate solution (72) and second order approximate solution (75) compared with the results of numerical simulations in the case of small damping β. Parameters: $\delta = 0$, $\mu = 1$, $\omega = 0.3$, $\varepsilon = 0.2$, $\beta = 0.01$. Steady state averaged variables Q and Z are given by expressions in (71) for the first approximation while for the second approximation they are obtained numerically ($Q = 2.0348$, $Z = 2.6838$) from the second order averaged equations (69), (70).

excited in two directions have been studied by an approximate method of separate motions in [33]. The similar problem of the spinner mounted loosely on a pivot with a prescribed bi-directional motion has been treated numerically and experimentally in [39].

Here we derive the exact solutions in the case of a circular trajectory of the waist center and approximate solutions in the case of an elliptic trajectory. We also check the condition of keeping contact with the waist during twirling.

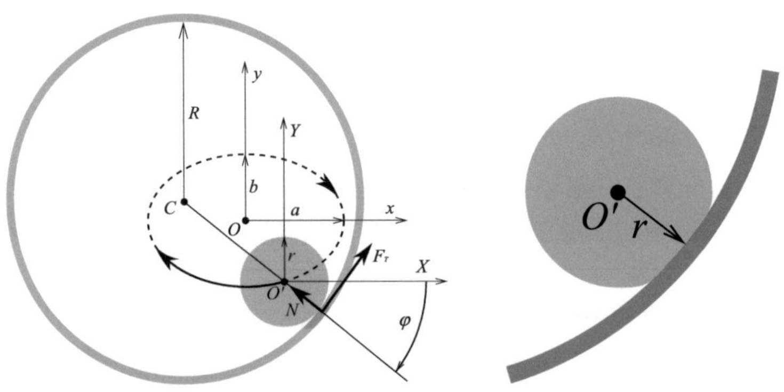

Figure 12. A hula-hoop with the radius R twirling with the angle φ around a circular waist (shaded) with the radius r. The center O' of the waist moves along the elliptic curve $x = a \sin \omega t$, $y = b \cos \omega t$ with the fixed center O. The hula-hoop acts on the waist with normal force N and tangential friction force F_T. There is also a rolling resistance due to the waist deformation (right).

4.1. Main relations

We assume that the center O' of a gymnast's waist moves in time according to the elliptic law $x = a \sin \omega t$, $y = b \cos \omega t$ with the amplitudes a, b and the excitation frequency $\omega > 0$, Fig. 12.

The equations of motion in the waist-fixed coordinate system take the following form

$$I_C\ddot{\theta} + k\dot{\theta} = -F_T R - d\,N\,\mathrm{sign}(\dot{\theta})\,, \tag{76}$$

$$m(R-r)\ddot{\varphi} = m\left(\ddot{x}\sin\varphi + \ddot{y}\cos\varphi\right) + F_T\,, \tag{77}$$

$$m(R-r)\dot{\varphi}^2 = N + m\left(\ddot{x}\cos\varphi - \ddot{y}\sin\varphi\right)\,, \tag{78}$$

where θ is the rotation angle around center of mass C, $I_C = mR^2$ is the central moment of inertia of the hula-hoop, φ is the angle between axis x and radius CO', r is the radius of the waist, m and R are the mass and radius of the hula-hoop. Equation (76) describes change of angular momentum due to linear viscous damping with coefficient k, rolling drag (rolling resistance) with coefficient d, and the tangential friction force F_T between the waist and the hoop. Equations (77) and (78) describe the motion of the hula-hoop in the longitudinal and transverse directions to the radius CO', where N is the normal reaction force of the hula-hoop to the waist. Equations (77) and (78) contain additional inertial forces since the waist-fixed reference system is noninertial.

Assuming that slipping at the point of contact is absent we obtain the kinematic relation

$$(R-r)\,\dot{\varphi} = R\dot{\theta}\,. \tag{79}$$

We exclude from equations (76) and (77) the force F_T and with relation (79) obtain the equation of motion

$$\begin{aligned}\ddot{\varphi} + \frac{k}{2mR^2}\dot{\varphi} + \frac{d}{2R}\,\mathrm{sign}\dot{\varphi}\left(\dot{\varphi}^2 + \frac{\omega^2\left(a\sin\omega t\cos\varphi - b\cos\omega t\sin\varphi\right)}{R-r}\right)\\ + \frac{\omega^2\left(a\sin\omega t\sin\varphi + b\cos\omega t\cos\varphi\right)}{2\left(R-r\right)} = 0.\end{aligned} \tag{80}$$

From equation (78) we find the normal force and imply the condition $N > 0$ as

$$(R-r)\,\dot{\varphi}^2 + \omega^2\left(a\sin\omega t\cos\varphi - b\cos\omega t\sin\varphi\right) > 0 \tag{81}$$

which means that the hula-hoop during its motion keeps contact with the waist of the gymnast.

We introduce new time $\tau = \omega t$ and non-dimensional parameters

$$\gamma = \frac{k}{2mR^2\omega}, \quad \delta = \frac{d}{2R}, \quad \varepsilon = \frac{a-b}{4\left(R-r\right)}, \quad \mu = \frac{a+b}{4\left(R-r\right)}, \tag{82}$$

where γ and δ are the damping and rolling resistance coefficients, μ and ε are the excitation parameters. Relation between μ and ε determines the form of ellipse – the trajectory of the waist center. For $\varepsilon = \mu$ the trajectory is a line, and for $\varepsilon = 0$ it is a circle. Then equation (80) and inequality (81) take the form

$$\begin{aligned}\ddot{\varphi} &+ \gamma\dot{\varphi} + \delta\dot{\varphi}\left|\dot{\varphi}\right| + \mu\cos(\varphi-\tau) - 2\mu\delta\,\mathrm{sign}(\dot{\varphi})\sin(\varphi-\tau)\\ &= \varepsilon\cos(\varphi+\tau) - 2\varepsilon\delta\,\mathrm{sign}(\dot{\varphi})\sin(\varphi+\tau)\,,\end{aligned} \tag{83}$$

$$\dot{\varphi}^2 - 2\mu\sin(\varphi-\tau) + 2\varepsilon\sin(\varphi+\tau) > 0\,, \tag{84}$$

where the dot means differentiation with respect to the time τ.

4.2. Exact solutions

When the waist center moves along a circle ($a = b$, i.e. $\varepsilon = 0$) equation (83) takes the following form

$$\ddot{\varphi} + \gamma\dot{\varphi} + \delta\dot{\varphi}\,|\dot{\varphi}| + \mu\cos(\varphi - \tau) - 2\mu\delta\,\mathrm{sign}(\dot{\varphi})\sin(\varphi - \tau) = 0 \tag{85}$$

and has the exact solution [6]

$$\varphi = \tau + \varphi_0 \tag{86}$$

with the constant initial phase φ_0 given by the equation

$$\gamma + \delta + \mu\cos\varphi_0 - 2\mu\delta\sin\varphi_0 = 0. \tag{87}$$

Therefore, solution (86) exists only under the condition $|\gamma + \delta| \leq |\mu|\sqrt{1 + 4\delta^2}$, so we find from equation (87)

$$\varphi_0 + \arccos\left(\frac{1}{\sqrt{1+4\delta^2}}\right) = \pm\arccos\left(-\frac{\gamma+\delta}{\mu\sqrt{1+4\delta^2}}\right) + 2\pi n, \quad n = 0, 1, 2, \ldots \tag{88}$$

provided that $\mu \neq 0$. Solutions (86), (88) correspond to the rotation of the hula-hoop with the constant angular velocity equal to the excitation frequency ω.

4.2.1. Stability of the exact solutions

Let us investigate the stability of the obtained solutions. For this purpose we take the angle φ in the form $\varphi = \tau + \varphi_0 + \eta(\tau)$ where $\eta(\tau)$ is a small quantity, and substitute it into equation (85). Taking linearization with respect to η and with the use of (87) we obtain a linear equation

$$\ddot{\eta} + (\gamma + 2\delta)\,\dot{\eta} - \mu\,(\sin\varphi_0 + 2\delta\cos\varphi_0)\,\eta = 0. \tag{89}$$

According to Lyapunov's theorem on the stability based on a linear approximation [17] solution (86), (88) is asymptotically stable if all the eigenvalues of linearized equation (89) have negative real parts. From Routh-Hurwitz criterion [17] we obtain the stability conditions as

$$\gamma + 2\delta > 0, \quad \mu\,(\sin\varphi_0 + 2\delta\cos\varphi_0) < 0. \tag{90}$$

Without loss of generality we assume $\mu > 0$ since the case $\mu < 0$ can be reduced to the previous one by the time transformation $\tau' = \tau + \pi$ in equation (83). The second condition in (90) can be written as $\sin(\varphi_0 + \arccos(1/\sqrt{1+4\delta^2})) < 0$. Thus, from conditions (90), relation (88) and due to the assumption $\mu > 0$ we find that for $0 < \gamma + 2\delta < \delta + \mu\sqrt{1+4\delta^2}$ solution (86) with

$$\varphi_0 = -\arccos\left(-\frac{\gamma+\delta}{\mu\sqrt{1+4\delta^2}}\right) - \arccos\left(\frac{1}{\sqrt{1+4\delta^2}}\right) + 2\pi n, \quad n = 0, 1, 2, \ldots \tag{91}$$

is asymptotically stable, and solution (86) with

$$\varphi_0 = \arccos\left(-\frac{\gamma+\delta}{\mu\sqrt{1+4\delta^2}}\right) - \arccos\left(\frac{1}{\sqrt{1+4\delta^2}}\right) + 2\pi n, \quad n = 0, 1, 2, \ldots \tag{92}$$

is unstable.

4.2.2. Condition for hula-hoop's contact with the waist

Let us verify for the exact solutions (86), (88) the condition of twirling without losing contact (84) which takes the form

$$\mu \sin \varphi_0 < \frac{1}{2}. \tag{93}$$

The second stability condition in (90) can be rewritten with the use of (87) as follows

$$\mu \sin \varphi_0 < \frac{2\delta\,(\gamma + \delta)}{1 + 4\delta^2}.$$

Thus, stable solution (86), (91) provides asymptotically stable twirling of the hula-hoop with the constant angular velocity ω without losing contact with the waist of the gymnast under the condition $\gamma\delta < 1/4$.

Without rolling resistance $\delta = 0$ condition (93) is always satisfied for the stable solution (86), (91). While for unstable solution (86), (92) we have $\mu \sin \varphi_0 = \sqrt{\mu^2 - \gamma^2}$ so condition (93) holds only if $\mu < \sqrt{1/4 + \gamma^2}$. The phase φ_0 of the stable solution belongs to the interval $[-\pi, -\pi/2] \mod 2\pi$, and for vanishing damping $\gamma \to +0$ the phase tends to $-\pi/2$. Below we will show that this phase inequality also holds for the approximate solutions. This is how to twirl a hula-hoop!

4.3. Approximate solutions

Let us find approximate solutions for the case of close but not equal amplitudes $a \approx b$. For the sake of simplicity from now on we will keep $\delta = 0$ and assume that $a \geq |b|$ which means $\varepsilon \geq 0$, $\mu \geq 0$. Taking ε as a small parameter we apply perturbation method assuming that the exact solution $\varphi_s(\tau)$ of (83) can be expressed in a series

$$\varphi_s(\tau) = \tau + \varphi_0 + \varepsilon\varphi_1(\tau) + o(\varepsilon). \tag{94}$$

After substitution of series (94) in (83) and grouping the terms by equal powers of ε we derive the following chain of equations

$$\varepsilon^0: \quad \gamma + \mu \cos(\varphi_0) = 0 \tag{95}$$

$$\varepsilon^1: \quad \ddot{\varphi}_1 + \gamma\dot{\varphi}_1 - \mu \sin(\varphi_0)\varphi_1 = \cos(\varphi_0 + 2\tau) \tag{96}$$

Taking solution of equation (95) for $\mu > 0$

$$\varphi_0 = -\arccos(-\gamma/\mu) + 2\pi n, \quad n = 1, 2, \ldots \tag{97}$$

corresponding to the stable unperturbed solution (86), (91) we write equation (96) as

$$\ddot{\varphi}_1 + \gamma\dot{\varphi}_1 + \sqrt{\mu^2 - \gamma^2}\,\varphi_1 = \cos(\varphi_0 + 2\tau)\,. \tag{98}$$

It has a unique periodic solution

$$\varphi_1(\tau) = C \sin(2\tau + \varphi_0) + D \cos(2\tau + \varphi_0)\,, \tag{99}$$

where constants C and D are defined as follows

$$C = \frac{2\gamma}{\mu^2+3\gamma^2-8\sqrt{\mu^2-\gamma^2}+16}, \quad D = \frac{-4+\sqrt{\mu^2-\gamma^2}}{\mu^2+3\gamma^2-8\sqrt{\mu^2-\gamma^2}+16}. \tag{100}$$

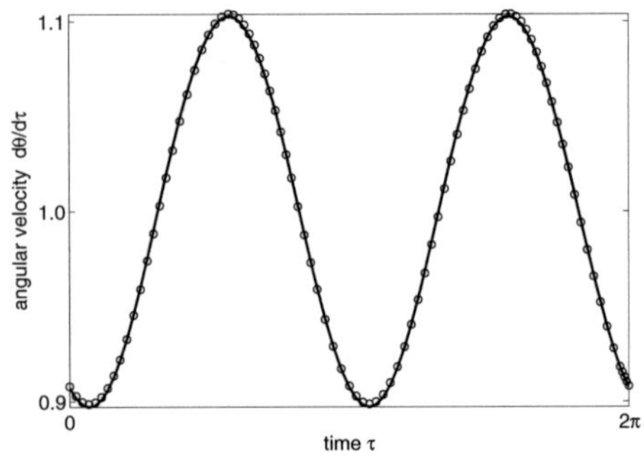

Figure 13. Comparison between the approximate analytical (solid line) and numerical (circles) results for parameters: $\mu = 1.2$, $\varepsilon = 0.2$, and $\gamma = 1$. Initial conditions for numerical calculations are $\theta(0) = -2.5389$, $\dot{\theta}(0) = 0.90802$.

We see that the approximate solution $\varphi(\tau) = \tau + \varphi_0 + \varepsilon\varphi_1(\tau)$ with (97), (99), (100) differs from the exact solution (86), (91) of the unperturbed system by small vibrating terms of frequency 2, see (99). Note that the approximate solutions were obtained with the assumption that the excitation amplitudes and damping are not small.

4.3.1. Stability of the approximate solutions

To find the stability conditions for solution (94), (97), (99) we take a small variation to the solution of (94) $\varphi = \varphi_s + u$ and substitute this expression into (83). After linearization with respect to u and keeping only terms of first order we obtain a linear equation

$$\ddot{u} + \gamma\dot{u} + (-\mu(\sin\varphi_0 + \varepsilon\varphi_1\cos\varphi_0) + \varepsilon\sin(2\tau + \varphi_0))\,u = 0\,, \tag{101}$$

where φ_0 is given by expression (97). Equation (101) can be written in the form of damped Mathieu-Hill equation as

$$\ddot{u} + \gamma\dot{u} + (p + \varepsilon\Phi(2\tau))\,u = 0\,, \tag{102}$$

where $p = -\mu\sin\varphi_0 = \sqrt{\mu^2 - \gamma^2}$, $\Phi(2\tau) = (\gamma C + 1)\sin(2\tau + \varphi_0) + \gamma D\cos(2\tau + \varphi_0)$. Then the stability condition (absence of parametric resonance at all frequencies $\sqrt{p}$) is given by the inequalities [17]

$$\varepsilon < \frac{2\gamma}{\sqrt{(\gamma C + 1)^2 + \gamma^2 D^2}} \tag{103}$$

with C and D defined in (100). This is the inequality to the problem parameters γ, ε and μ.

4.3.2. Condition for hula-hoop's contact with the waist

The condition of twirling without losing contact (84) takes the following form

$$\varepsilon < \frac{1 + 2\sqrt{\mu^2 - \gamma^2}}{2}\sqrt{\frac{\mu^2 + 3\gamma^2 - 8\sqrt{\mu^2 - \gamma^2} + 16}{\mu^2 + 8\gamma^2 - 12\sqrt{\mu^2 - \gamma^2} + 36}}. \tag{104}$$

Conditions (103), (104) imply restrictions to ε, i.e. how much the elliptic trajectory of the waist center differs from the circle.

4.3.3. Comparison with numerical simulations

In Fig. 13 the approximate analytical solution is presented and compared with the results of numerical simulation for the case when the excitation parameter μ and damping coefficient γ are not small.

4.4. Small excitation amplitudes and damping

It is interesting to consider the case when the excitation amplitudes and damping coefficient are small having the same order as ε. Then we introduce new parameters $\tilde{\mu} = \mu/\varepsilon$ and $\tilde{\gamma} = \gamma/\varepsilon$ and assume that the solution has the form

$$\varphi(\tau) = \rho\tau + \varphi_0(\tau) + \varepsilon\varphi_1(\tau) + o(\varepsilon), \tag{105}$$

where ρ is the angular velocity of rotation, and the functions $\varphi_0(\tau)$, $\varphi_1(\tau)$ are supposed to be bounded.

We substitute expression (105) into (83) and equating terms of the same powers of ε obtain the following equations

$$\varepsilon^0 : \ddot{\varphi}_0 = 0 \tag{106}$$

$$\varepsilon^1 : \ddot{\varphi}_1 = \cos(\varphi_0 + \tau + \rho\tau) - \tilde{\mu}\cos(\varphi_0 - \tau + \rho\tau) - \tilde{\gamma}\dot{\varphi}_0 - \tilde{\gamma}\rho. \tag{107}$$

From equation (106) we get that function $\varphi_0(\tau)$ can remain bounded only if it is constant ($\varphi_0(\tau) \equiv \varphi_0 = const$). Then equation (107) can have bounded solutions $\varphi_1(\tau)$ only when ρ takes the values -1, 0, 1. Thus, besides clockwise rotation we have also counterclockwise rotation $\rho = -1$, and no rotational solution $\rho = 0$. The letter is not interesting, so we omit it.

4.4.1. Clockwise rotation

For clockwise rotation $\rho = 1$, see Fig. 14 a), from equations (105), (106), and (107) we obtain in the first approximation the solution

$$\varphi_*(\tau) = \tau + \varphi_0 + \varepsilon\varphi_1(\tau), \tag{108}$$

$$\varphi_1(\tau) = -\frac{1}{4}\cos(\varphi_0 + 2\tau), \quad \cos\varphi_0 = -\frac{\gamma}{\mu}, \tag{109}$$

where function $\varphi_1(\tau)$ is found up to the addition of a constant, which we have set to zero for determinacy. Thus, we let only φ_0 contain a constant term of the solution. The first expression in (109) is the special case of (99), (100) when $\mu = \gamma = 0$.

To verify the stability conditions for solution (108) we use damped Mathieu-Hill equation (102) and for the case of small damping and excitation amplitudes get $\gamma > 0$, $\sin\varphi_0 < 0$, see [17]. These conditions are similar with inequalities (90) derived for the undisturbed exact solution. Thus, the stable solution (108) with $\varphi_0 = -\arccos(-\gamma/\mu) + 2\pi n$ exists for

$$0 < \gamma < \mu. \tag{110}$$

For solution (108) condition (84) of keeping contact in the first approximation reads

$$\varepsilon < \frac{1 + 2\sqrt{\mu^2 - \gamma^2}}{3}, \tag{111}$$

and holds true for sufficiently small ε.

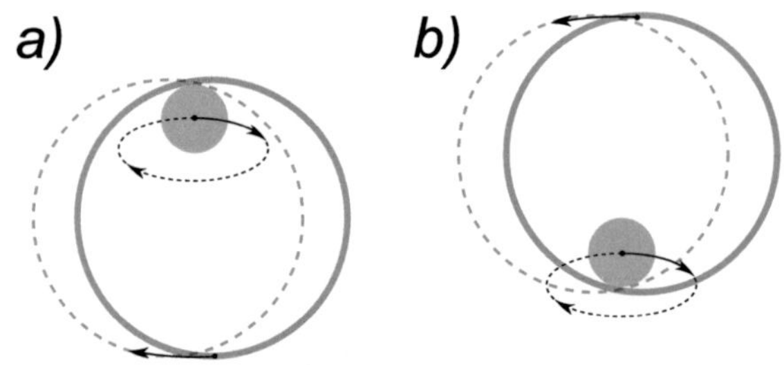

Figure 14. Stable twirling of the hula-hoop for the cases: a) direct twirling b) inverse twirling.

4.4.2. Counterclockwise rotation

For counterclockwise rotation $\rho = -1$ in Fig. 14 b) we obtain in the first approximation the solution

$$\varphi_*(\tau) = -\tau + \varphi_0 + \frac{\mu}{4}\cos(\varphi_0 - 2\tau), \quad \cos\varphi_0 = -\frac{\gamma}{\varepsilon}, \tag{112}$$

with the stability conditions $\gamma > 0$, $\sin\varphi_0 > 0$. Thus, the stable counterclockwise rotation (112) with $\varphi_0 = \arccos(-\gamma/\varepsilon) + 2\pi n$ exists for

$$0 < \gamma < \varepsilon. \tag{113}$$

For this case condition (84) takes the form similar to (111) and holds true for sufficiently small μ

$$\mu < \frac{1 + 2\sqrt{\varepsilon^2 - \gamma^2}}{3}. \tag{114}$$

4.4.3. Coexistence of clockwise and counterclockwise rotations

It follows from conditions (110), (113) that stable clockwise and counterclockwise rotations (108), (109) and (112) coexist if the following conditions are satisfied

$$0 < \gamma < \min\{\varepsilon, \mu\}. \tag{115}$$

Conditions (115) in physical variables take the form

$$0 < 2k\frac{R - r}{R^2\omega m} < a - |b|, \tag{116}$$

meaning that the trajectory of the waist should be sufficiently prolate. Coexisting clockwise and counterclockwise rotations are illustrated in Fig. 14.

4.4.4. Comparison with numerical simulations

In Fig. 15 the approximate analytical solutions for rotations in both directions are presented and compared with the results of numerical simulation for the case of small excitation parameters μ, ε and the damping coefficient γ. The values of μ, ε correspond to the dimensional parameters $a = 15cm$, $b = 10cm$, $r = 10cm$, $R = 50cm$.

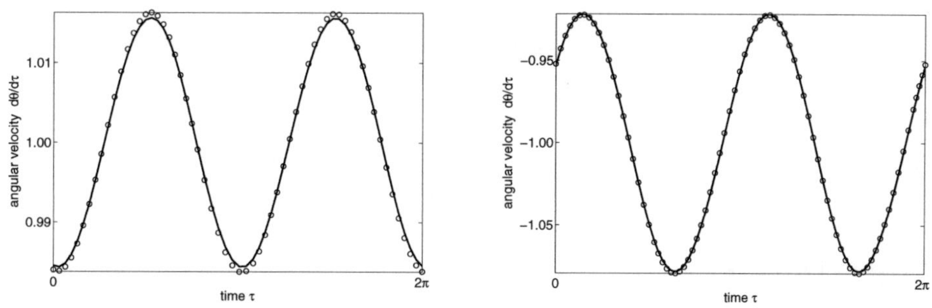

Figure 15. Comparison between the approximate analytical (solid line) and numerical (circles) results for small excitation amplitudes and damping coefficient for clockwise (left) and counterclockwise (right) rotations, for parameters: $\mu = 5/32$, $\varepsilon = 1/32$, and $\gamma = 1/40$. Initial conditions for numerical calculations are $\theta(0) = -1.7301$, $\dot{\theta}(0) = 0.98403$ (left) and $\theta(0) = 2.4696$, $\dot{\theta}(0) = -0.95212$ (right).

5. Conclusions

In section 2 we showed that the pendulum with periodically varying length exhibits diversity of behavior types. We recognized that the analytical stability boundaries of the vertical position of the pendulum and the frequency-response curve for limit cycles are in a good agreement with the numerical results. The second resonance zone appeared to be empty. The stability conditions of limit cycles are derived based on direct use of Lyapunov's theorem on stability of periodic solutions. We found numerically regular rotation, oscillation, and rotation-oscillation regimes with various periods and mean angular velocities of the pendulum including high-speed rotations and rotations with fractional relative velocities (it is rotation-oscillation regime when the pendulum makes regular sequence of rotations in both directions). We derived analytically the conditions for existence of regular rotation and oscillation regimes which agree with the numerical results. Domains for chaotic motions are found and analyzed numerically in the parameter space via calculation of Lyapunov exponents and bifurcation diagrams. Basins of attractions of different regimes of the pendulum motion were plotted and analyzed.

In section 3 we studied the planar rotational motion of the pendulum with the pivot oscillating both vertically and horizontally when the trajectory of the pivot is an ellipse close to a circle. The analysis of motion was based on the exact rotational solutions in the case of circular pivot trajectory and zero gravity. The conditions for existence and stability of such solutions were derived. Assuming that the amplitudes of excitations are not small while the pivot trajectory has small ellipticity the approximate solutions were found both for large and small linear damping. Comparison between approximate and numerical solutions was made for different values of the damping parameter demonstrating good accuracy of the method involved.

Finally, in section 4 we assumed that the waist of a sportsman twirling a hula hoop is a circle and its center moves along an elliptic trajectory close to a circle. We studied the system with both small and not small linear viscous damping as well as with some rolling resistance. For the case of the circular trajectory, two families of the exact solutions were obtained, similar to those in section 3. Both of them correspond to twirling of the hula-hoop with a constant angular speed equal to the speed of the excitation. We showed that one family of the solutions is stable, while the other one is unstable. These exact solutions allowed us to obtain the approximate solutions for the case of an elliptic trajectory of the waist. An interesting effect of inverse twirling was described when the waist moves in opposite direction to the hula-hoop rotation. It is shown that the approximate analytical solutions agree with the results of numerical simulation.

Acknowledgements

The authors express their gratitude to Angelo Luongo for his contribution to dynamics of a pendulum with variable length and fruitful discussions on nonlinear mechanics.

Author details

Belyakov Anton
Institute of Mechanics, Lomonosov Moscow State University (MSU), Moscow, Russia
Institute of Mathematical Methods in Economics, Vienna University of Technology, Vienna, Austria

Seyranian Alexander P.
Institute of Mechanics, Lomonosov Moscow State University (MSU), Moscow, Russia

Appendix. The standard method of averaging

The standard averaging method is a straightforward procedure and in the literature [9, 18] usually only two first approximations are written down. We use the averaging method down to the fifth approximation and feel obliged to present the resulting formulas.

The standard averaging method finds the transformation of the equation system with small and 2π time periodic right hand side

$$\dot{x} = f_1(x,t) + f_2(x,t) + \ldots + f_6(x,t) \tag{117}$$

into autonomous system (also called averaged system) which could be solved analytically

$$\dot{X} = F_1(X) + F_2(X) + \ldots + F_6(X) + \ldots, \tag{118}$$

where the lower index denotes the order of smallness with respect to one. A new variable X is presented in the following series

$$x = X + u_1(X,t) + u_2(X,t) + \ldots + u_5(X,t) + \ldots, \tag{119}$$

which is the approximate solution of system (117) and, hence, of the transformed system (118). The functions u_i and F_i can be found one by one after differentiating (119) with respect to time, substituting there expressions (117) and (118), expanding the functions f_i around X into Taylor's series, and collecting there terms of the same order.

Averaging operator is defined as follows

$$\langle\cdot\rangle = \lim_{T\to\infty}\frac{1}{T}\int_0^T \cdot|_{x=X}d\tau = \frac{1}{2\pi}\int_0^{2\pi}\cdot|_{x=X}d\tau.$$

We also define integral operator $\{\cdot\}$ with the following expression

$$\{f(x,\tau)\} = \int (f(x,\tau) - \langle f(x,\tau)\rangle)\, d\tau,$$

which is such an antiderivative that satisfies the condition $\langle\{f(x,\tau)\}\rangle = \{\langle f(x,\tau)\rangle\} = 0$. Latter condition is necessary to obviate an ambiguity. We define following vector product operators

$$fu_k = \sum_{i=1}^{n}\frac{\partial f}{\partial x_i}u_k^i, \quad fu_{k,m} = \sum_{i,j=1}^{n}\frac{\partial^2 f}{\partial x_i \partial x_j}u_k^i u_m^j, \quad \ldots$$

and so on, where i and j are the indices of vector components placed in u_k^i and u_m^j on the top not to confuse it with smallness order indices k and m; n is the length of vectors f, u_k and u_m. Hence, we can write recurrent expressions

$$F_1(X) = \langle f_1(x,t)\rangle, \tag{120}$$

$$u_1(X,t) = \{f_1(x,t)\} + U_1(X), \quad F_2(X) = \langle f_2(x,t) + f_1(x,t)u_1(X,t)\rangle, \tag{121}$$

$$u_2 = \{f_2 + f_1u_1 - u_1F_1\} + U_2, \quad F_3 = \left\langle f_3 + f_1u_2 + f_2u_1 + \frac{1}{2}f_1u_{1,1}\right\rangle, \tag{122}$$

$$u_3 = \left\{f_3 + f_1u_2 + f_2u_1 + \frac{1}{2}f_1u_{1,1} - u_1F_2 - u_2F_1\right\} + U_3, \tag{123}$$

$$F_4 = \left\langle f_4 + f_1u_3 + f_2u_2 + f_3u_1 + f_1u_{1,2} + \frac{1}{2}f_2u_{1,1} + \frac{1}{6}f_1u_{1,1,1}\right\rangle, \tag{124}$$

$$\begin{aligned} u_4 = \Big\{&f_4 + f_1u_3 + f_2u_2 + f_3u_1 + f_1u_{1,2} + \frac{1}{2}f_2u_{1,1} + \frac{1}{6}f_1u_{1,1,1} \\ &- u_1F_3 - u_2F_2 - u_3F_1\Big\} + U_4, \end{aligned} \tag{125}$$

$$\begin{aligned} F_5 = \Big\langle &f_5 + f_1u_4 + f_2u_3 + f_3u_2 + f_4u_1 + f_1u_{1,3} + f_2u_{1,2} \\ &+ \frac{1}{2}f_1u_{2,2} + \frac{1}{2}f_3u_{1,1} + \frac{1}{2}f_1u_{1,1,2} + \frac{1}{6}f_2u_{1,1,1} + \frac{1}{24}f_1u_{1,1,1,1}\Big\rangle, \end{aligned} \tag{126}$$

$$\begin{aligned} u_5 = \Big\{&f_5 + f_1u_4 + f_2u_3 + f_3u_2 + f_4u_1 + f_1u_{1,3} + f_2u_{1,2} \\ &+ \frac{1}{2}f_1u_{2,2} + \frac{1}{2}f_3u_{1,1} + \frac{1}{2}f_1u_{1,1,2} + \frac{1}{6}f_2u_{1,1,1} + \frac{1}{24}f_1u_{1,1,1,1} \\ &- u_1F_4 - u_2F_3 - u_3F_2 - u_4F_1\Big\} + U_5, \end{aligned} \tag{127}$$

$$F_6 = \Big\langle f_6 + f_1 u_5 + f_2 u_4 + f_3 u_3 + f_4 u_2 + f_5 u_1 + f_1 u_{1,4} + f_1 u_{2,3} + f_2 u_{1,3} + f_3 u_{1,2}$$
$$+ \frac{1}{2} f_2 u_{2,2} + \frac{1}{2} f_4 u_{1,1} + \frac{1}{2} f_1 u_{1,2,2} + \frac{1}{2} f_2 u_{1,1,2} + \frac{1}{6} f_3 u_{1,1,1} + \frac{1}{6} f_1 u_{1,1,1,2}$$
$$+ \frac{1}{24} f_2 u_{1,1,1,1} + \frac{1}{120} f_1 u_{1,1,1,1,1} \Big\rangle, \qquad (128)$$

where we similarly denote $u_k F_m = \sum_{i=1}^{n} \frac{\partial u_k}{\partial x_i} F_m^i$. Functions $U_k = U_k(X)$ can be chosen arbitrarily, so for convenience we set $U_k(X) \equiv 0$. Thus, knowing the functions F_i from expressions (120)-(128) we can write averaged system (118), which is simpler than the original system (117). If we solve averaged system (118) we are able to write the approximate solution (119) of system (117) substituting slow variables $X(t)$ into the functions $u_i(X,t)$ obtained from (121)-(127). Since the functions $u_i(X,t)$ are periodic with respect to time t the behavior of slow variables determines the behavior of the approximate solution. It means that we can study stability of the approximate solutions by stability of the regular solutions of averaged system (118).

6. References

[1] Seyranian A P (2004) The Swing: Parametric Resonance. Journal of applied mathematics and mechanics 68: 757-764.
[2] Seyranian A P, Belyakov A O (2008) Swing Dynamics. Doklady physics 53(7): 388-394.
[3] Belyakov A O, Seyranian A P, Luongo A (2009) Dynamics of the Pendulum with Periodically Varying Length. Physica D 238: 1589-1597.
[4] Belyakov A O, Seyranian A P (2010) On Nonlinear Dynamics of the Pendulum with Periodically Varying Length. Mechanics of machines, BulKToMM 18(87): 21-28.
[5] Belyakov A O (2011) On Rotational Solutions for Elliptically Excited Pendulum. Physics letters A 375: 2524-2530.
[6] Belyakov A O, Seyranian A P (2010) The Hula-Hoop Problem. Doklady physics 55: 99-104.
[7] Seyranian A P, Belyakov A O (2011) How to Twirl a Hula Hoop. American j. physics 79: 712-715.
[8] Kauderer H (1958) Nichtlineare Mechanik. Berlin: Springer.
[9] Bogolyubov N N, Mitropolsky Yu A (1961) Asymptotic Methods in the Theory of Non-Linear Oscillations. New York: Gordon and Breach.
[10] Panovko Ya G, Gubanova I I. (1987) Stability and Oscillations of Elastic Systems. Modern Concepts, Paradoxes and Mistakes. Moscow: Nauka.
[11] Magnus K (1976) Schwingungen. Eine Einfuhrung in die theoretische Behandlung von Schwingungensproblemen. Stuttgart: J.Teubner.
[12] Bolotin V V (1999) Vibrations in Engineering. A Handbook, Vol. 1. Oscillations of Linear Systems. Moscow: Mashinostroenie.
[13] Pinsky M F, Zevin A A (1999) Oscillations of a Pendulum with a Periodically Varying Length and a Model of Swing. International journal of non-linear mechanics 34: 105-109.

[14] Zevin A A, Filonenko L A (2007) Qualitative Study of Oscillations of a Pendulum with Periodically Varying Length and a Mathematical Model of Swing. Journal of applied mathematics and mechanics 71: 989-1003.

[15] Bozduganova V S, Vitliemov V G (2009) Dynamics of Pendulum with Variable Length and Dry Friction as a Simulator of a Swing. Mechanics of machines 17(82): 45-48.

[16] Seyranian A P (2001) Resonance Domains for the Hill Equation with Allowance for Damping. Doklady physics 46: 41-44.

[17] Seyranian A P, Mailybaev A A (2003) Multiparameter Stability Theory with Mechanical Applications. New Jersey: World Scientific.

[18] Volosov V M, Morgunov B I (1971) Averaging Method in the Theory of Nonlinear Oscillatoratory Systems. Moscow: MSU.

[19] Thomsen J J (2003) Vibrations and Stability. Advanced Theory, Analysis and Tools. Berlin: Springer.

[20] Awrejcewicz J, Krysko V A (2006) Introduction to Asymptotic Methods. Boca Raton, New York: Chapman and Hall, CRC Press.

[21] Nusse H E, Yorke A Y (1997) Dynamics: Numerical Explorations. New York: Springer.

[22] Szemplinska-Stupnicka W, Tyrkiel E, Zubrzycki A (2000) The Global Bifurcations that Lead to Transient Tumbling Chaos in a Parametrically Driven Pendulum. International journal of bifurcation and chaos 10: 2161-2175.

[23] Xu Xu, Wiercigroch M, Cartmell M P (2005) Rotating Orbits of a Parametrically-Excited Pendulum. Chaos solitons & fractals 23: 1537-1548.

[24] Lenci S, Pavlovskaia E, Rega G, Wiercigroch M (2008) Rotating Solutions and Stability of Parametric Pendulum by Perturbation Method. Journal of sound and vibration 310: 243-259.

[25] Awrejcewicz J, Kudra G, Wasilewski G (2007) Experimental and Numerical Investigation of Chaotic Regions in the Triple Physical Pendulum. Special Issue of Nonlinear Dynamics 50 (4): 755-766.

[26] Awrejcewicz J, Supeş B, Lamarque C-H , Kudra G, Wasilewski G, Olejnik P (2008) Numerical and Experimental Study of Regular and Chaotic Motion of Triple Physical Pendulum. International journal of bifurcation and chaos 18 (10): 2883-2915.

[27] Awrejcewicz J, Petrov A G (2008) Nonlinear Oscillations of an Elastic Two Degrees-of-Freedom Pendulum. Nonlinear dynamics 53 (1-2): 19-30.

[28] Seyranian A P, Yabuno H, Tsumoto K (2005) Instability and Periodic Motion of a Physical Pendulum with a Vibrating Suspension Point (Theoretical and Experimental Approach). Doklady physics 50(9): 467-472.

[29] Seyranian A A, Seyranian A P (2006) The Stability of an Inverted Pendulum with a Vibrating Suspension Point. Journal of applied mathematics and mechanics 70: 754-761.

[30] Horton B, Sieber J, Thompson J M T, Wiercigroch M (2011) Dynamics of the Nearly Parametricpendulum. International journal of non-linear mechanics 46: 436-442.

[31] Blekhman I I (1954) Rotation of an Unbalanced Rotor Caused by Harmonic Oscillations of its Axis. Izv. AN SSSR, OTN 8: 79-94. (in Russian)

[32] Blekhman I I (1979) Vibrations in Engineering. A Handbook. Vol. 2. Vibrations of Nonlinear Mechanical Systems. Moscow: Mashinostroenie. (in Russian)

[33] Blekhman I I (2000) Vibrational Mechanics. Nonlinear Dynamic Effects, General Approach, Applications. Singapore: World Scientific. 509 p.

[34] Akulenko L D (2001) Higher-order Averaging Schemes in the Theory of Non-linear Oscillations. Journal of applied mathematics and mechanics 65: 817-826.

[35] Trueba J L, Baltanáas J P, Sanjuáan M A F (2003) A Generalized Perturbed Pendulum. Chaos, solitons and fractals 15: 911–924.

[36] Fidlin A, Thomsen J J (2008) Non-Trivial Effects of High-Frequency Excitation for Strongly Damped Mechanical Systems. International journal of non-linear mechanics 43: 569–578.

[37] Caughey T K (1960) Hula-Hoop: an Example of Heteroparametric Excitation. American j. physics 28: 104–109.

[38] Horikawa T, Tsujioka Y (1987) Motion of Hula-Hoop and its Stability. Keio science and technology reports 40: 27-39.

[39] Wilson J F (1998) Parametric Spin Resonance for a Spinner with an Orbiting Pivot. Int. j. non-linear mech. 33: 189-200.

Chapter 5

Mathematical Modelling and Numerical Investigations on the Coanda Effect

A. Dumitrache, F. Frunzulica and T.C. Ionescu

Additional information is available at the end of the chapter

http://dx.doi.org/10.5772/50403

1. Introduction

Jets are frequently observed to adhere to and to flow around nearby solid boundaries. This general class of phenomena, which may be observed in both liquid and gaseous jets, are known as the Coanda effect. Flows deflected by a curved surface have caused great interest in last fifty years [1-4]. A major interest in the study of this phenomenon is caused by the possibility of using this effect to aircrafts with short takeoff and landing, for fluidic vectoring.

Flow control offers a multitude of opportunities for improving not only the aerodynamic performance, but also the safety and environmental impact of flight vehicles. Circulation control (CC) is one type of flow control which is currently receiving considerable attention. Such flow control is usually implemented by tangentially injecting a jet sheet over a rounded wing trailing edge. The jet sheet remains attached further along the curved surface of the wing due to the Coanda effect (i.e., a balance of pressure and centrifugal forces). This results in the effective camber of the wing being increased, producing lift augmentation.

At the beginning of the chapter we achieve an analytic solution that approximates a two-dimensional Coanda flow. The validity of the results is limited to cases $b/R \ll 1$, since in the tangential component of the momentum equation, the curvature was neglected ($y^* \ll 1$).

In many applications that use boundary layer control by tangential blowing, the solid surface downstream of the blowing slot is strongly curved and, in this case, the prediction of the jet involves both separation and a more accurate knowledge of the flow (radial and tangential pressure - velocity profiles) which can be done by CFD methods.

After the analytical approach, using the FLUENT code both external and internal flows are analyzed, with emphasis on the Coanda effect, in order to determine its advantages and limitations. Finally, we analyze the situations when bifurcations of the flow occur.

INTECH
open science | open minds

© 2012 The Author(s). Licensee InTech. This chapter is distributed under the terms of the Creative Commons Attribution License http://creativecommons.org/licenses/by/3.0), which permits unrestricted use, distribution, and reproduction in any medium, provided the original work is properly cited.

2. Similar solution for a Coanda flow

Jets are frequently observed to adhere to and flow round nearby solid boundaries. This general class of phenomena, which may be observed in both liquid and gaseous jets, is known as the Coanda effect. In recent years, great interest has been taken in flows deflected by a curved surface. Studying this phenomenon is very important due to the possibility of using the Coanda effect to aircrafts with short takeoff and landing, for fluidic vectoring.

This section deals with the steady two-dimensional, laminar and turbulent flow of an incompressible fluid that develops like a jet-sheet on a cylinder surface, i.e., a Coanda flow [5]. We show that this flow can be approximated well enough by similar solutions for both the laminar and the turbulent regime. Basically we use Falkner-Skan transformations of the momentum equations that can be reduced to one ordinary differential equation (ODE). These solutions are presented in this section for both the laminar and the turbulent flow. The results are given in the form of analytical expressions for the mass flow, thrust and jet-sheet thickness depending on the angle of deviation.

We also consider the possibility of the thrust augmentation yielded by the fluid entrainment of the jet flow. Thrust vectoring of aircraft which is the key technology for current and future air vehicles, can be achieved by utilizing the Coanda effect to alter the angle of the primary jet from an engine exhaust nozzle. Furthermore, the increased entrainment by the Coanda surface coupled with the primary jet fluid can augment the thrust, see e.g., [6].

The problem considered here is only a crude approximation of the physical phenomenon. However, we believe that the singular solutions that we develop pave the way towards a further, more accurate approach of the problem.

2.1. Mathematical model

Let us consider the steady two-dimensional flow of an incompressible fluid developed on a cylindrical surface like a jet-sheet. The boundary-layer type equations are written in a curvilinear coordinate system shown in Figure 1. Assuming that the width of the jet slot is small compared to the curvature radius of the cylinder, R, the boundary-layer approximations can be applied yielding the simplified equations of motion

$$\frac{1}{r}\frac{\partial V_\theta}{\partial \theta} + \frac{\partial V_r}{\partial r} + \frac{V_r}{r} = 0, \tag{1}$$

$$V_r \cdot \frac{\partial V_\theta}{\partial r} + \frac{V_\theta}{r} \cdot \frac{\partial V_\theta}{\partial \theta} = \frac{1}{\rho} \cdot \frac{\partial \tau}{\partial r}, \tag{2}$$

$$\rho \frac{V_\theta^2}{r} = \frac{\partial p}{\partial r}, \tag{3}$$

where the laminar shear stress is $\tau = \mu \frac{\partial V_\theta}{\partial r}$ and the turbulent shear stress, where the contribution of the laminar sublayer is neglected (omit the term $\mu \frac{\partial V_\theta}{\partial r}$), has the form $\tau = \mu_t \frac{\partial V_\theta}{\partial r}$, with μ_t the turbulent viscosity, assumed constant in a cross-section like shear layer, i.e., $\nu_t / \nu = \sigma \theta^c$.

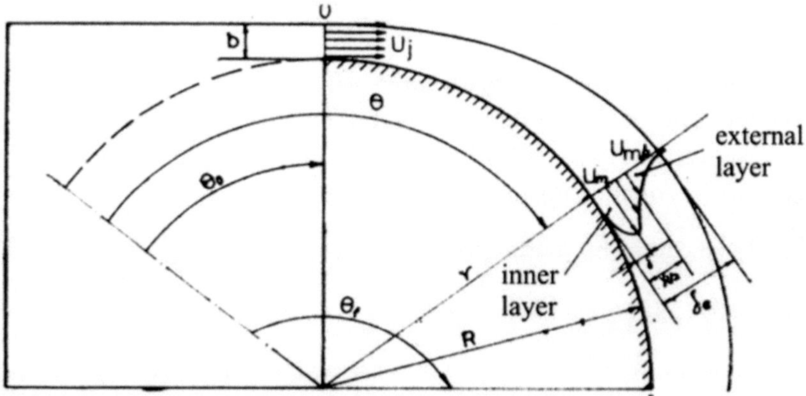

Figure 1. Coordinate system and notation.

The variables in equations (1)-(3) can be made dimensionless, as follows

$$V_r^* = \frac{V_r}{U_j},\ V_\theta^* = \frac{V_\theta}{U_j},\ r^* = \frac{r}{R},\ p^* = \frac{P}{\rho U_j^2},\ \mathrm{Re} = \frac{U_j R}{\nu},\ y^* = \frac{y}{R} = r^* - 1,\ \nu = \mu / \rho,\ y^* \ll 1,$$

where U_j is the velocity of the jet at the exit of nozzle $(\theta = \theta_0)$ (assumed constant in cross-section), Re is the Reynolds number based on the cylinder radius and y is the radial distance from the cylinder surface, i.e., $r = R$.

The dimensionless continuity equation (1) is satisfied by a stream function, chosen such that $V_r^* = \left(-\frac{1}{r^*}\right)(\partial\psi / \partial\theta)$ and $V_\theta^* = \frac{\partial \psi}{\partial r^*}$. Since y^* is much smaller than the unity, it may be neglected compared with the unity in the dimensionless equation (2). Introducing a modelling variable of the form

$$\eta = \mathrm{Re} \cdot y^* \cdot \frac{c+1}{\sigma} \theta^{(c+1)(a-1)} \tag{4}$$

and with the stream function chosen as

$$\psi = \theta^{a(c+1)} f(\eta), \tag{5}$$

equation (2) can be transformed into the following nonlinear ordinary differential equation:

$$f''' + a f f' + (1-2a) f'^2 = 0. \tag{6}$$

The choice of the constant a depends on the boundary conditions. By definition,

$$V_r^* = -\frac{1}{r^*}\cdot\frac{d\psi}{d\theta} \cong -\theta^{a(c+1)-1}\left[a(c+1)f + (c+1)(a-1)\eta f'\right],$$

$$V_\theta^* = \frac{\partial\psi}{\partial r^*} = \frac{\partial\psi}{\partial y^*} = \mathrm{Re}\frac{c+1}{\sigma}\theta^{(c+1)(2a-1)} f'.$$

Integrating equation (6) easily proves that the values $a = 1/3$ and $a = 1/2$ satisfy the boundary conditions of the free jet flow case and of the boundary layer on a flat plate with zero incident, respectively. For the Coanda type flow considered here, the boundary conditions attached to equation (6) are

$\eta = 0; f = 0, f'(0) = 0$, (non-slip condition),

$\eta \to \infty: f' = 0, f'' = 0$, (condition at the edge).

Integrating equation (6) from η to ∞, with the above conditions, yields

$$-f'' - aff' + (1-3a)\int_\eta^\infty f'^2 d\eta = 0. \tag{7}$$

Further integrating equation (7) by means of the integrant factor f', yields:

$$\frac{1}{2}f'^2 - af\int_\eta^\infty f'^2 d\eta + (1-4a)\int_\eta^\infty g f' d\eta = 0, \tag{8}$$

where $g(\eta) = \int_\eta^\infty f'^2 d\eta$.

Equation (8) written at the wall, i.e., $\eta = 0$, leads to $(1-4a)\int_0^\infty gf' d\eta = 0$. Since $\int_0^\infty gf' d\eta \neq 0$, then either $(1-4a) = 0$, or $a = 1/4$. This value satisfies the boundary conditions for the Coanda flow. Thus, the velocity components can be written as

$$V_r^* = -\frac{c+1}{4}\theta^{\frac{c+1}{4}-1}\left[1 - 3\eta f'\right], \tag{9}$$

$$V_\theta^* = \mathrm{Re}\frac{c+1}{\sigma}\theta^{-\frac{c+1}{2}} f'. \tag{10}$$

Hence, equation (6) is integrated for the value a corresponding to the Coanda flow. Writing equation (6) with $a = 1/4$ yields

$$f''' + \frac{1}{4} f f'' + \frac{1}{2} f'^2 = 0. \tag{11}$$

Successively integrating equation (11) with the integrant factors f and $f^{-3/2}$ and taking into account the appropriate boundary conditions firstly gives $f'' f + \frac{1}{4} f' f^2 - \frac{1}{2} f'^2 = 0$ and then

$$6f' + f^2 = f_\infty^{3/2} f^{1/2}, \tag{12}$$

where $f_\infty = \lim_{\eta \to \infty} f$. Choosing $F = \left(f / f_\infty \right)^{1/2}$ equation (12) becomes an equation with separable variables whose solution is written as

$$\eta f_\infty = 4\sqrt{3} actg \left(\frac{\sqrt{3}F}{2+F} \right) + 2 \ln \frac{F^2 + F + 1}{\left(F - 1 \right)^2}. \tag{13}$$

Thus, the quantities η and f can be expressed as explicit functions of the parameter F and so f becomes

$$f = f_\infty F^2. \tag{14}$$

We further obtain explicit relations of F for the quantities f', f'', f''' as follows:

$$f' = \frac{f_\infty^2}{6} \left(F - F^4 \right), \tag{15}$$

$$f'' = \frac{f_\infty^3}{72} \left(1 - F^3 \right) \left(1 - 4F^3 \right), \tag{16}$$

$$f''' = -\frac{f_\infty^4}{288} F^2 \left(1 - F^3 \right) \left(5 - 8F^3 \right). \tag{17}$$

The domain of values of F is from [0, 1], hence the independent variable η ranges from 0 to $\infty, \eta \in [0, \infty)$.

Thus, the problem is completely determined and the solution of equation (11) with given boundary conditions is represented by equations (13)-(17). Calculated values of ηf_∞, f / f_∞, f' / f_∞^2, f'' / f_∞^3 and f''' / f_∞^4 are presented in Figure 2

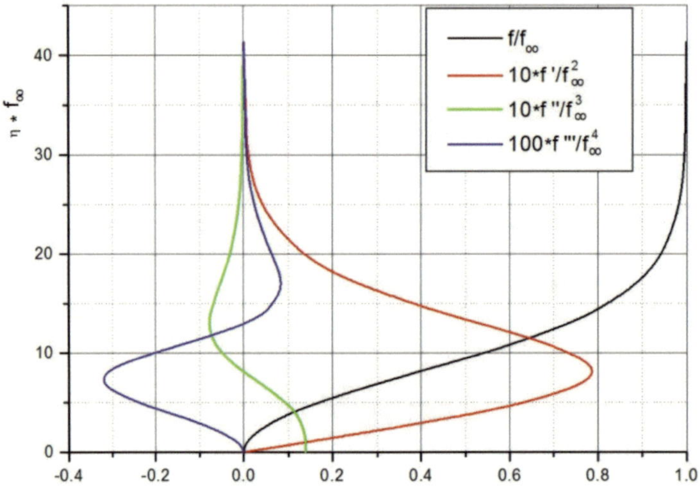

Figure 2. Variations of terms for the similar solution.

2.2. Results

Laminar flow. Since for a laminar flow, $\mu_t / \mu = 1$, then $\sigma = 1$ and $c = 0$. So, the components of the velocity become

$$V_r^* = -\frac{\theta^{-3/4}}{4}\left[f - 3\eta f'\right], \tag{18}$$

$$V_\theta^* = \mathrm{Re}\,\theta^{-1/2} f'. \tag{19}$$

If the mass flow and the momentum of the jet are given at output ($\theta = \theta_0$) then we are able to compute the value of θ_0. Since the boundary layer approximations are not valid near the origin of the jet, this value should be regarded as a virtual origin of the similar motion. Considering the mass flow per length unity of the slot $Q_m = \rho U_j b$, and b the width of the slot, then

$$Q_m = \rho U_j b = \int_0^\infty \rho V_\theta dy\Big|_{\theta_0} = \rho U_j R \theta_0^{1/4} f_\infty,$$

or

$$\theta^{1/4} = \left(\frac{b}{R}\right)\frac{1}{f_\infty}. \tag{20}$$

Now, let the momentum of the jet in the slot be $\rho U_j^2 b = Q_m U_j$. This assumption holds if the velocity does not depend on the length of the slot. This contradicts the assumption that the flows are similar at $\theta = \theta_0$. The jet model is similar to the material point model of solid mechanics, where only the physical size of the body (mass) is taken into account. According to this model, the theoretical jet with given momentum comes out from a zero width slot at $\theta = 0$ and achieves its mass flow in similar flow conditions at $\theta = \theta_0$. At the origin $\theta = 0$, the flow is singular. Equating the momentum of the physical jet with the one theoretically determined at $\theta = \theta_0$ yields $\rho U_j^2 b = \int_0^\infty \rho V_\theta^2 \mathrm{d}y_{|\theta_0}$ and further

$$b = R \frac{\mathrm{Re}}{\theta_0^{1/4}} \int_0^\infty f'^2 \mathrm{d}\eta. \tag{21}$$

The integral from equation (21) is easily computed yielding

$$\frac{b}{R} = \frac{\mathrm{Re}}{Q_0^{1/4}} \cdot \frac{f_\infty^3}{18}. \tag{22}$$

Substituting f_∞ from (20) into (22) leads to

$$\theta_0 = \left(\frac{b}{R}\right)^2 \frac{\mathrm{Re}}{18}. \tag{23}$$

Thus, f_∞ that does not appear in the expression of θ_0, depends only on the curvature wall and the jet characteristics, i.e.,

$$f_\infty = \left(\frac{b}{R}\right)^{1/2} \left(\frac{18}{\mathrm{Re}}\right)^{1/4}. \tag{24}$$

This means that f_∞ is a curvature parameter. In [7] an entrainment parameter A has been defined as a measure of the fluid amount involved in flow. The fluid entrainment is an important physical process because it determines the extent of the attachment region of the Coanda effect. The entrainment parameter is defined by

$$A = \frac{1}{U_j} \cdot \frac{\mathrm{d}}{R\mathrm{d}\theta} \int_0^\infty V_\theta \mathrm{d}y. \tag{25}$$

For the laminar flow, we obtain the dimensionless entrainment parameter as

$$A = \frac{f_\infty}{4} \cdot \theta^{-3/4}, \tag{26}$$

which shows that the entrainment attains its maximum at the jet origin.

Next, we calculate the thrust produced by the deflection of the jet with 90°. If F_t is the thrust per width unity, then $F_t = \int_0^\infty \rho V_\theta^2 \mathrm{d}y \Big|_{\theta=\theta_1} + \int_0^\infty (p - p_\infty) \mathrm{d}y \Big|_{\theta=\theta_1}$, where the first term represents the flow of the momentum and $\theta_1 = \theta_0 + 90^\circ$. Integrating equation (3) yields

$$p_\infty - p = \int_r^\infty \rho V_\theta^2 \frac{dr}{r} = \int_{y^*}^\infty \frac{\rho U_j^2 V_\theta^{*2}}{1+y^*} dy^* .$$

Since y^* is much smaller compared to the unity, we may neglect it in the denominator. Hence, the integral becomes $\int_\eta^\infty \rho U_j^2 \operatorname{Re} \theta^{-1/4} f'^2 d\eta$, which leads to

$$p_\infty - p = \rho U_j^2 \operatorname{Re} \frac{f_\infty^3}{18\theta^{1/4}} \left(1 - F^3\right)^2 .$$

Integrating once, we obtain the second term, which is the contribution of pressure, i.e.,

$$\int_0^\infty (p - p_\infty) dy \Big|_{\theta=\theta_1} = \frac{\rho U_j^2}{2} R \theta_1^{1/2} f_\infty^2 .$$

As the total momentum flux is $\rho U_j^2 R \operatorname{Re} \theta_1^{-1/4} \frac{f_\infty^3}{18}$, the expression of thrust force becomes

$$F_t = \rho U_j^2 R f_\infty^2 \left[\frac{\operatorname{Re} f_\infty}{18\,\theta^{1/4}} - \frac{\theta_1^{1/2}}{2} \right]. \tag{27}$$

By comparison with the non-deflected free jet, we introduce an enhancement factor of thrust defined by the ratio

$$T = \frac{F_t \big|_{\theta=\theta_1}}{F_t \big|_{\theta=\theta_0}} = \frac{F_t}{\rho U_j^2 b} = \frac{R}{b} \theta_1^{-1/4} \left[\frac{\operatorname{Re} f_\infty^3}{18} - \frac{f_\infty^2 \theta_1^{3/4}}{2} \right]. \tag{28}$$

Using equations (22), (23) and (24), T becomes

$$T = \left(\frac{\theta_0}{\theta_1} \right)^{1/4} - \frac{1}{2} \left[\frac{9\pi}{\operatorname{Re}} + \left(\frac{b}{R} \right)^2 \right]^{1/2} . \tag{29}$$

Since $\theta_1 = \theta_0 + \pi/2$, $9\pi/\operatorname{Re} << 1$ and $\left(\frac{b}{R} \right)^2 << 1$, one finds that $T < 1$, i.e., for the considered case there is no thrust increase, but only a change in its direction.

It defines the finite thickness of the jet sheet, δ, the same as the boundary layer, namely the value of y where the section $\theta > \theta_0$ has $\frac{V_\theta}{V_{\theta_{max}}} = 0.01$.

Since $V_\theta^* = \dfrac{V_\theta}{U_j} = \mathrm{Re} \cdot f'^{1/2}$, $\dfrac{V_\theta}{V_{\theta l\max}} = \dfrac{V_\theta^*}{V_{\theta_{\max}}^*} = \dfrac{f'}{f'_{\max}}$ the problem is to find the value of η for which $f'/f'_{\max} = 0.01$. By equation (15), $f'_{\max} = \left(f_\infty^2/8\right)\dfrac{1}{4^{1/3}}$ implying that $f'/f'_{\max} = \dfrac{4^{4/3}}{3}\left(F - F^4\right)$. For $f'/f'_{\max} = 0.01$ it follows that $F = 0.998421$ and by equation (13), we have $\eta f_\infty = 31.622$. Substituting these values in (4) finally yields $\delta^* = \dfrac{31{,}622}{\mathrm{Re}\, f_\infty}\theta^{3/4}$, where the angle θ is measured from the apparent origin (the point of zero sheet thickness). Denoting by $x = R\theta$ the distance along the cylinder surface, one notices that $\delta^* \sim x^{3/4}$ and $V_{\theta_{\max}}^* \sim x^{-1/2}$.

Turbulent flow. For turbulent flows it is first necessary to determine the eddy viscosity μ_t. Wall jets are important test cases for turbulence models because they contain a boundary-layer near the wall which interacts with a free shear layer (Figure 3). Thus, they grow much less than free jets. This reduction in the wall jet development is mainly due to the presence of the wall surface where the entrainment by the jet is inhibited on the side nearest to the surface. The velocity fluctuation damping is transmitted to the outer layer and, since the transfer of side momentum component is closely related to the side component of velocity fluctuations, the shear stress and the development of the jet are reduced. Conversely, a relatively high turbulence degree of the outer layer of the wall jet has an effect similar to the turbulence of free jet in the boundary layer case. Usually, the models of turbulence based on turbulent viscosity do not take into account the jet damping at the side wall and in this way, the empirical constants used for the free jet overestimate the development of the jet wall. For example, the $k-\varepsilon$ model with standard constants, given in [8], yields values of the width of the jet augmented by 30 %. Turbulence models based on the shear stress transport equations (6) can correctly predict the wall jet, with standard constants, only if the equations take into account the wall effect on the correlation pressure – strain velocity, see [8], [9]. If this influence is not taken into account the models yield 20% higher values. Use of the turbulence models based on transport equations is however complex and expensive. Hence, in this section (for analytical solution) we are limited to a simple algebraic model, of turbulent viscosity type, which estimates accurately enough the main features of the considered Coanda flow.

We assume that the turbulent viscosity for a moderate curvature of the flow is governed by the same laws as in the case without curvature. Hence

$$\nu_t = K \cdot V_{\theta m} \cdot y_{1/2} = \nu K \mathrm{Re}\, V_{\theta m}^* y_{1/2}^*, \tag{30}$$

where $V_{\theta m}$ is the maximum speed of flow, $y_{1/2}$ is the point of the outer layer where $\dfrac{V_\theta}{V_{\theta m}} = \dfrac{1}{2}$, and the mixing coefficient K is an empirical constant, same as in the case without curvature.

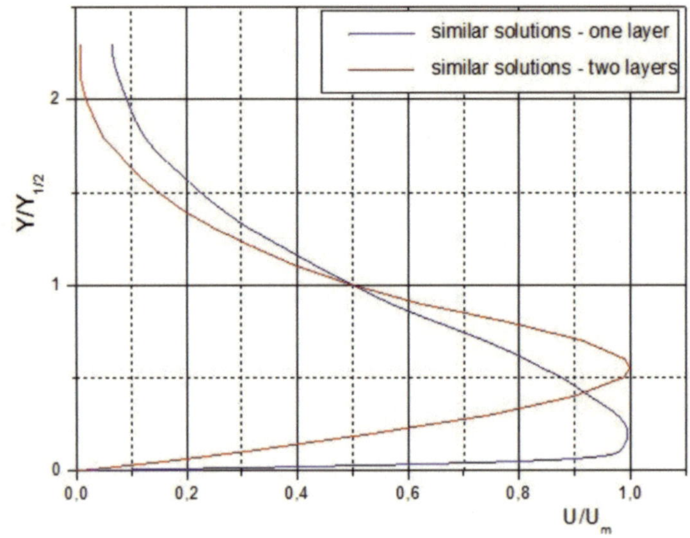

Figure 3. Similar velocity profiles.

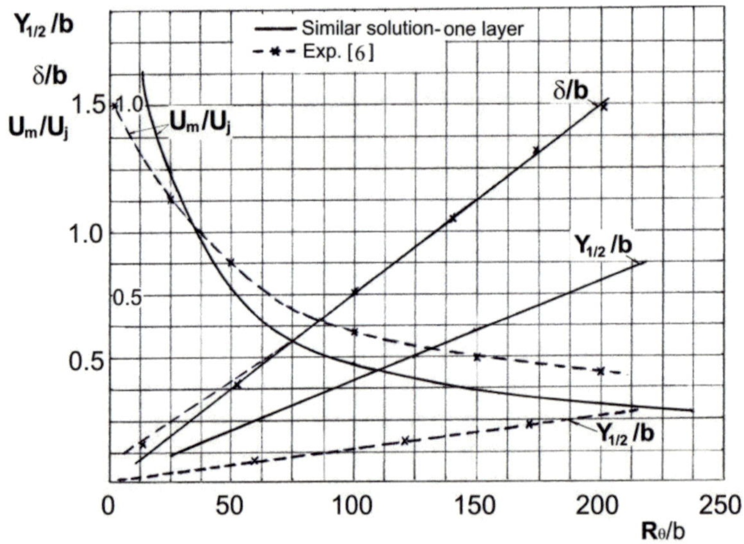

Figure 4. Governing laws for the wall jet flow.

It is easily seen that there are similar solutions of the motion equations (1), (2) and (3), if ν_t / ν has the form

$$\nu_t / \nu = \sigma \theta^c . \tag{31}$$

In the present model the viscosity does not take into account the outer intermittent flow, so that the assumption of constant value in the cross section leads to velocity profiles identical to those of laminar flows, though the general configuration of the flow development is different. Suppose in this case that a linear flow develops, then the dimensionless coordinate of equation (4) must be written as $\eta \sim y^* / \theta$, which yields $c = 1/3$. Since the rate of increase in width is $dy_{1/2} / dx = 0.075$, see [6], we obtain $\sigma = \mathrm{Re}\frac{4}{3}\frac{dy_{1/2} / Rd\theta}{\eta_{1/2}}$, which leads to the estimation $\sigma \cong \frac{\mathrm{Re}\, f_\infty}{148}$ (from the condition $\frac{V_\theta}{V_{\theta m}} = \frac{f'}{f'_{max}} = 0.5$ it follows that $F = 0.90396$ which is used to compute $\eta f_\infty \cong 14.8$). With these data we can now specify the development of the considered self-modelled flow (Figure 4):

- velocity components

$$V_r^* = -\theta^{-2/3} / 3\left[f - 3\eta f' \right], \tag{32}$$

$$V_\theta^* = \frac{4}{3}\frac{\mathrm{Re}}{\sigma} f'\theta^{-2/3}; \tag{33}$$

- position of virtual origin from the slot

$$\theta_0 = \frac{296}{27}\frac{b}{R}; \tag{34}$$

- curvature parameter

$$f_\infty = \frac{3}{2.37^{1/3}}\left(\frac{b}{R}\right)^{2/3}; \tag{35}$$

- entrainment parameter

$$A = \frac{f_\infty}{3}\theta^{-2/3}; \tag{36}$$

- augmentation factor of thrust

$$T = \left(\frac{\theta_0}{\theta_1}\right)^{1/3}\left[1 - \frac{1}{2}\frac{b}{R}\frac{\theta_1}{\theta_0}\right] < 1 \tag{37}$$

- finite thickness of jet sheet, defined as the value for which $V_\theta = V_{\theta\max} / 2$. For $\frac{V_\theta}{V_{\theta m}} = \frac{f'}{f'_{max}} = 0.5$, it follows that $F = 0.90396$ and $\eta f_\infty \cong 14.8$. Substituting in (4) we obtain the final result

$$y_{1/2}^* = 0.075\,\theta \quad \left(V_\theta = V_{\theta\max} / 2\right); \tag{38}$$

- decrease in the maximum flow velocity

$$\frac{V_{\theta m}}{U_j} = \left(18.5\frac{b}{\theta R}\right)^{2/3}; \tag{39}$$

- power laws for the wall jet flow

$$y_{1/2}^{*} \sim x,\ V_{\theta m}^{*} \sim x^{-2/3},\ \ x = R\theta. \tag{40}$$

In many applications that use boundary layer control by tangential blowing, the solid surface downstream of the blowing slot is strongly curved and, in this case, the prediction of the jet involves separation and a more accurate knowledge of the flow (radial and tangential pressure - velocity profiles) which can be done with CFD methods.

3. Control of the two-dimensional turbulent wall jet on a Coanda surface

Flow control refers to the ability to alter flows with the aim of achieving a desired effect. Examples include the delay of boundary layer separation and drag reduction, noise attenuation, improved mixing or increased combustion efficiency, among many other industrial applications. There are two possibilities to approach the problem of flow separation control: (1) passive control (vortex generators, flaps/slats, slots, absorbant surfaces and riblets) and (2) active control (mobile surface, planform control, jets, advanced controls - magnetodynamics).

The active control without additional net mass flow can be achieved by synthetic jets or small vibrating flap. A synthetic jet is a concept that consists of an orifice or neck driven by an acoustic source in a cavity, as in [10]. At sufficiently high levels of excitation by the acoustic source, a mean stream of flow has been observed to emanate from the neck. The excitation cycle increases the ability of the boundary layer to resist separation.

Another technique of flow control on the convex surfaces is to use passive devices, one of these being the slot mounted between lower-pressure and high-pressure points (near the separation point) on the upper surface. The tendency of equalization of the pressure will produce blowing-suction jets which maintain the boundary layer attached to the upper surface, see [11].

We investigate three issues related to flow control with applications to aerospace and wind energy: finding the appropriate turbulence model for the study of jets on convex surfaces, the passive control using a slot and the active control using a synthetic jet at medium frequencies on Coanda surfaces [12].

3.1. Coanda effect. Computational analysis

In this section the effect of the surface curvature (Coanda effect) on the development of a two-dimensional wall jet is numerically investigated. The main goals are providing a systematic survey of the performance of selected eddy-viscosity models in a range of curved flows and establishing more clearly their potential and limitations.

Reynolds averaged Navier-Stokes simulations (RANS) with different turbulence models have been employed in order to compute the two-dimensional turbulent wall jet flowing around a circular cylinder: (1) Spalart and Allmaras (SA - one turbulence model equation) [13], (2) Launder and Spalding $k-\varepsilon$ model [14], (3) Wilcox $k-\omega$ model [15] and (4) Menter $k-\omega$ SST model [16]. The predictions yielded by the simulations were compared to available experimental measurements from the literature. The surface curvature enhances the near-wall shear production of turbulent stresses and is responsible for the entrainment of the ambient fluid which causes the jet to adhere to the curved surface.

The particular configuration shown in Figure 5 is considered cylindrical. The wall jet properties have been reported by Neuendorf and Wygnanski [17] and provide the means to evaluate the simulation results (diameter $d = 2R = 0.2032$ m, nozzle height $b = 2.34$ mm and jet-exit velocity $U_j = 48$ m/s).

The computational grid used for these investigations consists of 900 x 220 nodes. For the turbulence models used in these calculations the laminar sublayer needed to be resolved. The y^+ values of the wall-next grid points were between 0.4 and 1, and the Δx^+ values were between 50 and 300. The grid resolution in the jet was between 40 and 180 times the local Kolmogorov length scale. A fully developed channel velocity profile was prescribed at the nozzle inflow (no near field), with a medium turbulence. The ambient was quiescent.

For some of these turbulence models the jet-velocity decay and jet-half-thickness versus the streamwise angle are plotted in Figure 6. The jet-half-thickness ($y_{1/2}$) represents the thickness where the jet velocity (U_j) is half of the maximum jet velocity (U_m) through the same section.

When the $k-\omega$ model was used in combination with the $k-\omega$ SST model, a close match of the jet-velocity decay with the measured data was achieved. However, even with this model, the downstream development of the jet-half-thickness was poorly predicted.

The shape of the normalized velocity profiles is best predicted by the $k-\varepsilon$ model (see Figure 7). Since the predicted half-thickness ($y_{1/2}$) is small for all models, the normalized velocity profiles do not match the experimental velocity profiles neither in the mild pressure region, nor in the adverse pressure region.

For the $k-\omega$ SST model, the separation location was slightly closer to the experimental data. When the $k-\varepsilon$ and Spalart-Allmaras models were used, the jet remained attached to the cylinder for more than 260 degrees (see Figure 8).

One weakness of the eddy-viscosity models is that these models are insensitive to streamline curvature and system rotation. Based on the work of Spalart and Shur [18] a modification of the production term has been derived, which allows the $k-\omega$ SST model to sensitize to the curvature effect.

The results obtained with the corrected (curvature correction – *c.c.*) $k-\omega$ SST turbulence model were presented in Figures 6, 7 and 8, respectively. The results are close to the experimental data up to about 120 degrees. For larger values, the development of the jet was poorly predicted.

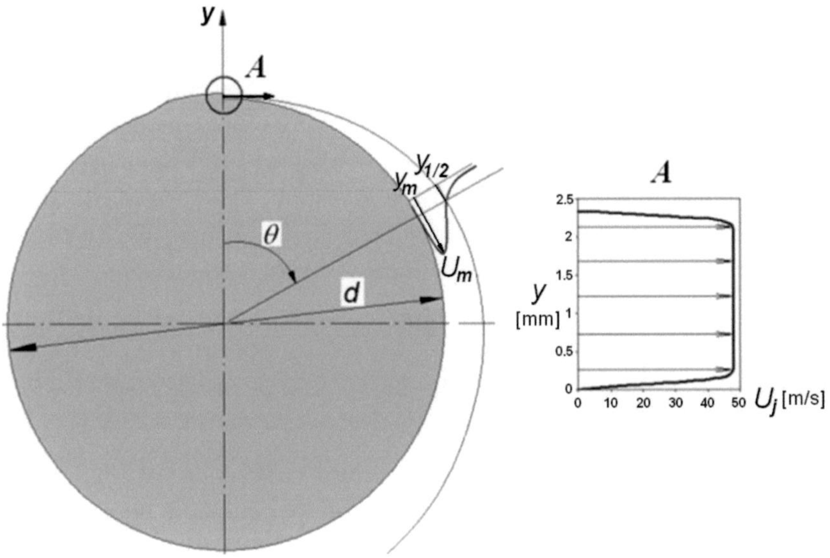

Figure 5. Configuration used in analysis.

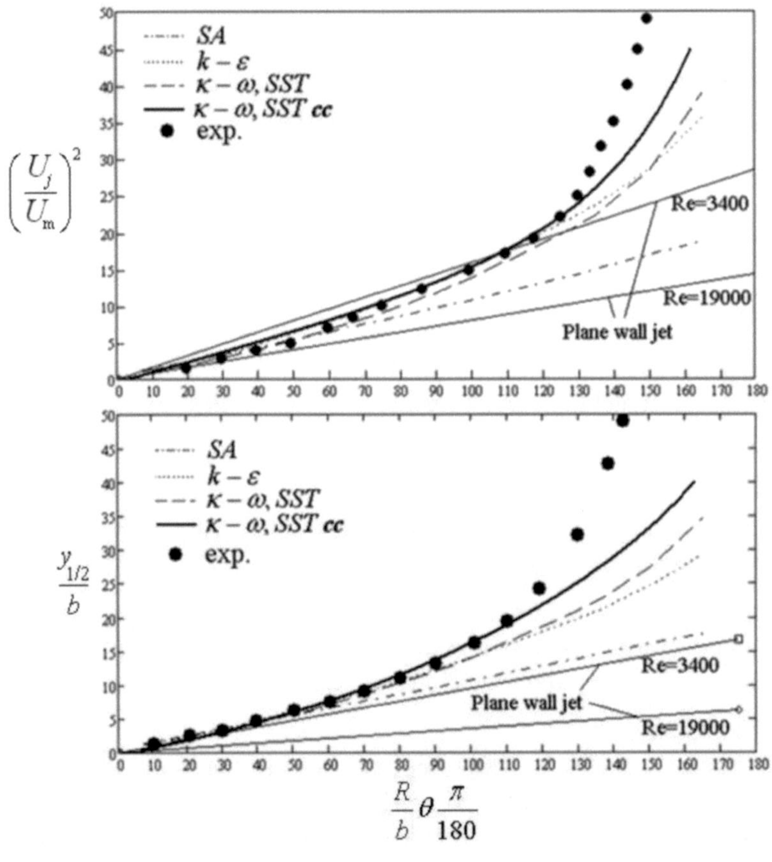

Figure 6. Jet velocity decay and jet-half-thickness (Exp.-Ref.[17]).

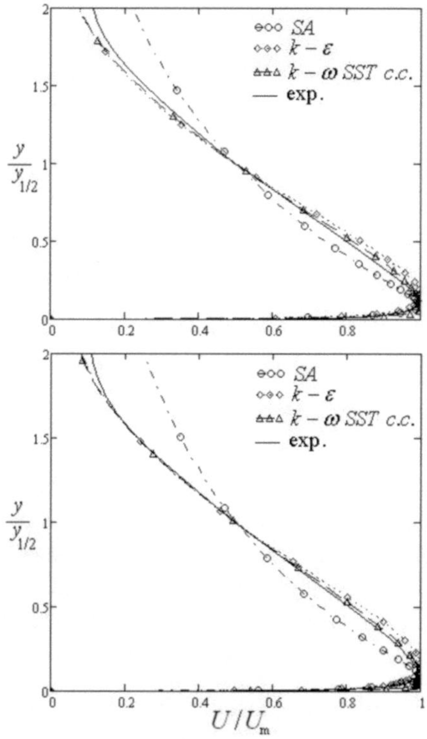

Figure 7. The shape of normalized velocity profiles at 90⁰ and 180⁰ [$y/y_{1/2}=f(U/U_m)$].

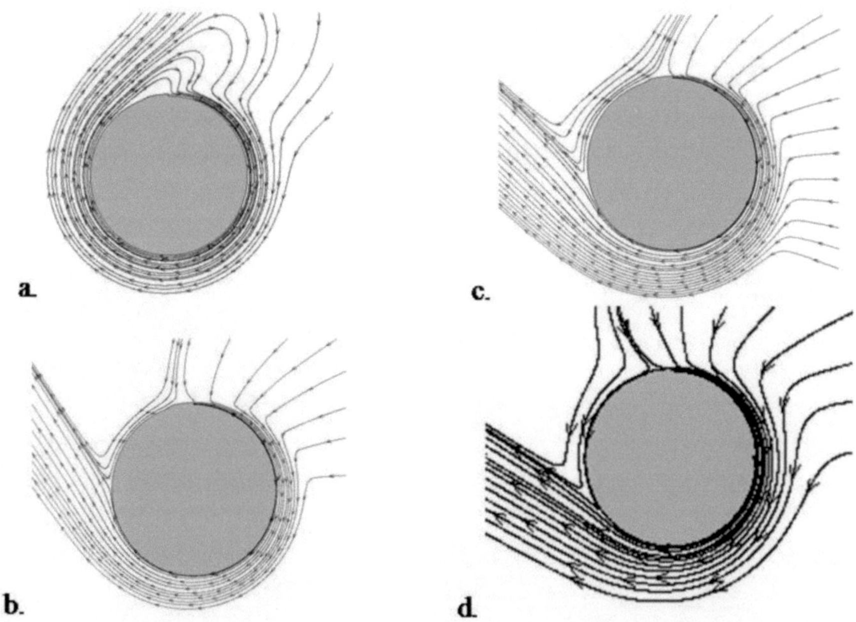

Figure 8. Streamline function: (a) Spalart-Allmaras turbulence model, (b) k- ε model (enhanced wall option), (c) k-ω SST model and (d) k-ω SST c.c. model.

3.2. Passive control using a slot

The first case study uses a simple convex surface and the second computational case uses the same convex surface with a slot between the over-pressure point and the under-pressure point on the surface (placed in separation boundary layer region). The tendency of equalizing the pressures leads to a blow in the first orifice of the slot, while in the second one the suction phenomenon occurs. The jet (U_j=25 m/s) is developed in a rectangular channel with 75 mm (height) x 20 cm (width) and passes over a convex surface (25 cm length). The shape of the surface is given by two elliptical fillet surfaces.

The experimental model has 11 pressure probes disposed on the median plane of the Coanda surface and connected to a digital pressure scanner.

For computations we use steady RANS with a *k-ω* SST *c.c.* turbulence model and the computation grid has 219,300 nodes. The suction-blowing phenomenon has a beneficial effect on keeping the boundary layer attached on 82% of the surface compared to the case without the slot when the boundary layer is attached to 58% of the surface. Figure 9 shows the velocity field in the computational domain and the pressure distribution on the surface for each of the aforementioned situations. The jet is deflected by 20 degrees from the original direction. Using a hydraulic resistance on the slot we can control the separation point of the jet and the jet orientation (the problem will be investigated in future work).

For an active control using synthetic jet concept [19], we use the same configuration as in the first case, but the configuration has an actuator with a lateral slot placed at the point of the detached boundary layer. The diaphragm oscillates in a sinusoidal way, with a frequency of 100 Hz and amplitude of 1 mm ($F^+ = f\,L/U_j = 1$). For simulation purposes, an unsteady RANS, *k-ω* SST turbulence model with curvature correction is used [20]. The computation grid has 160,000 nodes and the y^+ values of the wall-next grid points are between 0.05 and 1, and the Δx^+ values between 10 and 100. In this investigation the separation was not completely suppressed and the boundary layer was not enough energized by the generated vortices structures (see Figure 10). A small unsteady deviation on the jet of about 3 degrees was noticed (see Figure 11).

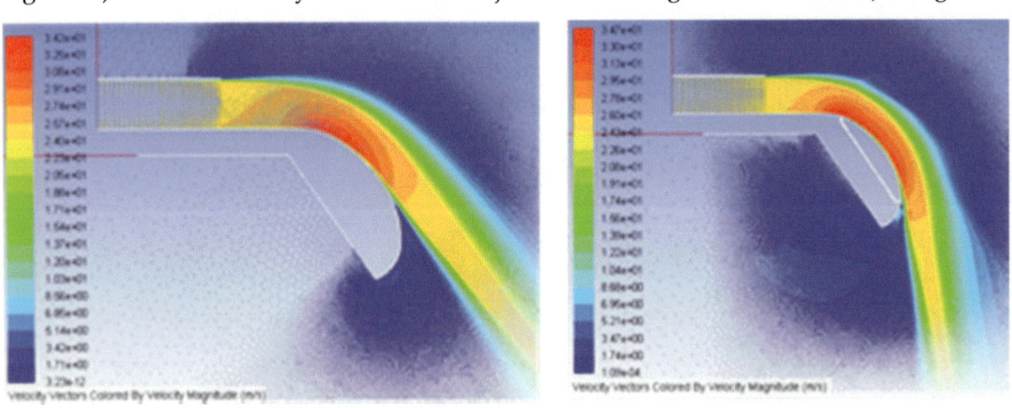

a. b.

Figure 9. Velocity vectors without (a) and with (b) slot.

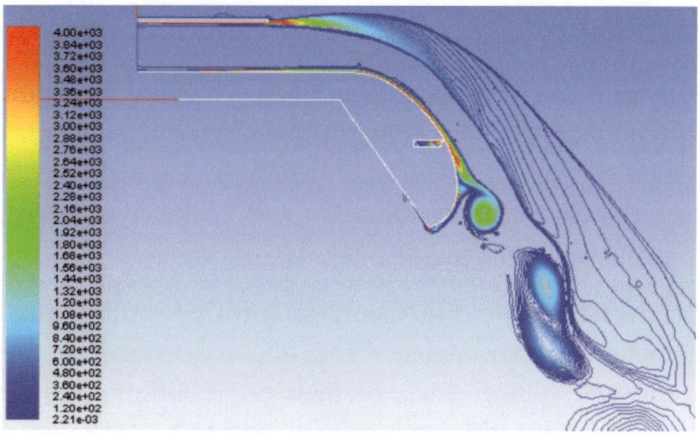

Figure 10. Contours of vorticity magnitudes (maximum expulsion).

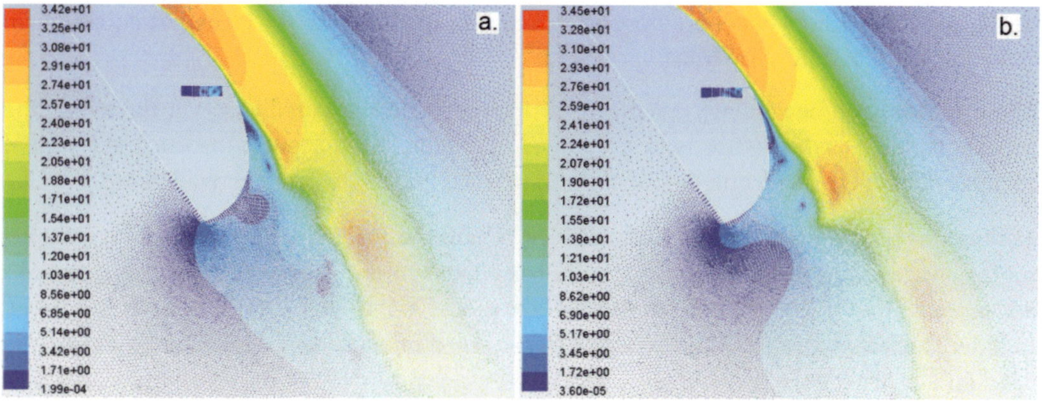

Figure 11. Velocity vectors at maximum ingestion (a), and maximum expulsion (b).

4. Numerical analyis of turbulent flow in a Coanda ejector

The task of this study is to investigate the influence of various geometric parameters and pressure ratios on the Coanda ejector performance.

The Coanda ejector is an axisymmetric device that uses the injected primary flow on the inner curved surface and entrains the secondary flow. The main purpose of the Coanda ejector is to provide a high ratio of the induced mass flow rate to the primary mass flow rate. A primary flow is supplied from a high pressure reservoir. The primary flow follows the curved contour of the ejector after a sonic throat, due to the Coanda effect, and expansion waves/compression waves are created depending on the pressure at the outlet section of the primary nozzle. The turbulent mixing of the primary flow with the ambient air near the entrance of the ejector transfers the momentum of the primary jet to the stagnant air in the ejector throat. The secondary, or induced flow is thus dragged by the turbulent shear stress along the viscous effects towards the ejector exit while being mixed with the primary flow by the persistence of a large turbulent intensity throughout the ejector. There are a few works [21, 22, 23] which examine the basic mechanism by which the secondary flow is induced by the ejector.

4.1. Mathematical model

Dimensionless Forms of Fluid Transport Equations. The fluid transport equations such as the mass (continuity), momentum, and energy conservation equations are used. We define: ρ_c, the characteristic (inlet) density of the fluid (kg/m^3), *U*, the characteristic (inlet) velocity of the fluid (m/s), t_c, the characteristic time (s), and *L*, the characteristic length, which is equal to the inlet diameter of ejector (m) [23, 24].

Then each term is converted to its dimensionless form by multiplying and dividing each term by their characteristic parameters, and then rearranging the equation to the dimensionless parameters. Since the geometrical configuration of the ejector is axisymmetric, the continuity equation and the momentum conservation equation have been used in axisymmetric coordinates.

In compressible fluids, the energy equation is used together with the transport equations in order to calculate fluid properties.

The equations can be spatially averaged to decrease computational cost, yet the averaging process yields a system with more unknowns than equations. Hence, the unclosed system requires a model (e.g., turbulence, or subgrid scale) to make the problem well posed.

Turbulence Closure Equations. The basic idea behind the SST model (see [16]) is to retain the robust and accurate formulation of the Wilcox model in the near wall region, and to take advantage of the free stream independence of the model in the outer part of the boundary layer. In order to achieve this, the $k-\varepsilon$ model is transformed into a $k-\omega$ formulation by means of a function that has the value one in the near wall region and zero away from the surface. The final form, the model parameters and the implementation are presented in detail in paper [16].

All the equations stated above are used to calculate fluid properties in a CFD code.

4.2. Numerical model and results

A numerical model of axisymmetric Coanda ejectors (Fig. 12a) have been built using the CFD software Fluent with a preprocessor, Gambit. The grid size was optimized to be small enough to ensure that the CFD flow results are virtually independent of the size, see [25]. The used grid is divided in a structured grid near the wall and an unstructured grid otherwise. The numerical results have been obtained for a total pressure value of 5 bar, imposed at the reservoir inlet. The computational domain includes the adjacent regions of the ejector with the physical opening boundaries condition. The flow is considered to be steady. We have used the following geometrical configurations (Fig. 12b): $e_1 =$ 0.25 mm, $R_1 = 7.5$ mm; $e_2 = 0.4$ mm, $R_2 = 37.5$ mm .

Figures 13a and 13b show the velocity vectors and the Mach number contours for the investigated axisymmetric Coanda ejector. The induced flow does not follow the path defined by the primary jet. The Mach contours clearly show the flow patterns of the primary and the induced flows and how they mix in the divergent portion of the ejector.

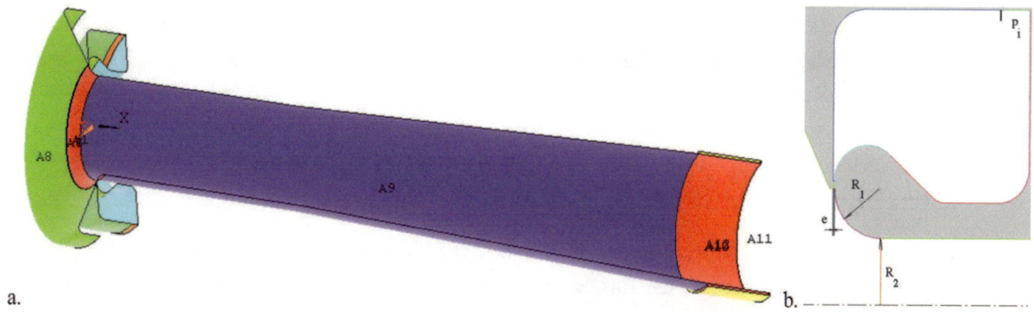

Figure 12. a) Geometry of the Coanda ejector 3D view ; b) detail of the throat gap (primary nozzle).

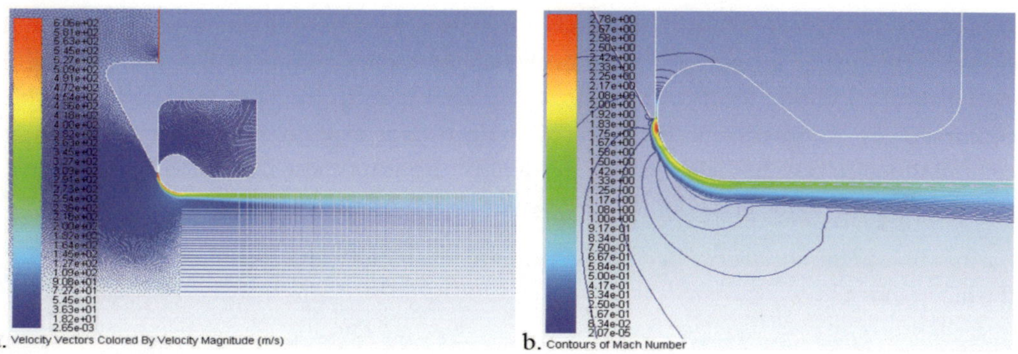

Figure 13. a) Velocity vectors; b) Mach number contour.

In Figures 14a and 14b the flow velocities at $x = 0$, and $x = 550$ are plotted versus the diameters of the Coanda ejector for various values of *e*. Note that the graph can be split into two parts: the first part characterized by a large velocity gradient with high velocities (the primary flow) and a second part (the induced flow) where the velocity gradient is small. The flat portion of the velocity profile indicates a mixed flow.

Also the flow velocities for two diameters of the Coanda ejector are analysed. Although cross sectional area increases when the diameter increases, the increment in mass flow rate is quite small.

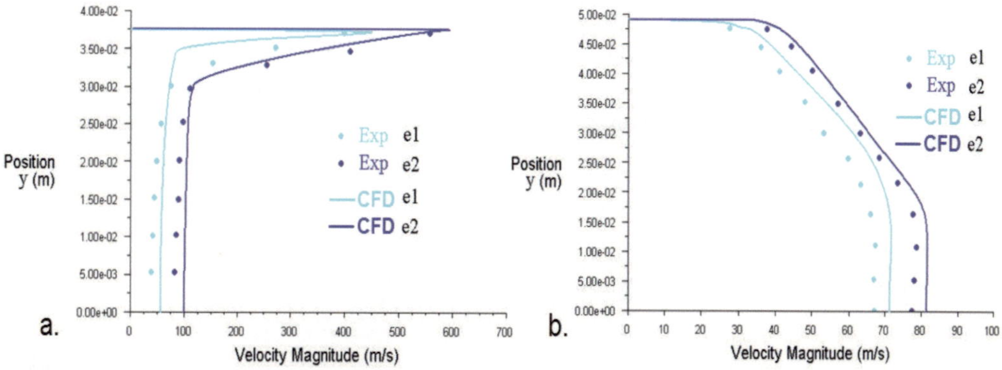

Figure 14. a) Velocity profiles at x = 0, and b) at x = 0.55 m - b

The optimization study of the Coanda ejector is attempted mainly based on the primary nozzle throat and the stagnation pressure ratio. Based on the computational results, it is seen that the throat gap and the stagnation pressure ratio are the two critical parameters which have great influence on the flow characteristics through the ejector and then on the performance of the Coanda ejector, see [24]. Based on these studies, the optimal configuration of a Coanda ejector might be obtained, in order to maximize the ratio of the mass flow rates.

By performing a computational study the effect of various geometric parameters on the performance of the Coanda ejector has been analyzed. The throat gap of the primary nozzle (e) has a strong influence on the ratio of mass flow rates of the induced flow and the primary flow and a critical control over the mixing length as well. For reduced throat gaps, the mixing length decreased, and this possibly indicates the rapid mixing layer growth in the ejector. The mixing layer was more developed for higher values of the diameters of the ejector throat. Validity limits of the calculation laws used in the numerical code have been confirmed by comparisons between numerical and experimental data. The present computational study has allowed us to identify the important parameters which have a strong influence on the behavior and performance of the Coanda ejector.

Further investigations are needed on the primary jet stability and its influence on the flow in the mixing area.

5. The pitchfork bifurcation flow in a symmetric 2D channel with contraction

An important application is the study of the incompressible flow in a symmetric 2D channel with contraction.

Experimental and numerical research (see the works [26 – 32]) were performed in order to evaluate the flow through the 2D channel, especially after contraction occurs. The experiments done by Cherdron and Sobey (see [27], [28]) show the preferential formation of a recirculating zone on one of the channel walls, at a given Reynolds number. For values larger than the critical value Re_{cr}, the flow through the channel loses its symmetry with respect to the channel summetry axis. This phenomenon is known as pitchfork bifurcation. Physically, in the fluid, a momentum transfer process occurs, causing the appearance of a pressure gradient across the channel. Such a pressure gradient may lead to an asymmetric flow. We refer to this phenomenon as the Coanda effect.

5.1. Physical model

The flow equations through the channel are characterized by the Navier-Stokes equations for a laminar, incompressible, stationary flow, given by

$$\begin{aligned} &\nabla\vec{V}=0 \\ &\rho(\vec{V}\cdot\nabla)\vec{V}=-\nabla p+\mu\nabla^2\vec{V} \end{aligned} \tag{41}$$

The geometry of the channel is shown in Figure 15 and it has a symmetry axis (y = 0). The contraction of the channel is given by the contraction ratio $k = D/d$. The inflow of the channel is at a coordinate $x = -L_1$ with respect to the contraction section, and has the following velocity profile

$$u(y) = 6\,V_{med}\left[0.25 - (y/D)^2\right] \tag{42}$$

where $V_{med} = \frac{1}{D}\int_{-D/2}^{D/2} u(y)dy$.

The Reynolds number is defined by the mean value of the velocity V_{med} and the maximum value of the height of the channel D, i.e., $\mathrm{Re} = \rho V_{med} D / \mu$, where ρ is the fluid (air) density and μ is the dynamical viscosity.

The outflow ($x = L_2$) is chosen sufficiently far from the contraction suction, such that the velocity gradient associated with the velocity profile is zero ("outflow" output condition).

On the solid boundaries (the walls of the channel) we impose a "no slip" condition. If, for numerical simulations, a half-channel is used, then the presence of the symmetry axis is imposed assuming a zero flux of all quantities across a symmetry boundary.

5.2. Numerical model

The "pressure-based" solver that is used has a SIMPLE-C algorithm implemented (see [33]), together with a multi grid technique for increasing the rate of convergence of stationary flow problems (Ansys Fluent). The spatial discretization is of second order accuracy, with a under-relaxation coefficient of 0.5 for both the pressure and the momentum. The solution has converged when the global L^2 - norm of the pressure and the velocity residuals is lower than 10^{-8}.

The Reynolds number sets were selected such that the flow in channel is laminar and stationary.

5.3. Numerical results

In the numerical simulations we use three channels with the contraction coefficients k = 2.4 and 8, repectively (see Table 1, and Figure 15). The fixed dimensions of the channel are D = 0.2 m, L_1 = 0.5 m and L_2 = 1 m.

k	$d = D/k$ (m)	Re	$\Delta s = \min(\Delta x, \Delta y)/D$	Grid nodes (full channel)
2	0.1	500…3600	2.50e-4	413,400
4	0.05	500…2000	1.25e-4	415,990
8	0.025	400…1700	1.00e-5	433,800

Table 1. Computation settings.

The recirculation zones that occur at the corners of the channel and beyond the contraction section have the lengths S_1, S_2 , and S_3^l, S_3^u, respectively (see Figure 15).

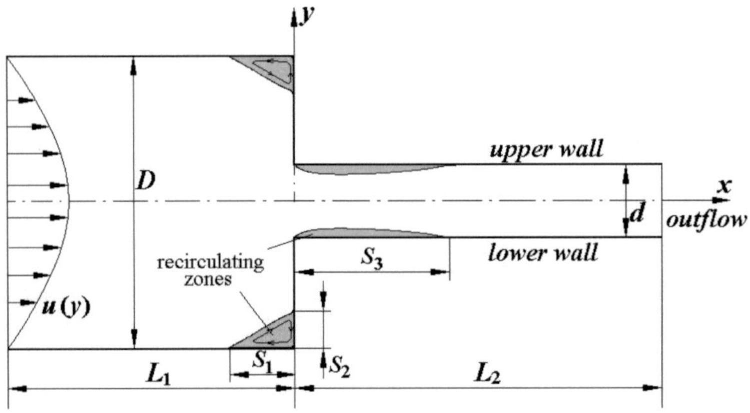

Figure 15. The geometry and the reference lenghts that characterize the recirculating zones (near corners and in downstream channel).

For each contraction coefficient *k*, two computation cases occur: a first case in which we have used half-channel ($y = 0$ is a symmetry boundary) and a second case in which the flow through the whole channel has been studied. The computational domain allows for generating orthogonal grids.

For a given value of *k*, without changing the value of the Reynolds number, the numerical simulations lead to similar values of the length(s) of the recirculation zone (S_1 and S_2) in the corner(s) of the upstream channel, in accordance with the results from the works of Hawken [30]. The relative error between the simulated values S_1 and S_2 (whole channel, upper and lower corner) and the values obtained from the half-channel simulations is under 1% (see Figure 16).

Figure 17 shows that, for the half-channel symmetric solution, the reattachment length S_3 is linear and monotonously increasing.

Concerning the full channel solution we make the following remarks:

- the use of a grid which is symmetric with respect to the symmetry axis of the channel and has symmetrically generated grid computing boundaries usually leads to a symmetric solution (see Figure 17b);
- if the grid is slightly asymmetric (the values of the symmetric coordinates are slightly perturbed or the number of discretization points is different for each symmetric boundary) or if the initial flow field is slightly perturbed, then the recirculation zone formed on the lower wall is different from the recirculation zone formed on the upper wall ($S_3^l \neq S_3^u$) (see Figure 17c). The results hereby have been obtained using four layers of cells, in the neighborhood of the lower wall, each cell splitted into two parts (keeping Δy and keeping the grid orthogonal).

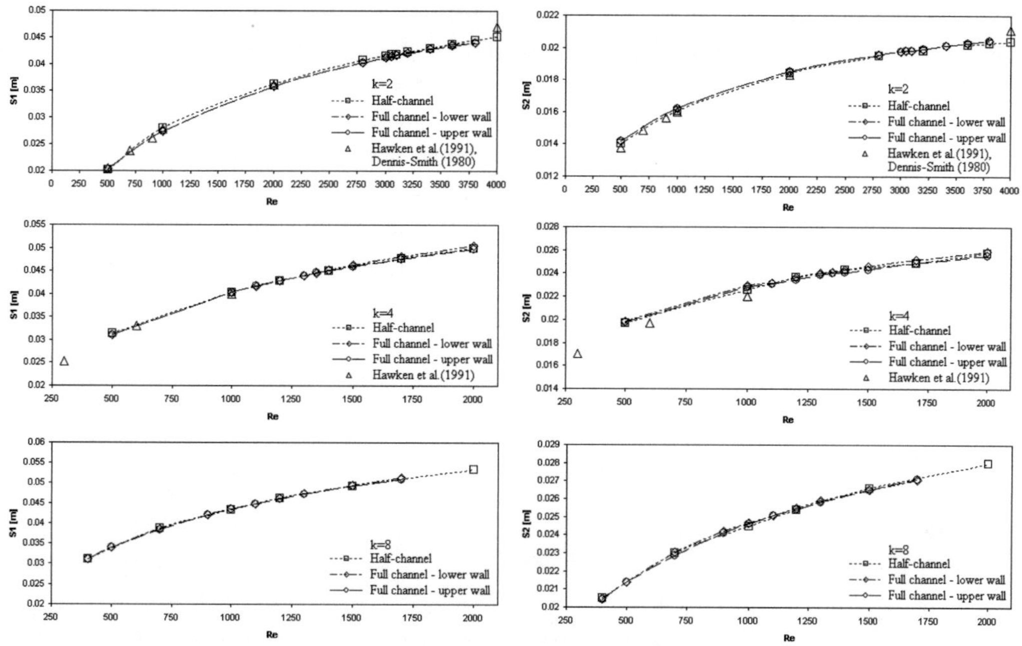

Figure 16. The computed separation (S_1) and reattachment lengths (S_2) for computed symmetric solution in a half-channel and full channel (lower and upper corner).

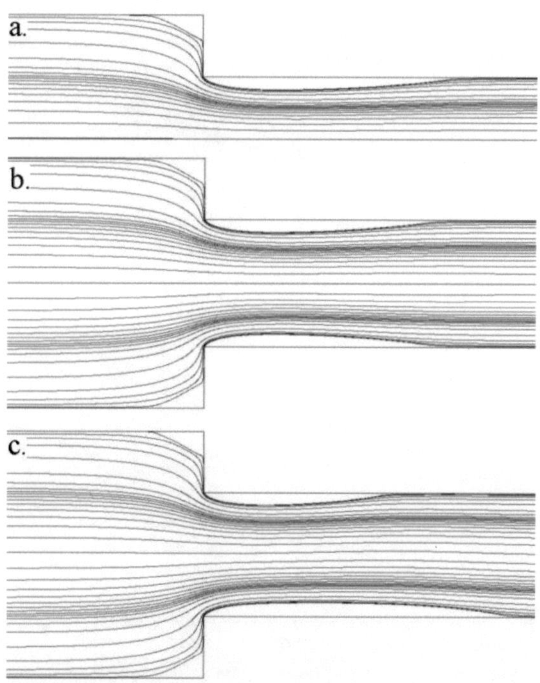

Figure 17. Pathlines for $k = 2$ and Re = 3600: (a) half-channel, (b) full-channel symmetric mesh and (c) full-channel with asymmetric orthogonal mesh.

Figure 18 shows the evolution of the reattachment length S_3 for both half-channel and full channel solutions, on the lower wall of downstream channel (S_3^l) and on the upper wall (S_3^u) respectively. Note that up to a certain value of the Reynolds number the full channel numerical solution is identical to the half-channel numerical solution (the relative error is under 1%). For values greater than the aforementioned Reynolds number, a longer recirculation zone occurs on one of the walls of the channel, e.g., the lower wall, in our case. The critical value of the Reynolds number that leads to the bifurcation of the solution lies in the range: (1) $3050 < \mathrm{Re}_{cr} < 3100$ for $k = 2$, (2) $1350 < \mathrm{Re}_{cr} < 1400$ for $k = 4$ and (3) $1050 < \mathrm{Re}_{cr} \approx 1100 < 1150$ for $k = 8$.

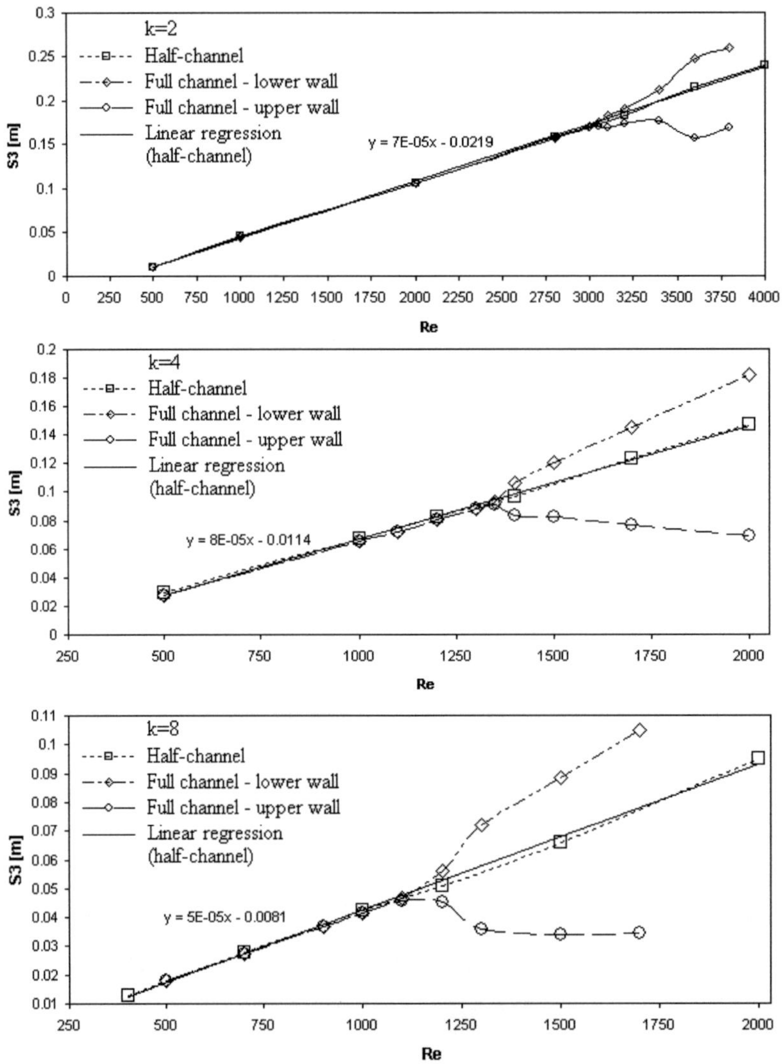

Figure 18. The plot of reattachment length S_3 after contracted section (i.e. downstream channel) as a function of Reynolds number .

The numerical simulations confirm the pitchfork bifurcation of the solution. The asymmetric solution given by the two different recirculation zones that occur on the upper and lower wall, respectively, can be stabilised up to a value of approximately 3800 of the Reynolds number for $k = 2$, 2000 for $k = 4$ and 1700 for $k = 8$, respectively.

Another application where the pitchfork bifurcation occurs is the study of two flows that go through a channel [34]. We consider the case when the velocity profile is described by equation (42), and assume that the flows have identical velocity profiles (see Figure 19). The flowing regime is characterized by small Reynolds numbers (< 25-30) such that the flows are laminar and stationary. The domain is discretized in 400 x 600 nodes [35]. When the Reynolds number ($\mathrm{Re} = \rho V_{med} D / \mu$) is larger than the critical value Re_{cr} the symmetry of the flow is lost, hence the flow becomes asymetric. Figure 20 shows the flow patterns for S/D = 10, S/H = 0.4 and L/H = 15, for Re = 15, 19 and 24, respectively. In Figure 20.b the jets unite into a single jet deflected towards one side-wall, which is then redirected to the opposite side-wall downstream. According to Figure 20.c the number of separation bubbles increase with the Re number, and the flow becomes unsteady.

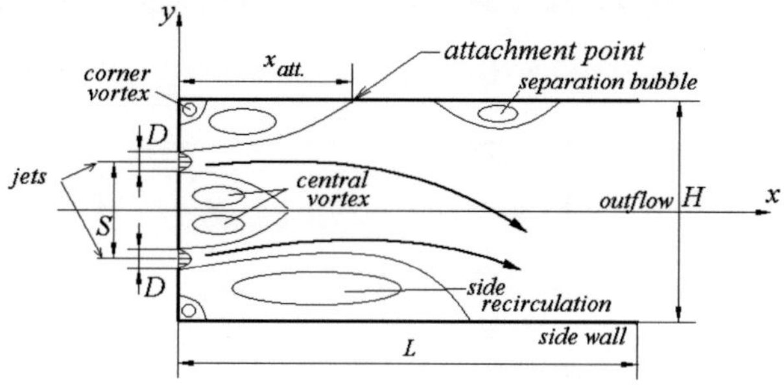

Figure 19. Twin-jet flow configuration.

Figure 21 shows the attachements point locations x_{att} for flows at various Reynolds numbers, for fixed ratio S/D = 10. This point aparently remains unchanged for S/H < 0.5, since the core region between jets is distant from side-walls and the walls do not influence it. For S/H ≥ 0.5, the walls are relatively closer to the jets, and the Coanda effects lead to the "attraction" of jets towards the walls with the merging point suddenly jumping to a further downstream location (observed for Re > 15).

The transonic airfoil buffet [36, 37] is a stability issue that leads to shock oscillations and large variations of the lift coefficient. The practical problem of the airplane buffet is given by the dynamic response of the elastic structure at the flow field [38].

The prediction of the onset and character of the unsteady transonic flow field is a great challenge. The transonic flow around an airfoil has been used as a model problem for understanding the unsteady forcing, phenomenon similar to airplane buffeting [39].

Many researchers analyze the problem using the Reynolds-averaged Navier–Stokes equations with adequate turbulence closure, which are a necessary approximation to cover the high Reynolds numbers at which transonic buffet occurs.

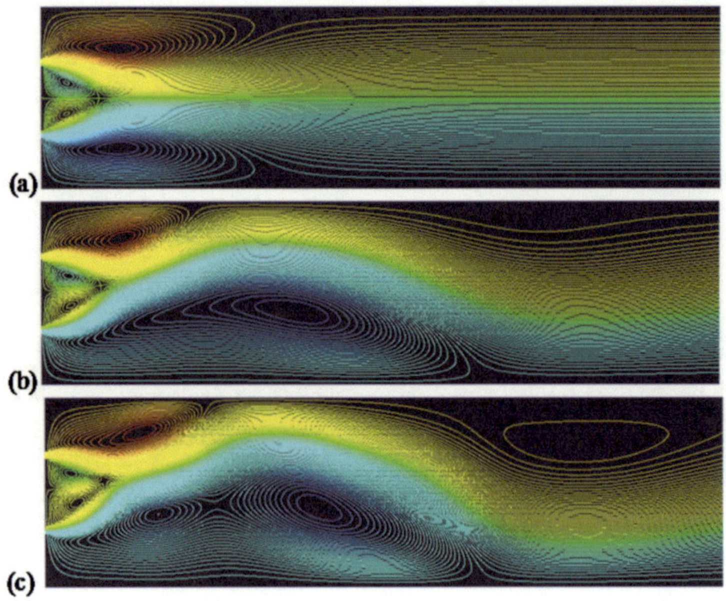

Figure 20. Streamlines for case S/D = 10, S/H = 0.4: (a) Re=15, (b) Re=19 and (c) Re=24.

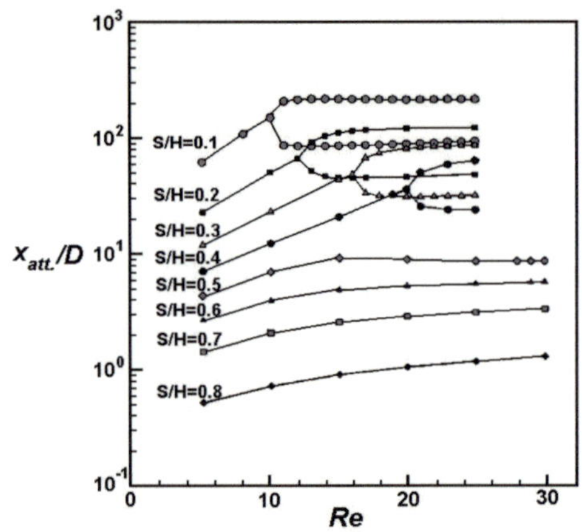

Figure 21. The attachment length for various Re and S/H for S/D = 10.

A simple exemplification for **bifurcation in transonic flow** over an particular airfoil is presented in the following section. The reference model can be found in ref . [40].

In figure 22, for the set of Mach incidence numbers $0.852 < M_{\infty} < 0.868$, one may notice the appearance of the solution bifurcation. The bifurcation is given by the relation $C_L = f(M_{\infty})$, resulting in four domains with different supersonic flow profiles:

- domains A and D, which correspond to the solution obtained by starting with a uniform flow at an angle of 0^0;
- domains B and C, which correspond to the solution obtained by starting with the initial solution at $\pm 1^0$ and M_{∞}=0.86.

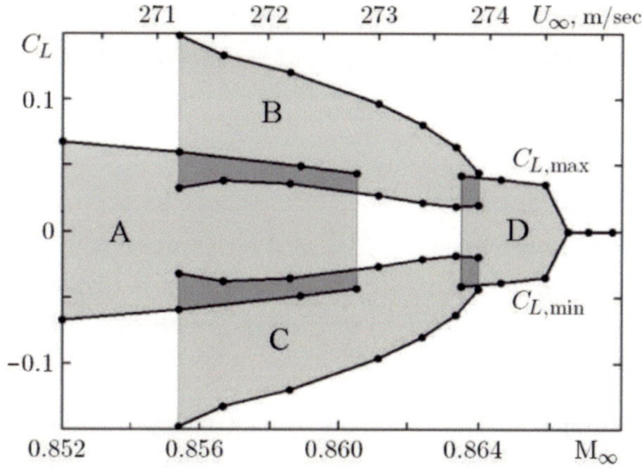

Figure 22. Extreme values of the lift coefficient C_L as function of the Mach number M_∞ for self-sustained flow oscillations about the particular symmetric airfoil (relative airfoil thickness $h = 0.09$) at 0^0 incidence and Re = 1.1 e+7: domains A, B,C and D describes the flow regimes with different location of supersonic regions [40].

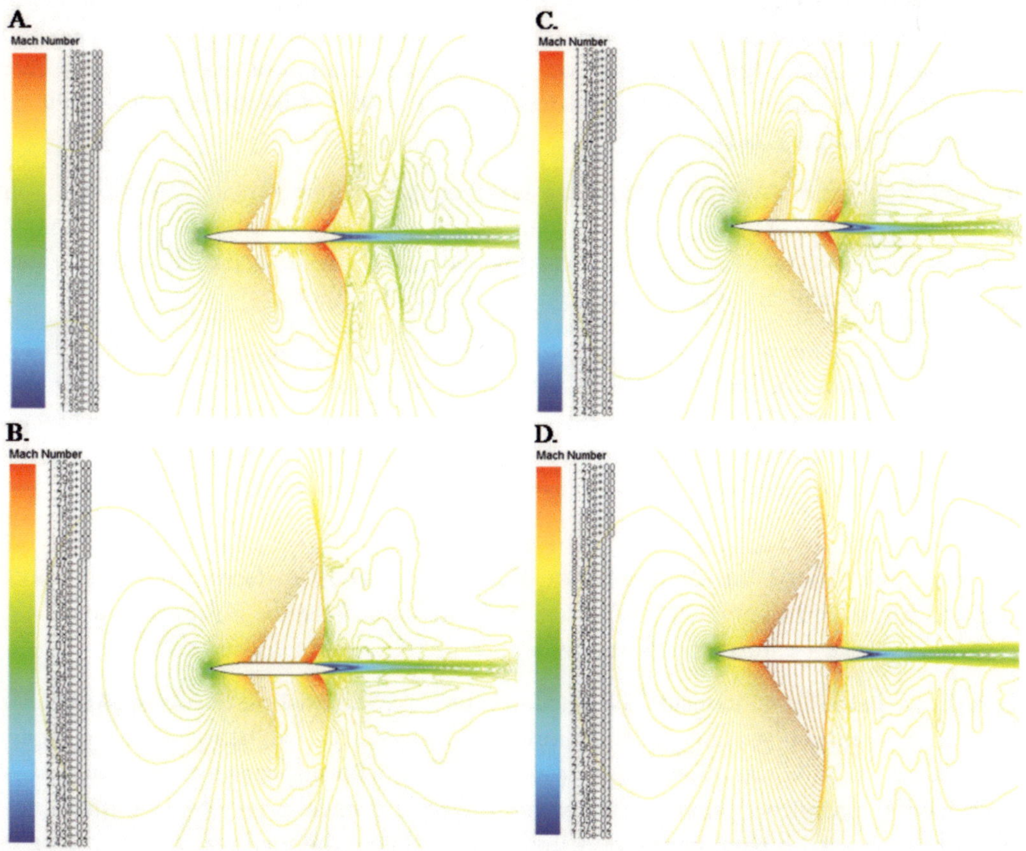

Figure 23. (A., B.,C., D.) - The flow regimes with different supersonic regions computational domains has 1.1 million cells (hybrid mesh with quadrilateral cells near the airfoil). Solver used: unsteady Reynolds-averaged Navier-Stokes implicit solver with SST k-ω turbulence model.

The nonlinear flow equations, the initial solution used for the numeric computations and the length of the airfoil midpart with a small or zero curvature are the principal factors for the onset of flow bifurcations [40, 41].

6. Conclusion

In the beginning of the chapter we have achieved an analytic solution that approximates a two-dimensional Coanda flow. The validity of the results is limited to the cases given by $b/R \ll 1$, since in the tangential component of momentum equation the curvature was neglected $\left(y^* \ll 1\right)$.

The validity of the laminar solution is for Reynolds numbers smaller than the critical value which is of the order 3×10^4. For the turbulent flow the similarity assumption has been reduced primarily to the general development along the flow, leaving the transverse distribution of shear stress.

In this model the edge of the intermittent flow has been neglected, so that assuming constant turbulent viscosity in cross section has lead to similar velocity profiles for both the laminar and turbulent flow. However, the general configuration of the flow development was different. In order to specify the development of this self-modeled flow only one empirical constant was need.

In many applications that use boundary layer control by tangential blowing, the solid surface downstream of blowing slot is strongly curved and, in this case, the prediction of jet involves separation and a more accurate knowledge of the flow (radial and tangential pressure - velocity profiles) which has been done by CFD methods.

The compressible Reynolds-averaged Navier–Stokes (RANS) equations have been solved for circulation control (CC) airfoil flows. Different turbulence models have been considered for closure, including the Spalart–Allmaras model with and without a curvature correction and the shear stress transport (SST) model of Menter. Numerical solutions have been computed with a structured grid solver. The effect of mesh density on the solutions has been examined.

We have investigated the characteristics of various Coanda surfaces, involving smooth curved surfaces and a polygonal curved surface with flap. Using the FLUENT code we have analyzed the distribution pressure and separation on the considered surfaces.

Further, we have taken interest in the detailed behavior of an existing Coanda ejector model, used in propulsion systems. For numerical investigations we have used an implicit formulation of RANS equations for axisymmetric flow with a shear stress transport $k-\omega$ (SST model) turbulence model. The numerical results have been obtained for a total pressure range of 1-5 bars, imposed at the reservoir inlet. The goal was to investigate the influence of various geometric parameters and pressure ratios on the Coanda ejector performance. The effect of various factors, such as the pressure ratio, primary nozzle and ejector configurations on the system performance has been evaluated based on the performance parameters. The mixing layer growth plays a major role in optimizing the performance of the Coanda ejector as it decides the ratio of secondary mass flow rate to primary mass flow rate and the mixing length.

Because single jet flows or multi-jet flows are extensively applied in conjunction with the Coanda surface, as confined or free jet flows, in the last part of the chapter we have provided further insight into complexities involving issues such as the variety of flow structure and the related bifurcation and flow instabilities.

We have considered two cases: i) the flow bifurcation in the symmetric planar contraction channel for different contraction ratio and Reynolds number (single jet) and ii) the flow structure as bifurcation phenomena involved in the confined twin-jet flow field, related to the parameters of jet momentum (Re), side-wall confinement and jet proximity effects.

Also was presented a simple exemplification for bifurcation in transonic flow over an particular airfoil.

Thus, we have determined the conditions and the limits within which one can benefit from the advantages of Coanda-type flows.

Author details

A. Dumitrache
Institute of Statistics and Applied Mathematics of the Romanian Academy, Bucharest, Romania

F. Frunzulica
Institute of Statistics and Applied Mathematics of the Romanian Academy, Bucharest, Romania
"POLITEHNICA" University of Bucharest, Faculty of Aerospace Engineering, Bucharest, Romania

T.C. Ionescu
Imperial College London, Dept. Electrical and Electronic Eng., Control & Power Group, London, UK

7. References

[1] Bourque C., Newman BG. Reattachment of a two – dimensional - incompressible jet of an adjacent flat plate. The Aeronautical Quarterly 1960; XI part 3 201-232.
[2] Newman BG. The deflection of plane jets adjacent boundaries – Coanda effect. In: Lachmann GV. (ed.) Boundary Layer and Flow Control Principles and Applications, vol. I, New York: Pergamon Press Inc.; 1961. p232-264.
[3] Kruka V., Eskinazi S. The wall – jet in a moving stream. Journal of Fluid Mechanics 1964; 20(4) 555-579.
[4] Williams JC., Cheng EH., Kim KH. Curvature effects in a laminar and turbulent free jet boundary. AIAA Journal 1971; 4 733-736.

[5] Dumitrescu H., Ion S., Dumitrache A. Similar solutions of a Coanda-type flow. St. Cerc. Mec. Apl., Bucharest, 1988; 47(6), 531-543 (in Romanian).
[6] Kind RJ. Calculation of the normal – stress distribution in a curved wall jet. The Aeronautical Journal of the Royal Aeronautical Society 1971; 343-348.
[7] Sawer RA. Two dimensional reattaching jet flow including the effect of curvature on entrainment. Journal of Fluid Mechanics 1963; 17 481-498.
[8] Ljuboja M., Rodi W. Calculation of turbulent wall jets with an algebraic Reynolds stress model. Journal of Fluids Engineering 1980; 102(3) 350-356.
[9] Reynolds AJ. Curgeri turbulente in tehnica. Bucuresti: Editura Tehnica; 1982. (in Romanian).
[10] Slomski JF., Chang PA. Large Eddy Simulation of a Circulation Control Airfoil. AIAA Paper 2006-3011 2006.
[11] Glezer A., Amitay M. Synthetic Jets. Annu. Rev. Fluid Mech. 2002; 34 503-529.
[12] Frunzulica F., Dumitrache A., Preotu O., Dumitrescu H. Control of two-dimensional turbulent wall jet on a Coanda surface. In: Brenn G., Holzapfel G.A., Schanz M., Steinbach O. (eds.) Proceedings in Applied Mathematics and Mechanics. Special Issue: 82nd Annual Meeting of the International Association of Applied Mathematics and Mechanics (GAMM), Graz 2011: H11(1), H651–652.
[13] Spalart PR., Allmaras SR. A one-equation turbulence model for aerodynamic flows. AIAA Paper 92-0439 1992.
[14] Launder BE., Spalding DB. The numerical computation of turbulent flows. International Journal for Numerical Methods in Fluids 1974; 15 127-136.
[15] Wilcox D. Simulation of transition with a two-equation turbulence model. AIAA Journal 1994; 32 1192-1199.
[16] Menter FR. Eddy viscosity transport equations and their relation to the $k-\omega$ model. ASME Journal of Fluids Engineering 1997; 119 876-884.
[17] Neuendorf R., Wygnanski I. On a turbulent wall jet flowing over a circular cylinder. Journal of Fluid Mechanics 1999; 381 1-25.
[18] Shur ML., Strelets MK., Travin AK., Spalart PR. Turbulence Modeling in Rotating and Curved Channels: Assessing the Spalart-Shur Correction. AIAA Journal 2000; 38(5) 784-792.
[19] Rizzetta DP., Visbal MR., Stanek MJ. Numerical investigation of synthetic jet flowfields. AIAA Paper 98-2910 1998.
[20] Fluent ANSYS 12. User Guide.
[21] Gregory-Smith DG., Senior P. The effect of base steps and axisymmetry on supersonic jets over Coanda surfaces. Intl. J. Heat and Fluid Flow 1994; 15 291-298.
[22] Guerriero V., Baldas L., Caen R. Numerical solution of compressible flow mixing in Coanda ejectors. In : Proc. the 8th Symposium on Fluid Control, Measurement and Visualization, 2005, Chengdu, China, 1–7.
[23] Kim HD., Lee JH., Segouchi T., Matsuo S. Computational analysis of a variable ejector flow. Journal of Thermal Science 2006; 15 140-144.

[24] Dumitrache A., Frunzulica F., Preotu O., Dumitrescu H. Numerical analysis of turbulent flow in a Coanda ejector. In: Brenn G., Holzapfel G.A., Schanz M., Steinbach O. (eds.) Proceedings in Applied Mathematics and Mechanics. Special Issue: 82nd Annual Meeting of the International Association of Applied Mathematics and Mechanics (GAMM), Graz 2011; H11(1), H647–648.

[25] Riffat SB., Everitt P. Experimental and CFD Modelling of an Ejector System for Vehicle Air Conditioning. Journal of the Institute of Energy 1999; 72(6) 41-47.

[26] Durst F., Schierholza WF., Wunderlich AM. Experimental and numerical investigations of plane channel flows with sudden contraction. ASME Journal of Fluids Engineering 1987; 109 376-383.

[27] Cherdron W., Durst F., Whitelaw J. Asymmetric flows and instabilities in symmetric channels with sudden expansion. Journal of Fluid Mechanics 1978; 84 13-31.

[28] Sobey IJ. Observation of waves during oscillating channel flow. Journal of Fluid Mechanics 1985; 151 395-426.

[29] Hunt R. The numerical solution of the laminar flow in a constricted channel at moderate high Reynolds number using Newton iteration. International Journal for Numerical Methods in Fluids, 1990; 11 247-259.

[30] Hawken DM., Townsend P., Webster MF. Numerical simulation of viscous flows in channels with a step. Computers & Fluids 1991; 20 59-75.

[31] Dennis SC., Smith FT., Steady flow through a channel with a symmetrical constriction in the form of a step. Proceedings of the Royal Society, London, Series A372 393-414.

[32] Drikakis D. Bifurcation in phenomena in incompressible sudden expansion flows, Phys. Fluids 1997; 9(1) 76-87.

[33] Van Doormaal JP., Raithby GD. Enhancement of the simple method for predicting incompressible fluid flows. Numerical Heat Transfer 1984; 7 147-163.

[34] Tzeng PY., Soong CY., Hsieh CD. A numerical investigation of side-wall confinement effect on two turbulent parallel plane jets. J. Chin. Soc. Mech. Eng. 1988; 19 167-178.

[35] Govaerts WJF. Numerical methods for bifurcation of dynamical equilibria. Gent: SIAM; 2000.

[36] Crouch J D., Garbaruk A., Magidov D., Travin A. Origin of transonic buffet on aerofoils. J. Fluid Mech. 2009; 628 357-369.

[37] Xiao Q., Tsai HM., Liu F. Numerical Study of Transonic Buffet on a Supercritical Airfoil. AIAA Journal 2006; 44(3) 620-628.

[38] Awrejcewicz J. Bifurcation and Chaos in Coupled Oscillators. World Scientific, Singapore, 1991.

[39] Awrejcewicz J., Krysko VA., Nonclassical Thermoelastic Problems in Nonlinear Dynamics of Shells. Springer-Verlag, Berlin 2003.

[40] Kuz'min G. Bifurcations of transonic flow past simple airfoils with elliptic and wedge-shaped noses. Journal of Applied Mechanics and Technical Physics 2010; 51(1) 16–21.

[41] Ivanova AV. Structural instability of inviscid transonic channel flow. Journal of Engineering d Thermophysics 2003; 76(6) 1262-1265.

Chapter 6

Universality of Transition to Chaos in All Kinds of Nonlinear Differential Equations

Nikolai A. Magnitskii

Additional information is available at the end of the chapter

http://dx.doi.org/10.5772/48810

1. Introduction

The basically finished universal theory of dynamical chaos in all kinds of nonlinear differential equations including dissipative and conservative, nonautonomous and autonomous nonlinear systems of ordinary and partial differential equations and differential equations with delay arguments is shortly presented in the paper. Consequence of the theory is an existence of the uniform universal mechanism of self-organizing in the huge class of the mathematical models having the applications in many areas of science and techniques and describing the numerous physical, chemical, biological, economic and social both natural and public phenomena and processes. All theoretical positions and results are received within last several years by extremely author and his pupils and confirmed with numerous examples, illustrations and numerical calculations.

The basis of this theory consists of the Feigenbaum theory of period doubling bifurcations in one-dimensional mappings (Feigenbaum, 1978), the Sharkovskii theory of subharmonic bifurcations of stable cycles of an arbitrary period up to the cycle of period three in one-dimensional mappings (Sharkovskii, 1964), the Magnitskii theory of homoclinic and heteroclinic bifurcations of stable cycles and tori in systems of differential equations and the Magnitskii theory of rotor type singular points of two-dimensional nonautonomous systems of differential equations with periodic coefficients of leading linear parts as a bridge between one-dimensional mappings and differential equations (Magnitskii & Sidorov, 2006; Magnitskii, 2007; Magnitskii, 2008; Magnitskii, 2008b; Magnitskii, 2010).

It is shown that this universal Feigenbaum-Sharkovskii-Magnitskii (FSM) bifurcation theory of transition to dynamical chaos takes place in all classical three-dimensional chaotic dissipative systems of ordinary differential equations including Lorenz hydrodynamic system, Ressler chemical system, Chua electro technical system, Magnitskii macroeconomic system and many others. It takes place also in well-known two-dimensional non-

INTECH
open science | open minds

© 2012 The Author(s). Licensee InTech. This chapter is distributed under the terms of the Creative Commons Attribution License http://creativecommons.org/licenses/by/3.0), which permits unrestricted use, distribution, and reproduction in any medium, provided the original work is properly cited.

autonomous and many-dimensional autonomous nonlinear dissipative systems of ordinary differential equations, such as Duffing-Holmes, Mathieu, Croquette and Rikitaki equations. It takes place also in nonlinear partial differential equations and differential equations with delay arguments, such as Brusselyator, Ginzburg-Landau, Navier-Stokes and Mackey-Glass equations, reaction-diffusion systems and systems of differential equations describing excitable and autooscillating mediums. Moreover, the same scenario of transition to chaos takes place also in conservative and, in particularly, Hamiltonian systems such as Henon-Heiles and Yang-Mills systems, conservative Duffing-Holmes, Mathieu and Croquette equation and many others.

Thus, the question is about discovery and description of the uniform universal mechanism of the arranging of surrounding us infinitely complex and infinitely various nonlinear world. And this nonlinear world is arranged under uniform laws, and these laws are laws of nonlinear dynamics, qualitative theory of nonlinear systems of differential equations and theory of bifurcations in such systems.

2. Dynamical chaos in nonlinear dissipative systems of ordinary differential equations

2.1. Two-dimensional systems with periodic coefficients

Consider a smooth family of two-dimensional real nonlinear non-autonomous systems of ordinary differential equations

$$\dot{u} = D(t,\mu)u(t) + H(u,t,\mu),\ H(0,t,\mu) \equiv 0, \tag{1}$$

with a $T(\mu)$ -periodic matrix $D(t,\mu)$ of the leading linear part depending on a scalar system parameter μ. Expansion of a function $H(u,t,\mu)$ on components of vector u begins with members of the second order. The Floquet theory states that the fundamental matrix solution $U(t,\mu)$ of the linear part of system of Eqs. (1) can be represented in the form $U(t,\mu) = P(t,\mu)e^{B(\mu)t}$, where $P(t,\mu)$ is some T -periodic complex matrix and $B(\mu)$ is some constant complex matrix whose eigenvalues are named as Floquet exponents. It is important that the real linear system can have various complex but not complex-conjugate Floquet exponents $\alpha_1(\mu)$ and $\alpha_2(\mu)$. Real parts $\mathrm{Re}\,\alpha_1(\mu) = \beta_1(\mu),\ \mathrm{Re}\,\alpha_2(\mu) = \beta_2(\mu)$, can be different but imaginary parts $\mathrm{Im}\,\alpha_2(\mu) = \mathrm{Im}\,\alpha_1(\mu) + 2\pi k$ can be equal or differ from each other on $2\pi k$. Singular point $O(0,0)$ of a two-dimensional non-autonomous real system of Eqs. (1) with periodic coefficients of its leading linear part is a **rotor** if corresponding linear system has complex Floquet exponents with equal imaginary and different real parts (Magnitskii, 2008, 2011; Magnitskii & Sidorov, 2006). Canonical form of a rotor is a linear system

$$\begin{aligned}\dot{u}_1 &= \frac{\beta_1 + \beta_2 + (\beta_1 - \beta_2)\cos\omega t}{2}u_1 + \frac{(\beta_1 - \beta_2)\sin\omega t - \omega}{2}u_2 \\ \dot{u}_2 &= \frac{(\beta_1 - \beta_2)\sin\omega t + \omega}{2}u_1 + \frac{\beta_1 + \beta_2 - (\beta_1 - \beta_2)\cos\omega t}{2}u_2\end{aligned} \tag{2}$$

with $2\pi/\omega$ - periodic coefficients. In Eqs. (2) β_1 and β_2 are arbitrary real constants.

2.1.1. FSM – scenario of transition to chaos

If real parts of Floquet exponents depend on parameter μ which is changing, then the Sharkovskii subharmonic cascade of bifurcations of stable limit cycles is realizing in system of Eqs. (1) in accordance with the Sharkovskii order (Magnitskii, 2008; Magnitskii & Sidorov, 2006):

$$\begin{aligned} 1 \lhd 2 \lhd 2^2 \lhd 2^3 \lhd \ldots \lhd 2^2 \cdot 7 \lhd 2^2 \cdot 5 \lhd 2^2 \cdot 3 \lhd \ldots \\ \ldots \lhd 2 \cdot 7 \lhd 2 \cdot 5 \lhd 2 \cdot 3 \lhd \ldots \lhd 7 \lhd 5 \lhd 3. \end{aligned} \tag{3}$$

The ordering $n \lhd k$ in (3) means that the existence of a cycle of period k implies the existence of a cycle of period n. So, if a system of Eqs. (1) has a stable limit cycle of period three then it has also all unstable cycles of all periods in accordance with the Sharkovskii order. So, the family of systems of Eqs. (1) can have irregular attractors only at infinitely many accumulation points of bifurcation values of the system parameter. Every such value is a limit of a sequence of values of some Feigenbaum subcascade of period doubling bifurcations in Sharkovskii cascade. Thus, any irregular attractor of the family of systems of Eqs. (1) with rotor type singular point is a **singular** attractor, as it is defined in (Magnitskii & Sidorov, 2006; Magnitskii, 2011). Simple singular attractor is almost stable non-periodic trajectory which is the limit of a sequence of periodic orbits of some Feigenbaum subcascade of period doubling bifurcations. Complex singular attractor exists only in bifurcation values corresponding to homoclinic or heteroclinic separatrix loops. For other values of the parameter μ the family of systems of Eqs. (1) has only regular attractors - asymptotically orbitally stable periodic trajectories, even of a very large period.

Obviously, the simplest singular attractor is the Feigenbaum attractor, i.e. the first non-periodic attractor existing in the family of systems of Eqs. (1) for $\mu = \mu_\infty$, where the value μ_∞ is the first limit of the sequence of bifurcation values μ for which period doubling bifurcations of the original cycle take place. Note that the Feigenbaum cascade of period doubling bifurcations is the beginning of the Sharkovskii subharmonic cascade. Note also, that the subharmonic cascade of bifurcations in accordance with the Sharkovskii order (3) does not exhaust the entire complexity of transition to chaos in two-dimensional nonlinear nonautonomous dissipative systems of ordinary differential equations with rotors. It can be continued at least by the Magnitskii homoclinic cascade of bifurcations of stable cycles converging to a homoclinic loop of the rotor type singular point.

As an example, we consider a simplest two-dimensional nonlinear non-autonomous system of Eqs. (1) with leading linear part (2) in which $\beta_1(\mu) = 2\mu,\ \beta_2(\mu) = 2\mu - 4$:

$$\begin{aligned} \dot{u}_1 &= 2(\mu - 1 + \cos(\omega t))u_1 + (2\sin(\omega t) - \omega/2)u_2 - u_2^2, \\ \dot{u}_2 &= (2\sin(\omega t) + \omega/2)u_1 + 2(\mu - 1 - \cos(\omega t))u_2. \end{aligned} \tag{4}$$

For $\omega = 4$ and for growth of the parameter μ stable cycles are generated in the system of Eqs. (4) in accordance with the Sharkovskii order (3) and then in accordance with the Magnitskii homoclinic order. These cycles of period two and three, one of the singular attractors and homoclinic cycles of periods four and five are presented in Fig.1 (Magnitskii & Sidorov, 2006). Thus, in this system full bifurcation FSM (Feigenbaum-Sharkovskii-Magnitskii) scenario of transition to dynamical chaos is realized.

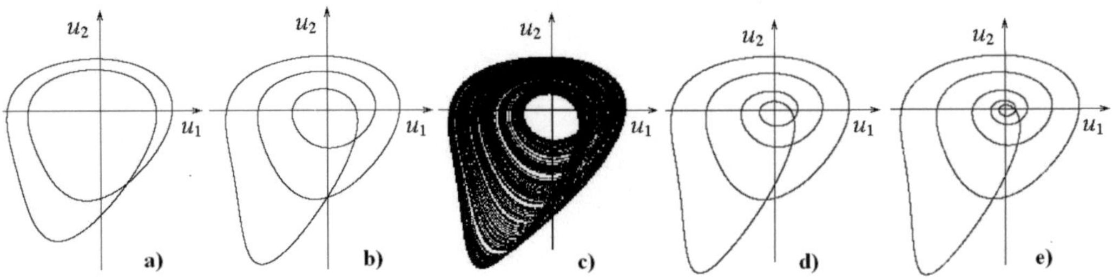

Figure 1. Stable cycles of period two (a) and period three (b), singular attractor (c) and homoclinic cycles of periods four (d) and five (e) in the system of Eqs. (4).

2.1.2. *Topological structure of singular attractors*

The problem which can be named as a main problem of chaotic dynamics of nonlinear systems of differential equations, is to find out how the boundary of the separatrix surface of the original singular cycle becomes more complex as the bifurcation parameter increases and how the onset of infinitely many regular and singular attractors of the system settle down on this separatrix manifold in accordance with a certain order (Sharkovskii order, homoclinic or heteroclinic order).

Note, that the simplest performance of a two-dimensional manifold in three-dimensional space on which all cycles in the Sharkovskii order and singular attractors can be placed without self-intersections was found in (Gilmore & Lefranc, 2002) in the form of branching manifold with the use of the Birman-Williams theorem and the principles of symbolic dynamics. However, such manifold must have a gluing, so that one can use it to explain the chaotic structure of semiflows but cannot generalize these results to flows, because this contradicts with the uniqueness theorem for solutions of differential equations. Hence, the representation given in (Gilmore & Lefranc, 2002) cannot be considered satisfactory.

We obtained a representation of the boundary of the separatrix surface of an original singular cycle of an arbitrary nonlinear dissipative system in a form of an infinitely folded two-dimensional heteroclinic separatrix manifold which Poincare section is named as **heteroclinic separatrix zigzag** (Magnitskii, 2010). It spanned by Moebius bands joining various cycles from the Feigenbaum period doubling cascade of bifurcations. From this consideration it becomes clear how and why cycles are arranged on this manifold in subharmonic and homoclinic order in the case of sufficiently strong dissipation, and why this order can be violated in systems with small dissipation and in conservative systems.

Rewrite the system of Eqs. (4) in the form of autonomous 4d-system

$$\begin{aligned} \dot{u} &= (2(\mu - b) + 2bp)u + (2bq - \omega / 2)v - v^2, \ \dot{p} = -\omega q \\ \dot{v} &= (2bq + \omega / 2)u + (2(\mu - b) - 2bp)v, \ \dot{q} = \omega p \end{aligned} \quad (5)$$

with $b = 1, u_2 = v$ and with the cycle $p^2 + q^2 = 1$. The parameter μ in system of Eqs. (5) is a bifurcation parameter, and the parameter b is responsible for dissipation. For small b and small μ the system is weakly dissipative, for large b and small μ it is strongly dissipative. Besides at $\mu = b$ the system of Eqs. (5) is conservative.

As a rule, all known dissipative systems of nonlinear differential equations are strongly dissipative, which has for many decades prevented one from studying the structure of their irregular attractors even with the use of most advanced computers. Last circumstance stimulated the development of numerous definitions of irregular attractors, ostensibly distinguished in their topological structure (strange, chaotic, stochastic, etc.). We illustrate this circumstance by the example of system of Eqs. (5) with strong dissipation for $b = 1, \omega = 4$, that is for the system of Eqs. (4). In this case, as the parameter $\mu > 0$ increases, system of Eqs. (5) has not only a complete subharmonic cascade of bifurcations in accordance with the Sharkovskii order, but also it has complete homoclinic cascade of bifurcations of cycles converging to the rotor homoclinic loop. The cause is clarified in Fig. 2a in which the Poincare section $(q = 0, p > 0)$ of the singular attractor of system of Eqs. (5) for $\mu = 0.12$ lying between cycles of period 5 and 3 in the Sharkovskii order is shown. The graph of the section almost coincides with the graph of one-dimensional unimodal mapping of a segment into itself, which has the above-listed cascades of bifurcations (Feigenbaum, 1978; Sharkovskii, 1964; Magnitskii & Sidorov, 2006). The projection of the manifold of the singular attractor onto the plane (p, u) corresponding to the section is shown in Fig. 2b.

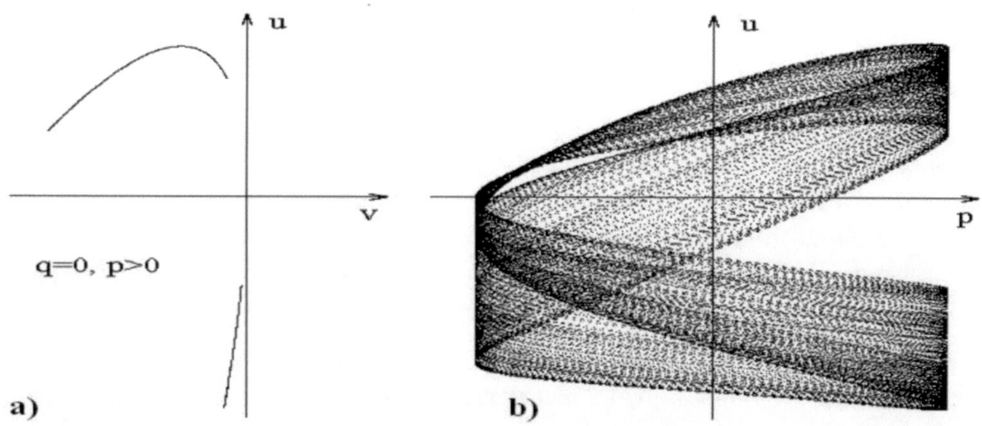

Figure 2. The projection of the Poincare section $(q = 0, p > 0)$ of solution of system of Eqs. (5) for $b = 1, \omega = 4, \mu = 0.12$ (a) and the projection of the manifold of the singular attractor onto the plane (p, u) corresponding to the section (b).

It seems that this is a two-dimensional strip whose lower part rotates around the original cycle, goes into its upper part in a revolution around it without twisting, and, in turn, the upper part goes into the lower part with twisting by 180 degrees in the next revolution. But in this case, to avoid contradiction with uniqueness theorem for solutions of systems of differential equations, two branches of the upper part should go into two branches of the lower part, which can be detected even under tenfold magnification (Fig.2a). Therefore, the upper part of the graph of the section in Fig. 2a should also consist of two branches, which makes its lower part to consist of four branches and so on. Consequently, the invariant manifold of the singular attractor shown in Fig. 2 should be a two-dimensional infinitely-sheeted folded surface. However, strong dissipation of the system in this case prevents correct understanding a topological structure of separatrix manifold of original singular cycle.

So, let us analyze the behavior of attractors of the system of Eqs. (5) with weak dissipation for $b = 0.05, \omega = 0.8$. A stable cycle of the double period, which is the boundary of the unstable Moebius band (an unstable two-dimensional manifold) of the original unit singular cycle $p^2 + q^2 = 1$, is generated in system of Eqs. (5) for small $\mu > 0$. It is an ordinary simple cycle of the period $4/\pi$ in the projection onto the two-dimensional subspace (u,v). Initially this cycle has two multipliers lying on the positive part of the real axis inside the unit circle and moving towards each other as the parameter μ grows. Then multipliers meet, become complex conjugated and continue to move on positive and negative half-circles inside an unit circle towards the negative part of the real axis. In this case, the unstable Moebius band of the original singular cycle becomes a complex roll around the stable cycle of the double period. The frequency of rotation of a trajectory on the roll around the stable cycle of the double period is specified by the frequency ω and also by imaginary parts of complex conjugated multipliers. Therefore, the approach of the multipliers to the negative part of the real axis leads to the flattening of the roll in one direction and to its degeneration into a stable Moebius band around the stable cycle of the double period.

Further multipliers of the cycle begin to move along the negative part of the real axis in opposite directions, which leads to appearance of two stable two-dimensional manifolds in the form of two transversal Moebius bands for the cycle of double period. Therefore, the cycle of double period becomes a singular stable cycle. Next, at the moment of intersection of the unit circle by one of the multiplies at the point -1, the cycle of double period becomes an unstable singular cycle, whose stable and unstable manifolds are two transversal Moebius bands. The boundary of its unstable manifold is a stable cycle of quadruple period. Thus, we came to an original situation, but for a singular cycle of double period.

The cascade of Feigenbaum period doubling bifurcations continues, up to infinity, the process of construction of a two-dimensional heteroclinic separatrix manifold, which consists of Moebius bands, joining the unstable singular cycles of the cascade. The self-similar separatrix figure obtained in the Poincare section is referred to as **Feigenbaum separatrix tree**. A nonperiodic stable trajectory passes through the top of the Feigenbaum tree and through the endpoints of all of its branches, and each neighborhood of that trajectory contains singular unstable cycles from the Feigenbaum cascade. This nonperiodic almost stable trajectory is the Feigenbaum attractor, which is the first and the simplest attractor in the infinite family of singular attractors.

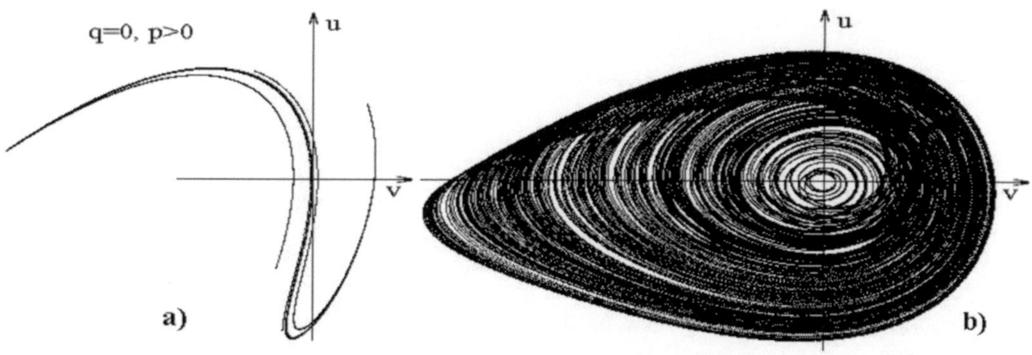

Figure 3. Projection of the Poincare section $(q=0, p>0)$ of solution of system of Eqs. (5) for $b=0.05, \omega=0.8, \mu=0.02$ (a) and the projection of the manifold of the singular attractor onto the plane (u,v) corresponding to the section (b).

Along with separatrix branches connecting unstable singular cycles of the Feigenbaum cascade, the Feigenbaum separatrix tree contains stable one-sided separatrix branches only entering cycles. These branches begin to close each other and form heteroclinic separatrix folds passing through various stable and unstable cycles from the Sharkovskii subharmonic cascade of bifurcations, which are generated during saddle-node bifurcations. New Feigenbaum separatrix trees are generated on the separatrices of the newly generated singular cycles, and the tops of these trees contain more complicated singular attractors. An infinitely folded separatrix two-dimensional manifold, which Poincare section is referred to as a **heteroclinic separatrix zigzag** is thereby generated. It is shown in Fig. 3a. In the case of weak dissipation Poincare section of solutions of the system (a heteroclinic separatrix zigzag) is already not close to the graph of the one-dimensional unimodal mapping, which leads to the violation of the Sharkovskii order in its right-hand side, i.e. cycles of periods 7, 5 and 3 may not exist in the system but may also be stable either simultaneously with cascades of bifurcations of some other cycle or without them. For example, system of Eqs. (5) for $b=0.05, \omega=1.5$ has simultaneously two stable cycles of periods one and three (for $\mu=0.0355$).

Thus, any unstable cycle of the system is unstable singular cycle joining neiboring separatrices of a heteroclinic zigzag. Any simple singular attractor is almost stable nonperiodic trajectory passing through vertices of some infinite Feigenbaum tree. Any trajectory of system from the attraction domain of the separatrix zigzag is first attracted to it along the nearest stable Moebius band, then approaches unstable sheets, goes along them, and tends either to a stable cycle or to a singular attractor depending on value of bifurcation parameter.

2.1.3. Some examples of classical two-dimensional nonautonomous systems

Consider three classical nonlinear ordinary differential equations of the second order with periodic coefficients such as Duffing-Holmes equation

$$\ddot{x} + k\dot{x} + \omega^2 x + \mu x^3 = f_0 \cos\Omega t, \tag{6}$$

modified dissipative Mathieu equation

$$\ddot{x} + \mu\dot{x} + (\delta + \varepsilon\cos\omega t)x + \alpha x^3 = 0, \tag{7}$$

and Croquette dissipative equation

$$\ddot{x} + \mu\dot{x} + \alpha\sin x + \beta\sin(x - \omega t) = 0. \tag{8}$$

All these equations are equivalent to two-dimensional nonlinear dissipative systems of ordinary differential equations with periodic coefficients and all of them have the same universal FSM scenario of transition to dynamical chaos (Magnitskii & Sidorov, 2006). For these equations, some important stable cycles and singular subharmonic attractors in accordance with the Sharkovskii order are presented in Fig. 4 - Fig. 6.

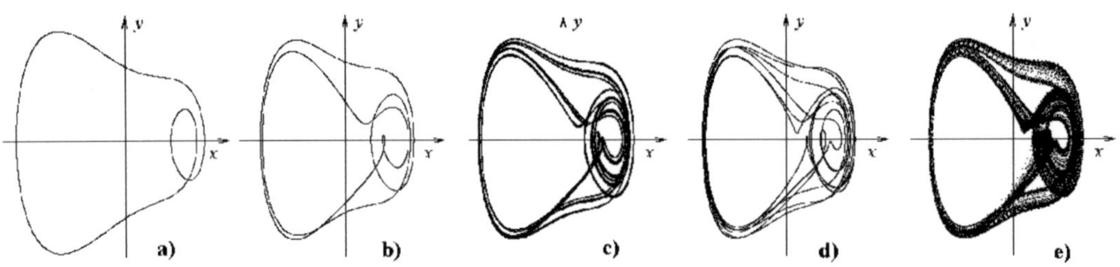

Figure 4. Original cycle (a), cycle of period two (b), Feigenbaum attractor (c), cycle of period six (d) from subharmonic cascade and more complex singular attractor (e) in the Duffing-Holmes equation (6).

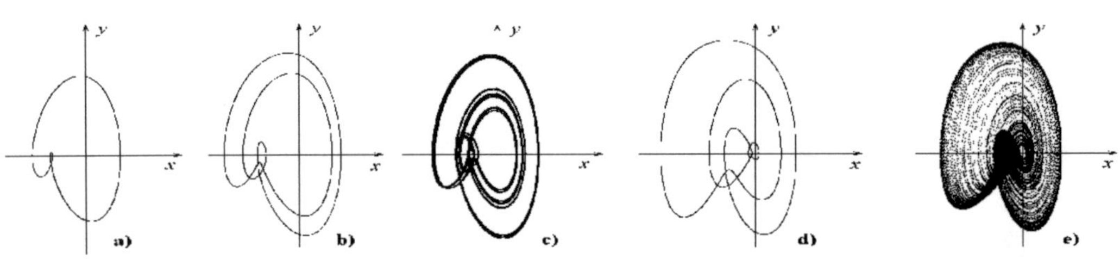

Figure 5. Original cycle (a), cycle of period two (b), Feigenbaum attractor (c), cycle of period three (d) from subharmonic cascade and more complex singular attractor (e) in the Mathieu equation (7).

Note that double period bifurcations were found also in (Awrejcewicz, 1989; Awrejcewicz 1991) for some other nonlinear ordinary differential equations of the second order with periodic coefficients.

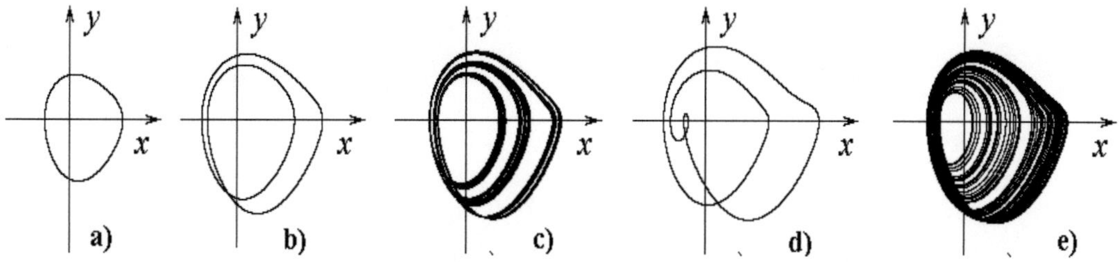

Figure 6. Original cycle (a), cycle of period two (b), Feigenbaum attractor (c), cycle of period three (d) from subharmonic cascade and more complex singular attractor (e) in the Croquette equation (8).

2.2. Three-dimensional autonomous systems

Consider a smooth family of three-dimensional nonlinear dissipative autonomous systems of ordinary differential equations

$$\dot{x} = F(x,\mu), \qquad x \in M \subset R^3, \quad \mu \in I \subset R, \quad F \in C^\infty, \tag{9}$$

depending on a scalar system parameter μ.

It is shown by the author in (Magnitskii & Sidorov, 2006; Magnitskii, 2008) that if a three-dimensional system of Eqs. (9) has a singular cycle of period T defined by complex Floquet exponents with equal imaginary parts (i.e. Moebius bands are its stable and unstable invariant manifolds), then by passing to a coordinate system rotating around the cycle, one can reduce such system to a two-dimensional nonautonomous system in coordinates, transversal to the singular cycle with zero rotor-type singular point corresponding to the cycle. So, all arguments listed in the previous section hold completely for autonomous three-dimensional systems with singular cycles.

2.2.1. *FSM – scenario of transition to chaos*

Therefore, three-dimensional autonomous system with singular cycle should have the same FSM scenario of transition to chaos as two-dimensional nonautonomous system with periodic coefficients and zero rotor-type singular point. As an example, consider the autonomous three-dimensional system

$$\begin{aligned}
\dot{x}_1 &= -\omega x_2 + x_1[((\mu-1)\sqrt{x_1^2+x_2^2} + x_1)(x_1^2+x_2^2-1) + (2x_2 - \omega/2)x_3],\\
\dot{x}_2 &= \omega x_1 + x_2[((\mu-1)\sqrt{x_1^2+x_2^2} + x_1)(x_1^2+x_2^2-1) + (2x_2 - \omega/2)x_3],\\
\dot{x}_3 &= (x_2 + \omega/4)(x_1^2+x_2^2-1) + 2(\mu-1-x_1)x_3.
\end{aligned} \tag{10}$$

For $\mu<1$ system of Eqs.(10) has the singular point $(0,0,\omega/8(\mu-1))$ and limit cycle $x_0(t,\mu)=(\cos\omega t, \sin\omega t, 0)^T$ with period $T=2\pi/\omega$ in the plane of variables (x_1,x_2). By changing the variables $x(t,\mu)=x_0(t,\mu)+Q(t,\mu)(0,u_1(t),u_2(t))^T$ with $2\pi/\omega$ -periodic matrix $Q(t)=\left(\dot{x}_0(t), x_0(t), (0,0,1)^T\right)$, one can reduce the system of Eqs. (10) to two-dimensional nonautonomous system with $2\pi/\omega$ -periodic coefficients and zero rotor-type singular point

$$\begin{aligned}\dot{u}_1 &= 2(\mu - 1 + \cos\omega t)u_1 + (2\sin\omega t - \omega/2)u_2 + h_1(u_1, u_2, t, \omega, \mu),\\ \dot{u}_2 &= (2\sin\omega t + \omega/2)u_1 + 2(\mu - 1 - \cos\omega t)u_2 + h_2(u_1, u_2, t, \omega, \mu),\end{aligned} \tag{11}$$

where

$$\begin{aligned}h_1 &= (\mu - 1 + \cos\omega t)((2u_1 + u_1^2)^2 + u_1^2) + ((2u_1 + 4)\sin\omega t - \omega/2)u_1 u_2,\\ h_2 &= (2\sin\omega t + (u_1 + 1)\sin\omega t + \omega/4)u_1^2 - +2u_1 u_2 \cos\omega t.\end{aligned}$$

Leading linear part of system of Eqs. (11) coincides with the linear part of system of Eqs. (4) with rotor. So, for $\mu < 0$ zero solution of system of Eqs. (11) and singular cycle $x_0(t, \mu)$ of system of Eqs. (10) are stable. For $\mu > 0$ all cascades of bifurcations in accordance with the theory FSM take place in both systems. Some cycles and singular attractors from these cascades are presented in Fig. 7, rotor and singular cycle separatrix loops are presented in Fig. 8. Thus, if parameter μ is changing, then the Sharkovskii subharmonic and Magnitskii

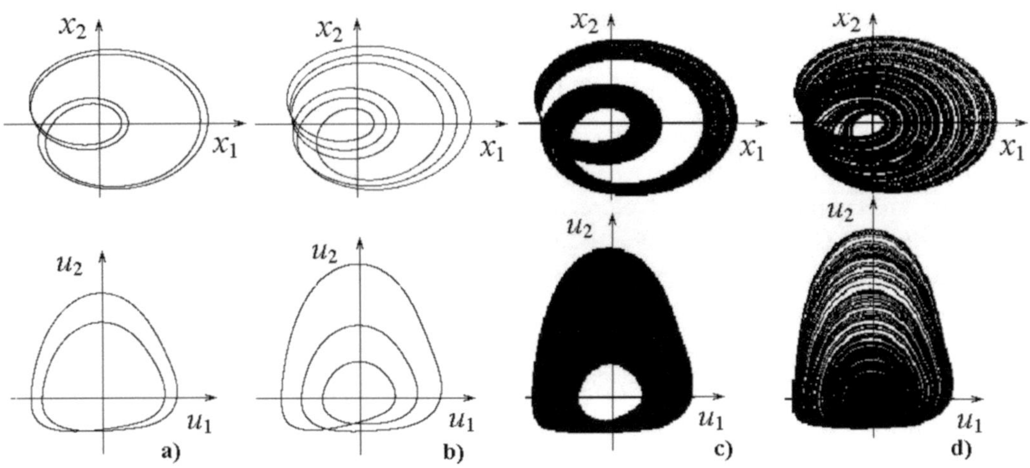

Figure 7. Projections of period four and six cycles and singular attractors of system of Eqs. (10) (above) and corresponding to them period two and three cycles and singular attractors of system of Eqs. (11) (below).

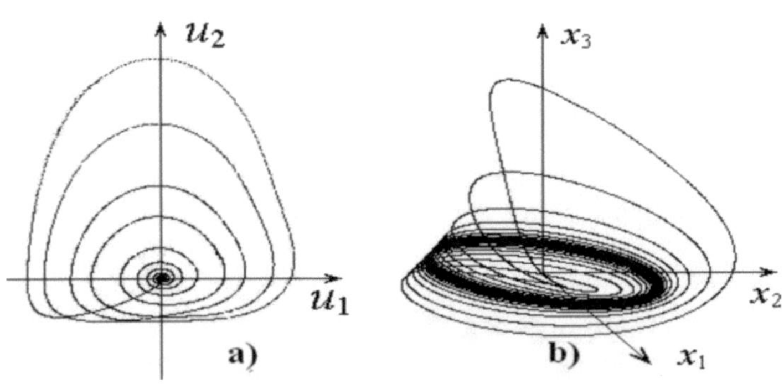

Figure 8. Rotor separatrix loop of system of Eqs. (11) (a) and corresponding to it separatrix loop of singular cycle of system of Eqs. (10) (b).

homoclinic cascades of bifurcations of stable limit cycles are realizing in any system of Eqs. (9) in accordance with the Sharkovskii and homoclinic orders. Cycle of period three is the last cycle in the Sharkovskii order. Therefore, to verify an existence of subharmonic cascade of bifurcations in any system one should to find a stable cycle of period three in this system or any stable homoclinic cycle.

2.2.2. *Topological structure of singular attractors*

The three-dimensional phase space of three-dimensional autonomous system containing the original singular cycle of period T is diffeomorphic to three-dimensional manifold of an autonomous four-dimensional system of the form of Eqs. (5), the first two equations of which have linear part of the form of Eqs. (2), and the remaining two equations with some condition define a motion on a plane along a simple cycle of period T. Therefore, the separatrix heteroclinic manifold constructed in previous Section for system of Eqs. (5) in the section $(q=0,p>0)$ (in the section of the singular cycle corresponding to the rotor) should be completely similar to the separatrix heteroclinic manifold of a three-dimensional autonomous system in the section of the original singular cycle.

As an example, consider the autonomous three-dimensional system:

$$\begin{aligned} \dot{x} &= -\omega y - \omega xz/2 - ((\mu-b)x+b)(1-x^2-y^2), \\ \dot{y} &= \omega x + 2(b-\omega y/4)z - (\mu-b)y(1-x^2-y^2) \\ \dot{z} &= 2(\mu-b-bx)z - (by+\omega/4)(1-x^2-y^2). \end{aligned} \tag{12}$$

System of Eqs. (12) has the periodic solution (the cycle) $x_0(t,\mu)=(\cos\omega t,\sin\omega t,0)^T$, which lies in the plane of the variables (x,y) and has the period $2\pi/\omega$. By linearizing system of Eqs. (12) on the cycle with respect to deviations y_1,y_2,y_3 from the cycle and by performing the change of variables $y(t)=Q(t)z(t)$ with $2\pi/\omega$- periodic matrix $Q(t)=(\dot{x}_0(t),x_0(t),(0,0,1)^T)$, one can obtain the following system of equations in the rotating variables transversal to the cycle:

$$\begin{aligned} \dot{z}_2 &= (2(\mu-b)+2b\cos\omega t)z_2 + (2b\sin\omega t-\omega/2)z_3, \\ \dot{z}_3 &= (2b\sin\omega t+\omega/2)z_2 + (2(\mu-b)-2b\cos\omega t)z_3. \end{aligned} \tag{13}$$

System of Eqs. (13) coincides with the linear part of system of Eqs. (4) considered in previous Section, and in addition, the coordinate tangent to the cycle has the form $\dot{z}_1=((-2b/\omega)\sin\omega t)z_2+((2b/\omega)\cos\omega t)z_3$, and does not influence the generation of the dynamics of solutions in a neighborhood of the cycle. Consequently, the heteroclinic separatrix manifold generated around the cycle of system of Eqs. (12) as the bifurcation parameter μ grows has the same structure as that of the heteroclinic separatrix manifold of a rotor-type singular point and should be similar to a heteroclinic separatrix zigzag in the Poincare section for small dissipation parameter b. The projection of Poincare section $(y=-0.1,x<0)$ of solution of system of Eqs. (12) is presented in Fig. 9a.

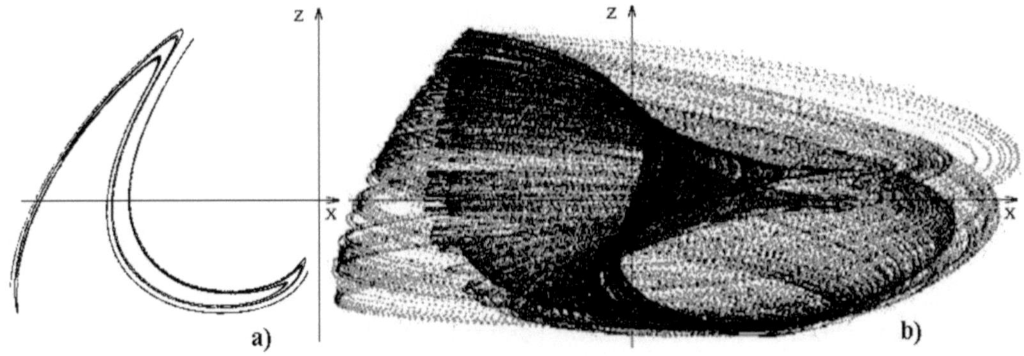

Figure 9. Projection of the Poincare section $(y = -0.1, x < 0)$ of solution of system of Eqs. (12) for $b = 0.08, \omega = 1, \mu = 0.051$ (a) and the projection of the manifold of the singular attractor onto the plane (x, z) corresponding to the section (b).

2.2.3. Some examples of classical tree-dimensional autonomous nonlinear systems

For instance let consider four classical tree-dimensional chaotic systems of nonlinear ordinary differential equations describing different natural and social processes:

the Lorenz hydrodynamic system

$$\dot{x} = \sigma(y - x), \quad \dot{y} = x(r - z) - y, \quad \dot{z} = xy - bz, \tag{14}$$

the Ressler chemical system

$$\dot{x} = -(y + z), \dot{y} = x + ay, \dot{z} = b + z(x - \mu), \tag{15}$$

the Chua electro technical system

$$\dot{x} = \mu[y - h(x)], \ \dot{y} = x - y + z, \ \dot{z} = -\beta y, \tag{16}$$

where $h(x)$ is a piecewise linear function; and the Magnitskii macroeconomic system

$$\dot{x} = bx((1 - \sigma)z - \delta y), \dot{y} = x(1 - (1 - \delta)y + \sigma z), \dot{z} = a(y - dx). \tag{17}$$

To demonstrate that the transition to chaos under variation of a system parameter in all these classical chaotic systems occurs in accordance with the described above unique FSM scenario, let show that all these systems have period three stable cycles in accordance with the Sharkovskii order (3). This stable period three cycles are presented in Fig. 10.

In Fig. 11 it is presented homoclinic cascade of bifurcations of stable cycles in the Lorenz system and the most complex separatrix contour in this system named as **heteroclinic butterfly** which is the limit of the heneroclinic cascade of bifurcations of stable heteroclinic cycles (Magnitskii & Sidorov, 2006; Magnitskii , 2008; Magnitskii , 2011).

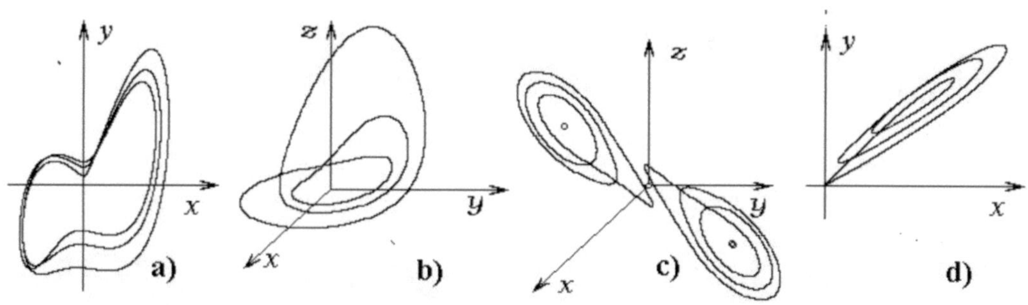

Figure 10. Cycles of period three in Lorenz (14) (a), Ressler (15) (b), Chua (16) (c) and Magnitskii (17) (d) systems.

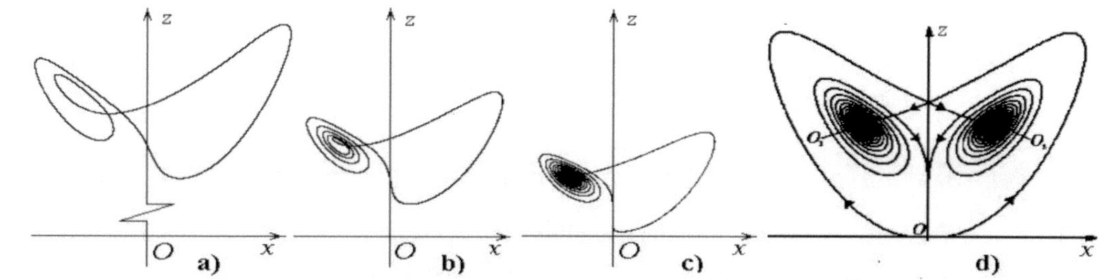

Figure 11. Homoclinic cascade of bifurcations of stable cycles (a)-(c) and heteroclinic butterfly separatrix contour (d) in the Lorenz system (14).

All these classical systems have also arbitrary stable cycles from the Sharkovskii subharmonic cascade of bifurcations and all singular attractors from this cascade. Moreover, these systems have also more complex complete or incomplete homoclinic or heteroclinic cascades of bifurcations which take place after Sharkovskii cascade and infinitely many homoclinic or heteroclinic singular attractors (Magnitskii & Sidorov, 2006; Magnitskii , 2008; Magnitskii , 2011).

In conclusion of this Section note that also very many other nonlinear three-dimensional autonomous systems of ordinary differential equations considered in the scientific literature have the same universal scenario of transition to dynamical chaos in accordance with the Feigenbaum-Sharkovskii-Magnitskii (FSM) theory. Among them there are systems of: Vallis, Anishchenko-Astakhov, Rabinovich-Fabricant, Pikovskii-Rabinovich-Trakhtengertz, Sviregev, Volterra-Gause, Sprott, Chen, Rucklidge, Genezio-Tesi, Wiedlich-Trubetskov and many others (Magnitskii , 2011; Magnitsky , 2007).

2.3. Many- and infinitely- dimensional autonomous systems

2.3.1. Transition to chaos through bifurcation cascades of stable cycles

At the beginning let us show that the scenario of transition to chaos through the Sharkovskii subharmonic and homoclinic cascades of bifurcations of stable cycles takes place also in many-dimensional dissipative nonlinear systems of ordinary differential equations. For example consider Rikitaki system

$$\dot{x} = -\mu x + yz, \dot{y} = -\mu y + xu, \dot{z} = 1 - xy - bz, \dot{u} = 1 - xy - cu, \quad (18)$$

modelling a change in dynamics of magnetic poles of the Earth. Some main cycles of subharmonic cascade of bifurcations in the system of Eqs. (18) and some singular attractors are presented in Fig. 12.

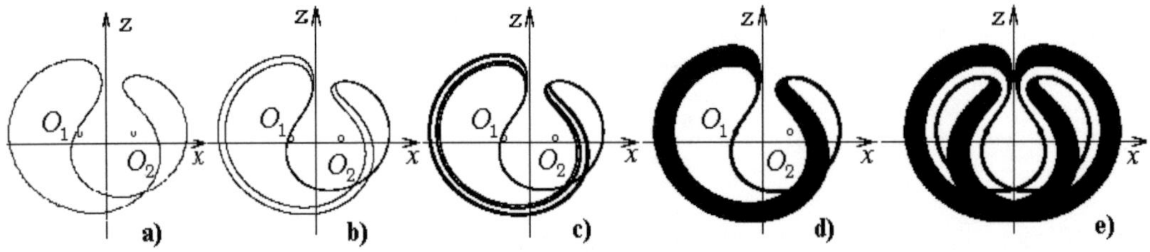

Figure 12. Projections of original singular cycle (a), cycle of period two (b), Feigenbaum attractor (c) and two more complex singular attractors in the Rikitaki system (d)-(e).

2.3.2. Transition to chaos through bifurcation cascades of stable two-dimensional tori

Besides described above mechanism of transition to chaos in accordance with subharmonic and homoclinic cascades of bifurcations of stable cycles, in many-dimensional dissipative nonlinear systems of ordinary differential equations there exists a scenario of transition to chaos through subharmonic and homoclinic cascades of bifurcations of stable two-dimensional or many-dimensional tori along any one or several frequencies simultaneously. The mechanism of this cascade of bifurcations has the same above considered FSM nature, and presently there not discovered really any other scenarios of transition to chaos in many-dimensional nonlinear systems of ordinary differential equations. Such a scenario of transition to chaos takes place in complex five-dimensional Lorenz system

$$\dot{X} = -\sigma X + \sigma Y, \quad \dot{Y} = -XZ + rX - aY, \quad \dot{Z} = -bZ + (X^* Y + XY^*)/2 \quad (19)$$

of two complex variables $X = x_1 + ix_2$ and $Y = y_1 + iy_2$ and one real variable Z. If values of parameters a, b, σ and $\mathrm{Re}\, r$ are fixed and the value of parameter $\mathrm{Im}\, r$ is decreasing, then at first a stable invariant torus is appearing from the stable cycle as a result of Andronov-Hopf bifurcation. After that the period two invariant torus is appearing from this original singular saddle torus as a result of double period bifurcation (Fig. 13). That is the beginning of Feigenbaum cascade of period doubling bifurcations. Then, after further decreasing of bifurcation parameter $\mathrm{Im}\, r$, all subharmonic cascade of bifurcations of stable two-dimensional tori with arbitrary period in accordance with the Sharkovskii order (3) takes place in the complex Lorenz system. Projections of sections of period one, two and three two-dimensional invariant tori and one of the toroidal singular attractor are presented in Fig. 13.

This example shows that the FSM (Feigenbaum-Sharkovskii-Magnitskii) scenario of transition to dynamical chaos in two-dimensional nonautonomous and three-dimensional autonomous systems of ordinary differential equations takes place also in many-dimensional systems. So, appearance of three-dimensional torus is not necessary condition for generation of chaotic dynamics in dissipative many-dimensional systems of differential equations.

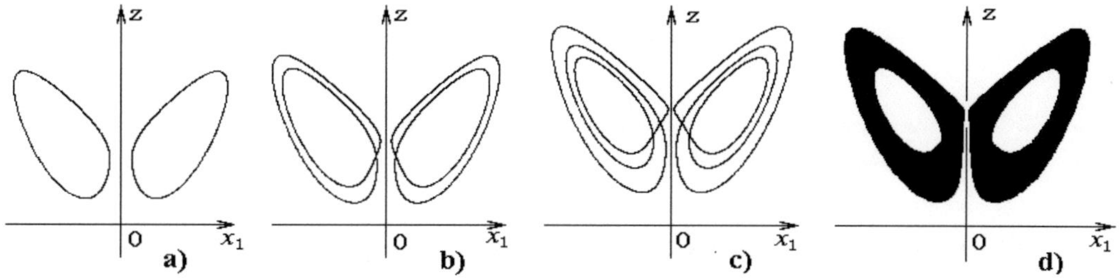

Figure 13. Projections of sections of two-dimensional invariant tori of period one (a), two (b) and three (c) and one of the toroidal singular attractor (d) in complex Lorenz system (19).

2.3.3. *Transition to chaos in nonlinear equations with delay argument*

Let us show now that the FSM scenario of transition to chaos is realized also in infinitely-dimensional nonlinear autonomous dissipative systems of ordinary differential equations, namely in nonlinear ordinary differential equations with delay arguments. For instance such scenario of transition to chaos through the Sharkovskii subharmonic cascade of bifurcations of stable cycles with arbitrary period in accordance with the Sharkovskii order (3) takes place in well-known Mackey-Glass equation (Mackey & Glass, 1977).

$$\dot{x}(t) = -ax(t) + \beta_0 \frac{\theta^n x(t-\tau)}{\theta^n + x^n(t-\tau)}. \tag{20}$$

In this equation the delay argument τ is a bifurcation parameter. When a value of parameter τ is small, then Mackey-Glass equation has unique stable stationary state. When τ is increasing, then at first a stable cycle is appearing in phase space of the equation from the stable stationary state as a result of Andronov-Hopf bifurcation. After that the period two stable cycle is appearing from this original singular cycle as a result of double period bifurcation. That is the beginning of the Feigenbaum cascade of period doubling bifurcations. Then, after further increasing of bifurcation parameter τ, all subharmonic cascade of bifurcations of stable cycles with arbitrary period in accordance with the Sharkovskii order takes place in the Mackey-Glass equation. Projections of some main stable cycles and singular attractors of the Mackey-Glass equation are presented in Fig. 14.

Thus we can make a conclusion that universal bifurcation Feigenbaum-Sharkovskii-Magnitskii theory describes transition to dynamical chaos in all nonlinear dissipative systems of ordinary differential equations. Scenario of transition to chaos consists of subharmonic and homoclinic (heteroclinic) cascades of bifurcations of stable cycles or stable two- or many-dimensional tori.

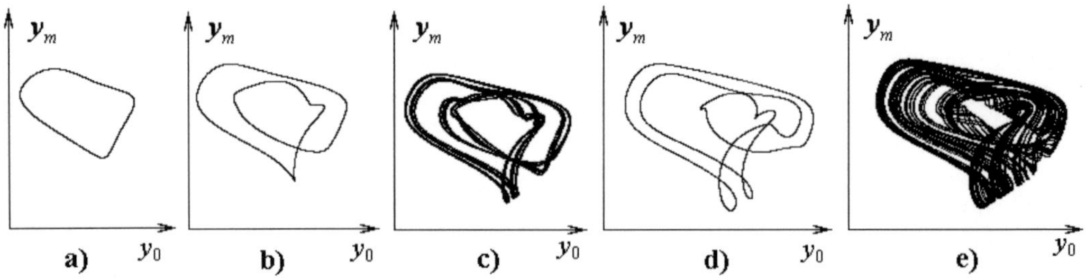

Figure 14. Projections of period one (a), two (b) and three (d) cycles, Feigenbaum attractor (c) and one of more complex singular attractor (e) in the Mackey-Glass equation (20).

3. Chaos in Hamiltonian and conservative systems

The modern classical theory of Hamiltonian systems reduces a problem of the analysis of dynamics of such system to the problem of its integralability, i.e. to a problem of construction of the canonical transformation reducing system to variables "action - angle" in which, as it is considered to be, movement occurs on a surface of n -dimensional torus and is periodic or quasiperiodic. Any nonintegrable nonlinear Hamiltonian system is considered as perturbation of integrable system, and the analysis of its dynamics is reduced to finding-out of a question on destruction or nondestruction some tori of nonperturbed system depending on value of perturbation.

In the present Section absolutely other bifurcation approach is considered for analysis of chaotic dynamics not only Hamiltonian, but also any conservative system of nonlinear differential equations. The method consists in consideration of approximating extended two-parametrical dissipative system of the equations, stable solutions (attractors) of which are as much as exact aproximations to solutions of original Hamiltonian (conservative) system. Attractors (stable cycles, tori and singular attractors) of extended dissipative system one can search by numerical methods with use the results of universal FSM (Feigenbaum-Sharkovskii-Magnitskii) theory, developed initially for nonlinear dissipative systems of ordinary differential equations and considered in detail in the previous Section of the chapter. It becomes clear what chaos is in Hamiltonian and simply conservative systems. And this chaos is not a result of destruction of some tori of nonperturbed system as it is considered to be in the modern literature, but, on the contrary, it is a result of bifurcation cascades of a birth of regular (cycles and tori) and singular attractors in extended dissipative system in accordance with the universal FSM theory when dissipation parameter tends to zero.

3.1. Bifurcation approach to analysis of Hamiltonian and conservative systems

The fact that the dynamics of any conservative system is a limit case of a dynamics of an extended dissipative system with weak dissipation as the dissipation parameter tends to zero was proved by author in (Magnitskii, 2008b; Magnitskii, 2011) and illustrated by numerous examples of Hamiltonian systems with one and a half, two and three degrees of freedom and by examples of simply conservative but not Hamiltonian systems. The stability domains of cycles of such a system with zero dissipation become tori of a conservative (Hamiltonian) system around its elliptic cycles into which the stable cycles themselves go. Complicated separatrix heteroclinic manifolds spanned by unstable singular cycles of the dissipative system become (for zero dissipation) even more complicated separatrix manifolds of the conservative (Hamiltonian) system along which the motion of a trajectory is treated as chaotic dynamics. Thus, it becomes clear why the order of the tori alternation in conservative (Hamiltonian) systems can differ from the Sharkovskii order existing in systems with strong dissipation.

3.1.1. Theoretical basis of bifurcation approach

Let's consider generally nonlinear conservative system of ordinary differential equations with a smooth right part

$$\dot{x} = f(x), x \in R^n, \ div\, f(x) = 0 \tag{21}$$

which variables are connected by some equation

$$H(x_1, ..., x_n) = \varepsilon. \tag{22}$$

Any Hamiltonian system is a special case of system of Eqs. (21)-(22) at even value of dimension n and at the given integral of movement (22) generating system of Eqs. (21). Movement in system of Eqs. (21) occurs in $n-1$-dimensional subspace, set by the equation (22).

Theorem. Let two-parametrical system of ordinary differential equations

$$\dot{x} = g(x, \varepsilon, \mu), \ x \in R^n, \tag{23}$$

possesses following properties: 1) the only solutions of system of Eqs. (21)-(22) are solutions of system of Eqs. (23) with initial conditions $H(x_{10}, ..., x_{n0}) = \varepsilon$ at $\mu = 0$; 2) at all $\mu > 0$ the system of Eqs. (23) is dissipative system on its solutions laying in neighborhoods of solutions of system of Eqs. (21)-(22). Then attractors of dissipative system of Eqs. (23) at small $\mu > 0$ are as much as exact approximations of solutions of conservative system of Eqs. (21)- (22) (see proof in (Magnitskii, 2008; Magnitskii, 2011)).

So, for application of the offered approach to the analysis of conservative and, in particular, Hamiltonian systems it is necessary to construct an extended dissipative system, satisfying the properties 1) and 2). Then for everyone $\varepsilon > 0$ one should to find numerically all stable solutions and their cascades of bifurcations according to the FSM scenario in extended dissipative system of Eqs. (23) when μ tends to zero, starting from the various initial conditions, satisfying the equality (22). Areas of stability of the found simple regular solutions (simple cycles) will generate at $\mu = 0$ regular solutions (tori) of original conservative system of Eqs. (21)-(22), and areas of stability of complex cycles and singular attractors and also heteroclinic separatrix manifolds will generate chaotic solutions. By the same method in the area of parameters $\varepsilon > 0$, $\mu \geq 0$ one can construct bifurcation diagrams of all bifurcations existing in two-parametrical extended dissipative system of Eqs. (23) and smoothly passing to bifurcations in conservative system of Eqs. (21)-(22) on the boundary $\mu = 0$.

3.1.2. *Subharmonic cascade of bifurcations in Hamiltonian and conservative systems*

From theoretical positions of bifurcation approach to the analysis of Hamiltonian and any conservative systems it follows, that at enough great values of parameter $\varepsilon > 0$ transition to chaotic dynamics in system of Eqs. (21)-(22) occurs according to universal FSM scenario and that bifurcation diagram of this scenario can be received by limiting transition at $\mu \to 0$ from similar bifurcation diagram of two-parametrical extended dissipative system of Eqs. (23). Let's illustrate this position by the example of classical conservative Croquette equation

$$\ddot{x} + \alpha \sin x + \beta \sin(x - \omega t) = 0, \tag{24}$$

modeling a magnet rotary fluctuations in an external magnetic field in absence of friction. It is easy to see, that the equation (24) is equivalent to two-dimensional conservative system with periodic coefficients (Hamiltonian system with one and a half degrees of freedom) and also to four-dimensional conservative (not Hamiltonian) autonomous system of the equations

$$\dot{x} = y, \;\; \dot{y} = -(\alpha + r)\sin x + z\cos x, \;\; \dot{z} = \omega r, \;\; \dot{r} = -\omega z \tag{25}$$

with a condition $H = z^2 + r^2 = \varepsilon^2$, $z_0 = z(0) = 0$. Extended dissipative system for the system of Eqs. (25) will be

$$\dot{x} = y, \;\; \dot{y} = -\mu y - (\alpha + r)\sin x + z\cos x, \;\; \dot{z} = \omega r, \;\; \dot{r} = -\omega z. \tag{26}$$

It is easy to check up numerically, that the two-parametrical system of Eqs. (26) with initial conditions $z_0 = z(0) = 0$, $r_0 = r(0) = \varepsilon$ has the subharmonic cascade of bifurcations at each value of parameter ε and at reduction of values of parameter μ . For each cycle of the cascade in a plane of parameters (ε, μ) it is possible to construct monotonously increasing bifurcation curve $\mu(\varepsilon)$ of births of the given cycle. Boundary values of such curves at $\mu = 0$ are bifurcation values of the subharmonic cascade of bifurcations in conservative Croquette system of the Eqs. (25) for parameter $\varepsilon > 0$ (see Fig. 15).

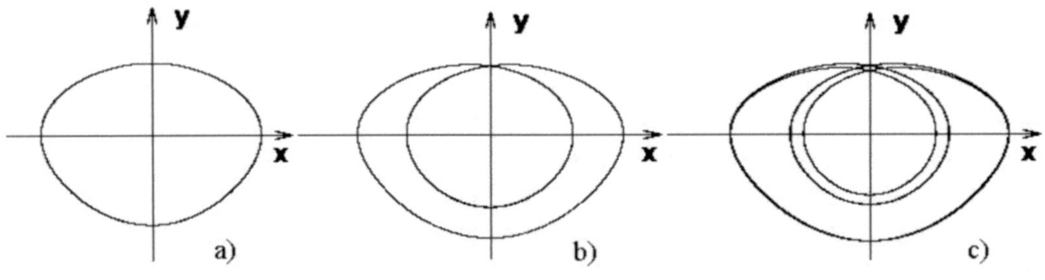

Figure 15. Projections on the plane (x, y) of the cycle (a) for $\varepsilon = 0.45$, period two cycle (b) for $\varepsilon = 0.48$ and period four cycle (c) for $\varepsilon = 0.497$ in conservative Croquette system of the Eqs. (25) for $\alpha = \omega = 1$.

3.1.3. *Heteroclinic separatrix manifolds*

Cascade of saddle-node bifurcations in extended dissipative system, consisting in a simultaneous birth of stable and saddle cycles, leads to formation in conservative (Hamiltonian) system of family of complex multiturnaround tori around of elliptic cycles and heteroclinic separatrix manifold which is tense on complex multiturnaround hyperbolic cycles of the system. In Poincare section it looks like a family of the hyperbolic singular points connected by separatrix contours. This picture at any shift in initial conditions passes into a family of so-called islands (points in Poincare section forming closed curves around of points of elliptic cycle).

At the same time, as follows from the theory, at enough great values of perturbation parameter $\varepsilon > 0$ in extended dissipative system there are cascades of bifurcations in accordance with scenario FSM. These cascades of bifurcations generate considered in the previous Section of the chapter infinitely folded heteroclinic separatrix manifolds having in Poincare section a kind of heteroclinic separatrix zigzag. These manifolds are tense on unstable singular cycles of FSM-cascade of dissipative system and they pass at zero dissipation in even more complex separatrix manifolds of conservative (Hamiltonian) system, movement of trajectories on which looks like as chaotic dynamics. Thus there is a stretching of an accordion of infinitely folded heteroclinic separatrix zigzag on some area of phase space of the conservative system. In the remained part of phase space elliptic cycles from the right part of subharmonic and homoclinic cascades can simultaneously coexist with tori around of them.

In Fig. 16a islands of solutions of conservative system of the Croquette Eqs. (25) are presented at $\varepsilon = 0.2$ in Poincare section $(z = 0, r = \varepsilon)$. Around of a picture presented in Fig. 16a there is not represented in figure an area of chaotic movement around of original separatrix contour of nonperturbed system connecting the points $(\pm\pi, 0)$. Development and complication of heteroclinic separatrix zigzag in the extended dissipative Croquette system of Eqs. (26) close to conservative system Eqs. (25) is presented in Fig. 16b,c for $\varepsilon = 0.55$. At reduction of values of dissipation parameter μ in system of Eqs. (26) the subharmonic cascade of bifurcations is observed. It generates the heteroclinic separatrix zigzag represented in Fig. 16b at $\mu = 0.1415$. At values of parameter $\mu < 0.138$ the accordion of heteroclinic separatrix zigzag starts to cover all phase space of the system merging with heteroclinic separatrix manifold which is tense on hyperbolic cycles from the cascade a saddle-node bifurcations.

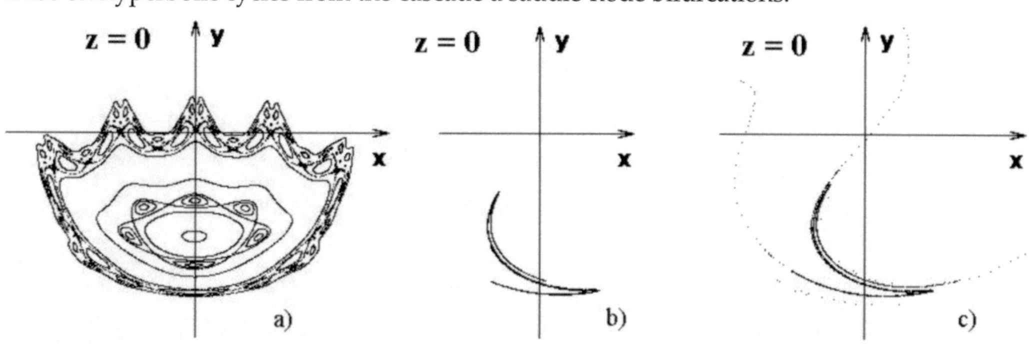

Figure 16. Projections on the plane (x, y) of Poincare section $(z = 0, r = \varepsilon)$ of solutions of conservative Croquette system of the Eqs. (25) for $\varepsilon = 0.2$ (a); development and complication of heteroclinic

separatrix zigzag in dissipative Croquette system of Eqs. (26), for $\varepsilon = 0.55$ and $\mu = 0.1415$ (b), $\mu = 0.138$ (c).

3.2. Hamiltonian systems with one and a half degrees of freedom

In modern scientific literature Hamiltonian systems with one and a half degrees of freedom refer to as nonautonomous conservative two-dimensional systems of ordinary differential equations with time-dependent Hamiltonian. Considered above Croquette system is an example of such a system. Let us analyze some other examples.

3.2.1. Hyperbolic nonautonomous concervative system

Consider nonautonomous conservative two-dimensional system of ordinary differential equations

$$\dot{x} = y,\ \ \dot{y} = (1 + \varepsilon \cos t)x - x^3. \tag{27}$$

Nonperturbed ($\varepsilon = 0$) system of Eqs. (27) has in the plane (x, y) two homoclinic separatrix loops of zero saddle singular point around singular points $O^{\pm} = (\pm 1, 0)$ which are centers of nonperturbed system. System of Eqs. (27) is equivalent to the perturbed four-dimensional conservative autonomous system

$$\dot{x} = y,\ \ \dot{y} = (1 + z)x - x^3,\ \dot{z} = r,\ \dot{r} = -z. \tag{28}$$

with conditions $H = z^2 + r^2 = \varepsilon^2$, $z_0 = z(0) = \varepsilon$. The system

$$\dot{x} = y,\ \ \dot{y} = (1 + z)x - x^3 - \mu y,\ \dot{z} = r,\ \dot{r} = -z. \tag{29}$$

is the extended dissipative system for conservative system of Eqs. (28). For large enough values of perturbation parameter (for example, $\varepsilon > 1.5$) conservative system of Eqs. (28) has a chaotic dynamics, because at reduction of values of parameter μ in dissipative system of Eqs. (29) there are subharmonic cascades of bifurcations in full accordance with the theory FSM.

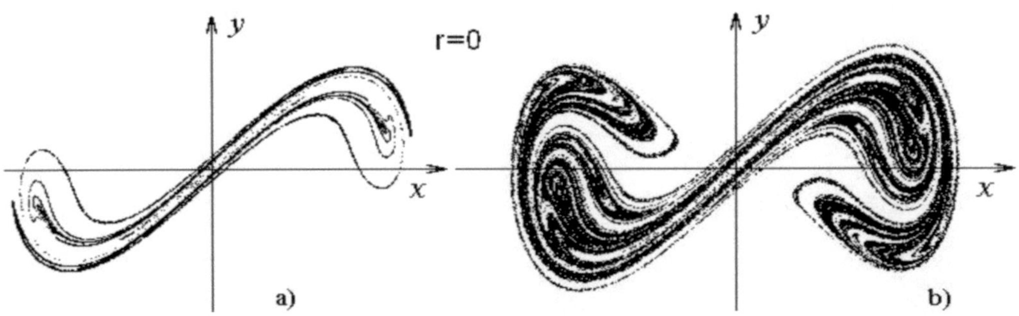

Figure 17. Projections on the plane (x, y) of Poincare section $(r = 0, z > 0)$ of solutions of dissipative system of Eqs. (29) for $\varepsilon = 1.5$ and $\mu = 0.25$ (a), $\mu = 0.04$ (b).

Then at $\mu \approx 0.251$ there is a merge of two tapes (separatrix manifolds) of singular attractors, accompanied formation of uniform heteroclinic separatrix zigzag. At the further reduction of values of parameter μ there is a development and complication of heteroclinic separatrix zigzag, accompanied a stretching of its accordion on all phase space of conservative system of Eqs. (28) at $\mu = 0$. In Fig. 17 accordions of infinitely folded heteroclinic separatrix zigzags in Poincare section of dissipative system of Eqs. (29) are shown at $\mu = 0.25$ and $\mu = 0.04$.

3.2.2. Standard example of a pendulum with oscillating point of fixing

Consider a standard example of a pendulum with vertically periodically oscillating point of fixing, that is a system with Hamiltonian

$$H(x,y,t,\varepsilon) = y^2 / 2 + (\omega^2 + \varepsilon \cos t)\cos x. \tag{30}$$

Let us write down the system of equations with Hamiltonian (30) in the form of four-dimensional conservative system of the equations

$$\dot{x} = y, \ \dot{y} = (\omega^2 + z)\sin x, \ \dot{z} = r, \ \dot{r} = -z \tag{31}$$

with conditions $H = z^2 + r^2 = \varepsilon^2$, $z_0 = z(0) = \varepsilon$. Let us consider alongside with system of Eqs. (31) the extended dissipative system of the equations

$$\dot{x} = y, \ \dot{y} = (\omega^2 + z)\sin x - \mu y, \ \dot{z} = r, \ \dot{r} = -z \tag{32}$$

and analyze numerically transition from solutions of dissipative system of Eqs. (32) to solutions of conservative system of Eqs. (31) at the fixed values of parameters $\varepsilon, \omega = 1$ when parameter μ tends to zero. It is convenient to analyze solutions of systems of Eqs. (31)- (32) in coordinates $(\sin x, y)$.

At value of perturbation parameter $\varepsilon = 2$ the conservative system of Eqs. (31) already possesses chaotic dynamics in sense of theory FSM. It is easy to be convinced of it if parameter μ in dissipative extended system of Eqs. (32) tends to zero. At $\mu \approx 0.38$ the double period bifurcation of each of original singular stable limit cycles $C^{\pm}$ occurs, that gives rise to cascades of Feigenbaum period doubling bifurcations. The given cascades of bifurcations come to the end with a birth of two singular Feigenbaum attractors at $\mu \approx 0.348$.

At further reduction of values of parameter μ the cascades of bifurcations of births of stable cycles with the periods according to the Sharkovskii order begin. Cycles of period five, for example, can be observed at $\mu \approx 0.3428$. At $\mu \approx 0.34$ two homoclinic cascades of bifurcations begin, then, as well as in other systems, there is a merge of two tapes of singular attractors (two infinitely folded heteroclinic separatrix manifolds) and then process of formation of new stable cycles proceeds on uniform infinitely folded heteroclinic separatrix surface. Development and complication of infinitely folded heteroclinic separatrix zigzag in dissipative extended system of Eqs. (32) accompanied a stretching of its accordion on the most part of phase space of conservative system of Eqs. (31) at reduction of values of parameter μ is shown in Fig. 18.

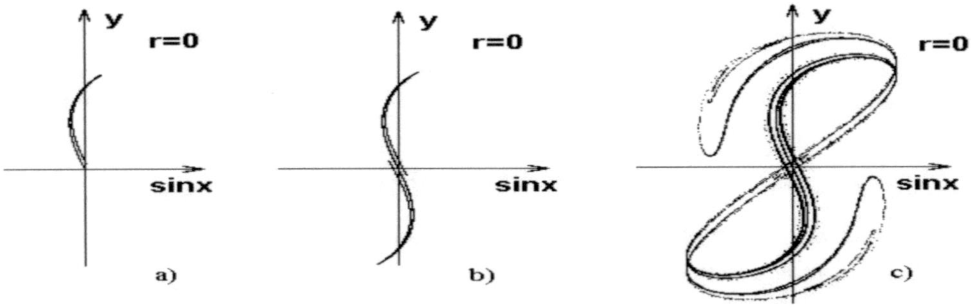

Figure 18. Projections on the plane $(\sin x, y)$ of Poincare section $(r = 0, z > 0)$ of solutions of dissipative system of Eqs. (32) for $\varepsilon = 2$ and $\mu \approx 0.337$ (a), $\mu \approx 0.33$ (b) and $\mu \approx 0.29$ (c).

3.2.3. *Conservative Duffing-Holmes equation*

Rewrite conservative Duffing-Holmes equation in the form of two-dimensional nonautonomous conservative system of the equations

$$\dot{x} = y,\ \dot{y} = \delta x - x^3 + \varepsilon \cos \omega t. \tag{33}$$

Nonperturbed $(\varepsilon = 0)$ system of Eqs. (33) has in the plane (x, y) two homoclinic separatrix loops of zero saddle singular point around singular points $O^{\pm} = (\pm\delta^{1/2}, 0)$ which are centers of nonperturbed system.

As other above considered systems, system of Eqs. (33) is equivalent to the perturbed four-dimensional conservative autonomous system

$$\dot{x} = y,\ \dot{y} = \delta x - x^3 + z,\ \dot{z} = \omega r,\ \dot{r} = -\omega z \tag{34}$$

with conditions $H = z^2 + r^2 = \varepsilon^2$, $z_0 = z(0) = \varepsilon$. The system

$$\dot{x} = y,\ \dot{y} = \delta x - x^3 + z - \mu y,\ \dot{z} = \omega r,\ \dot{r} = -\omega z \tag{35}$$

is the extended dissipative system for conservative system of Eqs. (34).

In the work (Dubrovsky, 2010) the two-parametrical bifurcation diagram of system of Eqs. (35) in space of parameters (ε, μ) is constructed. All cycles of the subharmonic cascade of bifurcations up to the cycle of period three, stable in dissipative system of Eqs. (35) at the some values of parameters $(\varepsilon, \mu > 0)$, are continued in a plane of parameters up to the value $\mu = 0$ (when the system becomes conservative) by the modified Magnitskii method of stabilization (Magnitskii & Sidorov, 2006). Thus, it is proved an existence of full subharmonic cascade of bifurcations of cycles of any period according to Sharkovskii order in conservative system of Duffing-Holmes equations (34). For large enough values of perturbation parameter ε conservative system of Eqs. (34) has also homoclinic cascade of bifurcations in full accordance with the FSM theory.

Note in conclusion of this item that the FSM scenario of transition to chaos takes place also in many other nonautonomous two-dimensional nonlinear conservative systems and, in particular, in classical generalized conservative Mathieu system

$$\dot{x} = y,\ \dot{y} = -(\delta + z)x - \alpha x^3,\ \dot{z} = \omega r,\ \dot{r} = -\omega z \tag{36}$$

which is equivalent to conservative generalized Mathieu equation (7) with $\mu = 0$ (Magnitskii , 2008b; Magnitskii , 2011)).

3.3. More complex Hamiltonian and conservative systems

In modern scientific literature Hamiltonian systems with two degrees of freedom refer to as autonomous Hamiltonian four-dimensional systems of ordinary differential equations, Hamiltonian systems with two and a half degrees of freedom refer to as nonautonomous conservative four-dimensional systems of ordinary differential equations with time-dependent Hamiltonian and Hamiltonian systems with three degrees of freedom refer to as autonomous Hamiltonian six-dimensional systems of ordinary differential equations. We consider examples of such systems and show that all such conservative systems satisfy the universal FSM theory of transition to chaos.

3.3.1. Hamiltonian systems with two degrees of freedom

Consider generalized Hamiltonian-Mathieu system with two degrees of freedom

$$\dot{x} = y,\ \ \dot{y} = -(\delta + z)x - x^3,\ \ \dot{z} = r,\ \ \dot{r} = -z - x^2 / 2 \tag{37}$$

wth Hamiltonian

$$H(x,y,z,r) = (\delta x^2 + y^2 + z^2 + r^2)/2 + zx^2/2 + x^4/4 = \varepsilon.$$

The system of Eqs. (37) contains additional composed $-x^2/2$ in the fourth equation of the conservative four-dimensional generalized Mathieu system of Eqs. (36). In this case extended dissipative system can have a kind of

$$\dot{x} = y,\ \ \dot{y} = -(\delta + z)x - x^3 - \mu y,\ \ \dot{z} = r,\ \ \dot{r} = -z - x^2/2 + (\varepsilon - H(x,y,z,r))r. \tag{38}$$

Let's consider a case $\delta = 0.5$ at which the cycle $z^2 + r^2 = \varepsilon^2$ $(x = y = 0)$ of Hamiltonian system of Eqs. (37) is an elliptic cycle at enough small ε. At $\varepsilon \approx 0.185$ period doubling bifurcation of the elliptic cycle occurs giving rise to various cascades of period doubling bifurcations and subharmonic cascades of bifurcations, generating infinitely folded heteroclinic separatrix manifolds both in extended dissipative system of Eqs. (38) and in Hamiltonian system of Eqs. (37) when $\mu \to 0$. Development and complication of infinitely folded heteroclinic separatrix zigzag in dissipative extended system of Eqs. (38) at $\varepsilon = 1$ accompanied a stretching of its accordion on all phase space of conservative system of Eqs. (37) at reduction of values of parameter $\mu \to 0$ is shown in Fig. 19.

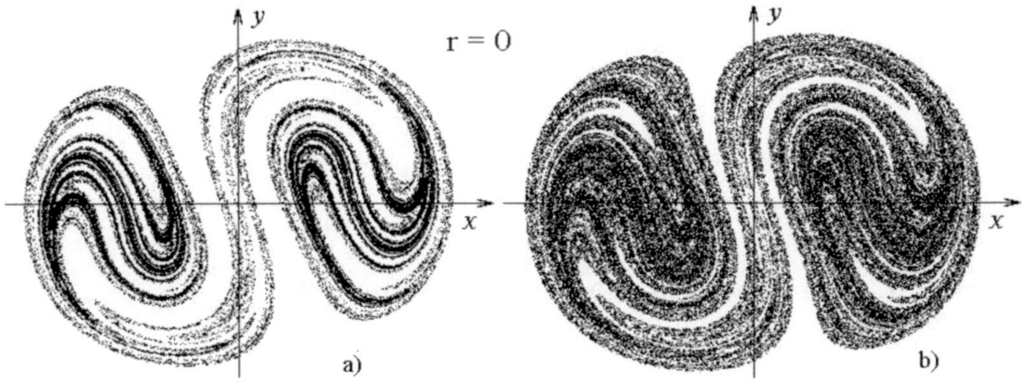

Figure 19. Projections on the plane (x, y) of the Poincare section $(r = 0, z > 0)$ of solutions of dissipative system of Eqs. (38) for $\delta = 0.5$, $\varepsilon = 1$ and $\mu \approx 0.029$ (a), $\mu \approx 0.01$ (b).

In conclusion of this item note that the FSM scenario of transition to chaos takes place also in classical Henon-Heiles system with Hamiltonian

$$H(x,y,z,r) = (x^2 + y^2 + z^2 + r^2)/2 + zx^2 - z^3/3$$

and in Yang-Mills-Higgs system (Magnitskii, 2008b; Magnitskii , 2009) with two degrees of freedom and with Hamiltonian

$$H = (\dot{x}^2 + \dot{z}^2)/2 + x^2 z^2/2 + \nu(x^2 + z^2)/2\,.$$

3.3.2. Hamiltonian systems with two and a half degrees of freedom

It is considered to be in modern literature that in case of systems with one and a half and two degrees of freedom, conservation of energy limits divergence of trajectories along all power surface, and in case of systems with two and a half and more degrees of freedom trajectories form in phase space uniform everywhere dense network named by Arnold web. Trajectories thus, as it is considered, for large enough time cover all power surface of system, approaching as much as close to its any point.

About inadequacy of the first part of this statement to the real situation all considered above examples of Hamiltonian systems with one and a half and two degrees of freedom testify. It follows from the established fact that chaotic dynamics in conservative systems is not consequence of tori resonances in nonperturbed systems, but is consequence of infinite cascades of bifurcations of births of new elliptic and hyperbolic cycles, not being cycles of nonperturbed systems. Thus the accordion of heteroclinic separatrix zigzag can be stretched on all phase space of perturbed conservative system (on all power surface), and this process is not connected in any way with tori of nonperturbed system.

Let's show now that the second part of the above mentioned statement does not correspond also to the real situation, and that in Hamiltonian systems with two and a half degrees of freedom trajectories are not obliged to cover all power surface even at the large perturbations. Thus, areas with regular, local chaotic and global chaotic dynamics can exist simultaneously on power surface of such systems even at large values of perturbation parameter.

Let's consider the system consisting from two nonlinear oscillators with weak periodic nonlinear connection. Hamiltonian of this system looks like

$$H=(\dot{x}^2+x^2+x^4/2+\dot{z}^2+z^2+z^4/2)/2+\varepsilon xz\cos t. \tag{39}$$

Hamiltonian (39) generates so called Hamiltonian system with two and a half degrees of freedom, i.e. four-dimensional system of ordinary differential equations with periodic coefficients

$$\dot{x}=y,\ \dot{y}=-x-x^3-\varepsilon z\cos t,\ \dot{z}=r,\ \dot{r}=-z-z^3-\varepsilon x\cos t. \tag{40}$$

Having designated $\varepsilon\cos t=u$ we shall receive from the system of Eqs. (40) the conservative six-dimensional autonomous system of ordinary differential equations

$$\dot{x}=y,\ \dot{y}=-x-x^3-zu,\ \dot{z}=r,\ \dot{r}=-z-z^3-xu,\ \dot{u}=v,\ \dot{v}=-u \tag{41}$$

with the condition $H=u^2+v^2=\varepsilon^2$, $u(0)=\varepsilon, v(0)=0$. In this case extended dissipative system can have a kind of

$$\dot{x}=y,\ \dot{y}=-x-x^3-zu-\mu y,\ \dot{z}=r,\ \dot{r}=-z-z^3-xu-\mu r,\ \dot{u}=v,\ \dot{v}=-u. \tag{42}$$

It is easy to see, that solutions of conservative system of Eqs. (41) with initial conditions $z_0=x_0, r_0=y_0$ are solutions of four-dimensional conservative system

$$\dot{x}=y,\ \dot{y}=-x-x^3-xu,\ \dot{u}=v,\ \dot{v}=-u \tag{43}$$

The right part of last system coincides with the right part of the considered above conservative generalized Mathieu system of Eqs. (36) with $\delta=1$. At large enough values of parameter ε (for example, $\varepsilon\geq 1.8$) conservative system of Eqs. (41) possesses chaotic dynamics even on solutions of system of Eqs. (43), as at reduction of values of parameter μ the subharmonic cascade of bifurcations of stable cycles exists in dissipative system of Eqs. (42) giving rise complex heteroclinic separatrix manifolds in four-dimensional subspace of solutions of conservative system of Eqs. (41) being solutions of system of Eqs. (43). However, chaotic dynamics of solutions of system of Eqs. (41) is local even inside this four-dimensional subspace of solutions and is limited by area of regular movements on two-dimensional tori (see in Fig. 20a). At the same time for solutions, not satisfying conditions $z_0=x_0, r_0=y_0$ or $z_0=-x_0, r_0=-y_0$ conservative system of Eqs. (41) has areas of complex global chaotic dynamics and areas of regular movement on three-dimensional tori even at such large values of perturbation parameter (see in Fig. 20b).

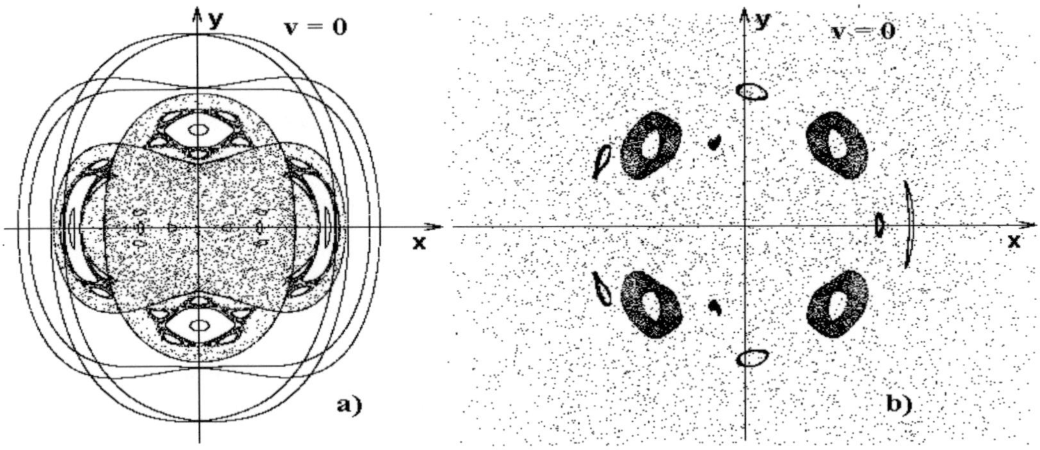

Figure 20. Projections of the section $v = 0$ of system of Eqs. (41) on the plane (x, y) for $\varepsilon = 1.8$, $z_0 = x_0, r_0 = y_0$ (a) and $z_0 \neq \pm x_0, r_0 \neq \pm y_0$ (b).}

So, in conservative system of Eqs. (41) even at enough large values of parameter ε there exist simultaneously areas of regular movement on two-dimensional tori around of basic cycles of the system, areas of regular movement on three-dimensional tori around of mentioned above two-dimensional tori, areas of local chaotic behaviour of trajectories of the system in four-dimensional subspace of five-dimensional phase space and areas of global chaotic behavior of trajectories of the system in the other part of phase space. All tori of the system are not tori of nonperturbed system, and are born as a result of various bifurcations in accordance with FSM theory. Global chaos in the system is not consequence of destruction of any mythical tori of nonperturbed system as this phenomenon is treated by the modern classical Hamiltonian mechanics and KAM (Kolmogorov-Arnold-Mozer) theory, and it is extreme consequence of complication of infinitely folded heteroclinic separatrix manifold of extended dissipative system of Eqs. (42) when dissipation parameter μ tends to zero (Magnitskii , 2011).

3.3.3. Hamiltonian system with three degrees of freedom

Let's consider a complex Hamiltonian system with three degrees of freedom

$$\dot{x} = y,\ \dot{y} = -(\delta + z)x - x^3,\ \dot{z} = r,\ \dot{r} = -z - x^2/2 - u^2/2,\ \dot{u} = v,\ \dot{v} = -(\gamma + z)u - u^3 \qquad (44)$$

with Hamiltonian

$$H(x,y,z,r,u,v) = (\delta x^2 + y^2 + z^2 + \gamma u^2 + v^2)/2 + z(x^2 + u^2)/2 + (x^4 + u^4)/4 = \varepsilon.$$

Extened dissipative two-parametrical system in this case can look like

$$\begin{gathered} \dot{x} = y,\ \dot{y} = -(\delta + z)x - x^3 - \mu y,\ \dot{z} = r, \dot{r} = -z - x^2/2 - u^2/2 \\ +(\varepsilon - H)r,\ \dot{u} = v,\ \dot{v} = -(\gamma + z)u - u^3 - \mu v. \end{gathered} \qquad (45)$$

The system of Eqs. (44) is interesting to those, that character of its dynamics contradicts practically to all propositions of the modern classical theory of Hamiltonian systems. In system of Eqs. (44) there exist simultaneously areas of regular movement on two-dimensional tori around of basic cycles of the system, areas of regular movement on three-dimensional tori around of mentioned above two-dimensional tori, areas of local chaotic behavior of trajectories of the system in four-dimensional subspace of a five-dimensional power surface and areas of global chaotic behavior of trajectories of the system in other part of a power surface even at enough large value of the perturbation parameter ε. All tori of the system are not tori of so-called nonperturbed system, but they are born as a result of various bifurcations. Global chaos in the system is not consequence of destruction of any mythical tori of nonperturbed system as this phenomenon is treated by the modern classical Hamiltonian mechanics and KAM theory. It is extreme consequence of complication of infinitely folded heteroclinic separatrix manifold of extended dissipative system of Eqs. (45) when dissipation parameter μ tends to zero. Corresponding heteroclinic separatrix zigzags in projections to the plane (x,y) of the section $r=0$ of solutions of extended dissipative system of Eqs. (45) at $\varepsilon=3$, $u_0=x_0, v_0=y_0$ and $\mu=0.125$, $\mu=0.095$ and $\mu=0.005$ are presented in Fig. 21 (see (Magnitskii , 2008b; Magnitskii , 2011)).

Thus we can make a conclusion that universal bifurcation Feigenbaum-Sharkovskii-Magnitskii theory describes also transition to dynamical chaos in nonlinear conservative and, in particular, Hamiltonian systems of ordinary differential equations at large enough values of perturbation parameter. Note that for small values of perturbation parameter the key role in complication of dynamics of any conservative system is played by nonlocal effect of duplication of hyperbolic and elliptic cycles and tori in a neighborhood of separatrix contour (or surface) of nonperturbed system opened and analyzed by the author in (Magnitskii , 2009b; Magnitskii , 2011).

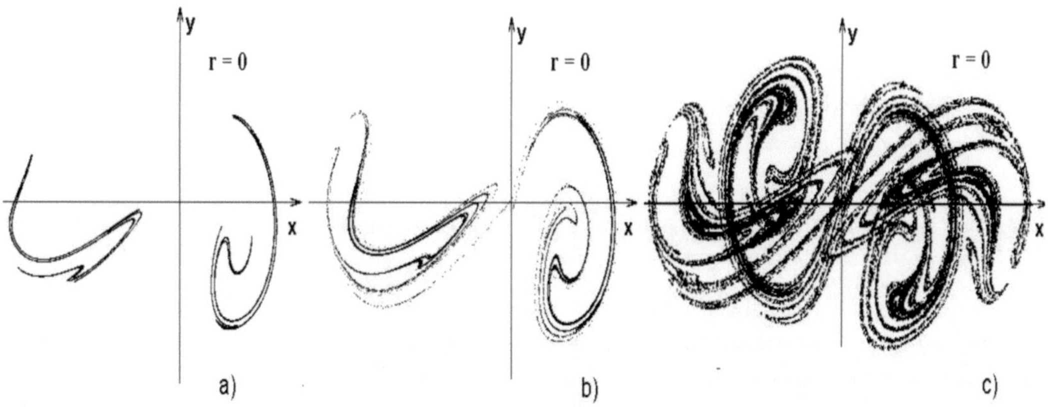

Figure 21. Development and complication of heteroclinic separatrix zigzag in dissipative system of Eqs. (45) for $\varepsilon=3$, $u_0=x_0, v_0=y_0$ and $\mu=0.125$ (a), $\mu=0.095$ (b) and $\mu=0.005$ (c).

4. Spatio-temporal chaos in nonlinear partial differential equations

4.1. Diffusion chaos in reaction-diffusion systems

Wide class of physical, chemical, biological, ecological and economic processes is described by reaction-diffusion systems of partial differential equations

$$u_t = D_1 u_{xx} + f(u,v,\mu), \quad v_t = D_2 v_{xx} + g(u,v,\mu), \quad 0 \le x \le l, \tag{46}$$

depending on scalar or vector parameter μ. Such system is very complex system. Behavior of its solutions depends on coefficients of diffusion and their ratio, length of space area and edge conditions. As a rule, there exists a value of the parameter μ_0, such that for all $\mu < \mu_0$ reaction-diffusion system has a stable stationary and space homogeneous solution (U,V), denoted as thermodynamic branch. When $\mu > \mu_0$, then thermodynamic branch loses its stability and after that reaction-diffusion system can have quite different solutions such as periodic oscillations, stationary dissipative structures, spiral waves and nonstationary nonperiodic nonhomogeneous solutions. Last solutions are known as diffusion or spatio-temporal chaos.

4.1.1. Diffusion chaos in the brusselator model

Considered on a segment $[0,l]$ the system of the brusselator equations offered for the first time by the Brussels school of I. Prigoging as a model of some self-catalyzed chemical reaction with diffusion

$$u_t = D_1 u_{xx} + A - (\mu + 1)u + u^2 v, \quad v_t = D_2 v_{xx} + \mu u - u^2 v. \tag{47}$$

It is easily to see, that stationary spatially-homogeneous solution (a thermodynamic branch) of the system of Eqs. (47) is the solution $u = A, v = \mu / A$. Therefore the first boundary problem for brusselator should satisfy the boundary conditions

$$u(0,t) = u(l,t) = A, \ v(0,t) = v(l,t) = \mu / A.$$

A more detailed analysis shows (Hassard *et al.*, 1981; Magnitskii & Sidorov, 2006) that at $\mu > \mu_0$ stable periodic spatially inhomogeneous solutions of the system of Eqs. (47) have the following asymptotic representations for small $\varepsilon = (\mu - \mu_0)^{1/2}$:

$$u(x,t) = A + \varepsilon \cos \omega t \cdot \sin\frac{\pi x}{l} + O(\varepsilon^2), \ v(x,t) = \frac{\mu}{A} + \varepsilon\gamma \cos \omega t \cdot \sin\frac{\pi x}{l} + \varepsilon\delta \sin \omega t \sin\frac{\pi x}{l} + O(\varepsilon^2),$$

where $\omega = \omega(\varepsilon) = \omega_0(1 + O(\varepsilon^2))$, γ, δ are some constants, and a kind of spatial harmonics is defined by boundary conditions of a problem. Points of a segment make fluctuations with identical frequency and a constant gradient of a phase. The effect of "wave" running on a segment is created. In the case of the second boundary value problem on a segment with the free ends, the periodic solutions born at $\mu > \mu_0$ will be spatially homogeneous (Magnitskii & Sidorov, 2006).

Let us show that the further complication of solutions of brusselator equations (47) at growth of values of parameter μ occurs according to the universal FSM theory both for the first and the second boundary value problems. In the beginning we shall consider the first boundary value problem on the segment $[0,\pi]$ for brusselator system with diffusion coefficients $D_1 = 0.15, D_2 = 0.3$. The Feigenbaum period doubling cascade of bifurcations of stable limit cycles and then the Sharkovskii subharmonic cascade of bifurcations exist in infinitely-dimensional phase space of solutions of the problem. Some main cycles and singular attractors of these cascades of bifurcations are presented in Fig. 22.

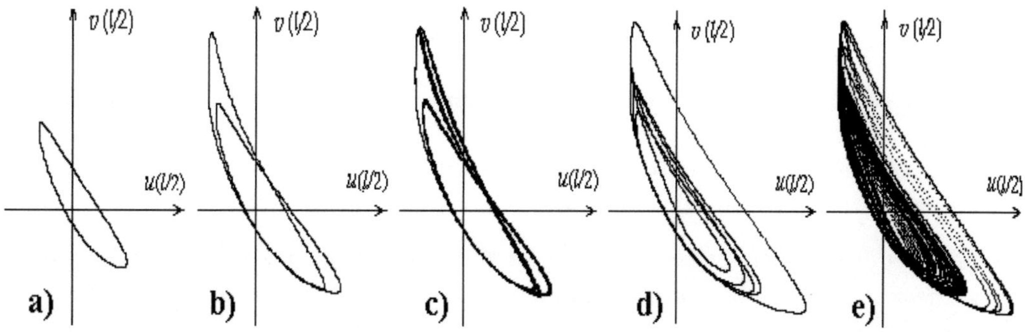

Figure 22. Singular cycle (a), cycle of the double period (b), the Feigenbaum attractor (c), cycle of period five (d) and one of the singular attractors in the first boundary value problem for the brusselator equations (47).

In the second boundary value problem singular toroidal attractors were found out in the brusselator equations for the parameter values $A = 4, l = \pi$ and for coefficients of diffusion $D_1 = 0.1, D_2 = 0.02$. At these fixed values of parameters a two-dimensional stable invariant torus is born from the stable limit cycle in the infinitely-dimensional phase space of the system of Eqs. (47). This torus begins the Feigenbaum period doubling cascade of bifurcations of stable tori on internal frequency generating by the end of cascade the Feigenbaum singular toroidal attractor (see Fig. 23).

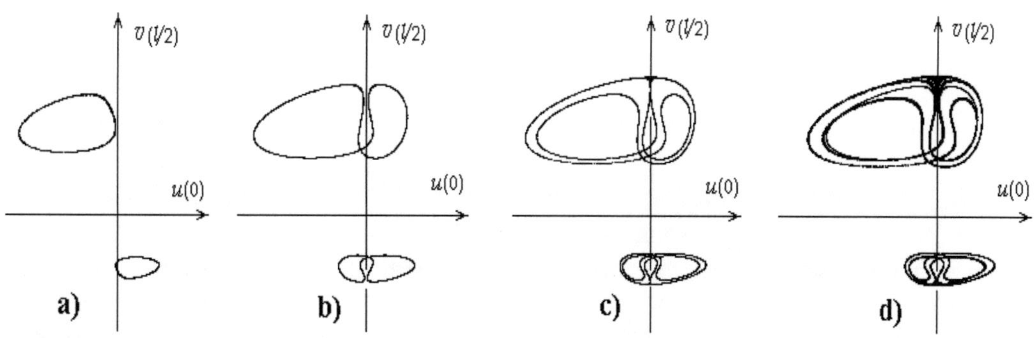

Figure 23. Projections of the section $u(l/2) = 0$ on the plane $(u(0), v(l/2))$ of two-dimensional torus (a), two-dimensional torus of double period on internal frequency (b), two-dimensional torus of period 4 (c) and the Feigenbaum singular toroidal attractor (d) in the second boundary value problems for the brusselator equations (47).

4.1.2. *Running waves, impulses and diffusion chaos in excitable mediums*

Special case of reaction-diffusion systems is the case of systems of the FitzHugh-Nagumo equations describing nonlinear processes occurring in so-called excitable mediums. Examples of such processes are distribution of impulse on a nervous fiber and a cardiac muscle and also various kinds of autocatalytic chemical reactions. The basic property describing a class of excitable mediums is slow diffusion of one variable in comparison with other variable in system of reaction-diffusion (46). Therefore the system of the FitzHugh-Nagumo equations can be written down in the following general form

$$u_t = Du_{xx} + f(u,v,\mu),\ \ v_t = g(u,v,\mu). \tag{48}$$

It is well-known, that in system of Eqs. (48) in one-dimensional spatial case there can be switching waves, running waves and running impulses, dissipative spatially nongomogeneous stationary structures, and also diffusion chaos - irregular nonperiodic nonstationary structures named sometimes as biological (or chemical) turbulence.

The analysis of solutions of system of Eqs. (48) on a straight line can be carried out by replacement $\xi = x - ct$ and transition to three-dimensional system of ordinary differential equations

$$\dot{u} = y,\ \dot{y} = -(cy + f(u,v,\mu))/D,\ \ \dot{v} = -g(u,v,\mu)/c, \tag{49}$$

where the derivative undertakes on a variable ξ. Thus the switching wave in system of Eqs. (48) is described by separatrix of the system (49) going from its one singular point into another singular point, running wave and running impulse of system of Eqs. (48) are described by limit cycle and separatrix loop of a singular point of the system (49).

Let's show, that diffusion chaos in the system of FitzHugh-Nagumo equations (48) is described by singular attractors of the system of ordinary differential equations (49) in accordance with the Feigenbaum-Sharkovskii-Magnitskii (FSM) theory. For this purpose consider the system of Eqs. (48)- (49) with nonlinearities

$$f(u,v,\mu) = -(u-1)(u-\delta v)/\varepsilon,\quad g(u,v,\mu) = arctg(\alpha u) - v, \tag{50}$$

where parameter ε is a small parameter. Note, that system of Eqs. (48) with polynomial function $f(u,v,\mu)$ and function $g(u,v,\mu)$ having at everyone v final limiting values at $u \to \pm\infty$, describes some kinds of autocatalytic chemical reactions (Zimmermann et al., 1997). It is easy to see that system of Eqs. (49)-(50) has singular point $O(0,0,0)$ for any values of parameters. Besides that, for $\alpha > 1/\delta$ system of Eqs. (49)-(50) has two more singular points $O_{\pm}(\pm u_*, 0, \pm u_*/\delta)$, where value u_* is a positive solution of the equation $\delta\, arctg(\alpha u_*) = u_*$.

A case of greatest interest is, naturally, a case when bifurcation parameter is the parameter c, not entering obviously in system of the Eqs. (48) and being the value of velocity of perturbations distribution along an axis x. For $c < \sqrt{1 + (\alpha\delta - 1)/(1-\varepsilon)}$ the limit stable cycle is born from a zero singular point as a result of Andronov-Hopf bifurcation. The singular point O becomes a saddle - focus. At the further reduction of values of parameter c the cascade of Feigenbaum double period bifurcations of stable limit cycles takes place in system of Eqs. (49)-(50) up to formation of the first singular attractor - Feigenbaum attractor. At the further reduction of values of parameter c in system of Eqs. (49)-(50) the full subharmonic cascade of bifurcations of stable cycles is realized according to the Sharkovskii order and then incomplete homoclinic cascade of bifurcations of stable cycles is realized. Last cycles converge to the homoclinic contour - the separatrix loop of the saddle-focus O (Fig. 24).

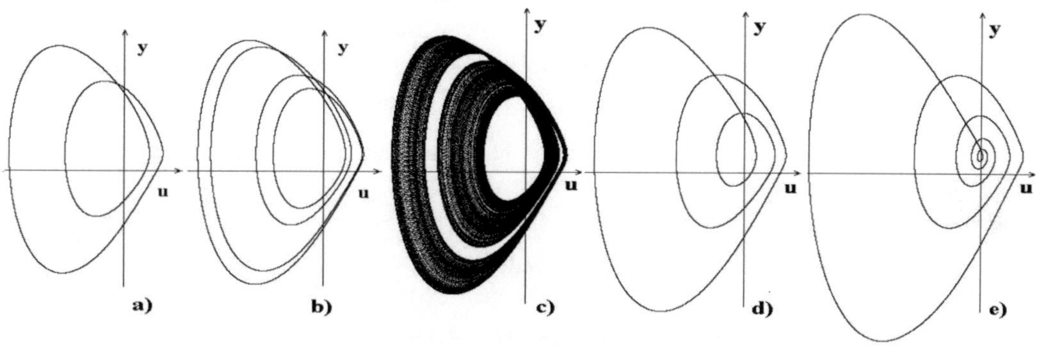

Figure 24. Projections of period two cycle (a), period four cycle (b), singular attractor (c), period three cycle (d) and homolcinic period four cycle (e) in the system of Eqs. (49).

Obtained result means, that the system of the FitzHugh-Nagumo equations (48) with fixed values of parameters can have infinite number of various autowave solutions of any period running along a spatial axis with various velocities and also infinite number of various regimes of spatio-temporal (diffusion) chaos.

4.1.3. Cycles and chaos in distributed market economy

Another, essentially different example of formation of spatio-temporal chaos in the nonlinear mediums is the distributed model of a market self-developing economy offered by the author and developed then in (Magnitskii & Sidorov, 2006). The model is a system of three nonlinear differential equations, two of which describe the change and intensity of motion (diffusion) of capital and consumer demand in a technology space under the influence of change of profit rate. The last is described by the third ordinary differential equation.

Self-development of market economy is characterized by spontaneous growth of capital and its movement in the technology space in response to differences in profitability. The model describes formation of social wealth, including production, distribution, exchange, and consumption. A distinctive feature of the model is that distribution of profitability (profit rates) determines the direction and the intensity of motion (diffusion) of capital and its spontaneous growth through generation of added value. Three economic agents having their own interests take part in economic processes that are employers, workers and government. In the model, being based on rigorous rules of Karl Marx's theory of added value, self-development of a market economy involves movement and spontaneous growth of capital of employers, which is the result of creation of added value by workers in the circulation process of capital under government control.

We show that the market economy system can exist in periodic or chaotic regimes only. Periodic regime can have any period in accordance with the theory FSM and any chaotic regime (economic crisis) can be described by some complex cycle or singular attractor.

The model assumes an unstructured closed economic system that is developing in a finite-dimensional Euclidean space R^n, called the technology space. Each point $c \in R^n$ corresponds to a certain production technology of some commodity and its coordinate $c_i, i = 1,...,n$ is the consumption of resource i per unit output.

System of market self-developing economy has the form (Magnitskii & Sidorov, 2005; Magnitskii & Sidorov, 2006):

$$\begin{aligned}
\frac{\partial x(t,c)}{\partial t} &= -div(d_1(c,x,z)grad\, z) + bx((1-\sigma)z - \delta y), \\
\frac{\partial y(t,c)}{\partial t} &= -div(d_2(c,y,z)grad\, z) + x(1-(1-\delta)y + \sigma z), \\
\frac{\partial z(t,c)}{\partial t} &= a(y - dx).
\end{aligned} \qquad (51)$$

where $x(t,c)$ is a normalized distribution of capital density, $y(t,c)$ is a normalized distribution of total consumer demand density and $z(t,c)$ is a distribution of profit rate at time t in the technology space; δ is government portion of added value (taxis, custom duties, etc.), σ is employers personal consumption portion of added value and a,b,d are structural economic parameters. Note that the system of Eqs. (51) is a particular case of systems with multicomponent diffusion, where the activator (the variable providing positive feedback) is the capital and the inhibitor (the variable suppressing capital growth) is the consumer demand.

System of equations describing the variation of macroeconomic variables can be similarly reduced to the form

$$\dot{x}(t) = bx((1-\sigma)z - \delta y),\ \dot{y}(t) = x(1-(1-\delta)y + \sigma z),\ \dot{z}(t) = a(y - dx). \qquad (52)$$

Parameters δ and σ are bifurcation parameters in system of Eqs. (52). Increase in values of parameter σ as well as reduction of values of parameter δ generate the Feigenbaum cascade of period-doubling bifurcations and then the Sharkovskii subharmonic cascade and chaotic dynamics in system of Eqs. (52) (cycle of period three is presented in Fig.10d). These results gave us possibility to draw the first important conclusion: uncontrolled growth of personal consumption of the employers as well as low government demand for consumer goods (government orders, government support to business, etc.) lead to various crisis phenomena and destroy the economic system.

Consider now the second boundary-value problem for the system of Eqs. (51) on an interval and the thermodynamic branch of this problem

$$(x^*, y^*, z^*) = \left(\frac{1-\sigma}{d(1-\delta-\sigma)}, \frac{1-\sigma}{1-\delta-\sigma}, \frac{1-\delta}{1-\delta-\sigma} \right)$$

Linearize the considered problem in the neighborhood of the thermodynamic branch, one can obtain that it is stable only when $d_1 \geq d_2 / d$ (Magnitskii & Sidorov, 2006). Thus, we can draw the second important conclusion: high inertia of the capital, slowing down its response to changes in profit rates and consumer demand, also makes the economic system unstable and lead to its destruction.

4.2. Spatio-temporal chaos in autooscillating mediums

It is well-known that any solution of the reaction-diffusion system (46) in a neighborhood $\mu > \mu_0$ of the thermodynamic branch can be approximated by some complex-valued solution $W(r,\tau) = u(r,\tau) + iv(r,\tau)$ of the Kuramoto-Tsuzuki (or Time Dependent Ginzburg-Landau) equation (Kuramoto & Tsuzuki, 1975):

$$W_\tau = W + (1+ic_1)W_{rr} - (1+ic_2)|W|^2 W, \tag{53}$$

where $r = \varepsilon x$, $\tau = \varepsilon^2 t$, $0 \leq r \leq R$, $\varepsilon = \sqrt{\mu - \mu_0}$, c_1, c_2 - some real constants. It is evident that for arbitrary phase ϕ the equation (53) has a space homogeneous solution $W(\tau) = \exp(-i(c_2\tau + \phi))$. Hence, each element of the medium (53) makes harmonious oscillations with frequency c_2 and this solution is stable in some area of parameters c_1 and c_2. Such mediums refer to as **autooscillating mediums**.

4.2.1. Transition to chaos in Kuramoto-Tsuzuki (Ginzburg-Landau) equation

In other area of parameters c_1 and c_2 the Kuramoto-Tsuzuki (Ginzburg-Landau) equation (53) has a stable automodel solution $W(r,\tau) = F(r)\exp(i(\omega\tau + a(r)))$. If $a(r) = kr$ then oscillations of the next elements occur with a constant phase lag, that corresponds to movement on space of a phase wave. In a two-dimensional case the equation (53) has also solutions in a kind of leading centers - sequences of running up concentric phase waves, and spiral waves. But equation (53) has also nonperiodic nonhomogeneous solutions in some areas of parameters - spatio-temporal or diffusion chaos.

From an opinion of most of researchers analysis of such solutions can be successfully fulfilled by using the Galerkin small-mode approximations for reducing the equation (53) to a nonlinear three-dimensional chaotic system of ordinary differential equations. As it was shown in (Magnitskii & Sidorov, 2005b; Magnitskii & Sidorov, 2006), all irregular attractors of reductive three-dimensional system are also singular attractors, and transition to chaos in this system occurs also in accordance with the Feigenbaum-Sharkovskii-Magnitskii (FSM) theory.

But further investigations of solutions of the Kuramoto-Tsuzuki (Ginzburg-Landau) equation (53) directly in its phase space showed that in reality subharmonic cascade of bifurcations of stable two-dimensional tori with arbitrary period in accordance with the Sharkovskii order in every frequency and in two frequencies simultaneously takes place in this equation.

It was considered the second boundary value problem on a segment $[0,l]$ for equation (53) and it was constructed four-dimensional subspace $(u(0),v(0),u(l/2),v(l/2))$ of infinitely-dimensional phase space of the problem. Then for different values of bifurcation parameters c_1 and c_2 the section of four-dimensional subspace has been carried out by the plane $u(l/2)=0$ and there were considered projections of this section on the plane $(u(0),v(l/2))$. Such method of the analysis of phase space of solutions of Kuramoto-Tsuzuki (Ginzburg-Landau) equation (53) appeared extremely fruitful and has enabled to find in the equation all cascades of bifurcations of two-dimensional tori in accordance with the theory FSM (see Figs. 25-26).

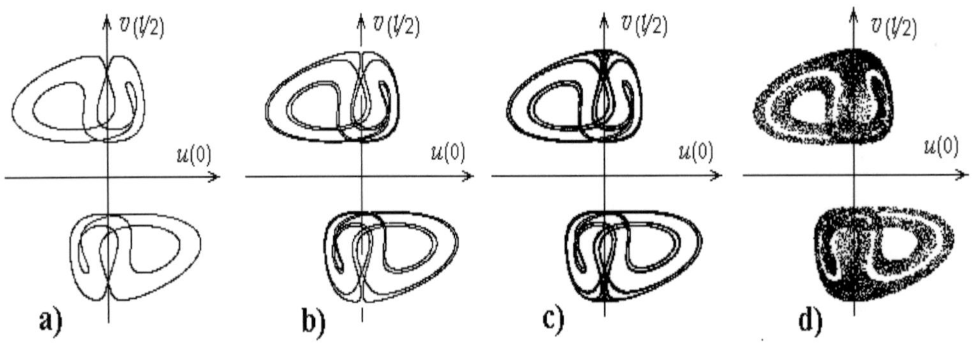

Figure 25. Bifurcation cascade on internal frequency in the equation (53). Projections of section $u(l/2)=0$ on the plane $(u(0),v(l/2))$ of two-dimensional invariant tori: period four torus (a), period eight torus (b), Feigenbaum toroidal singular attractor (c) and more complex toroidal singular attractor (d).

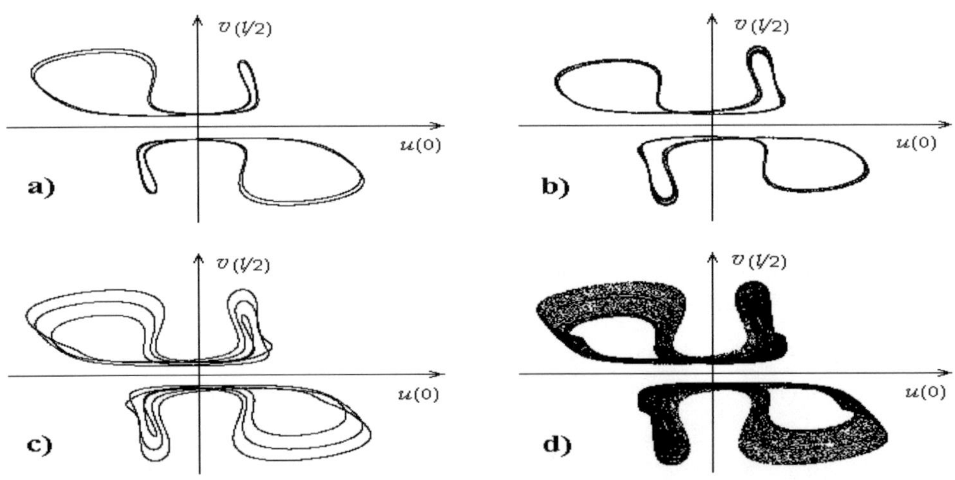

Figure 26. Bifurcation cascade on external frequency in the equation (53). Projections of section $u(l/2)=0$ on the plane $(u(0),v(l/2))$ of two-dimensional invariant tori: period two torus (a),

Feigenbaum toroidal singular attractor (b), period three torus (c) and more complex toroidal singular attractor (d).

Note that in monograph (Magnitskii & Sidorov, 2006) one can find full bifurcation diagram of existence of various subharmonic cascades of bifurcations of two-dimensional invariant tori in the second boundary value problem for the Kuramoto-Tsuzuki (Ginzburg-Landau) equation (53) in the space of parameters (c_1, c_2).

4.2.2. Running waves and chaos in autooscillating mediums

For the analysis of running waves and spatio-temporal chaos in autooscillating active mediums we apply the method used in the Section 4.1.2 for the analysis of mechanisms of formation of running waves, impulses and diffusion chaos in nonlinear excitable mediums. Let's show, that in case of autooscillating active mediums role of cascades of bifurcations of limit cycles converging to a separatrix loop of singular point is plaid by cascades of bifurcations of two-dimensional tori of four-dimensional system of ordinary differential equations converging to singular two-dimensional homoclinic structure, being the Cartesian product of a singular limit cycle on a separatrix loop of singular point. Thus the four-dimensional system has infinite number of subharmonic and homoclinic toroidal singular attractors, generating spatio-temporal chaos in original autooscillating system of partial differential equations. The solutions of four-dimensional system specifying movement on the singular homoclinic structure, tend to the periodic singular solution at $\xi \to \pm\infty$. Thus, formation of running waves and spatio-temporal chaos in autooscillating active mediums also is described by the universal bifurcation Feigenbaum-Sharkovskii-Magnitskii theory.

Rewrite the Cauchy problem on a straight line for the Kuramoto-Tsuzuki (Ginzburg-Landau) equation with complex-valued function $W(x,t) = u(x,t) + iv(x,t)$ as system of two parabolic equations with real variables $u(x,t)$ and $v(x,t)$

$$\begin{gathered} u_t = u + u_{xx} - c_1 v_{xx} - (u - c_2 v)(u^2 + v^2),\ v_t = v + c_1 u_{xx} + v_{xx} - (c_2 u + v)(u^2 + v^2), \\ -\infty < x < \infty,\ u(x,0) = u_0(x),\ v(x,0) = v_0(x),\ 0 \le t < \infty. \end{gathered} \tag{54}$$

We shall search a solution of system of Eqs. (54) as a running wave $u(x,t) = u(x - ct),\ v(x,t) = v(x - ct)$. Let's enter an automodel variable $\xi = x - ct$ and write down the system of (54) as the system of two ordinary differential equations of the second order

$$-c\dot{u} = u + \ddot{u} - c_1\ddot{v} - (u - c_2 v)(u^2 + v^2),\ -c\dot{v} = v + c_1\ddot{u} + \ddot{v} - (c_2 u + v)(u^2 + v^2), \tag{55}$$

where the derivative undertakes on a variable ξ. Resolving the system of Eqs. (55) concerning the second derivatives $\ddot{u}$ and $\ddot{v}$ and passing to phase variables $u, \dot{u} = z, v, \dot{v} = r$ we shall receive four-dimensional system of ordinary differential equations

$$\dot{u} = z,\quad \dot{z} = (-u - cz - c_1(v + cr) + ((c_1c_2 + 1)u + (c_1 - c_2)v)(u^2 + v^2)) / (1 + c_1^2),$$
$$\dot{v} = r,\quad \dot{r} = (-v - cr + c_1(u + cz) + ((c_1c_2 + 1)v + (c_2 - c_1)u)(u^2 + v^2)) / (1 + c_1^2), \tag{56}$$

The greatest interest, as well as in the case of excitable mediums, represents presence in the system of (56) cascades of bifurcations on parameter c, not entering obviously to system of the equations (54) and being the value of velocity of perturbation distribution along a spatial axis x. This case means, that the system of the Kuramoto-Tsuzuki (Ginzburg-Landau) equations (54) with the fixed parameters c_1 and c_2 can have infinite number of various autowave solutions of any period running along a spatial axis with various velocities, and also infinite number of various regimes of spatio-temporal chaos.

Let's illustrate the last statement with an example of system of Eqs. (56) with the fixed values of parameters $c_1 = 2$ and $c_2 = -0.1$. At these values of parameters the singular periodic solution

$$u = k\cos(\alpha\xi),\ v = k\sin(\alpha\xi),\ \alpha = (c + \sqrt{c^2 - 4c_2(c_1 - c_2)}) / (2(c_1 - c_2)),\ k = \sqrt{1 - \alpha^2}$$

of the system of Eqs. (56) is a stable cycle for $c > 1.306$. At smaller values of parameter c a stable two-dimensional torus is born from the singular cycle as a result of Andronov-Hopf bifurcation. At the further reduction of values of parameter c in system of Eqs. (56) the Feigenbaum cascade of period doubling bifurcations of stable two-dimensional tori on external frequency is realized. Then in system of Eqs. (56) the full subharmonic cascade of bifurcations of stable two-dimensional tori is realized according to the Sharkovskii order and then Magnitskii homoclinic cascade of bifurcations of stable tori is realized converging to the singular homoclinic structure being the Cartesian product of the original singular limit cycle on the separatrix loop of the singular point. Projections of Poincare section $(u = 0, z < 0)$ of some basic two-dimensional tori and singular toroidal attractors on the plane (r, v) are presented in Fig. 27.

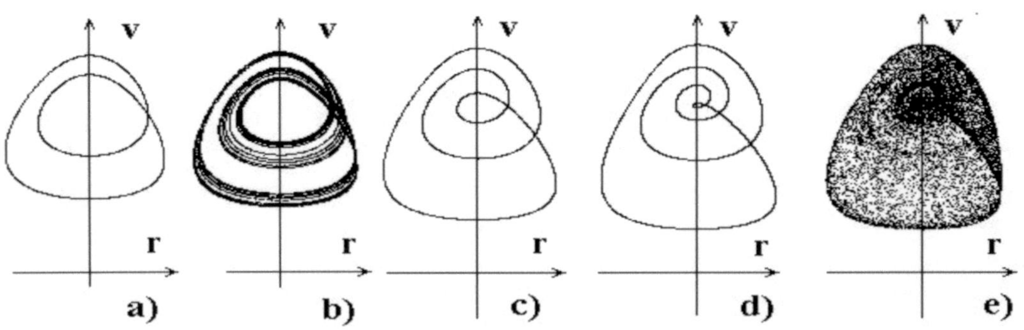

Figure 27. Projections of Poincare section $(u = 0, z < 0)$: period two torus (a), toroidal singular Feigenbaum attractor (b), period three torus (c), period four torus from homoclinic cascade (d) and more complex singular toroidal attractor (e) in the system of Eqs. (56).

4.2.3. *Spiral waves and chaos in two-dimensional autooscillating mediums*

Let's consider the second boundary value problem for Kuramoto-Tsuzuki (Ginzburg-Landau) equation in spatially two-dimensional area:

$$\begin{aligned} &W_t = W + (1+ic_1)(W_{xx} + W_{yy}) - (1+ic_2)|W|^2 W,\ 0 \le x \le l,\ 0 \le y \le l, \\ &W(x,y,0) = W_0(x,y),\ W_x(0,y,t) = W_x(l,y,t) = W_y(x,0,t) = W_y(x,l,t) = 0 \end{aligned} \tag{57}$$

with complex-valued function $W(x,y,t) = u(x,y,t) + iv(x,y,t)$. Well-known that solutions of the problem of Eqs. (57) can be plane waves, concentric phase waves (peasmakers) and also spiral waves, that are functions of a kind

$$W = R(r)e^{i(\omega t + a(r) + m\phi)},\ x = r\cos\phi,\ y = r\sin\phi.$$

Solutions with $m = 1$ correspond to one-coil spiral waves, with $m > 1$ - many-coils spiral waves. Spiral waves can be represented on a plane (x,y) by two kinds of areas, in one of which (shaded) $u(x,y,t) = \mathrm{Re}\,W(x,y,t) \ge 0$, and in another (not shaded) $u(x,y,t) = \mathrm{Re}\,W(x,y,t) < 0$. It is known also, that in some areas of change of values of parameters (c_1,c_2) the quantity of spiral waves starts to increase, that results finally in their destruction and to a forming in the active autooscillating medium, described by the equation (57), chaotic or turbulent regimes.

We show, that the mechanism of formation of spiral waves and turbulent regimes (spatio-temporal chaos) in the boundary value problem (57) for two-dimensional Kuramoto-Tsuzuki (Ginzburg-Landau) equation is subharmonic and homoclinic cascades of bifurcations of two-dimensional and many-dimensional tori in infinitely-dimensional phase space of variables $(u(x,y),v(x,y))$ that also satisfy the universal bifurcation Feigenbaum-Sharkovskii-Magnitskii (FSM) theory.

Detailed numerical analysis of the problem with initial conditions

$$\mathrm{W}_0 = u_0 + iv_0 = 0.1\sum_{m,n=0}^{4} \cos\frac{\pi m x}{l}\cos\frac{\pi n y}{l}[1 + i/(m+1)]$$

was carried out in the paper (Karamisheva, 2010) (see also (Magnitskii, 2011)) by the method of Poincare sections of finite-dimensional subspaces of infinitely-dimensional phase space. It was shown that for $c_1 = 0.5, l = 2$ spiral waves in the plane (x,y) appear at $c_2 < -0.65$ (see Fig. 28a for $c_2 = -0.68$). Then for four pairs of points (x_1,y_1) and (x_2,y_2), laying near the centers of four spiral waves, projections of sections $u(x_1,y_1) = 0$ on the plane of coordinates $(v(x_1,y_1),u(x_2,y_2))$ were constructed. The projection corresponding to a neighborhood of the center of the bottom spiral wave is represented in Fig. 28b. Thus, the Fig. 28 specifies that stable two-dimensional invariant torus is an image of a simple one-coil spiral wave in phase space of solutions of the problem (57).

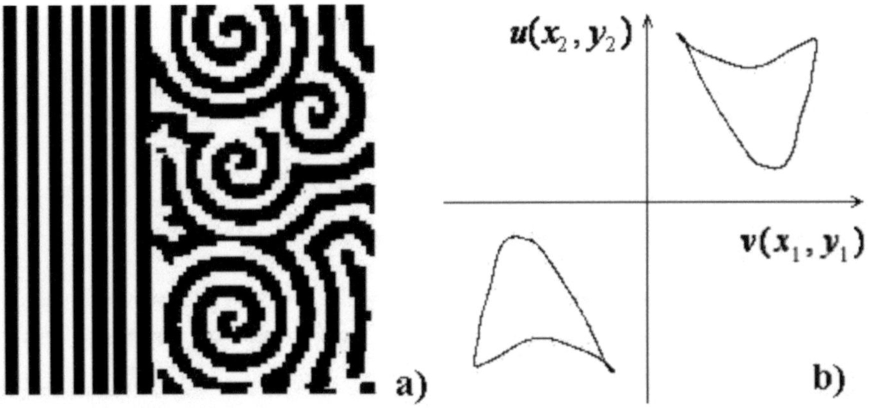

Figure 28. Spiral waves in the plane (x,y) at $c_1 = 0.5$, $c_2 = -0.68$ (a) and projection of the section $u(x_1,y_1) = 0$ of four-dimensional subspace of phase space near center of bottom spiral wave.

At the reduction of negative values of parameter c_2 there is a complication of structure of spiral waves and solutions corresponding to them in phase space of a boundary value problem (57). In Fig. 29a the picture of spiral waves on a plane (x,y) is shown at value $c_2 = -0.7$, and in Fig. 29b the projection of one of two parts of section $u(x_1,y_1) = 0$ on a plane of coordinates $(v(x_1,y_1), u(x_2,y_2))$ for two points from a neighborhood of the center of a spiral wave of the greatest radius from Fig. 29a is shown in the increased scale. It is visible, that in phase space of solutions complex two-dimensional torus of the period three from Sharkovskii subharmonic cascade corresponds to a neighborhood of the center of this spiral wave. In Fig. 29c the projection of section $u(x_1,y_1) = 0$ in a neighborhood of the other spiral wave located in a right bottom corner in Fig. 29a is presented. The projection represents the shaded ring area. But the second section by the plane $u(x_2,y_2) = -28$ of three-dimensional space of points received after carrying out the first section, gives in coordinates $(v(x_1,y_1), v(x_2,y_2))$ two closed curves. These curves testify the existence of three-dimensional torus in phase subspace of solutions in a neighborhood of the center of the second spiral wave.

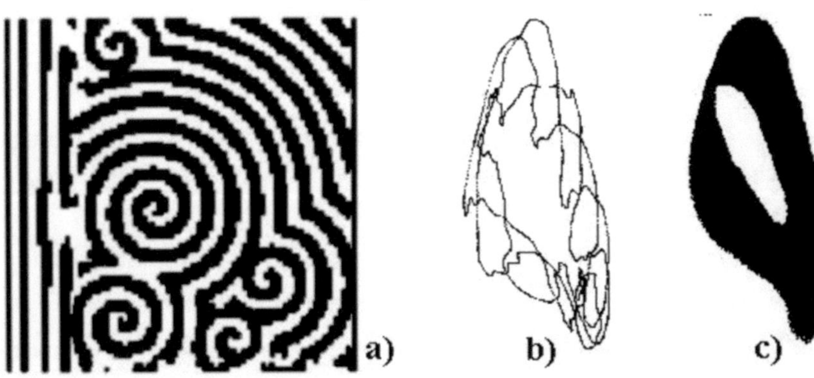

Figure 29. Spiral waves in the plane (x,y) at $c_1 = 0.5$, $c_2 = -0.7$ (a) and projections of parts of sections of four-dimensional subspace of phase space of solutions of the problem (57) in neighborhoods of two spiral waves (b), (c).

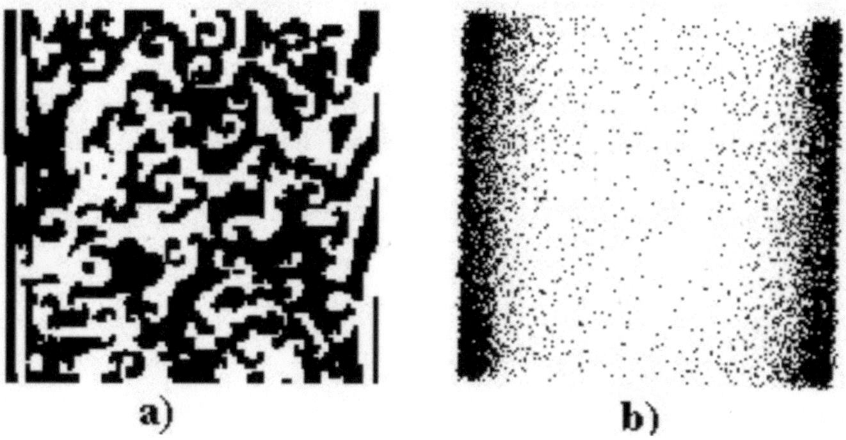

Figure 30. Spatio-temporal chaos at $c_1 = 0.5$, $c_2 = -0.9$ in the plane (x, y) (a) and in projection of section of four-dimensional subspace of phase space of solutions of the problem (57) (b).

At values of parameters $c_1 = 0.5$, $c_2 = -0.9$ already there are no stable spiral waves on a plane (x, y), and in projections of section $u(x_1, y_1) = 0$ of anyone four-dimensional subspace of phase space of solutions the continuous spatio-temporal chaotic regime is observed (Fig. 30).

5. Conclusion

In the chapter it is proved and illustrated with numerous analytical and numerical examples that there exists a uniform universal bifurcation mechanism of transition to dynamical chaos in all kinds of nonlinear systems of differential equations including dissipative and conservative, ordinary and partial, autonomous and non-autonomous differential equations and differential equations with delay arguments. This mechanism is working for all nonlinear continuous models describing both natural and social phenomena of a macrocosm surrounding us, including various physical, chemical, biological, medical, economic and sociological processes and laws. And this universal mechanism is described by the Feigenbaum-Sharkovskii-Magnitskii theory - the theory of development of complexity in nonlinear systems through subharmonic and homoclinic cascades of bifurcations of stable limit cycles or stable two-dimensional or many-dimensional invariant tori.

Notice, that theory FSM is also applicable for solutions of Navier-Stokes equations, i.e. it solves a problem of turbulence describing various bifurcation scenarios of transition from laminar to turbulent regimes in spatially three-dimensional problem of motion of a viscous incompressible liquid (Evstigneev *et al.*, 2009a,b; Evstigneev *et al.*, 2010; Evstigneev & Magnitskii, 2010). The solution of this super complex problem is presented in the separate chapter in the present book. Similar scenarios with classical Feigenbaum scenario and

Sharkovskii windows of periodicity where recently found also in (Awrejcewicz *et al.*, 2012} for initial-boundary value problems in continuous mechanical systems such as flexible plates and shallow shells. As to the processes occurring in a microcosm they, in opinion of the author, also can be successfully described by nonlinear systems of differential equations and their bifurcations. The first results in this direction are received by the author in (Magnitskii, 2010b; Magnitskii, 2011b; Magnitskii, 2012) where the basic equations and formulas of classical electrodynamics, quantum field theory and theory of gravitation are deduced from the nonlinear equations of dynamics of physical vacuum (ether).

Author details

Nikolai A. Magnitskii
Institute for Systems Analysis of RAS, Moscow, Russia

Acknowledgement

Paper is supported by Russian Foundation for Basic Research (grants 11-07-00126-a) and programs of Russian Academy of Sci. (projects 1.4, 2.5).

6. References

Feigenbaum, M. (1978). Quantitative universality for a class of nonlinear transformations. *J. Stat. Phys.*, Vol. 19, pp. 25--52.

Sharkovskii, A. (1964). Cycles coexistence of continuous transformation of line in itself. *Ukr. Math. Journal*, Vol. 26, 1, pp. 61-71.

Magnitskii, N. & Sidorov, S. (2006). *New Methods for Chaotic Dynamics*. World Scientific, Singapore, 360p.

Magnitskii, N. (2007). Universal theory of dynamical and spatio-temporal chaos in complex systems. *Dynamics of Complex Systems.*, Vol. 1, 1, pp. 18-39 (in Russian).

Magnitskii, N. (2008). Universal theory of dynamical chaos in nonlinear dissipative systems of differential equations. *Commun. Nonlinear Sci. Numer. Simul.*, Vol.13, pp. 416-433.

Magnitskii, N.(2008b). New approach to analysis of Hamiltonian and conservative systems. *Differential Equation,* Vol.44 , No.12, pp. 1682-1690.

Magnitskii, N. (2010). On topological structure of singular attractors. Differential Equations, Vol.46, No.11, pp.1552-1560.

Magnitskii, N. (2011). *Theory of dynamical chaos.* URSS , Moscow, 320p. (in Russian)

Gilmore, R. & Lefranc, M. (2002). *The topology of chaos.* Wiley, NY, 495p.

Awrejcewicz, J. (1989). *Bifurcation and Chaos in Simple Dynamical Systems.* World Scientific, Singapore, 126p.

Awrejcewicz, J. (1989). *Bifurcation and Chaos in Coupled Oscillators.* World Scientific, Singapore, 245p.

Magnitsky, Y. (2007). Regular and chaotic dynamics in nonlinear Weidlich-Trubetskov systems. *Differential Equations,* Vol.43, 12, pp. 1618-1625.

Hassard, B., Kazarinoff, N. & Wan, Y. (1981). *Theory and applications of Hopf bifurcation.* Cambridge Univ. Press, Cambridge, 311p.

Mackey, M. & Glass, L. (1977). Oscillations and chaos in physiological control systems. *Science,* Vol.197, pp. 287-289.

Dubrovsky, A. (2010). Nature of chaos in conservative and dissipative systems of Duffing-Holmes oscillator. *Differential Equations,* Vol.46, 11, pp. 1652-1656.

Magnitskii, N. (2009). Chaotic dynamics of homogeneous Yang-Mills Fields with two degrees of freedom. *Differential Equations,* Vol.45, 12, pp.1698-1703.

Magnitskii, N. (2009b). On nature of chaotic dynamics in neighborhood of separatrix of conservative system. *Differential Equations,* Vol.45, 5, pp.647-654.

Zimmermann M., et al. (1997). Pulse bifurcation and transition to spatio-temporal chaos in an excitable reaction-diffusion model. *Physica D.,* Vol.110, pp. 92-104.

Magnitskii, N. & Sidorov, S. (2005). Distributed model of a self-developing market economy. *Computational Mathematics and Modelling.,* Vol. 16, 1, pp. 83-97.

Kuramoto, Y. & Tsuzuki, T. (1975). On the formation of dissipative structures in reaction-diffusion systems. *Progr. Theor. Phys.,* Vol.54, 3, pp. 687-699.

Magnitskii, N. & Sidorov, S. (2005b). On transition to diffusion chaos through subharmonic cascade of bifurcations of two-dimensional tori. *Differential Equations,* Vol.41, 11, pp. 1550-58.

Karamisheva, T.(2010). Spiral waves and diffusion chaos in the Kuramoto-Tsuzuki equation. *Proc. ISA RAS. Dynamics of nonhomogeneous systems.,* Vol.53, 14, pp. 31-45.

Evstigneev, N., Magnitskii, N. & Sidorov, S. (2009). On nature of laminar-turbulent flow in backward facing step problem. *Differential equations,* Vol. 45, 1, pp.69-73.

Evstigneev, N., Magnitskii N. & Sidorov, S. (2009b). On the nature of turbulence in Rayleigh-Benard convection. *Differential Equations,* Vol.45, 6, pp.909-912.

Evstigneev, N., Magnitskii, N., Sidorov, S. (2010). Nonlinear dynamics of laminar-turbulent transition in three dimensional Rayleigh-Benard convection. *Commun. Nonlinear Sci. Numer. Simul.,* Vol.15, pp. 2851-2859.

Evstigneev, N. & Magnitskii, N. (2010). On possible scenarios of the transition to turbulence in Rayleigh-Benard convection. *Doklady Mathematics,* Vol. 82, No.1, pp. 659-662.

Awrejcewicz J., Krysko V.A., Papkova I.V., Krysko A.V. (2012). Routes to chaos in continuous mechanical systems. Part 1,2,3. *Chaos Solitons and Fractals.* (to appear).

Magnitskii, N. (2010b). *Mathematical theory of physical vacuum.* New Inflow, Moscow, 24p.

Magnitskii, N. (2011b). Mathematical theory of physical vacuum. *Commun. Nonlinear Sci. Numer. Simul.* Vol.16, pp. 2438-2444.

Magnitskii, N. (2012). Theory of elementary particles based on Newtonian mechanics. In *"Quantum Mechanics/Book 1"*- InTech, pp. 107-126.

Chapter 7

Non-Linearity in Structural Dynamics and Experimental Modal Analysis

Ulrich Fuellekrug

Additional information is available at the end of the chapter

http://dx.doi.org/10.5772/48812

1. Introduction

During the mechanical design and development of technical systems for power plants, in civil engineering, aerospace or mechanical engineering increasing demands are made concerning the performance, weight reduction and utilization of the material. The consequence is that the dynamical behaviour and the occurrence of vibrations of the load carrying parts, the so-called primary structures are becoming more and more important. It has to be avoided that undesired vibrations can disturb or even jeopardize the intended operation. Thus, the analysis of the dynamics and vibrations of structures is an important task. To perform dynamic analyses and to draw conclusions for possibly needed changes of the mechanical design several steps have to be carried out.

First, the dynamic analysis requires computational models which may be setup with the Finite Element Method (FEM) or other adequate techniques. Second, the computational models have to be validated because otherwise no reliable theoretical predictions are possible which can be used for optimizing the mechanical design. For the validation of the computational models it is required to perform experiments on components, prototypes or the structures themselves. In many cases the structures can be considered as linear and thus linear structural dynamics methods and approaches can be applied for modelling and validation. However, in some cases non-linear effects are important and have to be taken into account (Awrejcewicz & Krysko, 2008). If this is the case, it is not sufficient to include non-linearities only in the models. Also the experimental validation has to be able to identify, characterize and quantify non-linearities (Awrejcewicz, Krysko, Papkova, & Krysko, 2012), (Krysko, Awrejcewicz, Papkova, & Krysko, 2012), (Awrejcewicz, Krysko, Papkova, & Krysko, 2012).

Let us first consider the dynamic equations of structures with non-linearities, then take a look at experimental dynamic identification and modal analysis before we develop basic ideas for identifying non-linearities of structures.

INTECH
open science | open minds

© 2012 The Author(s). Licensee InTech. This chapter is distributed under the terms of the Creative Commons Attribution License http://creativecommons.org/licenses/by/3.0), which permits unrestricted use, distribution, and reproduction in any medium, provided the original work is properly cited.

2. Equations of motion for structures

Most large and complex technical structures, or at least large parts of them, can be considered as elastomechanical systems. That means, the dynamic behaviour and the vibration characteristics are determined by the quantity and distribution of masses, stiffness and damping. In principle, all of these structures are assembled by continuous parts. However, an analysis of continuous structures is only possible if the geometry is rather simple. Beams, plates and shells can be analysed by using ordinary or partial differential equations. However, the coupling of the basic elements, which are described by differential equations, becomes difficult and impossible due to complicated boundary conditions if the number of elements is high. Under practical considerations it is appropriate to discretize the structures. Discrete points have to be defined at all suitable locations and the dynamic motions are described by motions of these discrete points. If a computational analysis with e.g. the Finite Element Method (FEM) is performed the nodal points are such discrete points. If an experimental analysis is carried out, suitable points have to be defined. Here, it is essential to select all structural points which are required to describe the dynamic motions with sufficient accuracy. The displacements, velocities and accelerations of the selected discrete points can then be assembled in the vectors $\{u\},\{\dot{u}\}$ and $\{\ddot{u}\}$, the so-called displacement, velocity and acceleration vectors.

The equations of motion can be setup with different methods. As most general method, *Hamilton's* principle of least action (Williams, 1996), (Szabo, 1956), (Landau & Lifschitz, 1976) can be utilized. *Hamilton's* principle states that the time integral

$$\Im = \int_{t_1}^{t_2} (L+W)dt, \tag{1}$$

which contains *Lagrange's* function L and the work of non-conservative forces W, reaches a stationary value for the actual dynamic motions of the structure. The meaning of *Hamilton's* principle is that from all possible dynamic motions between two fixed states at points in time t_1 and t_2 the actual dynamic motions are those which cause a stationary value of the time integral of Eq. (1). Thus, arbitrary variations of Eq. (1) have to vanish and this leads to a method for setting up equations of motion (Williams, 1996).

Lagrange's L function consists of the kinetic and potential energy of the structure and can be written as

$$L = E_{kin} - E_{pot}. \tag{2}$$

The work of non-conservative forces $\{F\}$ can be computed with the displacements at discrete points $\{u\}$ at two fixed states at points in time t_1 and t_2 to

$$W = \int_{\{u(t_1)\}}^{\{u(t_2)\}} \{F\}^T d\{u\}. \tag{3}$$

Let us now separate notionally the structure in a complete linear part and some non-linear elements. In this case the kinetic energy of the linear part can be written with the physical mass matrix $\left[M\right]$ and the velocities $\{\dot{u}\}$ as follows

$$E_{kin} = \frac{1}{2}\{\dot{u}\}^T\left[M\right]\{\dot{u}\}. \tag{4}$$

In a similar way the potential energy of the structure's linear part is given by the physical stiffness matrix $\left[K\right]$ and the elastic deformations $\{u\}$ to

$$E_{pot} = \frac{1}{2}\{u\}^T\left[K\right]\{u\}. \tag{5}$$

The work of the non-conservative forces consists firstly of the work of the external forces $\{F_{ext}\}$ and the related deformations $\{u\}$

$$W_{ext} = \{u\}^T\{F_{ext}\}. \tag{6}$$

The damping of the elastomechanical structure can be taken into account by assuming discrete dampers, separating notionally the damping elements from the structure and considering the damping forces as external forces. Following this, the work of the damping forces is given for the structure's linear part by the physical damping matrix $\left[C\right]$ and the velocities $\{\dot{u}\}$ to

$$W_c = -\{u\}^T\left[C\right]\{\dot{u}\}, \tag{7}$$

where the minus sign indicates that the damping forces act into the opposite direction of the related velocities.

At next, the non-linear part of the structure has to be taken into account. Here, all non-linear elements are considered as discrete elements, are notionally separated from the linear structure and it is assumed that the forces between the structure and the non-linear elements depend only from the deformations and velocities at the connection points

$$\{F_{nl}\} = \{F_{nl}(\{u\},\{\dot{u}\})\}. \tag{8}$$

Thus, the non-linearties of the structure can be considered as the effect of external forces. Following this, the work of the non-linear forces is given by

$$W_{nl} = -\int_{\{u(t_1)\}}^{\{u(t_2)\}} \{F_{nl}(\{u\},\{\dot{u}\})\}\, d\{u\} \tag{9}$$

where the minus sign indicates that the non-linear forces act into the opposite direction of the related deformations and velocities.

The work of the non-conservative forces can now be written as

$$W = W_{ext} + W_c + W_{nl} = \{u\}^T \{F_{ext}\} - \{u\}^T [C]\{\dot{u}\} - \int_{\{u(t_1)\}}^{\{u(t_2)\}} \{F_{nl}(\{u\},\{\dot{u}\})\} d\{u\}. \tag{10}$$

The variation of Eq. (1)

$$\delta \Im = \delta \int_{t_1}^{t_2} (L + W) dt = \int_{t_1}^{t_2} (\delta E_{kin} - \delta E_{pot} + \delta W) dt = 0 \tag{11}$$

leads with Eqs. (4), (5) and (10) to

$$[M]\{\ddot{u}\} + [C]\{\dot{u}\} + [K]\{u\} + \{F_{nl}(\{u\},\{\dot{u}\})\} = \{F\}, \tag{12}$$

which is the well-known basic equation of linear structural dynamics extended by a term accounting for non-linearities.

3. Dynamic identification and modal analysis

With the purpose to validate analytical models of complex technical structures it is required to perform measurements on components or prototypes and to identify the dynamic properties. The most important dynamic properties are the modal parameters. Their identification is the essential goal of experimental modal analysis (Maia & Silva, 1997), (Ewins, 2000).

3.1. Modal parameters

To explain the basic ideas, let us first assume that the structure undergoing a modal identification test is linear and that the damping matrix is proportional to the mass and stiffness matrix. In this case Eq. (12) simplifies to

$$[M]\{\ddot{u}\} + [C]\{\dot{u}\} + [K]\{u\} = \{F\}, \tag{13}$$

where it is assumed

$$[C] = \beta_1 [M] + \beta_2 [K]. \tag{14}$$

The eigenvalues und eigenvectors of the undamped structure are determined by the eigenvalue problem

$$\left(\omega_{0r}^2 [M] + [K]\right)\{\phi\}_r = \{0\} \tag{15}$$

and are of great practical importance. The values ω_{0r} are the so-called eigenfrequencies and $\{\phi\}_r$ are the eigenvectors of the undamped structure. A fundamental property of the eigenvectors $\{\phi\}_r$ is the fact that the matrix of eigenvectors, the so-called modal matrix, diagonalises the mass and stiffness matrices

$$[\phi]^T[M][\phi]=\begin{bmatrix}\{\phi\}_1^T\\ \{\phi\}_2^T\\ \vdots\\ \{\phi\}_n^T\end{bmatrix}[M]\left[\{\phi\}_1\quad\{\phi\}_2\quad\cdots\quad\{\phi\}_n\right]=\begin{bmatrix}m_1&0&0&0\\0&m_2&0&0\\0&0&\ddots&0\\0&0&0&m_n\end{bmatrix},\tag{16}$$

$$[\phi]^T[K][\phi]=\begin{bmatrix}\{\phi\}_1^T\\ \{\phi\}_2^T\\ \vdots\\ \{\phi\}_n^T\end{bmatrix}[K]\left[\{\phi\}_1\quad\{\phi\}_2\quad\cdots\quad\{\phi\}_n\right]=\begin{bmatrix}k_1&0&0&0\\0&k_2&0&0\\0&0&\ddots&0\\0&0&0&k_n\end{bmatrix}.\tag{17}$$

The terms $m_1, m_2, \ldots, m_n$ are the so-called modal mass

$$m_r=\{\phi\}_r^T[M]\{\phi\}_r,\tag{18}$$

and in analogy, the terms $k_1, k_2, \ldots, k_n$ are the so-called modal stiffness

$$k_r=\{\phi\}_r^T[K]\{\phi\}_r.\tag{19}$$

In addition it is valid

$$\omega_{0r}=\sqrt{\frac{k_r}{m_r}},\tag{20}$$

$$\zeta_r=\frac{c_r}{2\sqrt{k_r m_r}},\tag{21}$$

where

$$c_r=\{\phi\}_r^T[C]\{\phi\}_r=\{\phi\}_r^T\left(\gamma_1[M]+\gamma_2[K]\right)\{\phi\}_r.\tag{22}$$

and

$$\omega_r=\omega_{0r}\sqrt{1-\zeta_r^2}.\tag{23}$$

Using the above modal parameters it can be shown that the dynamic responses of a structure (13) due to an impulse or a release from any initial condition are

$$\{u(t)\}=\sum_{r=1}^{n}\left(A_r\sin\omega_r t+B_r\cos\omega_r t\right)e^{-\zeta_r\omega_r t}\{\phi\}_r.\tag{24}$$

This equation reveals that the free decay vibrations are determined by a superposition of eigenvectors with damped harmonic vibrations at the respective eigenfrequencies. The contribution of each eigenvector depends on A_r and B_r, i.e. the initial conditions at time $t = 0$. The time history of the vibrations is determined by the eigenfrequency ω_r for the harmonic part and by the modal damping value ζ_r as well as the eigenfrequency ω_r for the decay part.

Also it can be shown that the steady state dynamic responses of a structure to a harmonic excitation with frequency ω

$$\{F(t)\} = \{\hat{F}\} e^{i\omega t} \tag{25}$$

is

$$\{u(t)\} = \{\hat{u}\} e^{i\omega t} = \sum_{r=1}^{n} \{\phi\}_r \frac{\{\phi\}_r^T \{\hat{F}\}}{m_r \left(\omega_{0r}^2 - \omega^2 + i2\zeta_r \omega_{0r} \omega\right)} e^{i\omega t}. \tag{26}$$

This equation shows that the steady state harmonic vibrations are defined by a superposition of eigenvectors with frequency dependent amplification or attenuation factors. The contribution of each eigenvector depends on the so-called modal force $\{\phi\}_r^T \{\hat{F}\}$, the modal mass m_r and the relationship of the excitation frequency ω to the respective eigenfrequency ω_{0r}. Near the resonance frequencies, where ω approaches ω_{0r}, i.e. $\omega \approx \omega_{0r}$, the modal damping ζ_r becomes important and limits the vibration amplitudes to finite values.

Considering Eqs. (24) and (26) shows that the complete dynamic behaviour of a complex structure is determined by a set of modal parameters $\omega_{0r}, \{\phi\}_r, m_r, \zeta_r$. Thus, the experimental identification of these parameters is of great practical importance and allows a detailed insight into the dynamic behaviour.

3.2. Experimental modal analysis

Since the 1970s numerous methods for experimental modal analysis have been developed (Maia & Silva, 1997), (Ewins, 2000), (Fuellekrug, 1988). In addition to the classical Phase Resonance Method (PhRM) a large number of Phase Separation Techniques (PhST) operating in the time or frequency domain has been developed and can be applied nowadays as a matter of routine during modal identification tests.

For the practical performance of high quality modal identification tests several concerns have to be accounted for. First, in many cases several hundred sensors are required to achieve a sufficient resolution of the spatial motions of all structural parts. Second, the excitation requires several large exciters which have to be operated simultaneously in order to excite all vibration modes. Third, the results have to be of high quality and accuracy since they are used for the verification and validation of analytical models. Therefore it has to be assured that all modes in the requested frequency range are identified and that the accuracy of the modal parameters is as high as possible.

All these demands lead to the fact that a highly sophisticated concept for the modal identification is required (Gloth, et al., 2001). During the modal identification testing of large complex structures also the possible non-linear behaviour has to be investigated. Usually, linear dynamic behaviour of the structure is assumed in the applied modal identification methods. However, in practice most of the investigated and tested structures exhibit some non-linear behaviour. Such non-linear behaviour can occur for example as a result of free play and different connection categories (e.g. welded, bolted) within joints or e.g. from hydraulic systems in control surfaces of aircraft.

4. Non-linear modal identification

The classical procedure for the modal identification is to perform normal-mode force appropriation with the Phase Resonance Method (PhRM). The structure is harmonically excited by means of an excitation force pattern appropriated to a single mode of vibration. However, the exclusive application of the Phase Resonance Method (PhRM) is time-consuming. Thus, an improved test concept is required which combines Phase Resonance Method (PhRM) with Phase Separation Techniques (PhST).

The core of such an optimized test concept applied e.g. to aircraft as Ground Vibrations Tests (GVT) is to combine consistently Phase Separation Techniques and the Phase Resonance Method with their particular advantages (Gloth, et al., 2001), see Figure 1. After the setup the GVT starts with the measurement of Frequency Response Functions (FRFs) in optimized exciter configurations. Second, the FRFs are analysed with Phase Separation Techniques. Hereafter the Phase Resonance Method is applied for selected vibration modes, e.g. for modes that indicate significant deviations from linearity, for modes known to be important for flutter calculations (if an aircraft is tested), or for modes which significantly differ from the prediction of the finite element analysis. Optimal exciter locations and amplitudes can be calculated from the already measured FRFs in order to accelerate the time-consuming appropriation of the force vector. The calculated force vector is applied and the corresponding eigenvector is tuned. Once a mode is identified, the classical methods for identifying modal damping and modal mass are applied. Also, a linearity check by simply increasing the excitation level is performed. During this linearity check, a possible change of the modal parameters with the force level can be investigated, see (Goege, Sinapius, Fuellekrug, & Link, 2005).

The identified eigenvectors are compared with the prediction of the finite element model and by themselves during the measurement in order to check the completeness of the data and its reliability. Multiply identified modes are sorted out. Additional exciter configurations have to be used and certain frequency ranges need to be investigated if not all expected modes are experimentally identified or if the quality of the results is not sufficient.

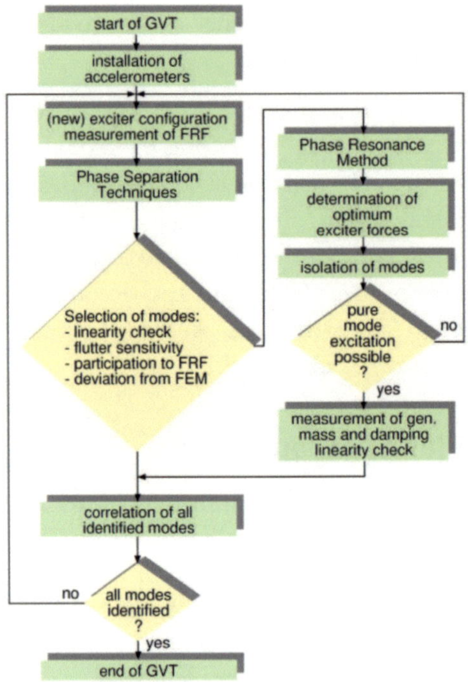

Figure 1. Test concept for modal identification of complex structures in a Ground Vibration Test (GVT)

4.1. Detection and identification of non-linearities

The above test concept allows the identification of non-linearities if some conditions are fulfilled: The response to harmonic excitation should be dominated by the excitation frequency and the mode shapes of the associated linear system should remain nearly unchanged at different force levels.

In order to characterize the non-linearities of a large complex structure, it is first required to detect the non-linearities. This can be done by simply increasing the force level. However, more detailed investigations are beneficial. The book (Worden & Tomlinson, 2001) gives a broad survey of non-linearities in structural dynamics. The detection, identification and modeling is described in great detail. Numerous suitable methods are presented and elucidated. The article (Gloth & Goege) proposes some methods for the fast detection of non-linearities within the described advanced modal survey test concept.

The step following the detection is the identification of the non-linearities. For complex lightly damped structures with weak non-linearities, the mode shapes can be divided into different groups as shown in (Wright, Platten, Cooper, & Sarmast, 2001):

- Linear proportionally damped modes, which are well separated in frequency.
- Linear proportionally damped modes, which are very close or identical in frequency.
- Linear non-proportionally damped modes, which are usually fairly close in frequency (significant damping coupling)

- Uncoupled modes, which are influenced by non-linear effects.
- Coupled modes, which are influenced by non-linear effects.

Most of the modes of real structures behave linear so that an identification using the classical linear methods and the test concept described above is still possible. Nevertheless, some modes show significant non-linear behaviour, which makes it impossible to adopt linear theory. A solution to this problem is a non-linear identification which can be based on the Masri-Caughey approach (Masri & Caughey, 1979), the force-state mapping (Crawley & Aubert, 1986) and a variant of it (Al_Hadid & Wright, 1989). The idea and basics of the non-linear resonant decay method (NLRDM) (Wright, Platten, Cooper, & Sarmast, 2001), (Platten, Wright, Cooper, & Sarmast, 2002), (Wrigth, Platten, Cooper, & Sarmast, 2003), (Platten, Wrigth, Worden, Cooper, & Dimitriadis, 2005), (Platten, Wrigth, Dimitriadis, & Cooper, 2009) appear to be an appropriated method for applying it to large and complex structures.

4.2. Basic equations for non-linear modal identification

In this section the theoretical background of the non-linear analysis of structures is outlined. The basic equations are established and a way for the modal identification in case of single non-linear modes and coupled non-linear modes is described.

The equations of motion for an elastomechanical system with linear and non-linear restoring forces are given according to Eq. (12) by

$$\left[M\right]\{\ddot{u}\}+\left[C\right]\{\dot{u}\}+\left[K\right]\{u\}+\left\{F_{nl}\left(\{u\},\{\dot{u}\}\right)\right\}=\{F_{ext}\}, \tag{27}$$

where, as above, $\left[M\right]$, $\left[C\right]$ and $\left[K\right]$ are the mass, damping and stiffness matrices, and $\{\ddot{u}\}$, $\{\dot{u}\}$ and $\{u\}$ are the vectors of physical displacements, velocities and accelerations. The non-linear restoring forces are given by $\left\{F_{nl}\left(\{u\},\{\dot{u}\}\right)\right\}$, and $\{F_{ext}\}$ is the vector of the external excitation forces.

The equations of motion Eq. (27) can be transformed from physical to modal space by using the modal matrix $\left[\phi\right]$ of the associated linear undamped system

$$\left\{u(t)\right\}=\sum_{r=1}^{n}\{\phi\}_r q_r(t)=\left[\phi\right]\{q(t)\} \tag{28}$$

where $\{q(t)\}$ is the vector of (generalized) modal coordinates, which represent modal degrees of freedom (DoF). Substituting the modal expansion of Eq. (28) into the equations of motion and pre-multiplying by the transposed of the modal matrix $\left[\phi\right]^T$ yields

$$\left[\phi\right]^T\left[M\right]\left[\phi\right]\{\ddot{q}\}+\left[\phi\right]^T\left[C\right]\left[\phi\right]\{\dot{q}\}+\left[\phi\right]^T\left[K\right]\left[\phi\right]\{q\}+\left[\phi\right]^T\left\{F_{nl}\left(\{u\},\{\dot{u}\}\right)\right\}=\left[\phi\right]^T\{F_{ext}\}. \tag{29}$$

This equation can be rewritten as

$$[m]\{\ddot{q}\}+[c]\{\dot{q}\}+[k]\{q\}+\{\delta\}=\{f_{ext}\}, \tag{30}$$

where $[m]$, $[c]$ and $[k]$ are the (generalized) modal mass, damping and stiffness matrices. The modal mass and stiffness matrices $[m]$, $[k]$ are diagonal since the real normal modes $\{\phi\}_r$ of the associated undamped system are orthogonal with respect to the physical mass and stiffness matrices $[M]$ and $[K]$. In case of so-called proportional damping also the modal damping matrix $[c]$ is diagonal. $\{\delta(t)\}$ is the vector of modal non-linear restoring forces, which includes stiffness and damping non-linearities, and $\{f_{ext}(t)\}$ is the vector of (generalized) modal excitation forces.

If the damping is proportional Eq. (30) simplifies to

$$m_r\ddot{q}_r+c_r\dot{q}_r+k_rq_r+\delta_r=f_r,\quad r=1,2,\ldots,n. \tag{31}$$

In case of $\delta_r(t)\equiv 0$, the dynamic equation for mode r is in the form of a single degree of freedom system. When the Phase Resonance Method according to the above described test concept is used, the excitation forces are appropriated to the specific mode and the whole structure vibrates in the linear case as a single DoF system. However, if non-linearities are present the modal DoF r may be coupled with other modal DoF. This is because the vector of the non-linear modal restoring forces $\{\delta(t)\}$ is, according to Eqs. (30) and (29), a function of all physical displacements and velocities

$$\{\delta(t)\}=[\phi]^T\{F_{nl}(\{u\},\{\dot{u}\})\}. \tag{32}$$

And thus, in the general case, the non-linear modal restoring forces $\delta_r(t)$ can be a function of all modal coordinates

$$\delta_r=\delta_r(q_1,q_2,\ldots,q_n;\dot{q}_1,\dot{q}_2,\ldots,\dot{q}_n). \tag{33}$$

The basic idea of the non-linear modal identification is to use time domain data of the modal DoF and to perform a so-called direct parameter estimation (DPE) in the modal space (Worden & Tomlinson, 2001) as well as to apply ideas of the non-linear resonant decay method (NLRDM) (Wright, Platten, Cooper, & Sarmast, 2001), (Platten, Wright, Cooper, & Sarmast, 2002), (Wrigth, Platten, Cooper, & Sarmast, 2003), (Platten, Wrigth, Worden, Cooper, & Dimitriadis, 2005), (Platten, Wrigth, Dimitriadis, & Cooper, 2009).

When the excitation forces are appropriated the whole structure vibrates in the linear case as a single DoF system. Thus, the analysis in modal space offers an effective way of identifying the non-linear damping and stiffness properties. Such a non-linear identification requires the previous identification of the linear modal parameters mass m_r, damping c_r and stiffness k_r. Also, it is required to determine the time histories of the modal coordinates $q_r(t)$ and the modal forces $f_r(t)$.

The rearrangement of Eq. (31) delivers

$$\delta_r(t)=-m_r\ddot{q}_r(t)-c_r\dot{q}_r(t)-k_rq_r(t)+f_r(t). \tag{34}$$

Here, m_r, c_r and k_r are experimentally identified e.g. from vector polar plot curve fit, evaluation of real part slopes or from the complex power method, see (Niedbal & Klusowski, 1989). The modal coordinate $\ddot{q}_r(t)$ is calculated from the physical acceleration responses $\{\ddot{u}(t)\}$ and the modal matrix $[\phi]$ by solving Eq. (28) e.g. with least squares:

$$\{\ddot{q}(t)\} = \left([\phi]^T [\phi]^{-1}\right)[\phi]^T \{\ddot{u}(t)\}. \tag{35}$$

$[\phi]$ in Eq. (35) represents the experimental modal matrix. This modal matrix contains the eigenvectors in the frequency band of interest, which were previously determined from linear modal analysis. If significant modal responses for other than the investigated mode of vibration are observable, coupling terms between the investigated modes and other modes exist.

The modal velocities $\dot{q}_r(t)$ and modal displacement responses $\ddot{q}_r(t)$ of the mode can be obtained by an integration of the modal acceleration responses. Prior to the integration, a band-pass filtering of the data is required in order to avoid a drift of the time domain signals. The modal excitation force $f_r(t)$ is calculated from measured eigenvectors and excitation forces according to

$$f_r(t) = \{\phi\}_r^T \{F(t)\}. \tag{36}$$

With the purpose of identifying the non-linear parameters it is required to use an analytical expression which is able to describe the non-linear behaviour. If the modal DoF r is non-linear in the stiffness and depends only on the modal coordinate $q_r(t)$, a polynomial function like

$$\delta_r(t) = \alpha_1 q_r(t) + \alpha_3 q_r^3(t) + \alpha_5 q_r^5(t) + \ldots \tag{37}$$

can be used. The coefficient α_1 describes the linear part of the stiffness and α_3, α_5,. characterize the cubic and higher polynomial parts of the stiffness.
The coefficients α_i of the function can be computed by writing Eq. (37) for several time steps t_j

$$\begin{Bmatrix} \delta_r(t_1) \\ \delta_r(t_2) \\ \vdots \\ \delta_r(t_\ell) \end{Bmatrix} = \begin{bmatrix} q_r(t_1) & q_r^3(t_1) & q_r^5(t_1) & \cdots \\ q_r(t_2) & q_r^3(t_2) & q_r^5(t_2) & \cdots \\ \vdots & \vdots & \vdots & \cdots \\ q_r(t_\ell) & q_r^3(t_\ell) & q_r^5(t_\ell) & \cdots \end{bmatrix} \begin{Bmatrix} \alpha_1 \\ \alpha_3 \\ \alpha_5 \\ \vdots \end{Bmatrix}. \tag{38}$$

The vector on the left hand side of the equation can be computed from Eq. (34) by inserting values for the modal parameters, m_r, c_r, k_r and time domain data at time steps t_j with $j = 1,2,\ldots,\ell$ of the modal coordinates $\ddot{q}_r(t_j), \dot{q}_r(t_j), q_r(t_j)$, and the modal force $f_r(t_j)$. The matrix on the right hand side is formed by time domain data of the modal coordinate $q_r(t_j)$.

The solution of Eq. (38) with least squares or any other appropriate method delivers the coefficients $\alpha_1, \alpha_3, \alpha_5, \ldots$. Care is needed for the appropriate number of time steps in Eq. (38) because too few or too many time steps can cause problems.

The quality of the non-linear identification can be checked by comparing the restoring force $\delta_r(t)$ of Eq. (34), which is based on measured data, and the recalculated restoring force, which is computed from Eq. (37) with the identified coefficients α_i. However, in cases of weak non-linearities (small non-linear restoring forces) the deviations may be high, although the agreement for the modal coordinates is very good. For this reason it is better to compare the modal accelerations of the measurement $\ddot{q}_r(t)$ with the recalculated modal accelerations $\ddot{\tilde{q}}_r(t)$, which are computed from the rearranged Eq. (34)

$$\ddot{q}_r(t) = \frac{1}{m_r}\left(f_r(t) - c_r\dot{q}_r(t) - k_r q_r(t) - \delta_r(t)\right), \tag{39}$$

where $\delta_r(t)$ is computed from Eq. (37). A qualitative comparison can be performed by visualizing the time histories of $\ddot{q}_r(t)$ and $\ddot{\tilde{q}}_r(t)$. In addition, a quantitative comparison can be obtained by the root mean square (RMS) values of the measured acceleration signal and the deviation between measured and recalculated signals.

4.3. Single mode identification

If the non-linearity in the modal DoF r is solely caused by displacements $q_r(t)$ and velocities $\dot{q}_r(t)$ of the same DoF r, the problem of non-linear identification is reduced to a single DoF problem.

To model stiffness non-linearities a polynomial with even and odd powers of the displacements $q_r(t)$ can be used

$$\delta_{k,r}(t) = \sum_{i=0}^{i_{\max}} \alpha_i q_r^{\,i}(t). \tag{40}$$

The involvement of terms with even powers in Eq. (40) allows for possible non-symmetric characteristics of the overall restoring force. If only terms with odd powers were employed, the overall restoring force would be completely anti-symmetric. Of course, the number of terms i and the associated coefficients α_i determine whether the overall force $\delta_{k,r}(t)$ always acts into the opposite direction of the respective displacements and has really the physical meaning of a restoring force.

In a quite similar way, the damping non-linearities can be modelled by the function

$$\delta_{c,r}(t) = \sum_{i=0}^{i_{\max}} \gamma_i \dot{q}_r^{\,i}(t). \tag{41}$$

Here as well, the involvement of terms with even powers in Eq. (41) allows for possible non-symmetric characteristics of the restoring forces. If only terms with odd powers would be employed, the overall restoring force would be completely anti-symmetric.

If stiffness and damping non-linearities occur together the functions of Eqs. (40) and (41) can be combined. In some cases it may also be appropriate to use mixed terms with displacements and velocities.

By modelling the non-linearities with functions of Eq. (40), Eq. (41) or an appropriate combination the non-linear identification is reduced to the estimation of the coefficients α_i and γ_i. The computation of the coefficients is in all cases based on an equation like Eq. (38).

The article (Goege, Fuellekrug, Sinapius, Link, & Gaul, 2005) describes in detail the identification of the non-linear parameters for a single mode of vibration. In addition, the paper shows a way of characterizing the identified non-linearities. The Harmonic Balance is used, and on the basis of the identified non-linear parameters α_i and γ_i the dependency of eigenfrequency ω_r and damping ζ_r versus the excitation level can be calculated and visualised in graphs, the so-called modal characterizing functions.

4.4. Coupled mode identification

In the case of coupled modes the function of Eq. (40) has to be extended by the contribution of other modal coordinates $q_s(t)$. If two modes r and s are coupled with respect to the stiffness, the polynomial function

$$\delta_{k,r}(t) = \sum_{i=0}^{i_{\max}} \sum_{j=0}^{j_{\max}} \alpha_{ij} q_r^{\,i}(t) q_s^{\,j}(t) \tag{42}$$

can be used. As above, the involvement of terms with even powers in Eq. (42) allows for possible non-symmetric characteristics of the restoring forces.

To model damping non-linearities the polynomial function

$$\delta_{c,r}(t) = \sum_{i=0}^{i_{\max}} \sum_{j=0}^{j_{\max}} \gamma_{ij} \dot{q}_r^{\,i}(t) \dot{q}_s^{\,j}(t) \tag{43}$$

can be used. For more general cases the functions of Eqs. (42) and (43) can be combined. In some cases it may also be appropriate to use mixed terms with displacements and velocities. If three or more modes are non-linearly coupled the functions of Eqs. (42) and (43) can be extended accordingly. Also, the identification is not generally restricted to polynomial functions. Any other function may be used where it is appropriate. The important fact is that the function has to contain parameter coefficients, which can be computed from measured data by using a suitable identification equation.

The estimation of the coefficients of the functions in Eq. (42) or Eq. (43) always leads to the solution of an over-determined set of linear equations like

$$\{\Delta\} = [Q]\{\alpha\}, \tag{44}$$

where $\{\Delta\}$ contains the values of the non-linear restoring forces $\delta(t)$ at discrete time steps (computed according to Eq. (34)), $[Q]$ is comprised by time domain data of the modal coordinates $q_r(t), \dot{q}_r(t), q_s(t), \dot{q}_s(t)$ and vector $\{\alpha\}$ is assembled by the unknown coefficients α_{ij}, γ_{ij}. The solution of Eq. (44) can be obtained by using least squares. However, also other appropriate parameter estimation methods can be applied.

4.5. Summarization of steps for non-linear modal identification

The steps for performing a non-linear modal identification according to the above theory can be summarized as follows:

- Identify the linear modal characteristics of the tested structure with the Phase Resonance Method or Phase Separation Techniques.
- Detect the modes that behave non-linear.
- Excite the non-linear modes with appropriated exciter forces at different force levels and use harmonic or sine sweep excitation. Measure time domain signals of forced vibrations alone, or signals of forced and free decay vibrations.
- Compute the participation of the modal coordinates according to Eq. (35) and check if modes are coupled.
- Perform single mode non-linear identification for the uncoupled modes.
- Perform coupled non-linear mode identification for the coupled modes.
- Check the quality of the identification by comparing the measured and recalculated modal signals.

5. Illustrative analytical example

In this section the non-linear identification is applied to an analytical vibration system with 3 DoF. The purpose is to illustrate the principles of and to demonstrate the applicability.

The vibration system is shown in Figure 2. The non-linearity consists of a non-linear spring with a cubic characteristic ($F_{nl} = k_{nolin} \cdot u_2^3$). The non-linear spring is attached parallel to the medial spring k_2. The eigenfrequencies of the associated linear undamped system are located at *2.845 Hz*, *3.774 Hz* and *8.954 Hz*. The modal matrix of the associated linear undamped vibration system is

$$[\phi] = \left[\{\phi\}_1 \;\; \{\phi\}_2 \;\; \{\phi\}_3\right] = \begin{bmatrix} 1 & -1 & -0.216 \\ 0.432 & 0 & 1 \\ 1 & 1 & -0.216 \end{bmatrix}, \tag{45}$$

where the columns of the modal matrix are the eigenvectors $\{\phi\}_r, r = 1,2,3$ with components at the 3 masses m_1, m_2, m_3. Due to the position of the non-linear spring, only modes 1 and 3 behave non-linear while mode 2 is completely linear. The reason is that mode 2 has no deformation at the attachment point of the non-linear spring k_{nolin}.

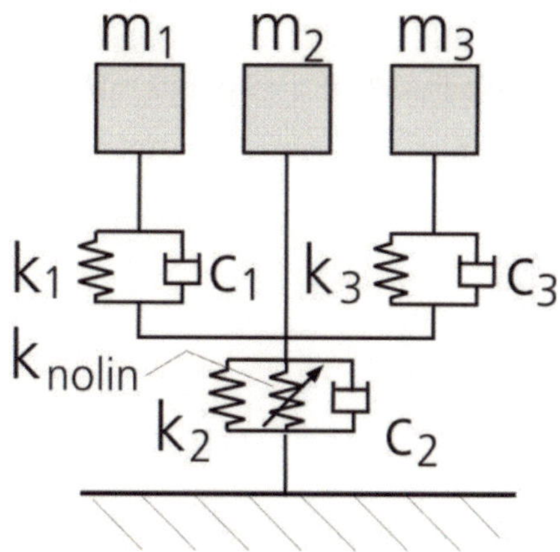

Figure 2. Vibration system with 3 Dof

For the simulation of 'measured' data the vibration system is excited with two single forces at mass 2 and mass 3. As excitation signal a sine sweep is used, which runs in *10 s* linearly from *2 Hz* to *12 Hz*.

For the non-linear analysis the *10 s* of the sine sweep excitation and *10 s* of the following free decay vibrations are used. The time domain integration of the 'measured' acceleration signals is realized by applying a digital band-pass filter to the accelerations and by integrating them once. The resulting velocities are also digitally band-pass filtered and then integrated to obtain displacements. Thus, no drift occurs during time domain integration. The force signals are also digitally high-pass filtered twice with the purpose to retain the correct phase relationship between the input and the output of the system.

Figure 3 shows the structural displacement responses following the above sine sweep excitation. *20 s* of the time histories of the modal coordinates $q_1(t)$, $q_2(t)$ and $q_3(t)$ are displayed. Figure 4 shows the mode participation of the modal coordinates as scaled root

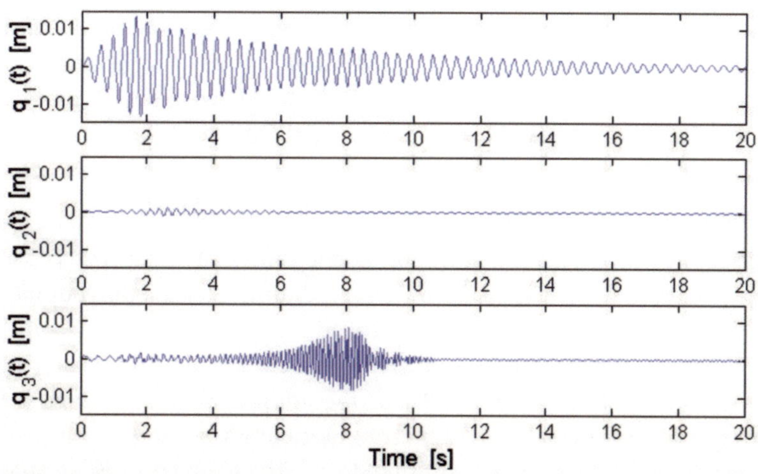

Figure 3. Time histories of the modal displacements

mean square (RMS) values. From the figures it can be seen that the above sine sweep excites clearly the modal DoF *1* and *3*, whereas DoF *2* responds only very weakly.

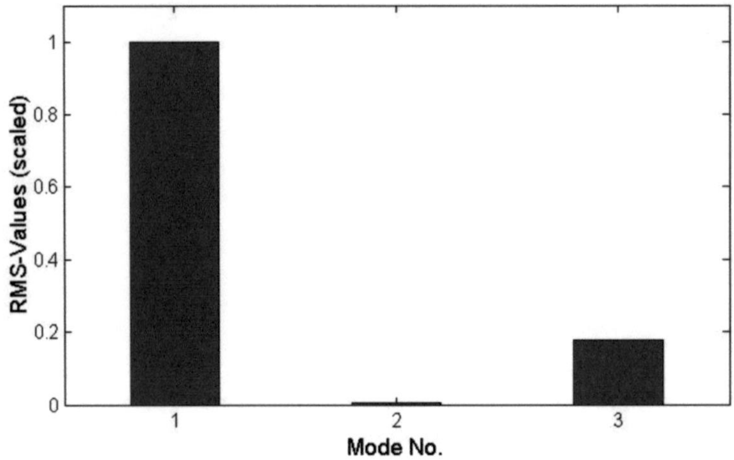

Figure 4. Mode participation

With the purpose to investigate the influence of measurement noise and errors in the data, a random signal with an RMS-value of *5 %* is added to the clean signals of excitation forces and responses prior to the non-linear identification. For the modal parameters m_r, c_r and k_r which are required for the computation of the non-linear restoring forces $\delta_r(t)$ according to Eq. (34), the correct values are used. Also, for the eigenvectors $\{\phi\}_r$ the correct data are used. A careful modal analysis at an appropriate excitation level should be able to deliver such accurate data of the underlying linear system.

In the following the simulated *20 s* time histories of $q_1(t)$, $q_2(t)$ and $q_3(t)$ are used for the non-linear modal identification. First, single mode identification on a trial basis is performed. The polynomial function of Eq. (40) with $i_{\max}$ increasing from *1* to *5* is employed. The result is always the same: the deviations between the 'measured' and recalculated signals remain high. Also, it shows that there are effects which cannot be accounted for with single mode non-linear identification. Figure 5 shows as an example the measured and recalculated restoring force of the modal DoF $r = 1$ for $i_{\max} = 5$. By the way, it is interesting that the usage of too much coefficients α_i causes no problems. The apparently unnecessary coefficients are computed to *0*.

Since the single mode non-linear identification is not sufficient, as next step coupled mode identification is performed. For the coupled mode identification the polynomial function of Eq. (42) is used. The number of terms is increased from $i_{\max} = j_{\max} = 1$ to $i_{\max} = j_{\max} = 3$. The deviations between the 'measured' and recalculated signals disappear completely for $i_{\max} = j_{\max} = 3$ and if the clean signals (without the additional random noise) are utilized. Again, it shows that the usage of too many coefficients α_i causes no problems. The apparently unnecessary coefficients are computed to *0*. The analysis of noisy signals leads to deviations. However, the deviations are not much higher than the noise itself. E.g. in the

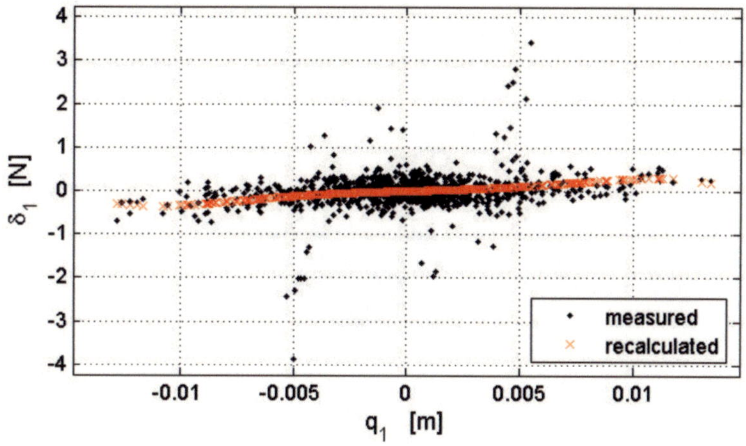

Figure 5. Non-linear restoring force of modal DoF 1 for single mode identification

case of *5 %* noise the deviations between the 'measured' and recalculated modal coordinates $q_1(t)$ and $q_3(t)$ amount to *7.4 %* and *7.7 %* respectively. It is apparent that no smaller deviation than *5 %* will be possible. Thus, the deviations are acceptable and indicate a good identification.

Figure 6 shows as an example the restoring force $\delta_1(q_1, q_3)$ identified from signals with *5 %* noise for mode $r=1$. In the figure the 'measured' and recalculated restoring forces at all time steps are plotted as points and crosses. Also, the interpolated restoring surface is depicted. The interpolated restoring surface is computed at a grid of 25×10 data points. The grid is spanned between the minimum and maximum values of q_1 and q_3. The values of the restoring surface are obtained by using Eq. (42) with the identified coefficients α_{ij} and inserting the values at the grid points for q_1 and q_3. The figure shows clearly that the restoring force $\delta_1(q_1, q_3)$ depends on both modal coordinates. If modal coordinate q_3 would be set to zero (or any other fixed value), which is assumed during single mode identification, just a trim curve would be identified. However, this trim curve is not able to describe the complete non-linear behaviour.

6. Experimental example

In this section an example of the application of the method in practice is shown. The method is exemplarily applied to an aileron mode of a large transport aircraft (Goege, Fuellekrug, Sinapius, Link, & Gaul, 2005), (Goege & Fuellekrug, 2004) (Goege, 2004).

6.1. Test structure and test performance

A modal identification test is performed as a Ground Vibration Test on an aircraft using the modal identification concept described above. The test duration was about two weeks and the aircraft was tested in two configurations. A total number of 352 accelerometers was employed to measure the mode shapes of the structure with a sufficient spatial resolution.

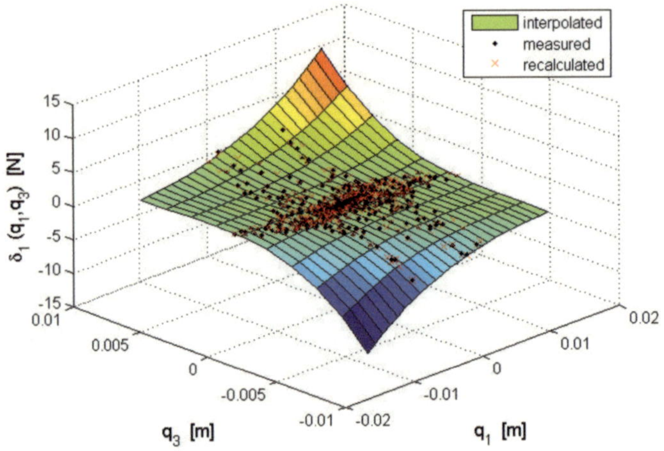

Figure 6. Restoring force of modal DoF 1 for coupled mode identification

The transport aircraft is dynamically characterized by a high modal density. During the GVT about 73 modes were identified. Most of the modes were linear. Only few modes exhibit non-linear behaviour. One mode with significant non-linear behaviour is the aileron mode.

6.2. Non-linear analysis

At first, the modal characteristics of the aileron mode were identified with the Phase Resonance Method at a level of the modal force of *10 N*. The aileron mode was excited with one single exciter, which was located at the aileron. Figure 7 displays schematically the test setup for the excitation of the aileron.

Figure 7. Excitation of the aileron mode of a large transport aircraft

Next, the level of the modal force was increased in several steps up to *121 N*. At each force level the aileron mode was measured with the Phase Resonance Method. Significant non-linear characteristics were observed: The resonance frequency of the aileron mode was changing over the load level by approximately *27 %*.

For the detailed non-linear analysis short parts of the time domain signals with harmonic steady-state excitation at the linear resonance frequency were measured. About 16 cycles of vibration were recorded. The modal accelerations were computed from the measured signals of the 352 accelerometers according to Eq. (35). The acceleration signals were filtered and integrated to obtain velocities and displacements as described above.

The analysis of the modal displacements was performed in the same way as for the simulated example. Figure 8 shows the RMS-values of the modal displacements for the lowest and highest excitation level. It shows that for the highest excitation level only the aileron mode $r = 60$ itself responds. However, for the lowest excitation level, a significant response of the bending mode of one winglet (mode $r = 71$) is also observed. This is not surprising because the motions of an aileron are in principle capable of exciting wing bending modes and thus motions of a winglet. The coupling of the aileron mode with all other modes is comparatively small.

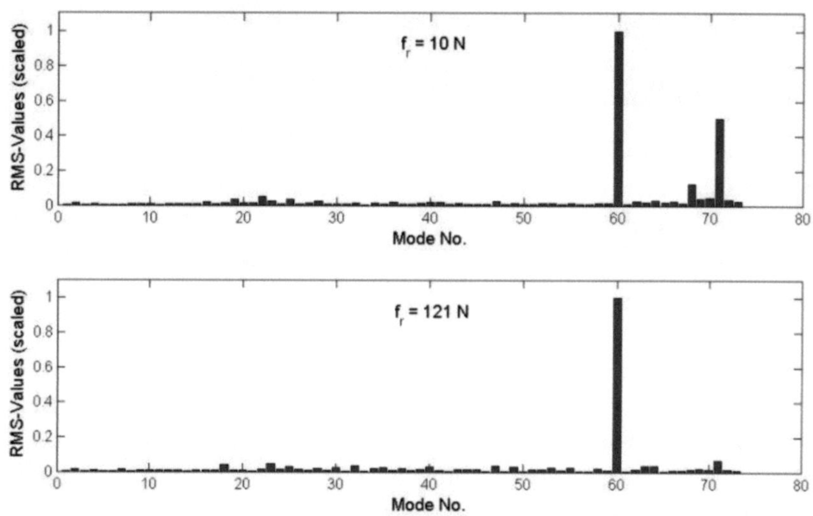

Figure 8. Mode participation for two different force levels

Figure 9 displays the restoring force of mode $r = 60$. The restoring force was calculated according to Eq. (34). In this equation the modal parameters, which were identified with the Phase Resonance Method on the highest level, are inserted together with the measured modal displacement q_{60} and the modal force f_{60}. The restoring function shows a hysteresis behaviour and may indicate a clearance non-linearity. This is imaginable because the structure vibrates at the lowest force level with only small amplitudes, which are close to the production tolerances of the aileron/wing attachment.

Under consideration of the observed mode coupling it makes sense to perform two types of non-linear identification: single mode identification for mode $r = 60$ and coupled mode identification for the two modes $r = 60$ and $r = 71$.

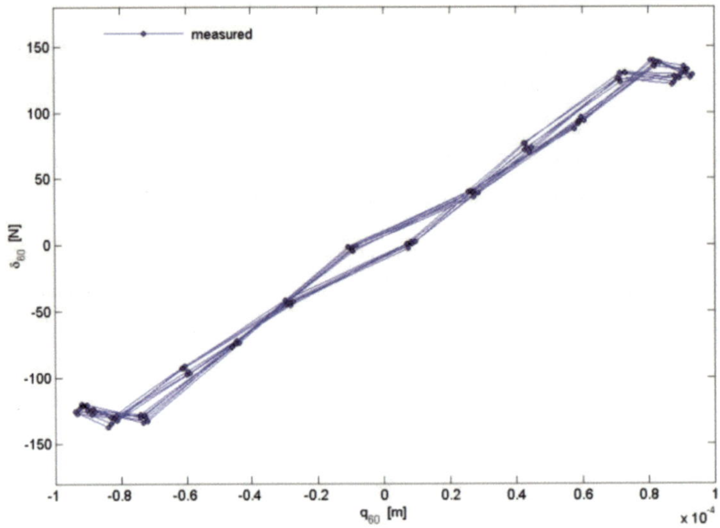

Figure 9. Measured restoring forces of the aileron mode

6.3. Single mode non-linear identification

For the single mode non-linear identification the polynomial functions of Eqs. (40) and (41) with the modal displacements and velocities of the aileron mode $r = 60$ are employed. The powers q_{60} and $\dot{q}_{60}$ are increased from $i_{\max} = 1$ to $i_{\max} = 3$. The selection of terms is performed in the way that several analysis runs with different terms on a trial and error basis are performed. The goal is to minimize the deviations between the measured and recalculated restoring forces and the measured and recalculated modal signals. Terms are included if they appear necessary to model the non-linear behaviour. They are excluded if they do not reduce the deviations. It turns out that the curve fit on the basis of Eqs. (40) and (41) is not completely successful. Thus, additional anti-symmetric terms on a trial and error basis are introduced. A stiffness term with the second power of q_{60}, namely $q_{60} \times |\dot{q}_{60}|$ reduces clearly the deviations and is therefore additionally included.

Table 1 shows the identified parameters which contribute clearly. The low value of the linear damping parameter γ_i constitutes only small changes with respect to the linear term, which is identified with the Phase Resonance Method and is a priori included in Eq. (34). No non-linear damping terms are detected.

Inserting the identified parameters in Eq. (37), the restoring force is calculated and displayed in Figure 10 together with the measured restoring force. It shows that the measured und recalculated restoring force match really well. Nevertheless, some deviations occur at the minima and maxima of the functions. The RMS-value of the deviation amounts to *0.90 %*.

Term of non-linearity	Single mode identification
q_{60}	$\alpha_1 = 3.204 \times 10^5$
$q_{60} \times \lvert q_{60} \rvert$	$\alpha_2 = 4.662 \times 10^{10}$
q_{60}^3	$\alpha_3 = -3.814 \times 10^{14}$
$\dot{q}_{60}$	$\gamma_1 = 7.0$

Table 1. Parameters for single mode identification

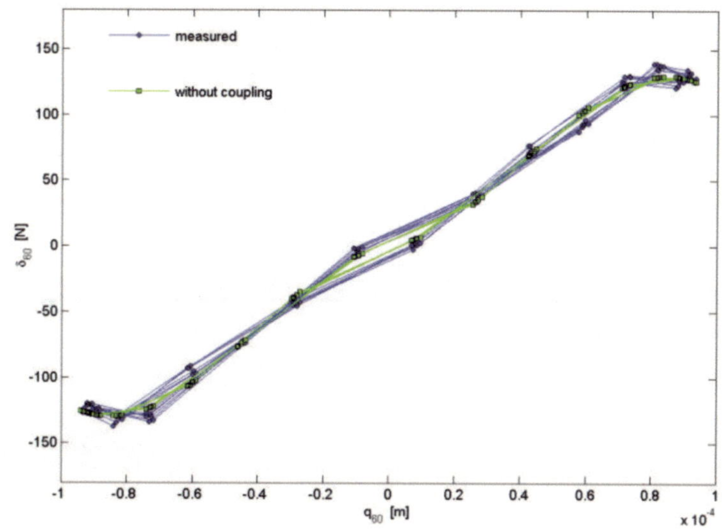

Figure 10. Measured and recalculated restoring forces (without coupling)

6.4. Coupled mode non-linear identification

For the coupled mode identification the polynomial functions of Eqs. (42) and (43) with the modal displacements and velocities of the aileron mode $r = 60$ and the winglet mode $r = 71$ are employed. The powers of q_{60}, q_{71} and $\dot{q}_{60}$, $\dot{q}_{71}$ are increased from $i_{\max} = j_{\max} = 1$ to $i_{\max} = j_{\max} = 3$. The selection of terms is performed in the same way as above. Several analysis runs with different terms on a trial and error basis are performed. Terms are only included if they contribute clearly to reduce the deviations between measured and recalculated modal signals.

Table 2 shows the identified parameters which contribute clearly. For mode $r = 60$ itself three stiffness and one damping parameter are identified again. In addition, two coupled stiffness and five coupled damping terms are identified. A significant difference to single mode identification for the four identified parameters of mode $r = 60$ is observed. The reason is that the analytical model has changed and that the coupled mode identification requires additional terms until the measured restoring forces are fitted with good accuracy. The different terms in the stiffness series compensate partly for each other. Thus, the physical meaning of the polynomial coefficients is limited. The polynomial coefficients may be considered rather as 'numbers' which enable a good fit to the measured data. The main criterion are the restoring functions.

Term of non-linearity	Coupled mode identification
q_{60}	$\alpha_1 = 2.264 \times 10^6$
$q_{60} \times \lvert q_{60} \rvert$	$\alpha_2 = -6.425 \times 10^9$
q_{60}^3	$\alpha_3 = -4.001 \times 10^{13}$
$\dot{q}_{60}$	$\gamma_1 = 1.190 \times 10^2$
$q_{60} \times q_{71}$	$\alpha_4 = 1.397 \times 10^9$
$q_{60}^2 \times q_{71}$	$\alpha_5 = 2.041 \times 10^{13}$
$\dot{q}_{60} \times \dot{q}_{71}$	$\gamma_2 = 1.806 \times 10^4$
$\dot{q}_{60}^2 \times \dot{q}_{71}$	$\gamma_3 = -1.458 \times 10^7$
$\dot{q}_{60} \times \dot{q}_{71}^2$	$\gamma_4 = -1.247 \times 10^6$
$\dot{q}_{60}^2 \times \dot{q}_{71}^3$	$\gamma_5 = 2.323 \times 10^{11}$
$\dot{q}_{60}^3 \times \dot{q}_{71}^3$	$\gamma_6 = -5.266 \times 10^{11}$

Table 2. Parameters for coupled mode identification

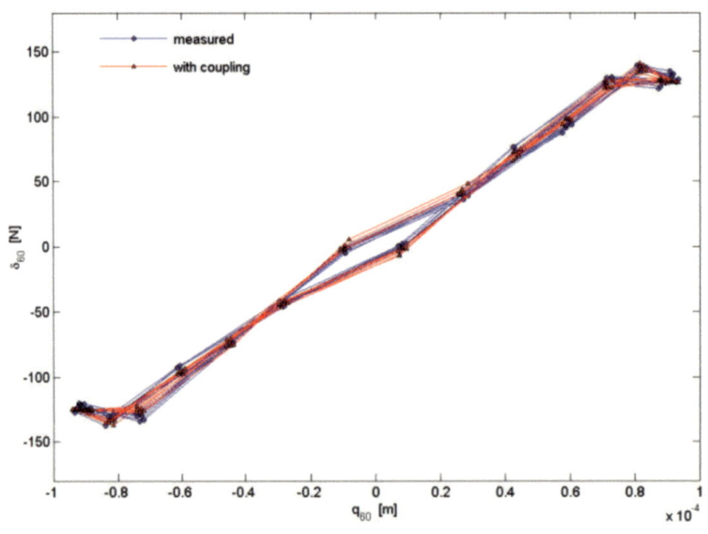

Figure 11. Measured and recalculated restoring forces (with coupling)

Figure 11 shows the measured and recalculated restoring forces. A nearly perfect agreement can be seen, even at the minima and maxima of the functions. The quantitative assessment via RMS-value delivers a deviation of *0.15 %*. In order to show the influence of the coupling terms, the restoring stiffness forces δ_{60} are visualized as surfaces in Figure 12.

The restoring surfaces are computed at a grid of data points which is spanned by the minimum and maximum values of q_{60} and q_{71}. The restoring force surface of the single mode nonlinear identification is computed at the grid points from Eq. (41) with the parameters of Table 1. This surface is depicted as black mesh. The restoring force surface of

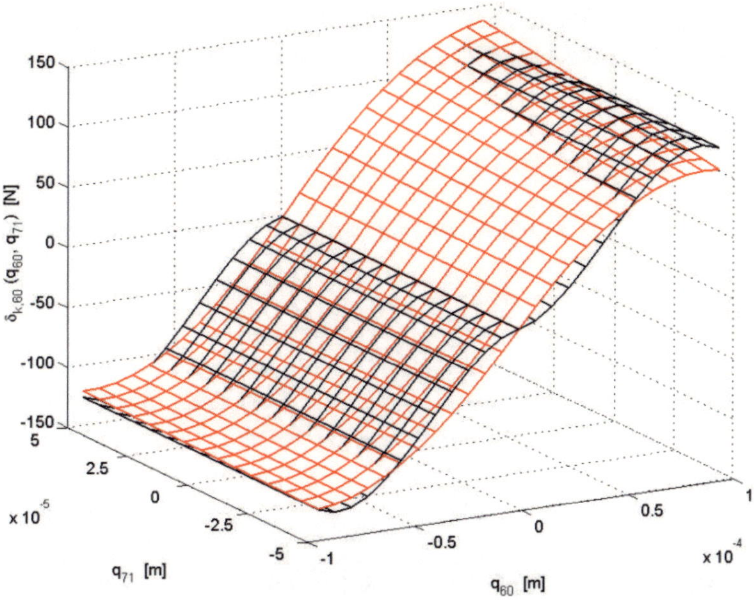

Figure 12. Restoring surface of single mode identification (black) and coupled mode identification (red)

the coupled mode identification is computed at the grid points from Eq. (42) with the parameters of Table 2. This surface is depicted as red mesh. Both surfaces exhibit a clear difference. Thus, the restoring force δ_{60} depends clearly from both modal coordinates q_{60} and q_{71}. Single mode identification with the black mesh restoring surface is not able to describe the complete non-linear behaviour.

7. Conclusion

This book chapter derives first the basic dynamic equations of structures with non-linearities and considers the experimental modal identification. Then the theoretical basis for non-linear identification is explained and a test strategy for non-linear modal identification, which can be used within a test concept for modal testing, is described. The basic idea is to use modal force appropriation, to employ equations in modal space and to identify the modal non-linear restoring forces. This is realized by computing the coefficients of applicable functions for the restoring forces from time domain data. The required steps for single mode and coupled mode non-linear identification are developed and discussed in detail. The identification is then illustrated by an analytical example where it could be shown that the method is able to identify the non-linear coupled modes of vibration. A second example taken from a modal identification test on a large transport aircraft shows the application of the approach in practice.

The non-linear identification may be further developed by using other functions for the restoring forces or to extend it to a higher number of modal DoF. Also, it can be elaborated whether and how it would be possible to derive the required linear modal parameters from applying Phase Separation Techniques. Thus, the experimental effort of applying the Phase Resonance Method could be avoided leading to a reduced test duration.

Author details

Ulrich Fuellekrug
Institute of Aeroelasticity, Deutsches Zentrum fuer Luft- und Raumfahrt (DLR), Germany

8. References

Al_Hadid, M. A., & Wright, J. R. (1989, Vol. 3, No. 3). Developments in the Force-State Mapping Technique for Non-Linear Systems and the Extension to the Localisation of Non-Linear Elements in a Lumped Parameter System. *Mechanical Systems and Signal Processing*, pp. 269-290.

Awrejcewicz, J., & Krysko, V. A. (2008). *Chaos in Structural Mechanics.* Berlin: Springer-Verlag.

Awrejcewicz, J., Krysko, V. A., Papkova, I. V., & Krysko, A. V. (2012, 45). Routes to chaos in continuous mechanical systems. Part 1: Mathematical models and solution methods. *Chaos Solitons & Fractals*, pp. 687-708.

Awrejcewicz, J., Krysko, V. A., Papkova, I. V., & Krysko, A. V. (2012, 25). Routes to chaos in continuous mechanical systems. Part 3: The Lyapunov exponents, hyper, hyper-hyper and spatial-temporal chaos. *Chaos Solitons & Fractals*, pp. 721-736.

Crawley, E. F., & Aubert, A. C. (1986, Vol. 24, No. 1). Identification o f Nonlinear Structural Elements by Force-State Mapping. *AIAA Journal*, pp. 155-162.

Ewins, D. J. (2000). *Modal Testing: Theory, Practice and Application.* Baldock, Hertfordshire, England: Research Studies Press Ltd.

Fuellekrug, U. (1988, Vol. 27-1). Survey of Parameter Estimation Methods in Experimental Modal Analysis. *Journal of the Society of Environmental Engineers*, pp. 33-44.

Gloth, G., & Goege, D. (2004). Handling of Non-Linear Structural Characteristics in Ground Vibration Testing. *Proceedings of the International Conference on Noise and Vibration Engineering (ISMA)*, (pp. 2129-2143). Leuven (Belgium).

Gloth, G., Degener, M., Fuellekrug, U., Gschwilm, J., Sinapius, M., Fargette, P., & Levadoux, B. (2001, Vol. 35, No. 11). New Ground Vibration Testing Techniques for Large Aircraft. *Sound and Vibration*, pp. 14-18.

Goege, D. (2004). *Schnelle Identifikation und Charaktersisierung von Linearitaetsabweichungen in der experimentellen Modalanalyse grosser Luft- und Raumfahrtstrukturen.* Forschungsbericht DLR_FB_2004-36.

Goege, D., & Fuellekrug, U. (2004). Analyse von nichtlinearem Schwingungsverhalten bei großen Luft- und Raumfahrtstrukturen. *VDI Schwingungstagung, VDI-Berichte Nr. 1825*, (pp. 123-155). Wiesloch (Germany).

Goege, D., Fuellekrug, U., Sinapius, M., Link, M., & Gaul, L. (2005, Vol. 46, No. 5). INTL - A Strategy for the Identification and Characterization of Non-Linearities within Modal Survey Testing. *AIAA Journal*, pp. 974-986.

Goege, D., Sinapius, M., Fuellekrug, U., & Link, M. (2005, Vol. 40). Detection and Description of Non-Linear Phenomena in Experimental Modal Analysis Via Linearity Plots. *International Journal of Non-Linear Mechanics*, pp. 27-48.

Krysko, V. A., Awrejcewicz, J., Papkova, I. V., & Krysko, A. V. (2012, 45). Routes to chaos in continuous mechanical systems. Part 2: Modelling transitions from regular to chaotic dynamics. *Chaos Solitons & Fractals*, pp. 709-720.

Landau, L. D., & Lifschitz, E. M. (1976). *Lehrbuch der theoretischen Physik - Mechanik.* Akademie Verlag.

Maia, N. M., & Silva, J. M. (1997). *Theoretical and Experimental Modal Analysis.* New York: John Wiley & Sons Inc.

Masri, S. F., & Caughey, T. K. (1979, Vol. 46). A Nonparametric Identification Technique for Nonlinear Dynamic Systems. *Journal of Applied Mechanics*, pp. 433-447.

Niedbal, N., & Klusowski, E. (1989, Vol. 13, No. 3). Die Ermittlung der generalisierten Masse und des globalen Daempfungsbeiwertes im Standschwingungsversuch. *Zeitschrift für Flugwissenschaften und Weltraumforschung*, pp. 91-100.

Platten, M. F., Wright, J. R., Cooper, J. E., & Sarmast, M. (2002). Identifcation of Multi-Degree of Freedom Non-Linear Simulated and Experimental Systems. *Proceedings of the International Conference on Noise and Vibration Engineering (ISMA*, (pp. 1195-1202). Leuven (Belgium).

Platten, M., Wrigth, J., Dimitriadis, G., & Cooper, J. (2009, Vol. 23, No. 1). Identification of multi degree of freedom non-linear systems using an extended modal space model. *Mechanical Systems and Signal Processing*, pp. 8-29.

Platten, M., Wrigth, J., Worden, K., Cooper, J., & Dimitriadis, G. (2005). Non-Linear Identification Using a Genetic Algorithm Approach for Model Selection. *Proceedings of the 23rd International Modal Analysis Conference (IMAC-XXIII).* Orlando, FL (USA).

Szabo, I. (1956). *Höhere Technische Mechanik.* Berlin/Göttingen/Heidelberg: Springer Verlag.

Williams, J. H. (1996). *Fundamentals of Applied Dynamics.* New York: John Wiley & Sons, Inc.

Worden, K., & Tomlinson, G. R. (2001). *Nonlinearity in Structural Dynamics -Detection, Identification and Modelling-.* Bristol and Philadelphia: Insitute of Physics Publishing.

Wright, J., Platten, M., Cooper, J., & Sarmast, M. (2001). Identification of Multi-Degree of Freedom Weakly Non-linear Systems using a Model based in Modal Space. *Proceedings of the International Conference on Strucutral System Identification*, (pp. 49-68). Kassel (Germany).

Wrigth, J., Platten, M., Cooper, J., & Sarmast, M. (2003). Experimental Identification of Continuous Non-Linear Systems Using an Extension of Force Appropriation. *Proceedings of the 21st International Modal Analysis Conference (IMAC-XXI)*. Kissimmee, FL (USA).

Chapter 8

Numerical Reproducing of a Bifurcation in the Stress Distribution Obtaining Process in Post-Critical Deformation States of Aircraft Load-Bearing Structures

Tomasz Kopecki

Additional information is available at the end of the chapter

http://dx.doi.org/10.5772/48069

1. Introduction

Modern aviation structures are characterised by widespread application of thin-shell load-bearing systems. The strict requirements with regard to the levels of transferred loads and the need to minimise a structure mass often become causes for accepting physical phenomena that in case of other structures are considered as inadmissible. An example of such a phenomenon is the loss of stability of shells that are parts of load-bearing structures, within the range of admissible loads.

Thus, an important stage in design work on an aircraft load-bearing structure is to determine stress distribution in the post-critical deformation state. One of the tools used to achieve this aim is nonlinear finite elements method analysis. The assessment of the reliability of the results thus obtained is based on the solution uniqueness rule, according to which a specific deformation form can correspond to one and only one stress state. In order to apply this rule it is required to obtain numerical model's displacements distribution fully corresponding to actual deformations of the analysed structure.

An element deciding about a structure's deformation state is the effect of a rapid change of the structure's shape occurring when the critical load levels are crossed. From the numerical point of view, this phenomenon is interpreted as a change of the relation between state parameters corresponding to particular degrees of freedom of the system and the control parameter related to the load. This relation, defined as the equilibrium path, in case of an occurrence of mentioned phenomenon, has an alternative character, defined as bifurcation. Therefore, the fact of taking a new deformation form by the structure corresponds to a sudden change to the alternative branch of the equilibrium path [1-4].

INTECH
open science | open minds

© 2012 The Author(s). Licensee InTech. This chapter is distributed under the terms of the Creative Commons Attribution License http://creativecommons.org/licenses/by/3.0), which permits unrestricted use, distribution, and reproduction in any medium, provided the original work is properly cited.

Therefore, a prerequisite condition for obtaining a proper form of the numerical model deformation is to retain the conformity between numerical bifurcations and bifurcations in the actual structure. In order to determine such conformity it is required to verify the results obtained by an appropriate model experiment or by using the data obtained during the tests of the actual object. It is often troublesome to obtain reliable results of nonlinear numerical analyses and it requires an appropriate choice of numerical methods dependent upon the type of the analysed structure and precise determination of parameters controlling the course of procedures.

Due to the number of state parameters, the full equilibrium path should be interpreted as hyper-surface in state hyperspace, satisfying the matrix equation for residual forces:

$$\mathbf{r}(\mathbf{u},\Lambda)=\mathbf{0}, \tag{1}$$

where $\mathbf{u}$ is the state vector containing structure nodes' displacement components corresponding to current geometrical configuration, Λ is a matrix composed of control parameters corresponding to current load state, and $\mathbf{r}$ is the residual vector containing uncompensated components of forces related to current system deformation state. The set of control parameters may be expressed by a single parameter that is a function of the load. Equation (1) takes then the following form:

$$\mathbf{r}(\mathbf{u},\lambda)=\mathbf{0}, \tag{2}$$

called a monoparametric equation of residual forces.

The prediction-correction methods of determining the consecutive points of the equilibrium path used in modern programs contain also a correction phase based on the satisfaction of an additional equation by the system, called an increment control equation or constraints equation:

$$c(\Delta\mathbf{u}_n,\Delta\lambda_n)=0, \tag{3}$$

where the increments:

$$\Delta\mathbf{u}_n=\mathbf{u}_{n+1}-\mathbf{u}_n \;\; and \;\; \Delta\lambda_n=\lambda_{n+1}-\lambda_n \tag{4}$$

correspond to the transition from n-th state to $n+1$-th state.

The graphic interpretation of the increment control equation is presented in Figure 1.

In order to find out whether there is full conformity between the character of actual deformations and their numerical representation it would be required to compare the combinations of the relevant state parameters in all the phases of the course of the phenomenon considered herein. Because of the complication of such a comparative system, the deformation processes are represented in practice by applying substitute characteristics called representative equilibrium paths. They define the relations between a control parameter related to load and a selected, characteristic geometric value related to a structure's deformation, an increment of which corresponds to a change of the value of all or some state parameters [5-9].

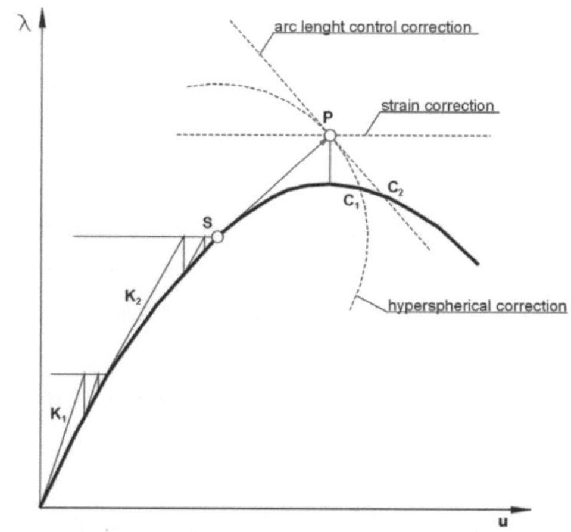

Figure 1. Graphic presentation of various correction strategies for the representative system of one degree of freedom

In case of a large number of state parameters it is not possible at all to represent the character of bifurcation by applying a representative equilibrium path. Sometimes, changes of state parameters resulting from local bifurcation may show the lack of perceptible influence on the representative value, which results in non-occurrence of any characteristic points on the representative path. In general, however, these changes cause a temporary drop in the control parameter value (Figure 2).

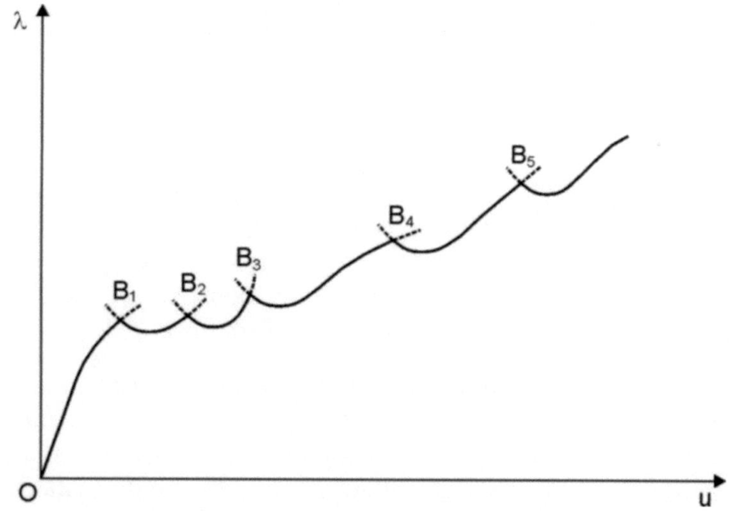

Figure 2. Bifurcation points on a representative equilibrium path (u - representative geometric value, λ – control parameter related to load).

So both the experiment itself and nonlinear numerical analyses may result only in a representative equilibrium path. In that case, the problem of the numerical representation of bifurcation comes down to the preservation of conformity of the representative equilibrium path obtained by a numerical method with the one obtained experimentally, where a sine qua non for the application of the solution uniqueness rule is to recognise the similarity of the post-critical deformation forms of the experimental and numerical models as sufficient [10,11].

An additional problem occurred during the experimental determination of the equilibrium path, resulting from the lack of abilities of recording the said temporary, little drops in load, arising from local bifurcations, causing changes of the values of some state parameters. In the majority of experiments, the load of the tested model is achieved by force control, e.g. using a gravitational system, or displacement, by means of various types of load-applying devices (Figure 3).

Figure 3. A stand for testing thin-shell structures subject to torsion: left – a version with a loading system controlling the displacement (turnbuckle), right – a version with a system controlling by force (gravitational).

However, even in case of devices with high level of technical advancement, in general it is not possible to register precisely short-lasting force changes, occurring from the beginning of a bifurcation phenomenon to the moment of reaching the consecutive deformation form by the model. Therefore, the representative equilibrium path obtained as a result of the experiment is of smooth characteristics, and its formation is based on measuring points corresponding to the consecutive deformation states determined.

In case of nonlinear numerical analysis in the finite element approach, the accuracy of the obtaining of the representative equilibrium path may be much more accurate. The existing commercial programs usually offer the results of all the increment steps, followed during the calculation process, and thus they also allow observing slight fluctuations of the control parameter. The only limitation here is exclusively the value of the incremental step itself. In spite of this, due to the lack of possibilities of relating the results obtained to the relevant detailed changes of the experimental characteristics, it seems appropriate to determine the numerical representative equilibrium path of the same level of simplification as in the case of the experiment.

2. Analyses of example structures

The comparative analysis of such representative equilibrium paths is not, however, a method that allows a complete enough verification of the reliability of the results of numerical calculations. An example of a problem in which the calculated results have been deemed incorrect despite the seeming full conformity of the representative equilibrium paths is a thin-shell open cylindrical structure with edges strengthened by stringers, working in the conditions of constrained torsion (Figure 4). This type of systems is quite often used in aviation structures. They form areas of cockpits and large cut-outs, e.g. in cargo airplanes and they are usually adjacent to much stiffer fragments of the structures [12,13].

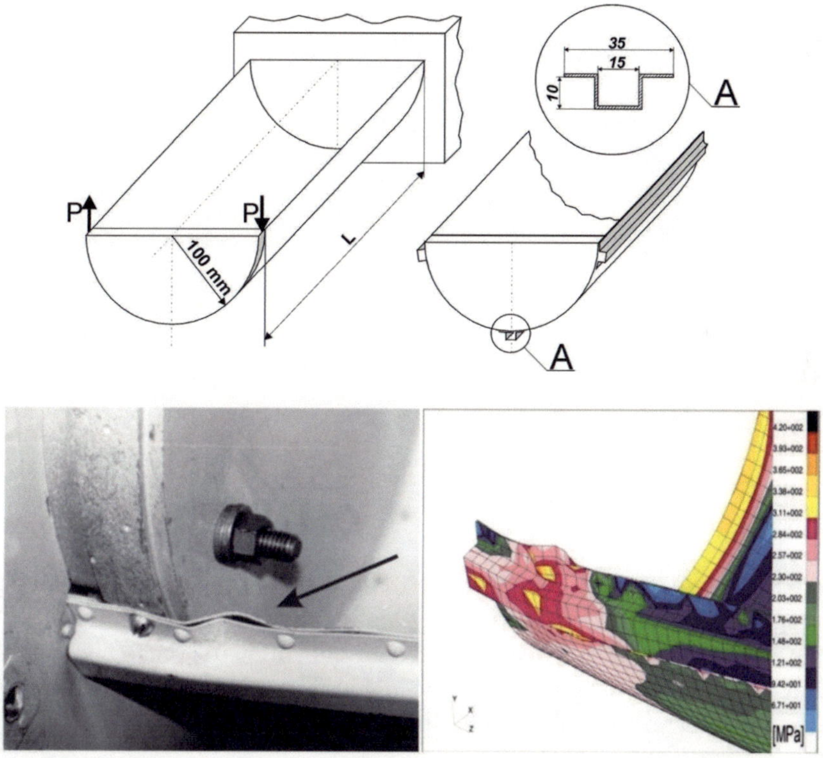

Figure 4. A schematic view of the tested structure and comparison of the deformation forms obtained experimentally and numerically

The area adjacent directly to the closing frame turned out to be crucial in the problem under consideration. The stringer strengthening the edge of the structure was buckled, and the experimental model sustained plastic deformation. The relation between the total torsion angle of the examined structure and the torque moment constituting the load was adopted as the representative equilibrium path. In spite of the seeming conformity of the obtained characteristics, a different character of deformation was observed in the critical area (Figure 3). The divergence seems to result here from the symmetric character the bifurcation phenomenon initiating deformation. Different deformations correspond to two possible variants of the actual equilibrium path, the differentiation of which is not possible in case of using a simplified representative path. So, in spite of the conformity of the characteristics, the effective stress distributions obtained numerically cannot be considered as reliable.

The need for an analysis of stress distributions in case of structures similar to the one presented above occurs quite rarely due to the commonly adopted principle, pursuant to which beam structures after buckling are considered as damaged [14]. The considerations relating to shells used in the aviation industry, e.g. semi-monocoque structure elements are of much greater practical significance.

Examples of such a system are open cylindrical shells, which were subjected to a cycle of tests, during which it was assumed that stringers are characterised by a sufficient margin of stiffness and they do not lose stability (Figure 5).

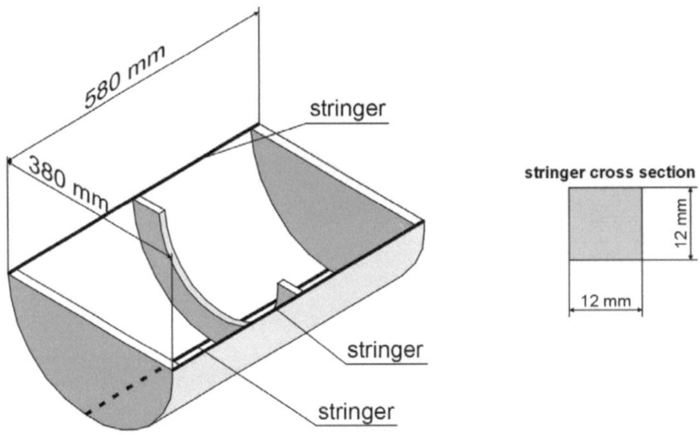

Figure 5. Geometry of the model

So the models made of polycarbonate were strengthened with longitudinal members with large, rectangular cross-section and relatively high values of geometric moments of inertia. Each of the examined systems was subjected to constrained torsion, using a test stand presented in Figure 3. The tests aimed at developing the methodology of determining stress distributions in a structure's shell in post-critical deformation states.

In the conditions of torsion the stress state created in the shell of such system, may be interpreted as an incomplete tension field. As a result, even if there are no geometrical imperfections, the shell loses its stability. On the other hand, the post-critical deformation increment causes significant stress redistribution. The experiments repeated a number of times showed that the final form of post-critical deformations of such systems, occurring at sufficiently high load values, is always the same in spite of the alternative character of the course of the structure state changing process (Figure 6).

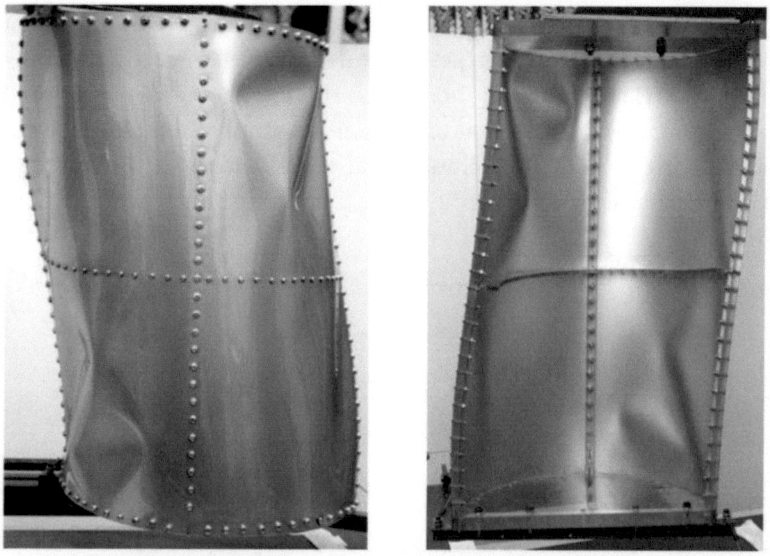

Figure 6. A post-critical deformation form of a two-segment structure at maximum load (the total torsional torque equals 180Nm)

The fact that local bifurcations following increases in loads occur with some scatter of locations and stress levels makes the nonlinear numerical analysis particularly troublesome in this case. It is practically impossible to develop a FEM model allowing to reproduce accurately the entire process of the structure's state changes, using commercial software, due to the nature of the functioning of algorithms for choosing the variants of the equilibrium path at the bifurcation points and the impossibility of the user's interference in the form of those algorithms. In this situation it seems appropriate to focus only on obtaining a numerical solution consistent with the experiment results at assumed load values.

The selection of an appropriate combination of numerical methods and parameters controlling the course of the analysis seems particularly vital in this case, likewise the proper representation of the model's stiffness. Even small mistakes in this respect result in the occurrence of incorrect forms of deformation (Figure 7).

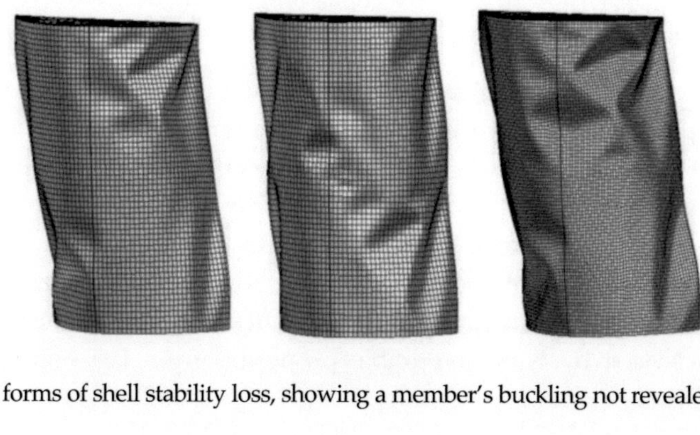

Figure 7. Incorrect forms of shell stability loss, showing a member's buckling not revealed in the experiment

It should be emphasised that it seems very risky to rely in the design process on the results of nonlinear numerical analysis of similar structures without appropriate verification in an experiment, if only a relatively cheap model experiment. In practice, multiple repetition of the analysis and systematic comparison of its results with the results of the experiment are required to obtain correct results of the numerical representation of a structure's state in the conditions of post-critical loads (Figure 8).

Figure 8. Accepted as satisfactory form of post-critical deformation obtained as a result of nonlinear analysis (left) and a deformation form obtained as a result of experimental tests at identical load (right)

Examples of aviation load-bearing structures, being much less troublesome from the point of view of nonlinear numerical analyses, working in the conditions of post-critical deformations are semi-monocoque box structures with flat walls. This type of solutions is often applied in modern military aircrafts. They usually constitute compact load-bearing structures of fuselages and internal wing areas, most frequently ogival or delta-shaped.

A representative element of this type of structures that was subjected to a detailed experimental and numerical analysis was a multi-segment multi-member thin-shell structure, working in the conditions of dominant torsion (Figure 9).

Several variant of the presented structure were examined, including the ones with openings of various shapes, corresponding to all types of functional and service cut-outs.

Striving to preserve the local character of post-critical deformations of the segments of skin, similar structures possess relatively thin shells at considerable stiffness of the framing. So, a characteristic feature of the structure's shell is a low level of critical load. Even at little levels of shearing stress, the waving of the skin occurs within particular segments. At the same time, as proved by the experiments carried out, to reach the global loss of stability corresponding to the damage of the structure it is required to increase the load value considerably. Thus, in spite of the natural susceptibility to a local loss of skin stability, the structure is characterised by very favourable properties from the point of view of their application in aviation.

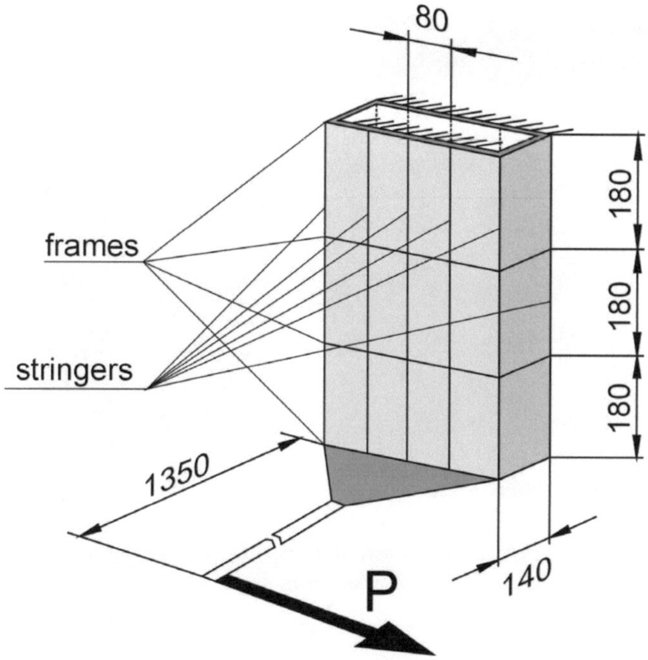

Figure 9. The examined representative case of a box structure

The process of the appearance of consecutive bifurcations is in this case of established and repeatable character. This results from the mere mechanism of tension fields formation in particular segments of the skin. Additionally, the character of the stress distribution in each of the structure's segments is similar. All these factors result in the fact that in contrast to the structure discussed earlier, nonlinear numerical analysis of semi-monocoque box structures is possible when using alternatively various numerical methods, and the successful course of the analysis does not require the application of correction strategies based on arc length control methods (Figure 1).

The numerical analysis of the examined structure allowed to obtain results of high degree of conformity with the results of the experiment (Figure 10).

The conformity between the courses of the representative equilibrium paths did not raise any objections either, and their form itself constitutes the confirmation of a gentle nature of the phenomenon (Figure 11).

While analysing the constructional solutions used in aviation load-bearing structures, the need should be emphasised for drawing attention to the area of working shells, in which local cracks occur due to cyclic loads. Such damages are usually dimensionless in nature, with edges touching each other. In case the skin is stretched along the normal to the direction of the crack, in the vicinity of the front of the crack there occur both tensile stresses and compressive stresses (crosswise), which creates conditions for the occurrence of a local buckling. In these conditions, in the weakened zone there occurs a phenomenon of a loss of shell stability within the area of a crack, referred to as wrinkling [11].

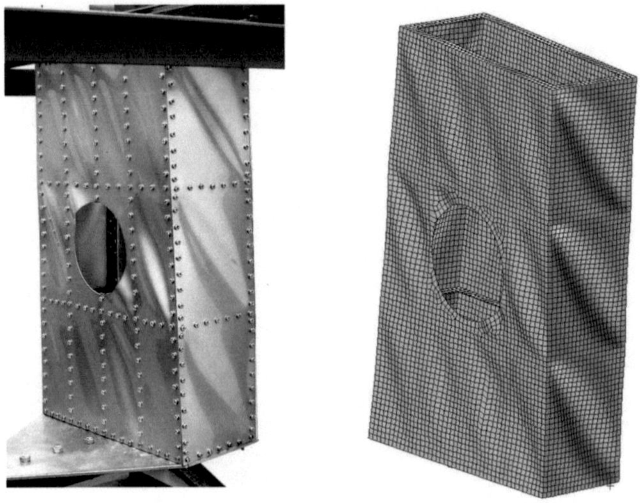

Figure 10. Comparison of the deformation form in a structure with a circular opening obtained during the experiment (left) and as a result of MES nonlinear analysis (right)

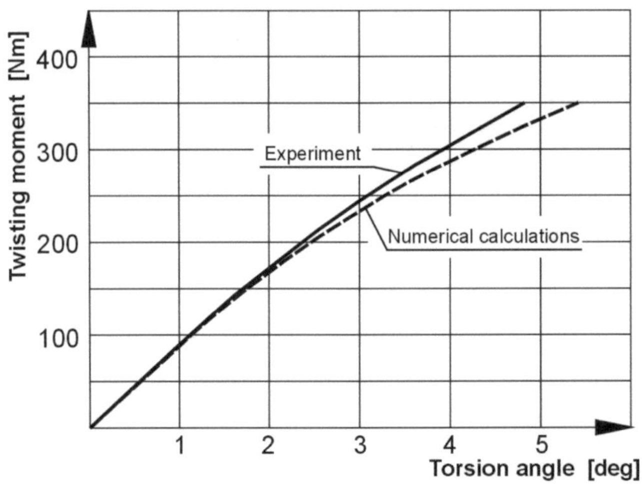

Figure 11. Comparison of the representative equilibrium paths – a structure with a circular opening

During the experimental and numerical analysis of the phenomenon the model shown in Figure 12 was used.

In order to examine the influence of a crack length on the nature of the phenomenon, a number of the models versions were analysed. The quantitative and qualitative character of deformations in the vicinity of the crack was determined by means of the Shadow moiré method (Figure 13).

Although during the experiment with many repetitions of load application, in the form of a force stretching the examined plate along the normal to the direction of the crack edges (Figure 11), the post-critical deformation was repeatable in nature, it turned out troublesome to obtain correct results of nonlinear numerical analyses. The deformation phenomenon is relatively gentle in nature, which results from the susceptibility of the structure to the slightest geometric imperfections near the crack.

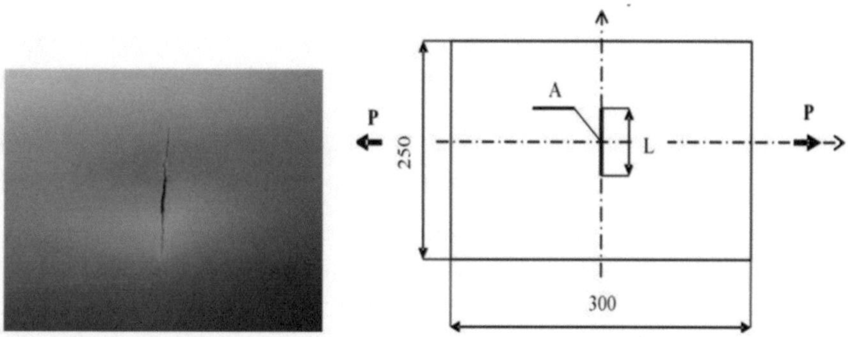

Figure 12. The model for experimental examination of the wrinkling phenomenon

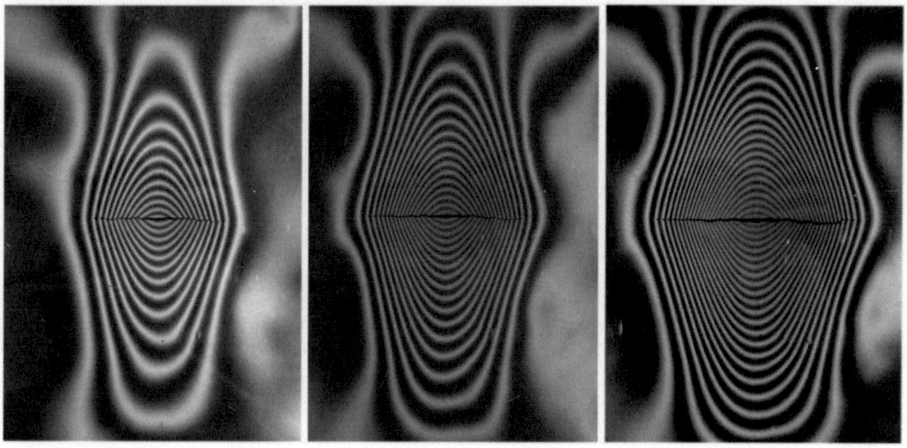

Figure 13. The Shadow moiré patterns for three crack lengths: 30mm, 50mm, 70mm

According to the expectations, an attempt to carry out a nonlinear numerical FEM analysis using an ideally flat plate model resulted in the settlement of the state of equilibrium of the structure which did not show any displacements in the perpendicular direction to the plate surface. In such a case, bifurcation point does not appear at all on the equilibrium path in the numerical space of the state.

The nature of the analysis undergoes a complete qualitative change in case of taking imperfection into consideration, either in the form of initial displacements of crack edges, or in the form of preload residual loads. The degree of the representation of the nature of the phenomenon depends mostly on the type of the finite element. As results from the numerical test performed, very few elements offered by commercial software manufacturers possess a sufficiently appropriate mathematical model of displacements to represent in full the phenomenon of wrinkling. They are usually elements of rich, nonlinear shape functions, the application of which in complex nonlinear tasks causes in turn the lack of effectiveness of algorithms for choosing the proper branches of the equilibrium path. They may result in e.g. the occurrence of the so-called "ping-pong effect", consisting in reversing the course of the analysis and its further advancement within the range of negative values of the control parameter.

From the numerical point of view, the mere complexity of nonlinear procedures forces the application of a possibly uncomplicated displacement model of the used finite elements. In case of elements of linear shape functions, recommended by the manufacturers of MSC MARC software for nonlinear analyses, the correct selection of the type and size of the initial imperfection turned out to be crucial for forcing the appropriate course of the phenomenon.

It turned out the most effective to apply continuous cross-bending load of constant intensity along both edges of a crack, possessing the value by several orders lower in relation to the force loading the plate. The post-critical deformation distributions obtained as a result of numerical analyses were compared with the experiment both in qualitative and quantitative terms, which was possible owing to the moiré patterns.

After numerous repetitions of the analyses and the selection of the most appropriate set of numerical methods, a satisfactory displacement distribution was obtained (Figure 14).

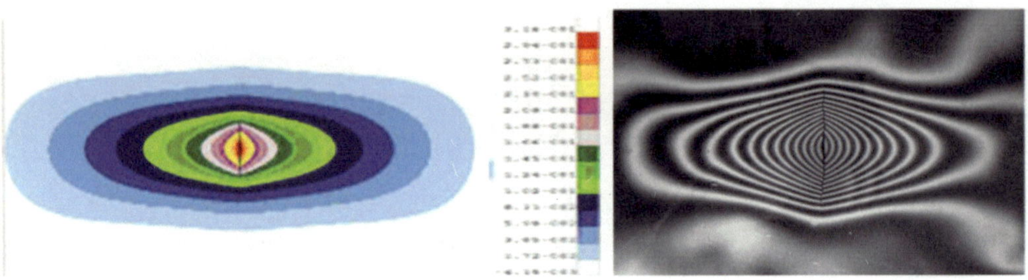

Figure 14. Comparison of deflection distributions in the post-critical state for a model with a crack 30 mm long

Owing to the examination of models with cracks of various lengths, it became possible to determine the nature of relation between the crack length and the resultant value of cross-bending load, constituting a factor causing initial imperfection (Figure 15).

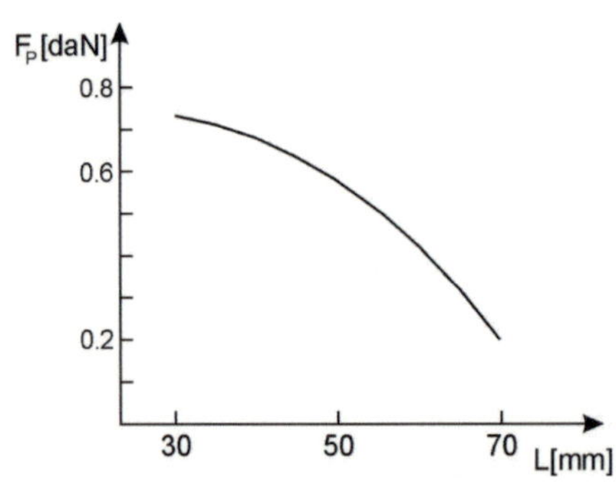

Figure 15. The relation of the values of total cross-bending load in the crack length function

The research results of various load-bearing structures confirm that the difficulty related to carrying out an appropriate nonlinear numerical analysis results from the nature of bifurcation. If the change in a structure's form is gentle in nature and it occurs in a small area, then the bifurcations related to it occur gradually, in relatively small subsets of state parameters. The numerical simulation of the process is then easy to perform and it may take place when using prediction-correction methods with simple correction based on state control. But if the deformation occurs in a larger area, and the change of the form is violent in nature, then the bifurcation corresponds to the simultaneous change of a great number of state parameters, and the determination through a numerical procedure of their appropriate combination, corresponding to the new state of static equilibrium, may be hindered or even impossible. In such a case it is necessary to apply matchings of prediction methods with correction strategies based on arc length control methods, such as the Riks correction or the Crisfield hyperspherical correction [15] (Figure 9).

Bifurcation changes of the forms of load-bearing structures, containing shells of considerable curvature, occur more violently if there is a higher relation of the square of the smaller of dimensions of the shell segment area limited by the adjacent members' frames to the value of the local radius of its curvature [16]. Thus, semi-monocoque structures of relatively low number of the framing elements are especially troublesome in nonlinear numerical representations.

An example of such a structure is a closed cylindrical shell presented in Figure 16.

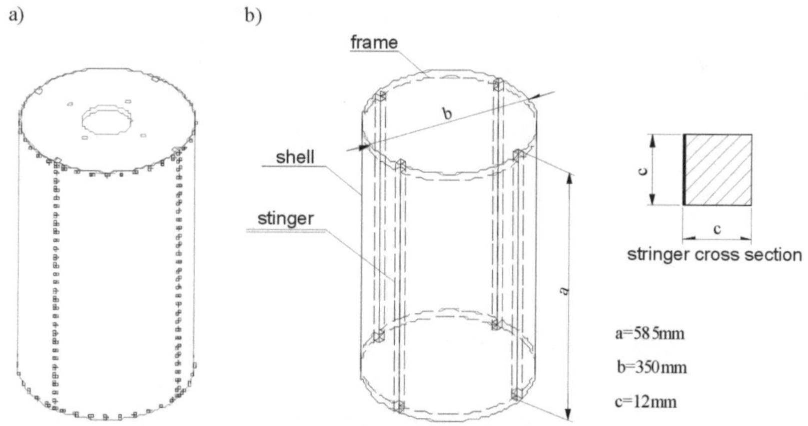

Figure 16. A schematic view of a complete cylindrical shell reinforced by four members (a) and a schematic view of a structure including dimensions (b)

The structure's framing consists of a minimum number of crosswise elements, i.e. two closing frames and four longitudinal members. The type of the structure itself corresponds to solutions commonly used in the aviation technology, e.g. the construction of a fuselage of an aircraft. It should be emphasised, however, the model subjected to examinations constitutes a special instance of a structure of purposefully minimised number of longitudinal members. The actual solutions are usually based on much more extended framings. The structure described corresponds to an isolated phase of a wider cycle of examinations aiming at determining direct dependences between the number of framing elements and post-critical deformation distributions.

The examined structure was subjected to constrained torsion using a modified version of the stand presented in Figure 2.

According to the expectations, post-critical deformations occurred in a violent way. Due to the gravitational way of load application, the measurement of the relation between torsion angle and the torque moment, assumed as the representative equilibrium path, corresponded to the steady states (Figure 21).

Using this mode of taking measurements, the representative characteristics does not reflect bifurcation points in an overt way, but attention should be drawn to the occurrence of its horizontal section. It corresponds to this phase of the experiment in which a sudden change occurred in the structure state with the simultaneous constant load level.

With regard to the symmetry, the deformed structure possessed four characteristic grooves in all the shell segments (Figure 17). During the experiment the surface geometry was registered using the projection moiré method. Atos scanner manufactured by a German company, GOM Optical Measuring Techniques was used as a registering device.

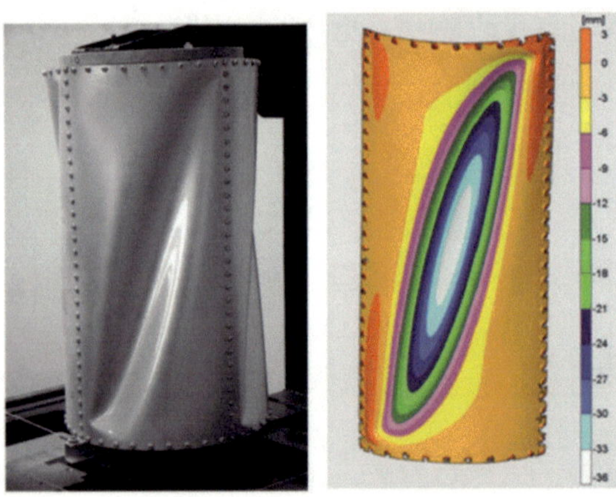

Figure 17. The advanced post-critical deformation of the examined structure (left) and the distribution of contour lines representing the size of the deformation, made using the projection moiré method (right)

The problem discussed belongs to one of the most troublesome from the point of view of a FEM nonlinear numerical simulation. A number of tests performed using the MSC MARC software revealed the lack of effectiveness of its procedures in case of this problem, with regard to determining the appropriate post-buckling state of a structure. The algorithms used in those procedures are characterised by inability to represent the symmetry of the phenomenon. With the idealised geometric form of the model, the obtaining of the new form of the structure after crossing the critical load value occurs only in one of the segments, in spite of the apparently correct, symmetrical initiation of stability loss. This proves the faults in the algorithms for choosing the appropriate variants of the equilibrium path in case of the appearance of changes in the state parameters combination in several of their independent subsets.

The situation was improved when shell imperfections were implemented, by applying normal forces to the skin, in the central points of particular skin (Figure 18).

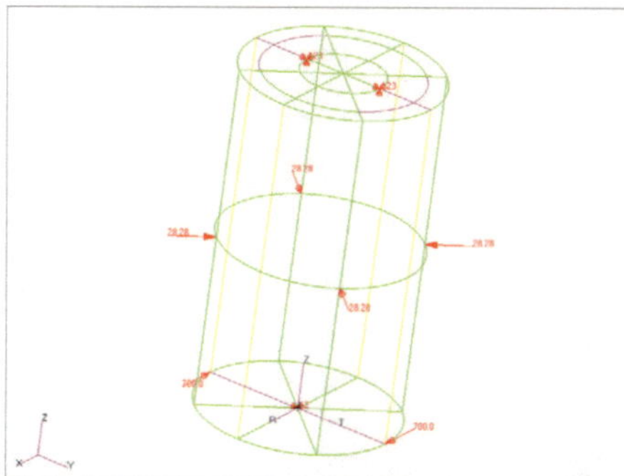

Figure 18. A geometrical model of a structure made in MSC PATRAN environment with boundary conditions and loading

However, even in the case of applying this type of forcing a form change, it was very difficult to obtain results that would fully correspond to the experimental results. Assuming the use of skin elements with linear shape functions, the appropriate density of the mesh turned out to be the key factor, but its excessive density caused incorrect forms of post-critical deformations (Figure 19).

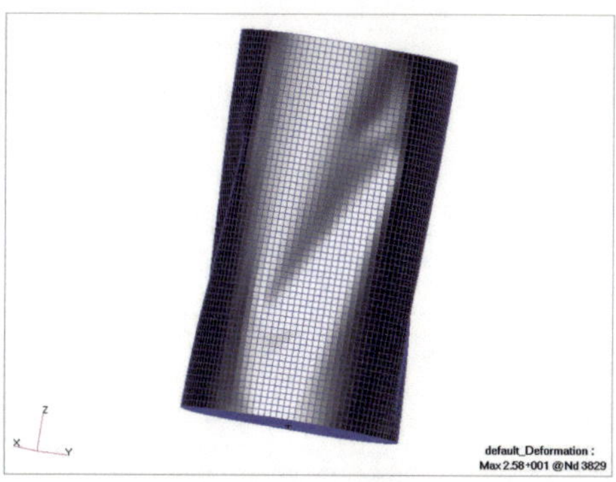

Figure 19. The incorrect form of deformation, obtained in case of too many elements

The better result, in case of application of beam elements as a representation of stringers, was obtained with the use of a relatively low density of mesh. This proves the rightness of the thesis, proved a number of times in many studies, pursuant to which the decrease in the general number of degrees of freedom, corresponding to the number of state parameters, in case of nonlinear procedures used in the available commercial programs, often brings benefits that considerably exceed the deficiencies of a mathematical description resulting from the decrease in the number of elements.

The best result was obtained only after the fundamental change of the concept of FEM model, when the different kind of finite elements was applied as a representation of stringers (thick shell element was used instead of a recommended beam element). However this solution, from the point of view of mathematical description is much less correct, it turned out much more effective in case of relatively low values of the total torsion angle of the structure.

The results of analysis of this FEM model version, obtained using secant prediction method and strain correction strategy (Figure1), are presented in Figure 20.

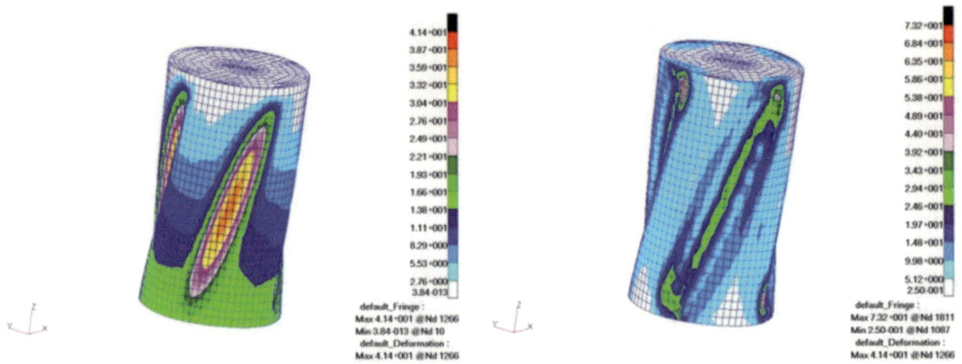

Figure 20. The deformation distribution (left) and reduced stress distribution acc. to Huber-Mises hypothesis (right) for 100% of the maximum load (stringers modelled with thick shell bilinear element)

The strain-correction strategy turned out most effective in case of significant, violent change of the form of deformations, when the representative equilibrium path contains relatively long "horizontal section".

The relation between the representative equilibrium paths is presented in Figure 21.

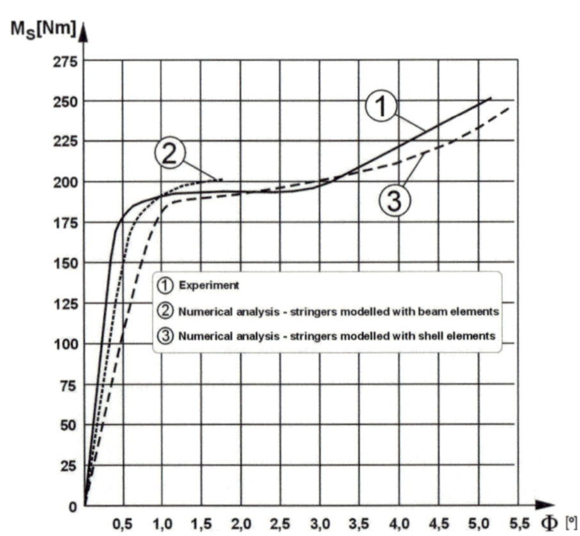

Figure 21. The presentation of the representative equilibrium paths

3. Conclusion

The presented examples of load-bearing structures represent only some of those many used in the modern aviation technology. But the criterion applied while selecting them as objects of experimental and numerical analyses was its representativeness for the most commonly met elements of constructions, in case of which the occurrence of a local stability loss is acceptable in the conditions of service load.

The fundamental conclusion that can be drawn from the presented results of the research is the absolute need for using experimental verifications with regard to FEM nonlinear numerical analyses of this type of structures. The more so that even in the cases in which the correctness of the results obtained seems unquestionable, they may be in fact burdened with errors resulting from the very limited reliability of the numerical procedures used in commercial programs.

Based on the nonlinear numerical analyses, related to the presented structures, frequently repeated many times, a general recommendation may also be formulated for the maximum possible limitation of the size of a task. Striving for increasing the accuracy of the calculations by increasing the density of finite elements mesh, applied successfully in linear analyses, may turn out ineffective in case of a nonlinear analysis and may lead to incorrect results or the lack of convergence of calculations.

The numerical representation of bifurcation, by virtue of the mere idea of the discrete representation of continuous systems, must be simplified in case of the finite elements method. In such a situation, based on the quoted examples, the need must be emphasised for obtaining the indispensable convergence of the experimental and obtained numerically relations between a selected geometric parameter characterising the essence of a structure's deformation and a selected value relating to the load, recognised as representative equilibrium paths. This convergence, in combination with the accepted as sufficient similarity of post-critical deformation forms, constitutes the grounds for accepting the reliability of stress distributions determined by means of numerical methods.

Author details

Tomasz Kopecki
Faculty of Mechanical Engineering and Aeronautics,
Rzeszów University of, Technology, Rzeszów, Poland

4. References

[1] Marcinowski J. (1999). *Nonlinear stability of elastic shells.* Publishing House of Technical University of Wrocław, Poland

[2] Felippa C. A. (1976): *Procedures for computer analysis of large nonlinear structural system in large engineering systems.* ed. by A. Wexler, Pergamon Press, London, UK

[3] Bathe K.J. (1996). *Finite element procedures*, Prentice Hall, USA

[4] Doyle J.F. (2001). *Nonlinear analysis of thin-walled structures.* Springer-Verlag, Berlin, Germany

[5] Andrianov J., Awrejcewicz J., Manewitch L.I. (2004) *Asymptotical Mechanics of thin-walled structures.* Springer, Berlin, Germany

[6] Awrejcewicz J., Krysko V.A., Vakakis A.F. (2004) N*onlinear dynamics of continuous elastic systems.* Springer, Berlin, Germany

[7] A.V. Krysko, J. Awrejcewicz, E.S. Kuznetsova, V.A. Krysko, *Chaotic vibrations of closed cylindrical shells in a temperature field*, International Journal of Bifurcation and Chaos, 18 (5), 2008, 1515-1529.

[8] A.V. Krysko, J. Awrejcewicz, E.S. Kuznetsova, V.A. Krysko, *Chaotic vibrations of closed cylindrical shells in a temperature field*, Shock and Vibration, 15 (3-4), 2008, 335-343.

[9] I.V. Andrianov, V.M. Verbonol, J. Awrejcewicz, *Buckling analysis of discretely stringer-stiffened cylindrical shells*, International Journal of Mechanical Sciences, 48, 2006, 1505-1515.

[10] Arborcz J. (1985). *Post-buckling behavior of structures. Numerical techniques for more complicated structures.* Lecture Notes In Physics, 228, USA

[11] Kopecki T. (2010). *Advanced deformation states in thin-walled load-bearing structure design work.* Publishing House of Rzeszów University of Technology, Rzeszów, Poland

[12] Niu M. C. (1988). *Airframe structural design.* Conmilit Press Ltd., Hong Kong10] Lynch C., Murphy A., Price M., Gibson A. (2004). *The computational post buckling analysis of fuselage stiffened panels loaded in compression.* Thin-Walled Structures, 42:1445-1464, USA

[13] Mohri F., Azrar L., Potier-Ferry M. (2002). *Lateral post buckling analysis of thin-walled open section beams.* Thin-Walled Structures, 40:1013-1036, USA

[14] Rakowski G., Kacprzyk Z. (2005). *Finite elements method in structure mechanics.* Publishing House of Technical University of Warszawa, Warszawa, Poland

[15] Ramm E. (1987). *The Riks/Wempner Approach – An extension of the displacement control method in nonlinear analysis*. Pineridge Press, Swensea, UK Aben H. (1979). *Integrated photoelasticity.* Mc Graw-Hill Book Co., London, UK

[16] Brzoska Z. (1965). *Statics and stability of bar and thin-walled structures.* PWN, Warszawa, Poland

Chapter 9

Nonlinear Plate Theory for Postbuckling Behaviour of Thin-Walled Structures Under Static and Dynamic Load

Tomasz Kubiak

Additional information is available at the end of the chapter

http://dx.doi.org/10.5772/48961

1. Introduction

A thin plate or thin-walled constructions are used in the sports industry, automotive, aerospace and civil engineering. As an example of such structural elements snowboard, skis, poles may be mentioned, as well as all kinds of crane girders, structural components of automobiles (car body sheathing or all longitudinal members), aircraft fuselages and wings, supporting structures of the walls and roofs of large halls and warehouses. All the above structures, as well as many others which can be regarded as a thin, exhaust carrying capacity not by exceeding the allowable stresses but by the stability loss. Therefore, not only critical load but also the postbuckling behaviour of thin-walled structures subjected to static and dynamic load is essential knowledge for designers. The use of more accurate mathematical models allows to explore the phenomena occurring after the loss of stability and to describe more precisely their behaviour. Engineers and designers need guidelines to construct as well as quick and easy software to use for analyse the behaviour of thin-walled structures. Therefore, the author of this chapter decided to explore this issue, propose a mathematical model and the method of analysis of orthotropic thin-walled structures subjected to static and dynamic load.

1.1. Static buckling

The buckling and postbuckling of thin-walled structures subjected to static load have been investigated by many authors for more than one hundred years. To the group of precursors of the investigation on the stability of thin-walled structures problem should be included following scientists: Euler [1], Timoshenko [2] and Volmir [3].

INTECH
open science | open minds

© 2012 The Author(s). Licensee InTech. This chapter is distributed under the terms of the Creative Commons Attribution License http://creativecommons.org/licenses/by/3.0), which permits unrestricted use, distribution, and reproduction in any medium, provided the original work is properly cited.

This chapter considered the thin plate or thin-walled structures composed of flat plates. Such structures have a various of buckling modes which can differ from one another both in quantitative (e.g., by the number of half-waves) and in qualitative (e.g., by global and local buckling) way.

The stability loss or buckling is a system transition from one equilibrium to another (the bifurcation point), or jump from the stable to the unstable equilibrium path (the limit point). Load resulting in the loss of stability is called the critical load. The behaviour of the structure subjected to load higher than the critical one can be described by a stable (the grow of displacement is caused by increased load) or unstable (displacements grow with decreasing load) postbuckling equilibrium path.

The postbuckling behaviour of the structures depends on their type. For example, the cylindrical shells subjected to axial compression change their equilibrium stage (buckling) by unstable bifurcation point or limit point. Long rods or columns subjected to axial compression have usually a sudden global buckling (bifurcation point of passage to the unstable postbuckling equilibrium path). Thin plates supported on all edges lose their stability having the local buckling mode and the stable postbuckling equilibrium path. Mentioned above type of buckling and postbuckling behaviour for given thin-walled structures are the same for ideal structures as for structures with geometrical imperfection. Columns made of thin prismatic plates could have the local buckling mode, global (flexural, torsional or distorsional) one or coupled. The structures after local buckling are able to sustain further load, because increasing the displacement is only possible by increasing the load value (stable postbuckling equilibrium path), further increasing the load leads to plasticity or reaching the new, this time unstable bifurcation point (global buckling). The dangerous form of stability loss is the interactive buckling (coupled buckling), which usually causes the structure transition to the unstable equilibrium path what leads to the destruction of the structure with load lower than the critical load corresponding to each mode separately. The interaction of different buckling modes occurs when the critical loads corresponding to the different buckling modes are close to each other.

A more comprehensive review of the literature concerning the interactive buckling analysis of an isotropic structure can be found for example in Ali and Sridharan [3], Benito and Sridharan [5], Byskov [6], Koiter and Pignataro [7], Kolakowski [8–10], Manevich [11], Moellmann and Goltermann [12], Pignataro et al. [13], Pignataro and Luongo [14, 15], Sridharan and Ali [16, 17]. The interactive buckling of orthotropic structures can be found for example in [18, 19].

1.2. Dynamic buckling

In literature a quantity of "pulse intensity" [20] or "pulse velocity" [21] is introduced. The analysis of dynamic stability of plates under in-plane pulse loading can be divided into three categories depending on pulse duration and magnitude of its amplitude. For pulses of high intensity the impact phenomenon is observed whereas for pulses of low intensity the problem becomes quasi-static. The phenomenon of dynamic stability and dynamic buckling are often confused with each other. In this chapter the dynamic buckling phenomenon is examined but the concept of dynamic stability is broader and applies also to the stability of motion, which for thin-walled structures can be found for example in [22, 23]. The dynamic buckling occurs when the loading process is of intermediate amplitude and the pulse duration is close to the period of fundamental natural flexural vibrations (in range of milliseconds). In such case the effects of dumping are neglected [24]. Damping neglecting is only possible for problem solved in elastic range [25].

It should be noted that dynamic stability loss may occur only for structures with initial geometric imperfections; therefore the dynamic bifurcation load does not exist. For the ideal structures (without geometrical imperfection) the critical buckling amplitude of pulse loading tends to infinity [26]. The dynamic buckling load should be defined on the basis of the assumed buckling criterion.

The precise mathematical criteria were formulated for structures having unstable postcritical equilibrium path or having limit point [26, 27]. But for the structures having stable postbuckling equilibrium path (thin plate, thin-walled beam-columns with minimal critical load corresponding to local buckling) the precise mathematical criterion have not been defined till now.

Therefore Simitses [27] suggested not to define the dynamic buckling for the structures with stable postbuckling behaviour, but rather it should be defined as a dynamic response to pulse loads.

It is a reason why in world literature a lot of criteria can be found. In the sixties of the twenty century Volmir [28] proposed a criterion for plates subjected to in-plane pulse loading. The Volmir criterion - considered the easiest to use - states that *the dynamic critical load corresponds to the amplitude of pulse force (of constant duration) at which the maximum plate deflection is equal to some constant value k (k - one half or one plate thickness)* [28].

In many publications the dynamic buckling load is determined on the basis of stability criterion of Budiansky and Hutchinson [26, 29, 30]. However, this criterion was formulated for shell structures but also it can be used for the plate structures [31-34]. Budiansky and Hutchinson noticed that in some range of the amplitude value, the deflection of structures grows more rapidly than in other. Budiansky and Hutchinson formulated the following criterion: *Dynamic stability loss occurs when the maximum deflection grows rapidly with the small variation of the load amplitude* [26].

In the end of 90's Ari-Gur and Simonetta [20] analysed laminated plates behaviour under impulse loading and formulated four own criteria of dynamic buckling, two of them of collapse-type conditions. One of them states: *Dynamic buckling occurs when a small increase in the pulse intensity causes a decrease in the peak lateral deflection* [20].

The failure criterion was proposed by Petry and Fahlbush [34], who suggest that for structures with stable postbuckling equilibrium path the Budiansky-Hutchinson criterion is conservative because it does not take into account load carrying- capacity of the structure.

Based on examples [35] it was noticed that for the thin-walled structures subjected to pulse loading, which lose their stability according to Budiansky-Hutchinson criterion or Volmir criterion, the maximal radius r_{max} calculated from characteristic root χ=a+jb (where j= $\sqrt{-1}$) of Jacoby matrix is equal or greater than unity in complex plane.

Therefore the criterion for thin-walled structures proposed by author [35] can be formulated as follows: *Thin-walled structures subjected to pulse loading of finite duration lose their stability even if one characteristic root χ=a+jb of Jacoby matrix find for every time moment from 0 to 1.5T_p lies in the complex plane outside the circle with radius equal to unity.*

Teter [36, 37] in his works analysed the long columns with longitudinal stiffeners and basing on the phase portrait for dynamic response of these structures defined the following criterion: *The dynamic buckling load for the tracing time of solutions has been defined as the minimum value of the pulse load such that phase portrait is an open curve.*

The dynamic buckling problem has been well known in the literature for over 50 years and was the subject of numerous works [20, 24, 26-34]. The extensive list of work dealing with dynamic buckling can be found for example in the book edited by Kowal-Michalska [38] or written by Simitses [39] or Grybos [40]. It seems that the analysis of dynamic buckling of thin-walled structures, especially structures with flat walls is not sufficiently investigated. There is a lack of both single-and multimodal analysis of dynamic buckling of columns with complex cross-sections made of thin flat walls. The author of this paper decided to fill this gap presenting a method for the analysis of the local (single mode) and interactive (coupled mode – local and global) buckling of thin-walled structures subjected to pulse loading.

It should be mentioned that the presented method can be used only if the structures are in the elastic range. The case of dynamic buckling in elasto-plastic range including the viscoplastic effect has been investigated by Mania and Kowal-Michalska [25, 41-43]. In world literature it is also possible to find the paper dealing with the dynamic buckling of thin-walled structures subjected to combined load [44]. Czechowski [45] modelled the girder subjected to twist and bending considering only one plate subjected to shear and compression. The general summary showing which parameters have an influence on dynamic buckling of plated structures can be found in [46, 47].

2. Thin orthotropic plate theory

The thin isotropic or orthotropic plates with constant or widthwise variable material properties are considered. The thin-walled beam-columns or girders composed of mentioned above plates are also analysed. In order to taken into account all buckling modes (global, local and their interaction) the plate model was adopted to the analysed structures.

2.1. Basic assumptions

The basic assumption for thin plate are given by Kirchhoff for linear and by von Kármán and Marquerre for nonlinear thin plate theory. They made their assumption for isotropic material; lots of authors extended these assumptions for orthotropic or even for orthotropic multilayer thin plate [18, 48]. The assumptions are as follows:

- the plate is homogeneous (for example orthotropic homogenisation is made for fibre composite – resin matrix and fibre-reinforcement)
- the plate is thin – other dimensions (length and width) are at least 10 times higher than plate thickness;
- the material of the plate subjects to Hooke's law;
- the plane stress state is considered for the plate – stress acting in the plate plane dominates the plate behaviour, stress acting in normal to plate plane direction are assumed to be zero;
- all strains (normal and shear) in plate plane are small compared to unity and they are linear;
- the strains of the plate to its normal direction are neglected (thickness of the plate do not change after deformation) – this assumption are made according to the Kirchhoff-Love hypothesis;
- straight lines normal to the mid-surface of the plate remain straight and normal to the mid-surface after deformation
- there is no interaction in normal direction between layers parallel to middle surface;
- deflections of the plate can be considerable in terms of nonlinear geometrical relations;

Additionally, it is assumed that principal axes of orthotropy are parallel to the edges of analysed structures (plate, beam, column, beam-column or girder).

2.2. Geometrical equations for thin plate

A plate model has been assumed for a thin plates and thin-walled beam-columns or girders. For easier explanation the plate (Figure 1a) or each *i*-th strip (Figure 1b) of the plate (or wall of the girder) or each *i*-th wall of the girder (Figure 1c) are called plate.

To describe the middle surface strains for each plate the following strain tensor have been assumed:

$$\begin{aligned} \varepsilon_{ix}^{m} &= u_{i,x} + \tfrac{1}{2}(w_{i,x}^2 + u_{i,x}^2 + v_{i,x}^2), \\ \varepsilon_{iy}^{m} &= v_{i,y} + \tfrac{1}{2}(w_{i,y}^2 + u_{i,y}^2 + v_{i,y}^2), \\ \gamma_{ixy}^{m} &= u_{i,y} + v_{i,x} + w_{i,x}w_{i,y} + u_{i,x}u_{i,y} + v_{i,x}v_{i,y}, \end{aligned} \tag{1}$$

where: u_i, v_i, w_i - displacements parallel to the respective axes x_i, y_i, z_i of the local Cartesian system of co-ordinates, whose plane x_iy_i coincides with the middle surface of the *i*-th plate before its buckling (Figure 1).

In the majority of publications devoted to structure stability, the terms $(u_{i,x}^2 + v_{i,x}^2)$, $(u_{i,y}^2 + v_{i,y}^2)$ and $(u_{i,x}u_{i,y} + v_{i,x}v_{i,y})$ are in general neglected for ε_{ix}^{m}, ε_{iy}^{m}, γ_{ixy}^{m} correspondingly, in (1) in the strain tensor components.

The change of the bending and twisting curvatures of the middle surface are assumed according to [48, 49] as follows:

$$\begin{aligned} \kappa_{ix} &= -w_{i,xx}, \\ \kappa_{iy} &= -w_{i,yy}, \\ \kappa_{ixy} &= -w_{i,xy}. \end{aligned} \tag{2}$$

The geometrical relationship given by equations (1) and (2) allow to consider both out-of-plane and in-plane bending of the plate.

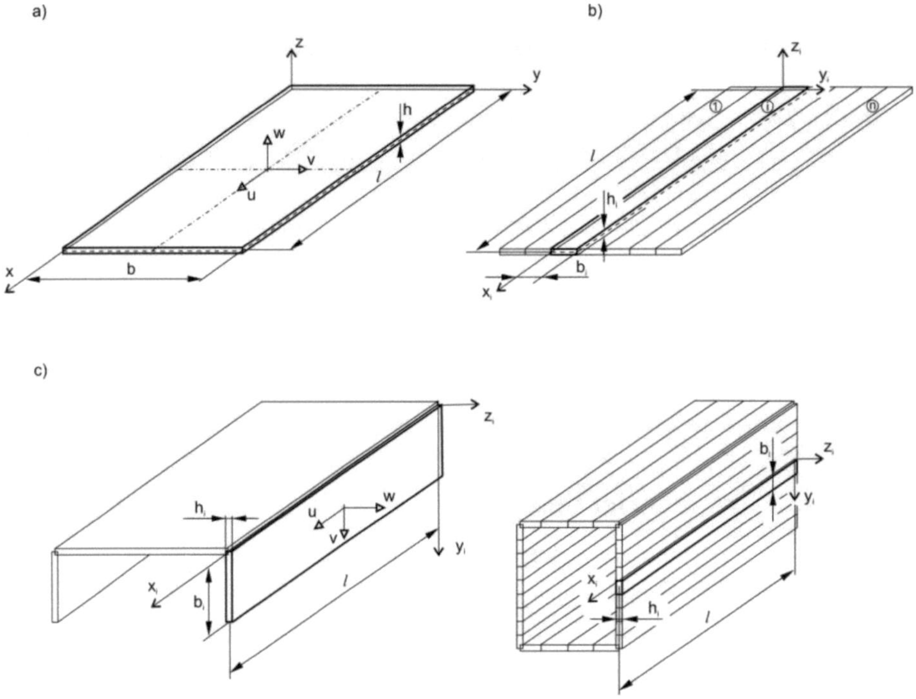

Figure 1. Possible models: plates, strips or wall with assumed dimension, coordinate systems and direction of deflections

2.3. Constitutive equations for orthotropy

Let's consider orthotropic plate with principal axes of ortothropy *1* and *2* parallel to plate edges (Figure 2).

As same as in previous paragraph let's consider *i*-th plate or strip of structures under analysis. The stress – strain relationship for orthotropic plate can be written in following form:

$$\begin{Bmatrix} \sigma_{i1} \\ \sigma_{i2} \\ \tau_{i12} \end{Bmatrix} = \begin{bmatrix} Q_{i11} & Q_{i12} & 0 \\ Q_{i21} & Q_{i22} & 0 \\ 0 & 0 & Q_{66} \end{bmatrix} \begin{Bmatrix} \varepsilon_{i1} \\ \varepsilon_{i2} \\ \gamma_{i12} \end{Bmatrix}, \tag{3}$$

where:

$$\begin{Bmatrix} \varepsilon_{i1} \\ \varepsilon_{i2} \\ \gamma_{i12} \end{Bmatrix} = \begin{Bmatrix} \varepsilon_{i1}^{m} \\ \varepsilon_{i2}^{m} \\ \gamma_{i12}^{m} \end{Bmatrix} + z \begin{Bmatrix} \kappa_{i1} \\ \kappa_{i2} \\ 2\kappa_{i12} \end{Bmatrix} \tag{4}$$

$$\begin{aligned} Q_{i11} &= \frac{E_{i1}}{1-\nu_{i12}\nu_{i21}}, \\ Q_{i12} &= Q_{21} = \nu_{i21}\frac{E_{i1}}{1-\nu_{i12}\nu_{i21}} = \nu_{i12}\frac{E_{i2}}{1-\nu_{i12}\nu_{i21}}, \\ Q_{i22} &= \frac{E_{i2}}{1-\nu_{i12}\nu_{i21}}, \\ Q_{i66} &= G_{12}. \end{aligned} \tag{5}$$

and E_{i1}, E_{i2} are Young modulus in longitudinal *1* and transverse *2* direction respectively, ν_{i12} is a Poisson ratio for which strains are in longitudinal direction *1* and stress in transverse direction *2*, G_{i12} is a shear modulus (Kirchhoff modulus) in *12* plane.

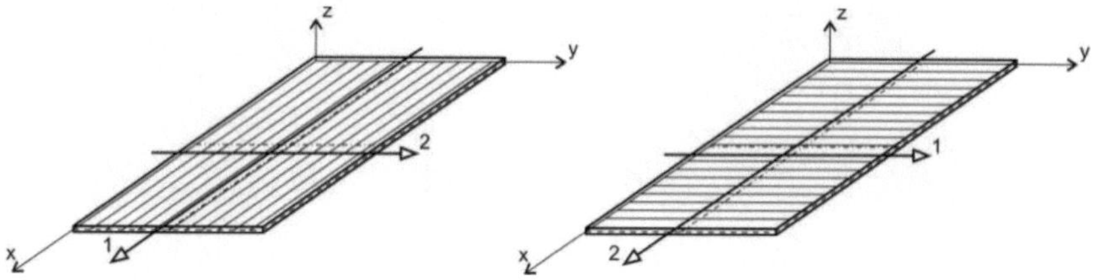

Figure 2. Plates or walls with principal axes of orthotropy

Young modulus and Poisson ratio occurring in (5) according to Betty-Maxwell theorem or according to symmetry condition of stress tensor should fulfil following relation:

$$E_{i1}\nu_{i21} = E_{i2}\nu_{i12}\,. \tag{6}$$

For isotropic plate (wall of beam-columns) the constitutive equations are as follows:

$$\begin{Bmatrix} \sigma_{ix} \\ \sigma_{iy} \\ \tau_{ixy} \end{Bmatrix} = \frac{E_i}{1-\nu_i^2} \begin{bmatrix} 1 & \nu_i & 0 \\ \nu_i & 1 & 0 \\ 0 & 0 & \dfrac{1-\nu_i}{2} \end{bmatrix} \left\{ \begin{Bmatrix} \varepsilon_{ix}^{m} \\ \varepsilon_{iy}^{m} \\ \gamma_{ixy}^{m} \end{Bmatrix} + z \begin{Bmatrix} \kappa_{ix} \\ \kappa_{iy} \\ 2\kappa_{ixy} \end{Bmatrix} \right\}. \tag{7}$$

2.4. Generalized sectional forces

Substituting stress-strain relation from previous subchapter, the sectional moments and forces:

- for *i*-th isotropic plate or wall of beam-column are expressed by:

$$\begin{Bmatrix} N_{ix} \\ N_{iy} \\ N_{ixy} \end{Bmatrix} = \frac{E_i h_i}{1-\nu_i^2} \begin{bmatrix} 1 & \nu_i & 0 \\ \nu_i & 1 & 0 \\ 0 & 0 & \frac{1-\nu_i}{2} \end{bmatrix} \begin{Bmatrix} \varepsilon_{ix}^m \\ \varepsilon_{iy}^m \\ \gamma_{ixy}^m \end{Bmatrix},$$
$$\begin{Bmatrix} M_{ix} \\ M_{iy} \\ M_{ixy} \end{Bmatrix} = D_i \begin{bmatrix} 1 & \nu_i & 0 \\ \nu_i & 1 & 0 \\ 0 & 0 & 1-\nu_i \end{bmatrix} \begin{Bmatrix} \kappa_{ix} \\ \kappa_{iy} \\ \kappa_{ixy} \end{Bmatrix}, \tag{8}$$

where: $D_i = \dfrac{E_i h_i^3}{12\left(1-\nu_i^2\right)}$

- for i-th orthotropic strip or wall are:

$$\begin{Bmatrix} N_{ix} \\ N_{iy} \\ N_{ixy} \end{Bmatrix} = \frac{h_i}{1-\nu_{ixy}\nu_{iyx}} \begin{bmatrix} E_{ix} & \nu_{iyx}E_{ix} & 0 \\ \nu_{ixy}E_{iy} & E_{iy} & 0 \\ 0 & 0 & \left(1-\nu_{ixy}\nu_{iyx}\right)G_{ixy} \end{bmatrix} \begin{Bmatrix} \varepsilon_{ix}^m \\ \varepsilon_{iy}^m \\ \gamma_{ixy}^m \end{Bmatrix},$$
$$\begin{Bmatrix} M_{ix} \\ M_{iy} \\ M_{ixy} \end{Bmatrix} = \begin{bmatrix} D_{ix} & \nu_{iyx}D_{ix} & 0 \\ \nu_{ixy}D_{iy} & D_{iy} & 0 \\ 0 & 0 & D_{ixy} \end{bmatrix} \begin{Bmatrix} \kappa_{ix} \\ \kappa_{iy} \\ \kappa_{ixy} \end{Bmatrix}, \tag{9}$$

where: $D_{ix} = \dfrac{E_{ix} h_i^3}{12\left(1-\nu_{ixy}\nu_{iyx}\right)}$, $D_{iy} = \dfrac{E_{iy} h_i^3}{12\left(1-\nu_{ixy}\nu_{iyx}\right)}$, $D_{ixy} = \dfrac{G_{ixy} h_i^3}{6}$.

2.5. Dynamic equations of stability for thin plate

Differential equations of motion of the plate were derived basing on Hamilton's principle. It states that the dynamics of a physical system is determined by a variation problem for a functional based on a single function, the Lagrangian, which contains all physical information concerning the system and the forces acting on it. In dynamic buckling problem the motion should be understand as the time dependent deflection.

The Hamilton's principles for conservative systems states that the true evolution (compatible with constrains) of the system between two specific states in specific time range (t_0, t_1) is a stationary point (a point where the variation is zero) of the action functional Ψ. Action functional Ψ for i-th plate is described by following equation:

$$\Psi = \int_{t_0}^{t_1} \Lambda dt = \int_{t_0}^{t_1} (K - \Pi) dt \tag{10}$$

where Λ is the Lagrangian function for the system, K is a kinetic energy of the system and Π is a total potential energy of the system.

The subscript i denoting i-th plate or strip in all equations in this subchapter is omitted – all equations are presented for one plate, which could be i-th plate, wall or strip of considered plate, beam-columns or girder (Figure 1).

Taking the action functional Ψ in form (9) the Hamilton's principle can be written as:

$$\delta\Psi = \delta\int_{t_0}^{t_1} \Lambda dt = \delta\int_{t_0}^{t_1} (K - \Pi) dt = 0 . \tag{11}$$

The total potential energy variation $\delta\Pi$ for i-th thin plate (or strip) can be written in form:

$$\delta\Pi = \delta Q - \delta W \tag{12}$$

where δQ is a variation of internal elastic strain energy:

$$\delta Q = \int_{\Omega} (\sigma_x \delta\varepsilon_x + \sigma_y \delta\varepsilon_y + \tau_{xy} \delta\gamma_{xy})\, d\Omega , \tag{13}$$

and Ω is the volume of the plate and S is its area, the volume can be expressed as $l{\cdot}b{\cdot}h$ or $S{\cdot}h$.

The variation of internal elastic strain energy for i-th plate or strip could be expressed by strain and sectional forces and moments in a following way:

$$\begin{aligned} \delta Q &= \delta Q^m + \delta Q^b = \\ &= \int_S (N_x \delta\varepsilon_x^m + N_y \delta\varepsilon_y^m + N_{xy} \delta\gamma_{xy}^m) dS - \int_S (M_x \delta w_{,xx} + M_y \delta w_{,yy} + 2M_{xy} \delta w_{,xy}) dS. \end{aligned} \tag{14}$$

The work W of external forces (neglecting the out-of plane load) done on i-th plate can be written as follows:

$$W = \int_0^b h[p^0(y)u + \tau_{xy}^0(y)v]dy + \int_o^\ell h[p^0(x)v + \tau_{xy}^0(x)u]dx, \tag{15}$$

where: $p^0(x)$, $p^0(y)$, $\tau^0_{xy}(x)$, $\tau^0_{xy}(y)$ are the prebuckling load applied to the middle surface of the considered plate (wall or strip)

For thin plates, it is assumed that the displacements u and v do not depend on rotation $w_{,x}$ and $w_{,y}$ and therefore do not depend on the coordinate z. This approach results in exclusion of rotational inertia [50] in the equation for kinetic energy, which for the i-th thin plate (strip) can be written as:

$$K = \frac{1}{2}\rho \int_{\Omega} \left((\dot{u})^2 + (\dot{v})^2 + (\dot{w})^2 \right) d\Omega \tag{16}$$

The Hamilton's principle, it is the variation of the action functional $\delta\Psi$ (10) for *i*-th thin plate (strip or wall) which after taking into consideration equations from (11) to (15) can be written as:

$$\delta\Psi = \int_{t_0}^{t_1} \left(\delta K - \delta Q^m - \delta Q^b + \delta W \right) dt = 0 \tag{17}$$

The Lagrangian function for the whole system is equal to the sum of the Lagrangian functions of all n plates of which the system was composed. To determine the variation of action $\delta\Psi$ for *i*-th plate, the following identity:

$$X\ \delta Y = \delta(XY) - Y\delta X \tag{18}$$

was used.

In the obtained equation, terms with the same variations were grouped, and then each of the obtained groups of terms (due to the mutual independence of variations) were equated to zero, giving:

- equilibrium equations:

$$\begin{aligned}
&\int_{t_0}^{t_1}\int_S \{[N_{x,x} + N_{xy,y} + (N_x u_{,x})_{,x} + (N_y u_{,y})_{,y} + (N_{xy} u_{,x})_{,y} + (N_{xy} u_{,y})_{,x}] - h\rho \ddot{u}\}\delta u dS dt = 0 \\
&\int_{t_0}^{t_1}\int_S \{[N_{xy,x} + N_{y,y} + (N_x v_{,x})_{,x} + (N_y v_{,y})_{,y} + (N_{xy} v_{,x})_{,y} + (N_{xy} v_{,y})_{,x}] - h\rho \ddot{v}\}\delta v dS dt = 0 \\
&\int_{t_0}^{t_1}\int_S \{[M_{x,xx} + M_{y,yy} + 2M_{xy,xy} + (N_x w_{,x})_{,x} + (N_y w_{,y})_{,y} + (N_{xy} w_{,x})_{,y} + (N_{xy} w_{,y})_{,x}] + \\
&-h\rho \ddot{w}\}\delta w dS dt = 0
\end{aligned} \tag{19}$$

- boundary conditions for lateral edges of the plate (x = const):

$$\begin{aligned}
&\int_{t_0}^{t_1}\int_0^b [N_x + N_x u_{,x} + N_{xy} u_{,y} - hp^0(y)]\delta u dy dt\Big|_{x=const} = 0 \\
&\int_{t_0}^{t_1}\int_0^b [N_{xy} + N_x v_{,x} + N_{xy} v_{,y} - h\tau^0_{xy}(y)]\delta v dy dt\Big|_{x=const} = 0 \\
&\int_{t_0}^{t_1}\int_0^b M_x \delta w_{,x} dy dt\Big|_{x=const} = 0 \\
&\int_{t_0}^{t_1}\int_0^b (M_{x,x} + 2M_{xy,y} + N_x w_{,x} + N_{xy} w_{,y})\delta w dy dt\Big|_{x=const} = 0
\end{aligned} \tag{20}$$

- boundary conditions for longitudinal edges of the plate (y = const):

$$\begin{aligned}
&\int_{t_0}^{t_1}\int_0^{\ell}[N_y + N_y v_{,y} + N_{xy} v_{,x} - hp^0(x)]\delta v dx dt\Big|_{y=const} = 0 \\
&\int_{t_0}^{t_1}\int_0^{\ell}[N_{xy} + N_y u_{,y} + N_{xy} u_{,x} - h\tau_{xy}^0(x)]\delta u dx dt\Big|_{y=const} = 0 \\
&\int_{t_0}^{t_1}\int_0^{\ell} M_y \delta w_{,y} dx dt\Big|_{y=const} = 0 \\
&\int_{t_0}^{t_1}\int_0^{\ell}(M_{y,y} + 2M_{xy,x} + N_y w_{,y} + N_{xy} w_{,x})\delta w dx dt\Big|_{y=const} = 0
\end{aligned} \tag{21}$$

- boundary condition for the plate corners (x = const and y = const):

$$\int_{t_0}^{t_1} 2M_{xy}\delta w dt\Big|_{x=const}\Big|_{y=const} = 0 \tag{22}$$

- initial conditions for t = const:

$$\begin{aligned}
&\int_S h\rho \dot{u}\,\delta u dS\Big|_{t=const} = 0 \\
&\int_S h\rho \dot{v}\,\delta v dS\Big|_{t=const} = 0 \\
&\int_S h\rho \dot{w}\,\delta w dS\Big|_{t=const} = 0
\end{aligned} \tag{23}$$

Above conditions are fulfilled for the entire structure, so if one apply the restrictions in moment of the initial t_0 and in moment of the final t_1 that the displacement variations are zero at all points of the structure. Then the system of equations (23) vanishes.

- already used the relationship between deformations and internal forces and moments (8) or (9):

$$\begin{aligned}
&\int_{t_0}^{t_1}\int_S \left(E_x h\varepsilon_x - N_x + \nu_{xy} N_y\right)\delta N_x dS dt = 0 \\
&\int_{t_0}^{t_1}\int_S \left(E_y h\varepsilon_y + \nu_{yx} N_x - N_y\right)\delta N_y dS dt = 0 \\
&\int_{t_0}^{t_1}\int_S \left(2Gh\varepsilon_{xy} - N_{xy}\right)\delta N_{xy} dS dt = 0
\end{aligned} \tag{24}$$

$$\int_{t_0}^{t_1}\int_S \left(\frac{E_x h^3}{12}\kappa_x - M_x + \nu_{xy} M_y \right) \delta M_x dS dt = 0$$
$$\int_{t_0}^{t_1}\int_S \left(\frac{E_y h^3}{12}\kappa_y + \nu_{yx} M_x - M_y \right) \delta M_y dS dt = 0 \tag{25}$$
$$\int_{t_0}^{t_1}\int_S \left(\frac{G h^3}{6}\kappa_{xy} - M_{xy} \right) \delta M_{xy} dS dt = 0$$

3. Solution method

To determine the critical loads, natural frequencies and the coefficients of the equation describing the postbuckling equilibrium path, the analytical-numerical method has been employed. The proposed method also allows analysing dynamic response of the structure subjected to pulse loading. Taking the time courses of deflections and applying the relevant dynamic buckling criteria it is possible to determine the dynamic critical load.

3.1. Equilibrium equations

The differential equations of equilibrium for orthotropic plate or strip directly from the equations (19) can be derived and have the form:

$$N_{x,x} + N_{xy,y} + \{(N_x u_{,x})_{,x} + (N_y u_{,y})_{,y} + (N_{xy} u_{,x})_{,y} + (N_{xy} u_{,y})_{,x}\} - h\rho \ddot{u} = 0$$
$$N_{xy,x} + N_{y,y} + \{(N_x v_{,x})_{,x} + (N_y v_{,y})_{,y} + (N_{xy} v_{,x})_{,y} + (N_{xy} v_{,y})_{,x}\} - h\rho \ddot{v} = 0 \tag{26}$$
$$M_{x,xx} + M_{y,yy} + 2M_{xy,xy} + (N_x w_{,x})_{,x} + (N_y w_{,y})_{,y} + (N_{xy} w_{,x})_{,y} + (N_{xy} w_{,y})_{,x} - h\rho \ddot{w} = 0$$

Above equilibrium equations after omitting the inertia forces $h\rho\ddot{u}$, $h\rho\ddot{v}$ and $h\rho\ddot{w}$ becomes the equilibrium equations for thin plates allowing analysis of both local and global buckling mode.

3.2. Boundary and initial condition

As the wave propagation effects have been neglected, the boundary conditions referring to the simply supported columns at their both ends, i.e. $x = 0$ and $x = l$, according to (20), are assumed to be:

$$\begin{aligned}
&\frac{1}{b_i}\int N_{ix}\left(x_i=0,y_i,t\right)dy_i=\frac{1}{b_i}\int N_{ix}\left(x_i=l,y_i,t\right)dy_i=N_{ix}^{(0)},\\
&v_i\left(x_i=0,y_i,t\right)=v_i\left(x_i=l,y_i,t\right)=0,\\
&w_i\left(x_i=0,y_i,t\right)=w_i\left(x_i=l,y_i,t\right)=0,\\
&M_{ix}\left(x_i=0,y_i,t\right)=M_{iy}\left(x_i=l,y_i,t\right)=0,
\end{aligned} \tag{27}$$

The condition written as a first of equations of (27) is satisfied for the prebuckling state and first-order approximation, the condition for deflection v (27) is satisfied for the first and second order of approximations, while the other two conditions are met for prebuckling state as well as for the first and second order of approximation. The condition of displacement in the y direction in the prebuckling state can be found for example in [51]. This approach allows to take into account the Poisson effect on the edges of the walls of the column. The boundary conditions described by equations (27) assume the lack of displacement possibility of points lying at the loaded edges in the transverse v and normal w directions to the surface in a wall or column. Furthermore, it is assumed that the moments M_{ix} (as a vector parallel to the edge of the plate or end edge of the column walls) are zero.

For structures with material properties varying widthwise the strip model was adopted what forces the boundary conditions modification in the second order approximations [51]. Modification consists of changing the first condition of (27) onto the following form:

$$\sum_{i=1}^{J}\frac{1}{b_i}\int_0^{b_i}N_{ix}^{(2)}dy_i\bigg|_{x=0;l}=0 \tag{28}$$

Summation is performed only for the J number of the strips, between which the angle $\varphi_{i,i+1}$ (Figure 3) is equal to zero.

To determine the boundary conditions on the longitudinal edges of plates or free edges of columns with open cross-sections the equations (20) were used. Whereas, directly from equations (23) result the following initial conditions:

$$\begin{aligned}
&\dot{u}_i(x_i,y_i,t=t_0)=\tilde{u}_i(x_i,y_i) \quad \text{and} \quad u_i(x_i,y_i,t=t_0)=\overline{u}_i(x_i,y_i),\\
&\dot{v}_i(x_i,y_i,t=t_0)=\tilde{v}_i(x_i,y_i) \quad \text{and} \quad v_i(x_i,y_i,t=t_0)=\overline{v}_i(x_i,y_i),\\
&\dot{w}_i(x_i,y_i,t=t_0)=\tilde{w}_i(x_i,y_i) \quad \text{and} \quad w_i(x_i,y_i,t=t_0)=\overline{w}_i(x_i,y_i),
\end{aligned} \tag{29}$$

where the following functions $\overline{u}_i,\ \overline{v}_i,\ \overline{w}_i,\ \tilde{u}_i,\ \tilde{v}_i,\ \tilde{w}_i$ are given for the initial moment $t = t_0$.

3.3. Interaction condition between adjacent plates

Static and kinematic junction conditions on the longitudinal edges of adjacent plates (Figure 3), according to (21), can be written as:

$$
\begin{aligned}
& u_{i+1}\big|^{-} = u_i\big|^{+}, \\
& w_{i+1}\big|^{-} = w_i\big|^{+}\cos(\phi) - v_i\big|^{+}\sin(\phi), \\
& v_{i+1}\big|^{-} = w_i\big|^{+}\sin(\phi) + v_i\big|^{+}\cos(\phi), \\
& w_{i+1,y}\big|^{-} = w_{i,y}\big|^{+}, \\
& M_{(i+1)y}\big|^{-} = M_{iy}\big|^{+}, \\
& N^{*}_{(i+1)y}\big|^{-} - N^{*}_{iy}\big|^{+}\cos(\phi) - Q^{*}_{iy}\big|^{+}\sin(\phi) = 0, \\
& Q^{*}_{(i+1)y}\big|^{-} + N^{*}_{iy}\big|^{+}\sin(\phi) - Q^{*}_{iy}\big|^{+}\cos(\phi) = 0, \\
& N^{*}_{(i+1)xy}\big|^{-} = N^{*}_{ixy}\big|^{+},
\end{aligned}
\tag{30}
$$

where:

$$
\begin{aligned}
& N^{*}_{ixy} = N_{iy} + N_{iy}v_{i,y} + N_{ixy}v_{i,x} \\
& N^{*}_{ixy} = N_{ixy} + N_{ixy}u_{i,x} + N_{iy}u_{i,y} \\
& M_{iy} = -\eta_i D_i(w_{i,yy} + \nu_i w_{i,xx}), \\
& Q^{*}_{iy} = -\eta_i D_i w_{i,yyy} - (\nu_i \eta_i D_i + 2D_{1i})w_{i,xxy} + N_{iy}w_{i,y} + N_{ixy}w_{i,x}, \\
& \phi = \phi_{i;i+1}.
\end{aligned}
\tag{31}
$$

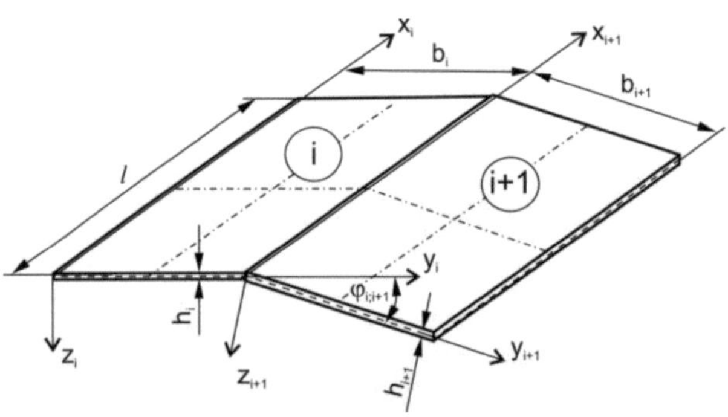

Figure 3. The geometrical dimensions and local coordinate systems adjacent plates

3.4. Buckling and postbuckling equilibrium paths

A non-linear stability problem has been solved by means of the Koiter's asymptotic theory. The displacement field $\bar{U}$, and sectional force field $\bar{N}$ have been expanded into the power series with respect to the parameter ξ, - the buckling linear eigenvector amplitude (normalised with the equality condition between the maximum deflection and the thickness of the first plate h_1).

$$\bar{U} = \lambda\bar{U}^{(0)} + \xi_i\bar{U}^{(i)} + \xi_i\xi_j\bar{U}^{(ij)} + ...$$
$$\bar{N} = \lambda\bar{N}^{(0)} + \xi_i\bar{N}^{(i)} + \xi_i\xi_j\bar{N}^{(ij)} + ... \quad (32)$$

It was assumed that the dimensionless amplitude of the initial deflections (imperfections) correspond to the considered buckling mode (for s-th buckling mode) is:

$$\bar{U} = \xi_s^*\bar{U}^{(i)}. \quad (33)$$

By substituting expansions (32) into equations of equilibrium (26) with neglected inertia terms (static buckling problem), junction conditions (30) and boundary conditions (27), the boundary problem of the zero (superscript $^{(0)}$ in Equations (32) and further), first (superscript $^{(i)}$) and second (superscript $^{(ij)}$) order has been obtained [18, 50, 52, 53]. The zero approximation describes the prebuckling state, whereas the first order approximation allows for determination of critical loads and the buckling modes corresponding to them, taking into account minimisation with respect to the number of half-waves m in the lengthwise direction. The second order approximation is reduced to a linear system of differential heterogeneous equations, which right-hand sides depend on the force field and the first order displacements only.

The most important advantage of this method is that it enables us to describe a complete range of behaviour of thin-walled structures from all global (i.e. flexural, flexural–torsional, lateral, distortional buckling and their combinations) to the local dynamic stability. In the solution obtained, the shear lag phenomenon, the effect of cross-sectional distortions and also the interaction between all the walls of structures are included.

Having found the solutions to the first and second order of the boundary problem, the coefficients a_{ijs}, b_{ijks} have been determined [18, 50, 52, 53]:

$$a_{ijs} = \frac{\boldsymbol{\sigma}^{(i)} * L_{11}(\mathbf{U}^{(j)},\mathbf{U}^{(s)}) + 0.5\boldsymbol{\sigma}^{(s)} * L_{11}(\mathbf{U}^{(i)},\mathbf{U}^{(j)})}{-\lambda_s\boldsymbol{\sigma}^{(0)} * L_2(\mathbf{U}^{(s)})},$$

$$b_{ijks} = \frac{2\boldsymbol{\sigma}^{(i)} * L_{11}(\mathbf{U}^{(jk)},\mathbf{U}^{(s)}) + \boldsymbol{\sigma}^{(ij)} * L_{11}(\mathbf{U}^{(k)},\mathbf{U}^{(s)})}{-\lambda_s\boldsymbol{\sigma}^{(0)} * L_2(\mathbf{U}^{(s)})}, \quad (34)$$

where: λ_s – is the critical load corresponding to the s-th mode, L_{11} is the bilinear operator, L_2 is the quadratic operator and $\boldsymbol{\sigma}^{(i)}$, $\boldsymbol{\sigma}^{(ij)}$ are the stress field tensors in the first and second order.

The postbuckling static equilibrium paths for coupled buckling can be described by the equation:

$$\left(1 - \frac{\lambda}{\lambda_s}\right)\xi_s + a_{ijs}\xi_i\xi_j + b_{ijks}\xi_i\xi_j\xi_k = \frac{\lambda}{\lambda_s}\xi_s^*; \quad (s = 1,...,N), \quad (35)$$

which for the uncoupled problem have the form:

$$\left(1-\frac{\lambda}{\lambda_{cr}}\right)\xi + a_{111}\xi^2 + b_{1111}\xi^3 = \xi^* \frac{\lambda}{\lambda_{cr}} \tag{36}$$

where λ_{cr} is the critical load value.

In a special case, i.e. for the so-called ideal structure without initial imperfections ($\xi^*=0$) and when the equilibrium path (a_{111}) is symmetrical, the postbuckling equilibrium path is defined by the equation:

$$\frac{\lambda}{\lambda_{cr}} = 1 + b_{1111}\xi^2 \tag{37}$$

3.5. Natural frequencies

Determination of the natural frequencies is similar to the determination of critical buckling load and the natural frequencies are found by solving the eigenvalue problem.

Natural frequencies of thin-walled structures were determined by solving a dynamic problem, which uses the approach proposed by Koiter in his asymptotic stability theory of conservative systems in the first-order approximation [52].

To determine the natural frequencies [55] of the structure the adopted equilibrium equations (26) contain cross-sectional inertia forces acting in the direction normal to the middle surface of the plate (column wall) and in the middle plane of plate (i.e. $h\rho\ddot{u} \neq 0$ and $h\rho\ddot{v} \neq 0$).

3.6. Lagrange equations

In the dynamic analysis (while finding the frequency of natural vibrations [55]), the independent non-dimensional displacement ξ and the load factor λ become a function dependent on time, and dynamic terms were added to equations describing postbuckling equilibrium path. Neglecting the forces associated with the inertia terms of prebuckling state and the second-order approximations, and taking into account the orthogonality conditions for the displacement field in the first $\bar{U}^{(i)}$ and second-order approximation $\bar{U}^{(ij)}$, the Lagrange equations can be written as [56]:

$$\frac{1}{\omega_s^2}\ddot{\xi}_s + \left(1-\frac{\lambda}{\lambda_s}\right)\xi_s + a_{ijs}\xi_i\xi_j + b_{ijks}\xi_i\xi_j\xi_k = \xi_s^* \frac{\lambda}{\lambda_s}; \ \left(s = 1,2,\ \ldots,N\right) \tag{38}$$

where ω_s is a natural frequency with mode corresponding to buckling mode; a_{ijs} and b_{ijks} are the coefficients (34) describing the postbuckling behaviour of the structure (independent of time); however the parameters of load λ and the displacement ξ are the functions of time *t*.

For the uncoupled buckling, i.e. the single-mode buckling (where index $s = N = 1$), the equations of motion may be written in the form:

$$\frac{1}{\omega_1^2}\ddot{\xi}_1 + \left(1-\frac{\lambda}{\lambda_1}\right)\xi_1 + a_{111}\xi_1^2 + b_{1111}\xi_1^3 = \xi_1^*\frac{\lambda}{\lambda_1}; \tag{39}$$

It is assumed that in the initial moment of time $t = 0$ the non-dimensional displacement ξ, as well as the velocity of displacement are equal to zero, i.e.:

$$\xi(t=0)=0 \;\text{ and }\; \dot{\xi}(t=0)=0\,. \tag{40}$$

The Runge-Kutta method [57] for solving the equation (39) requires the following substitutions:

$$\begin{aligned} &\dot{\xi} = \Gamma(t), \\ &\dot{\Gamma} = -\omega_1^2\left(1-\frac{\lambda(t)}{\lambda_1}\right)\xi - \omega_1^2 b_{111}\xi_1^2 - \omega_1^2 b_{1111}\xi_1^3 + \omega_1^2\frac{\lambda(t)}{\lambda_1}\xi^*, \end{aligned} \tag{41}$$

which lead to the system of two differential equations. ''Complete'' equations of motion (41) are solved with the numerical Runge–Kutta method of order 8 (5,3), thanks to Dormand and Price (with step-size control and density output).

4. Exemplary results of calculations

The exemplary results of numerical calculation are presented in this sub-chapter. All results are obtained using explained above proposed analytical-numerical method (ANM) based on the nonlinear orthotropic plate theory.

The material properties (E – Young modulus, ν – Poisson ratio, G=E/[2(1+ν)] – Kirchhoff modulus; ρ – density) for materials taken into account are presented in Table 1.

material type:	E [GPa]	ν	ρ [kg/m³]
steel	200	0.3	7850
aluminium	70	0.33	2950
epoxy resin	3.5	0.33	1249
glass fibre	71	0.22	2450

Table 1. Assumed material properties

The fibre composite material was modelled as orthotropic but for components (resin and fibre) the isotropic material properties (Table 1) was assumed. Necessary equations for material properties homogenization based on theory of mixture [57, 58] are as follows:

$$
\begin{aligned}
E_x &= E_m(1-f) + E_f f,\\
E_y &= E_m \frac{E_m\left(1-\sqrt{f}\right)+E_f\sqrt{f}}{E_m\left[1-\sqrt{f}\left(1-\sqrt{f}\right)\right]+E_f\sqrt{f}\left(1-\sqrt{f}\right)},\\
\nu_{yx} &= \nu_m\left(1-\sqrt{f}\right)+\nu_f\sqrt{f},\\
G &= G_m \frac{G_m\sqrt{f}\left(1-\sqrt{f}\right)+G_f\left[1-\sqrt{f}\left(1-\sqrt{f}\right)\right]}{G_m\sqrt{f}+G_f\left(1-\sqrt{f}\right)}.
\end{aligned}
\tag{42}
$$

where E_m and E_f are the Young's modulus of elasticity for matrix and fibre, respectively, G_m and G_f are the shear modulus for matrix (subscript m) and fibre (subscript *f*), ν_m and ν_f are the Poisson's ratios for matrix and fibre and $f = V_f/(V_m + V_f)$ is the fibre volume fraction.

For static buckling the critical buckling load and corresponding modes are presented as well as the postbuckling equilibrium paths.

For dynamic buckling the proposed by Budiansky and Hutchinson parameter called Dynamic Load Factor DLF is introduced. The DLF is defined as a ratio of pulse loading amplitude to static buckling load. The results are presented of nondimensional deflection ξ versus DLF. The critical dynamic load factor DLF_{cr} corresponding to dynamic buckling has been estimated using different criteria – the obtained results were compared.

For the proposed method the validation of the results was made by comparison with the other Authors [34] calculations (Figure 4) or with the results obtained with FEM [38]. The results presented in Figure 4 were obtained for thin (ratio length to thickness equals 200) aluminium square plate simply supported at all edges and subjected to sinusoidal pulse load. The time of pulse duration was equal to the period of natural vibration of the plate. The considered plate has a geometrical imperfection corresponding to buckling mode with amplitude equal to 0.05 of the plate thickness.

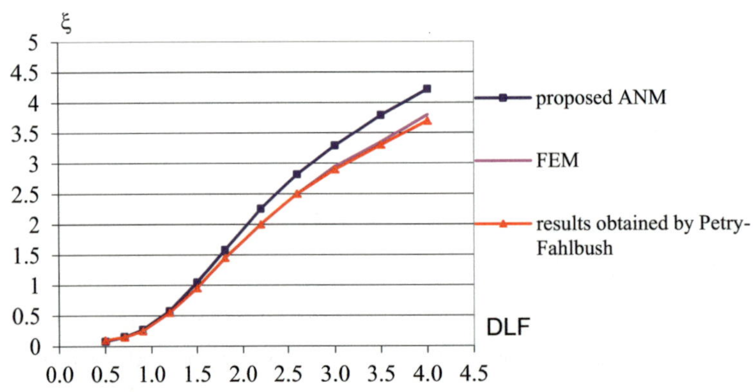

Figure 4. The results of different calculation comparison

4.1. Plates

The rectangular thin plates simply supported on loaded edges with different boundary conditions along the unloaded ones were considered (Figure 5). On the longitudinal edges five different boundary condition cases were taken into account. Following notations is used in Figure 5: s – simply supported edge, c – clamped edge, e – free edge.

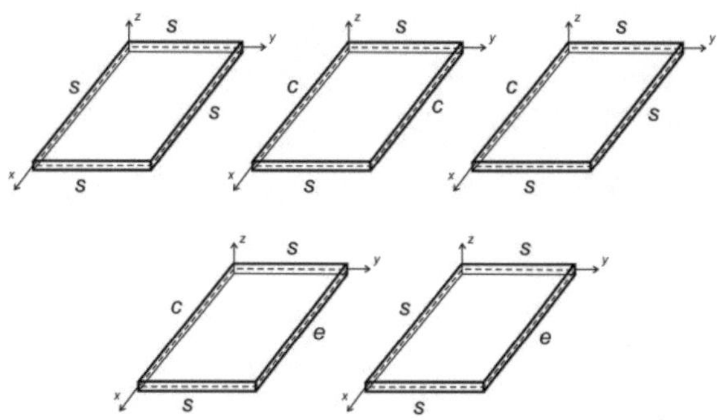

Figure 5. Analysed plates with different boundary conditions

material:	boundary condition	P_{cr} [kN]		ω [rad/s]	
		ANM	FEM	ANM	FEM
steel	ss	7.23	7.24	3016	3010
	cc	15.6 ($m = 1$) 13.9 ($m = 2$)	15.7 14.0	4423 ($m = 1$) 8363 ($m = 2$)	4423 8344
	se	2.53	2.54	1784	1784
	ce	2.99	2.99	1935	1935
	sc	10.38	10.41	3613	3607
composite $f = 0.5$	ss	0.54	0.54	1703	1709
	cc	0.93	0.95	2231	2237
	se	0.34	0.35	1351	1351
	ce	0.36	0.37	1389	1389
	sc	0.69	0.70	1916	1923

Table 2. Critical load P_{cr} and natural frequencies ω for analysed plates

Exemplary results were calculated for steel and epoxy glass composite (fibre volume factor $f = 0.5$) square plates subjected to rectangular compressive pulse loading. The buckling load for plate under analysis is presented in Table 2. The pulse duration T_P was equal to the period of natural vibration with mode corresponding to the buckling mode.

The dimensions of analysed plates were assumed as follows: the length (width) a= b= 100 mm and thickness $h = 1$ mm.

The geometrical imperfection was assumed in the shape corresponding to the buckling mode with amplitude $\xi^* = 0.01$, where ξ^* is an amplitude of deflection divided by the plate thickness. The Figure 6 presents postbuckling equilibrium paths for ideal flat composite plate (Figure 6a) and for plate with geometrical imperfection with amplitude $\xi^* = 0.01$ (Figure 6b).

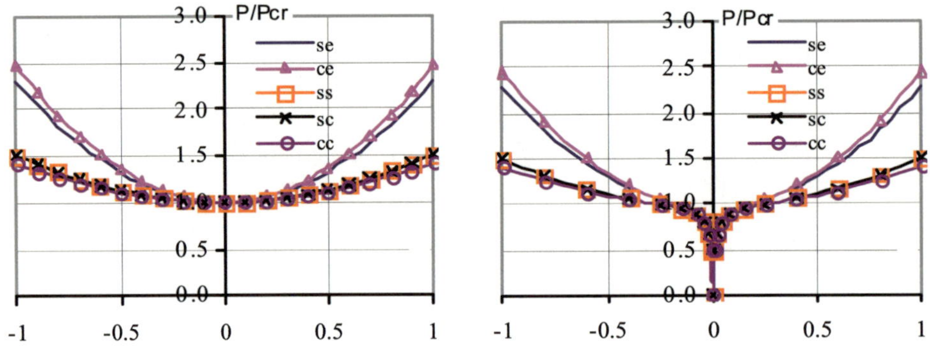

a) b)

Figure 6. Postbuckling equilibrium paths for square ideal plates (a) and plates with imperfection (b) with different boundary conditions on non-loaded edges

In the dynamic buckling case the results are shown as graphs presenting nondimensional deflection ξ or radius r calculated from real and imaginary part of maximal characteristic root of Jacoby matrix as a function of dynamic load factor DLF. The graphs mentioned above allow to find critical amplitude of pulse loading using the proposed criterion (PC) [35] and to compare the obtained results with Budiansky-Hutchinson (B-H) or Volmir (V) criteria. In brackets the notation used in Figures and Tables is given. The critical deflection according to Volmir criterion was assumed as $\xi_{cr} = 1$.

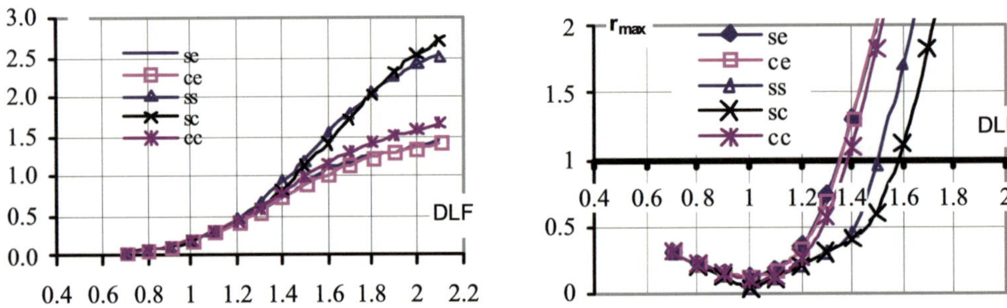

a) b)

Figure 7. Nondimensional deflection ξ (a) and maximum radius r_{max} (b) vs. DLF for square plates with different boundary conditions on non-loaded edges [35]

Basing on curves presented in Figure 7 the critical value of dynamic load factor can be found. The comparison of obtained critical DLF values using different criteria is presented in Table 3. All critical DLF values except the case denoted as *se* obtained from the proposed criterion (PC) are in the range obtained from Budiansky-Hutchinson criterion (B-H). For the case denoted as **se** the greatest differences between critical dynamic load factors from the proposed and Budiansky-Hutchinson criterion were obtained but these differences are less than 10%.

boundary conditions for unloaded edges	mode	B-H	V	PC
se	m=1	1.5÷1.6	1.55	1.35
ce	m=1	1.3÷1.4	1.58	1.35
ss	m=1	1.5÷1.6	1.43	1.51
sc	m=1	1.4÷1.8	1.46	1.58
cc	m=2	1.4÷1.5	1.51	1.39

Table 3. Comparison of DLF_{cr} obtained from different criteria for compressed plate

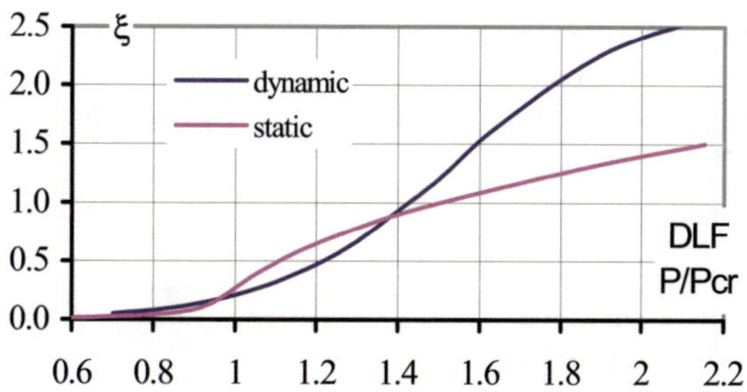

Figure 8. Nondimensional deflection ξ vs. DLF for simply supported square plate

The comparison for postbuckling behaviour of the rectangular plate simply supported at all edges subjected to static and dynamic load are obtained using proposed analytical-numerical method and presented in Figure 8.

4.2. Segments of the girders

As a next example the static and dynamic buckling of composite (epoxy glass composite with different volume fibre fraction f) girders with open cross-section (Figure 9) is presented. The assumed boundary conditions on loaded edges correspond to simply support. The calculation was carried out for short segment of girder with length to web width ratio $l/b_1 = 1$ and for the following dimensions of the cross-section: $b_1/h = 50$, $b_2/h = 25$ and $b_3/h = 12.5$.

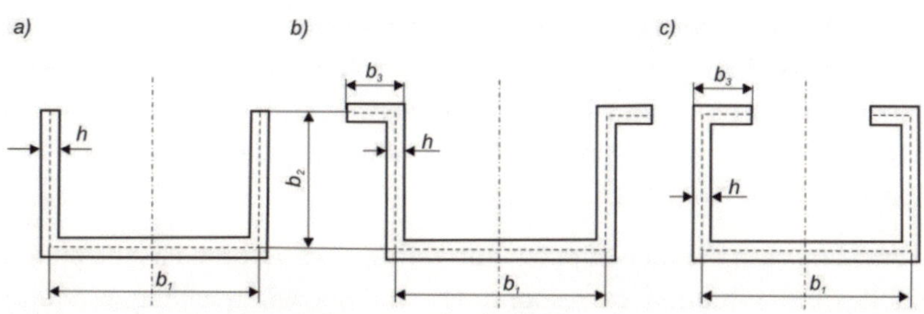

Figure 9. Cross-sections of analysed segment of the girders

The geometrical imperfection was assumed in the shape corresponding to the buckling mode with amplitude $\xi^* = 0.01$. The static buckling load and fundamental flexural natural frequency obtained with analytical-numerical method for girders made of composite with different fibre fraction are presented in Tables 4. The static critical buckling loads are presented in Table 4 and postbuckling equilibrium paths are presented in Figure 10.

volume fibre fraction f: / cross-section	critical load P_{cr} [N]		natural frequency ω [rad/s]	
	0.4	0.6	0.4	0.6
channel (Figure 9a)	1526	2281	1076	1076
channel with inner stiffeners (Figure 9c)	2821	4217	1308	1308
omega (Figure 9b)	2819	4214	1308	1308

Table 4. Critical load and natural frequencies for analysed girder's segments

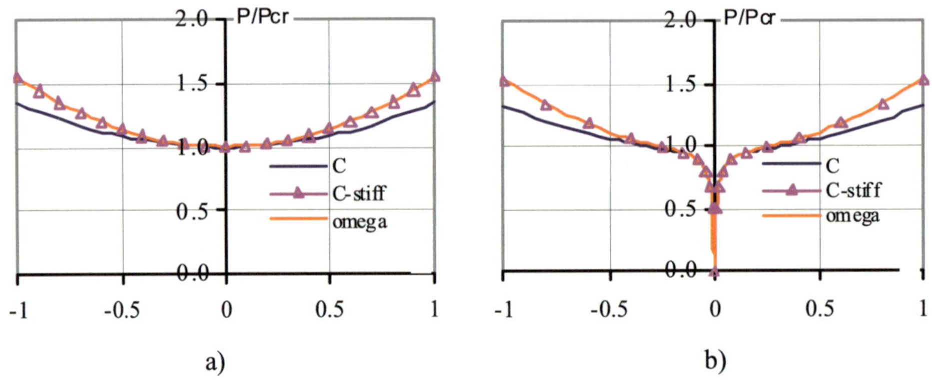

Figure 10. Postbuckling equilibrium paths for segment of girders

Buckling load and natural frequency for girder segment with omega and stiffened channel cross-section are similar – it is true only for local buckling case. For girder with channel cross-section the buckling was caused by flanges – this is a reason why for this cross-section the buckling load and natural frequency are smaller than for two others analysed cross-sections. Looking at obtained results (Table 4) it can be seen that increasing the volume fibre fraction f leads to an increasing the buckling loads as well as the natural frequencies. The postbuckling equilibrium paths for stiffened cross-section (channel with inner stiffeners and omega) overlap. The postbuckling path for channel cross-section lies below the equilibrium paths of girders with stiffened cross-section – it is obvious because the girders with stiffened cross-section have similar stiffness and the girder with channel cross-section is more flexible.

In dynamic buckling case the time of pulse duration T_p was assumed as the period of natural fundamental flexural vibration corresponding to the local buckling mode. Considered shapes of pulse loading are presented in Figure 11.

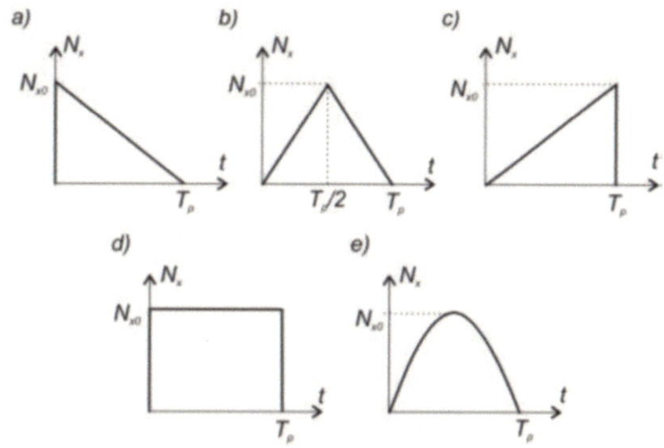

Figure 11. Shapes of pulse loading: three triangular impulses T1 (a), T2 (b), T3 (c), rectangular R (d) and sinusoidal S (e)

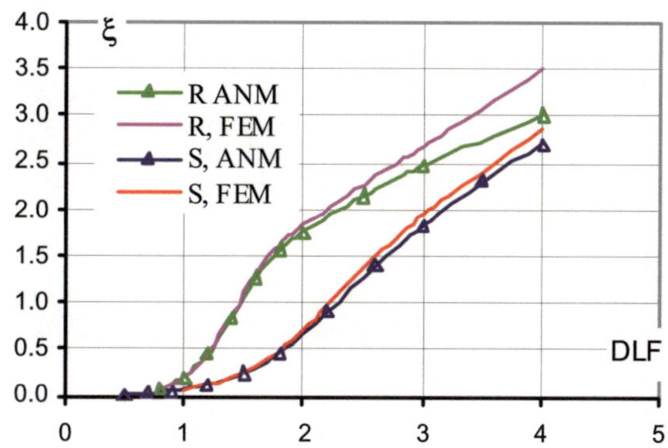

Figure 12. Dimensionless deflection vs. *DLF* for channel girder – results comparison obtained with analytical-numerical method ANM and finite element method FEM [56]

The results presented in Figure 12 were obtained with the proposed analytical-numerical method (ANM) and compared with FEM computations. They are similar for both assumed shapes of pulse loading (Figure 11): rectangular (R) and sinusoidal (S). Nevertheless, for rectangular shape pulse loading some small differences in deflection are visible for *DLF* greater than 2. It should be noted that obtained curves (Figure 12) from both methods allow to find the same critical dynamic load factor DLF_{cr} using Budiansky-Hutchinson or Volmir criterion – for rectangular pulse loading $DLF_{cr} \approx 1.4$ (both criteria) and for sinusoidal pulse loading $DLF_{cr} \approx 2.1$ (Budiansky-Hutchinson) or 2.3 (Volmir).

In Figure 13 the dynamic response comparison of girders made of composite material (f = 0.5) with different cross-section subjected to triangular T3 pulse was presented. The curves for omega cross-section and channel cross-section with inner stiffeners cover each other's.

Dynamic responses for girder with channel cross-section with inner stiffeners for different pulse loading are presented in Figure 14. The curves denoted by *S=R* were obtained for sinusoidal pulse loading with the same area as rectangular pulse loading (for the same pulse duration the amplitude was higher for sinusoidal pulse). The highest deflection was

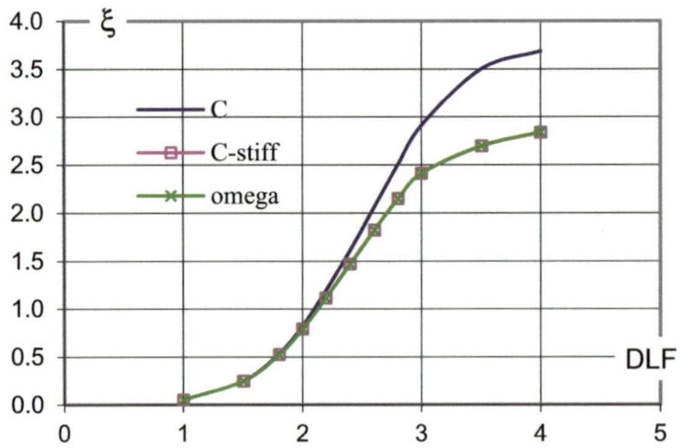

Figure 13. Nondimensional deflection ξ as a DLF function for girder with different channel cross-section subjected to T3 pulse loading [56]

obtained for pulse denoted by *S=R* because this pulse has the highest amplitude. The rest of compared pulses have the same amplitude and the same duration. Analysing the curves ξ(*DLF*) for rectangular, sinusoidal and three triangular impulses it can be seen that the highest increment of deflection for the smallest *DLF* takes place for rectangular pulse loading.

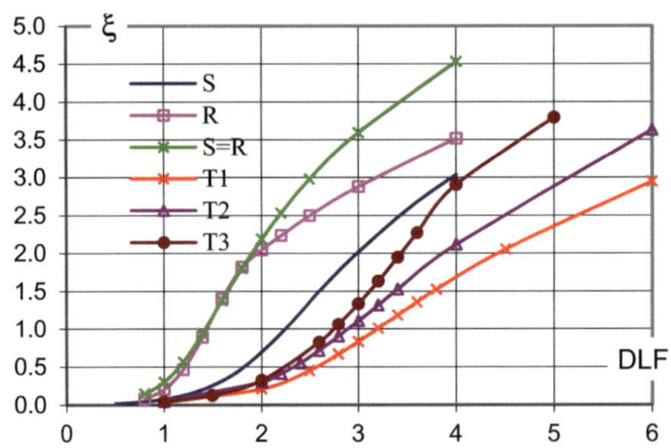

Figure 14. Dimensionless deflection vs. dynamic load factor for different shapes of pulse loading - channel with inner stiffeners, composite material $f = 0.7$ [56]

The comparison of DLF_{cr} obtained using Budiansky-Hutchinson criterion for girders with different cross-section made of composite materials ($f = 0.5$) are presented in Table 5. Only the average values from the obtained critical ranges are presented. The dynamic load factors for different cross-sections are in the same relation as buckling loads (Table 4) – the same or similar DLF_{cr} for cross-sections with stiffeners (omega and channel with inner stiffeners).

4.3. Columns

Next, the exemplary results for the dynamic interactive buckling of channel cross-section the columns are presented. The columns subjected to rectangular compressive pulse loading were analysed. The calculation was carried out for columns with various length to web width ratio $l/b_1 = 4$; 6 and 8 and for the following dimensions of the cross-section: width of the web to its thickness $b_1/h = 100$, width of the flange to its thickness $b_2/h = 50$.

analysed cross-section: / type of pulse	channel	omega	channel with inner stiffeners
S	2.0	2.1	2.1
R	1.6	1.6	1.6
S=R	1.4	1.4	1.4
T1	3.1	3.1	3.1
T2	2.3	2.5	2.5
T3	2.4	2.5	2.5

Table 5. The DLF_{cr} value for different shape of applied pulses

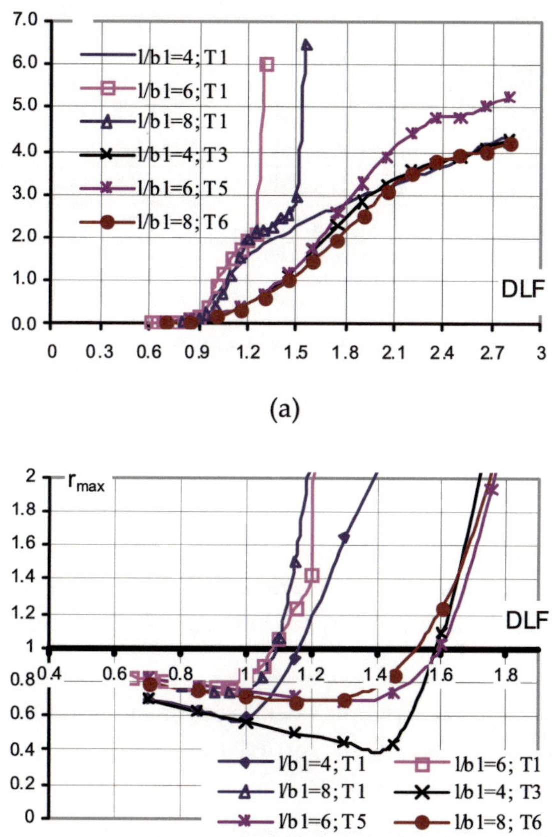

(a)

(b)

Figure 15. Nondimensional deflection ξ (a) and maximum radius r_{max} (b) as a function of DLF for channel beam-columns with different length ratio l/b_1 and pulse duration T_p [35]

The interaction between global buckling mode $m = 1$ and the first local buckling mode $m > 1$ was considered. The geometrical imperfection were assumed in the shape corresponding to the buckling mode with amplitude $\xi^* \equiv \xi_2^*$equals 1/100 wall thickness for local mode and ξ_1^* equal to length to one thousand wall thickness ($l/1000h$) for global mode. Time of pulse duration T_p was assumed as T_1 equal to the period of natural fundamental flexural vibration or T_m (where m is a number of half waves of local buckling mode) equal to the period of natural vibration with mode corresponding to the local buckling mode.

Figure 15 presents dimensionless deflection ξ as a function of dynamic load factor and maximal radius r_{max} calculated for maximal characteristic root of Jacoby matrix as a function of DLF.

From curves presented in Figure 15a the critical value of dynamic load factor based on Volmir (V) or Budiansky-Hutchinson (B-H) criterion can be found. The curves presented in Figure 15b help to find critical DLF value based on proposed criterion (PC) [35]. The obtained critical DLF's according to mentioned above criteria are presented in Table 6.

columns length ratio	pulse duration T_p [ms]	buckling modes	B-H	V	PC
$l/b_1 = 4$	T_1= 1.6	m=1; 3 local	1.0÷1.15	1.07	1.16
$l/b_1 = 6$	T_1= 1.9	m=1; 5 local	0.95÷1.0	1.01	1.08
$l/b_1 = 8$	T_1= 2.7	m=1; 6 local	1.1÷1.15	1.09	1.09
$l/b_1 = 4$	T_3= 0.8	m=1; 3 local	1.6÷1.75	1.40	1.58
$l/b_1 = 6$	T_5= 0.7	m=1; 5 local	1.6÷1.75	1.39	1.59
$l/b_1 = 8$	T_6= 0.8	m=1; 6 local	1.6÷2.05	1.43	1.51

Table 6. Comparison of DLF_{cr} obtained from different criteria for interactive buckling

The comparison of the obtained results shows that they are in good agreement. In all cases with pulse duration equal to the period of natural vibration of the form corresponding to local buckling mode the results obtained from the proposed criterion (PC) are between the results obtained from Volmir (V) and Budiansky-Hutchinson (B-H) criteria. For loading with time of pulse duration equal to the period of natural fundamental vibration T_1 the critical DLF values obtained using the proposed criterion (PC) are equal or a bit greater (about 6%) than the critical dynamic load factors from Budiansky-Hutchinson (B-H) or Volmir (V) criteria.

Should be pointed out that in the dynamic buckling problem also for the short columns the multimodal buckling analysis should be carried out. It has been proven on exemplary channel columns with following dimensions: $b_1/h = 100$, $b_2/h = 50$, $b_3/h = 25$ and $l/b_1 = 4$.

The problem has been calculated with the analytical-numerical method and the finite element method [56].

The dimensionless deflection ξ as a function of dimensionless time (time divided by pulse duration) for channel column is presented in Figure 16. The characteristic points are located in the middle cross-section of the columns and in the middle of the web (point 1 – Figure 16) and on the edge between the web and the flange (point 2 – Figure 16).

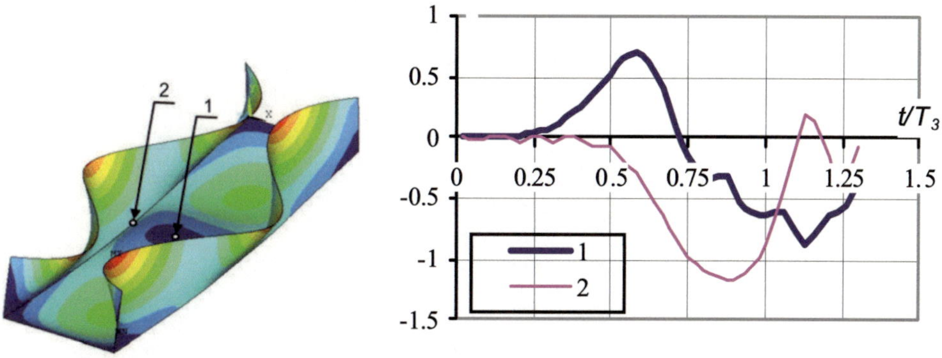

Figure 16. Deflection in time for channel columns subjected to rectangular pulse loading with the DLF = 1.6

Analysing the results presented in Figure 16, it can be said that the global buckling appears for channel cross-section columns. The column edge deflections are greater than deflections of the middle part of the web. The FEM results of calculations presented in Figure 16 have initiated the need for a multimodal buckling analysis also for short columns subjected to pulse loading.

Results for linear buckling and modal analyses obtained with proposed analytical-numerical method are presented in Table 7. As it will be presented below (see Figure 17) the four modes should take into consideration in ANM to obtain similar results to this obtained with FEM. The finite element method gives results (global mode) even in the case when only one buckling mode as the initial imperfection (for example, the local buckling mode m=3) has been taken into account [56].

Mode	σ_{cr} [MPa]	n [Hz]
local mode m = 3	53	614
primary local mode m = 1	123	312
secondary local mode m = 1′	972	880
global mode m = 1″	5122	2001

Table 7. Buckling stress and natural vibration for the channel column

A comparison between the results obtained with the analytical-numerical method and the finite element method on plots presenting a dimensionless deflection vs. a dynamic load factor for columns with channel cross-sections are shown in Figure 17. Some differences appear because the analytical-numerical model has only a few degrees of freedom in contrary to FE model, which has thousands DoF. However the curves presented in Figure 17 are different the critical DLF values estimated using the Budiasky-Hutchinson criterion is similar. From the ANM, the DLF_{cr} = 2.7 and from the FEM, the DLF_{cr} = 2.6.

Good agreement between the results obtained with ANM and FEM is possible because the interactive dynamic buckling problem has been solved in the analytical-numerical method. Four modes have been taken into account. The buckling stress and the natural frequency obtained with the analytical-numerical method for all the modes assumed in the multimodal analysis are listed in Table 7. The buckling modes taken into consideration in the multimodal analysis are presented in Figure 18, correspondingly.

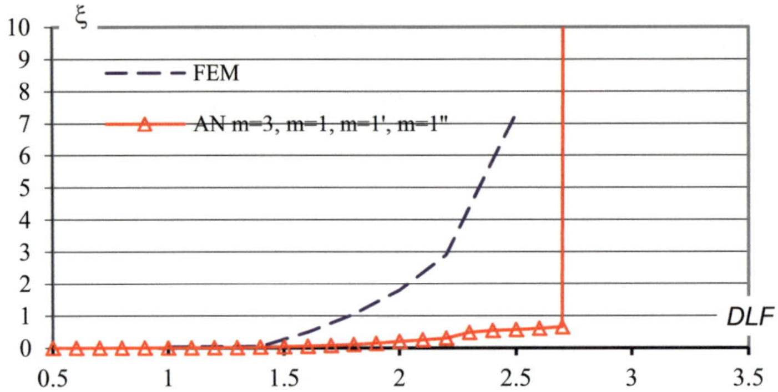

Figure 17. Dimensionless edge deflection ξ vs. the *DLF* for channel columns subjected to rectangular pulse loading $T_p = T_3 = 1.6$ ms [56]

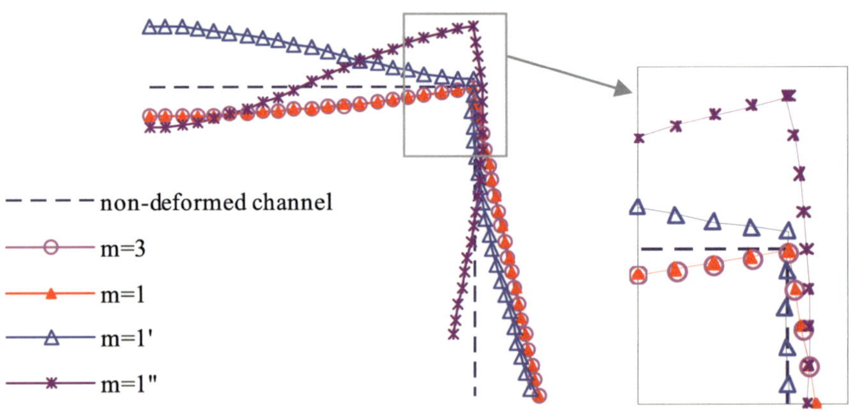

Figure 18. Buckling modes for the channel cross-section column

5. Conclusion

Taken into consideration the nonlinear thin plate theory for orthotropic material allows, as is presented in exemplary results of calculation, to analyse thin-walled structures composed of flat plates and subjected to static and dynamic load. The nonlinear orthotropic plate theory is the base for the proposed analytical-numerical method which allows to find buckling load with corresponding buckling mode, natural frequencies with corresponding modes and to analyse the postbuckling behaviour – drawing the postbuckling equilibrium paths for plate, segment of girders or columns made of isotropic, orthotropic or even composite materials. As it was shown not only static load can be considered but also dynamic load with intermediate velocity – the dynamic buckling can be analysed using assumed plate theory and the proposed method of solution.

The proposed analytical–numerical method gives almost the same results for eigenvalue problem (buckling loads, natural frequencies with corresponding modes) and similar results for dynamic buckling as the finite element method. However the dimensionless deflection versus dynamic load factor relation obtained with both (proposed and FEM) methods are not identical (especially for higher DLF value). These relations allow to find similar critical value of DLF taking into consideration one of the well-known criterion. The differences in the dimensionless deflection ξ appear because the numerical model in the FEM has more degrees of freedom than the model in the analytical–numerical method, but the results from the ANM are obtained in a significantly faster way than those from the finite element method.

Author details

Tomasz Kubiak
Department of Strength of Materials, Lodz University of Technology, Poland

Acknowledgement

This publication is a result of the research work carried out within the project subsidized over the years 2009-2012 from the state funds designated for scientific research (MNiSW - N N501 113636).

6. References

[1] Bernoulli J., Euler L. (1910) Abhandlungen uber das Gleichegewicht und die Schwingungen der Ebenen Elastischen Kurven, Wilhelm Engelmann, Nr 175, Leipzig.
[2] Timoshenko S.P., Gere J.M. (1961) Theory of Elastic Stability, McGraw-Hill Book Company, Inc. NewYork, Toronto, London.
[3] Volmir A.S. (1967), Stability of Deformation Systems, Moscow, Nauka, Fizmatlit /in Russian/.
[4] Ali M.A., Sridharan S. (1988) A Versatile Model for Interactive Buckling of Columns and Beam-Columns, International Journal of Solids and Structures, 24(5): 481–486.
[5] Benito R., Sridharan S. (1985) Mode Interaction in Thin-Walled Structural Members, Journal of Structural Mechanics, 12(4): 517–542.
[6] Byskov E. (1988) Elastic Buckling Problem with Infinitely Many Local Modes, Mechanics and Structures Machines, 15(4): 413–435.
[7] Koiter WT, Pignataro M. (1974) An Alternative Approach to the Interaction Between Local and Overall Buckling in Stiffened Panels, Buckling of structures Proceedings of IUTAM Symposium, Cambridge: 133–148.

[8] Kolakowski Z. (1989) Some Thoughts on Mode Interaction in Thin-Walled Columns Under Uniform Compression, Thin Wall Structures, 7(1): 23–35.

[9] Kolakowski Z. (1993) Interactive Buckling of Thin-Walled Beams with Open and Closed Cross-Sections, Thin Wall Structures, 15(3): 159–183.

[10] Kolakowski Z. (1993) Influence of Modification of Boundary Conditions on Load Carrying Capacity in Thin-Walled Columns in The Second Order Approximation. International Journal of Solids and Structures, 30(19): 2597–2609.

[11] Manevich A.I. (1988) Interactive Buckling of Stiffened Plate under Compression. Mekhanika Tverdogo Tela, 5: 152–159 /in Russian/.

[12] Moellmann H., Goltermann P. (1989) Interactive Buckling in Thin-Walled Beams, Part I: Theory, Part II: Applications. Internatioan Journal of Solids and Structures, 25(7): 715–728 and 729–749.

[13] Pignataro M., Luongo A., Rizzi N. (1985) On the Effect of the Local Overall Interaction on the Postbuckling of Uniformly Compressed Channels, Thin Wall Structures, 3(4): 283–321.

[14] Pignataro M., Luongo A. (1987) Asymmetric Interactive Buckling of Thinwalled Columns with Initial Imperfection, Thin Wall Structures, 5(5): 365–386.

[15] Pignataro M., Luongo A. (1987) Multiple Interactive Buckling of Thin-Walled Members in Compression, Proceedings of the International Colloqium on Stability of Plate and Shell Structures, Ghent: University Ghent: 235–240.

[16] Sridharan S, Ali M.A. (1985) Interactive Buckling in Thin-Walled Beam-Columns, Journal of Engineering Mechanics ASCE, 111(12): 1470–1486.

[17] Sridharan S., Ali M.A. (1986) An Improved Interactive Buckling Analysis of Thin-Walled Columns Having Doubly Symmetric Sections, International Journal of Solids Structures, 22(4): 429–443.

[18] Kolakowski Z., Kowal-Michalska K. (Eds.) (1999) Selected Problems of Instabilities in Composite Structures, Technical University of Lodz Press, A Series of Monographes, Lodz, 1999.

[19] Kolakowski Z., Kubiak T. (2007) Interactive Dynamic Buckling of Orthotropic Thin-Walled Channels Subjected to In-Plane Pulse Loading, Composite Structures, 81: 222–232

[20] Ari-Gur J., Simonetta, S.R. (1997) Dynamic Pulse Buckling of Rectangular Composite Plates, Composites Part B, 28: 301–308.

[21] Cui S., Hai H., Cheong H.K. (2001) Numerical Analysis of Dynamicbuckling of Rectangular Plates Subjected to Intermediate Velocity Impact, International Journal of Impact Engineering, 25 (2): 147–167.

[22] Awrejcewicz J., Krysko V.A. (2003) Nonclassical Thermoelastic Problems in Nonlinear Dynamics of Shells, Springer-Verlag, Berlin.

[23] Awrejcewicz J., Andrianov I.V., Manevitch L.I. (2004) Asymtotical Mechanics of Thin-Walled Structures . A Handbook, Springer-Verlag, Berlin.

[24] Kounadis A.N., Gantes C., Simitses G. (1997) Nonlinear Dynamic Buckling of Multi-DOF Structural Dissipative System under Impact Loading, International Journal of Impact Engineering, 19 (1): 63-80.

[25] Mania R.J. (2011) Viscoplastic Thin-Walled Columns Response to Pulse Load, Proc. ICTWS, Thin-Walled Structures, (ed.) Dubina D.: 415–430.

[26] Budiansky B. (1965) Dynamic Buckling of Elastic Structures: Criteria and Estimates, Report SM-7, NASA CR-66072.

[27] Simitses G.J. (1990) Dynamic Stability of Suddenly Loaded Structures, Springer Verlag, New York.

[28] Volmir S.A., (1972) Nonlinear Dynamics of Plates and Shells, Science, Moscow /in Russian/.

[29] Budiansky B., Hutchinson J.W. (1966) Dynamic Buckling of Imperfection-Sensitive Structures, Proceedings of the Eleventh International Congress of Applied Mechanics, Goetler H. (ed.), Munich: 636–651.

[30] Hutchinson J.W., Budiansky B. (1966) Dynamic Buckling Estimates, AIAA Journal, 4(3): 525–530

[31] Bisagni C. (2005) Dynamic Buckling of Fiber Composite Shells under Impulsive Axial Compression, Thin-Walled Structures 43(3): 499–514

[32] Gilat R., Aboudi J. (2002) The Lyapunov Exponents as a Quantitive Criterion for the Dynamic Buckling of Composite Plates, International Journal of Solid and Structures 39(2): 467–481

[33] Kubiak T. (2005) Dynamic Buckling of Thin-Walled Composite Plates with Varying Widthwise Material Properties, International Journal of Solid and Structures 42(20): 5555–5567

[34] Petry D., Fahlbusch G. (2000) Dynamic Buckling of Thin Isotropic Plates Subjected to In-Plane Impact, Thin-Walled Structures 38: 267–283.

[35] Kubiak T. (2007) Criteria of Dynamic Buckling Estimation of Thin-Walled Structures, Thin-Walled Structures 45 (10-11): 888–892.

[36] Teter A. (2010) Application of Different Dynamic Stability Criteria in Case of Columns with Intermediate Stiffeners, Mechanics and Mechanical Engineering, 14(2): 165–176.

[37] Teter A. (2011) Dynamic Critical Load Based on Different Stability Criteria for Coupled Buckling of Columns with Stiffened Open Cross-Sections, Thin-Walled Structures 49 (5): 589–595.

[38] Kowal-Michalska K. (ed.) (2007) Dynamic Stability of Composite Plate Structures, WNT, Warsaw.

[39] Simitses G. J., (1987) Instability of Dynamically Loaded Structures, Applied Mechanics Revision, 40 (10): 1403–1408.

[40] Grybos R. (1980) Statecznosc Konstrukcji pod Obciazeniem Uderzeniowym, PWN /in Polish/.

[41] Mania R.J. (2011) Dynamic Buckling of Orthotropic Viscoplastic Column, Thin-Walled Structures, 49(5): 591–588.

[42] Mania R., Kowal-Michalska K. (2006) Behaviour of Composite Columns of Closed Cross-Section under In-Plane Compressive Pulse Loading, Thin-Walled Structures, 45 (10-11): 125–129.

[43] Mania R.J., Kowal-Michalska K. (2010) Elasto-Plastic Dynamic Response of Thin-Walled Columns Subjected to Pulse Compression, Proc. of SSTA 2009, CRC Press: 183–186.

[44] Krolak M., Mania J.R. (eds.), (2011) Stability of Thin-Walled Plate Structures, Vol.1 of Statics, Dynamics and Stability of Structures, A Series of Monographs, TUL Press, Lodz.

[45] Czechowski L. (2008) Dynamic Stability of Rectangular Orthotropic Plates Subjected to Combined In-Plane Pulse Loading in the Elasto-Plastic Range, Mechanics and Mechanical Engineering, 12(4): 309–321.

[46] Kowal-Michalska K. (2010) About Some Important Parameters in Dynamic Buckling Analysis of Plated Structures Subjected to Pulse Loading, Mechanics and Mechanical Engineering, 14: 269–279.

[47] Kowal-Michalska K., Mania J.R. (2008) Some Aspects of Dynamic Buckling of Plates under In-Plane Pulse Loading, Mechanics and Mechanical Engineering, 12(2): 135–146.

[48] Kubiak T. (2007), Dynamic Coupling Buckling of Thin-Walled Columns, 998 Scientific Bulletin of TUL, Lodz */in Polish/*.

[49] Pietraszkiewicz W. (1989) Geometrically Nonlinear Theories of Thin Elastic Shells. Advances in Mechanics, 12 (1): 51–130.

[50] Wozniak Cz. (ed.), (2001) Mechanika Sprezysta Plyt i Powlok. Wydawnictwo Naukowe PWN, Warszawa */in Polish/*.

[51] Kubiak T. (2001) Postbuckling Behavior of Thin-Walled Girders with Orthotropy Varying Widthwise, International Journal of Solid and Structures, 38 (28-29): 4839–4856.

[52] Koiter W.T. (1963) Elastic Stability and Post-Buckling Behaviour, Proceedings of the Symposium on Nonlinear Problems, Univ. of Wisconsin Press, Wisconsic: 257–275.

[53] Krolak M. (ed.), (1990) Stany Zakrytyczne i Nosnosc Graniczna Cienkosciennych Dzwigarow o Scianach Plaskich, PWN, Warszawa – Lodz, */in Polish/*.

[54] Krolak M. (ed.), (1990) Statecznosc, Stany Zakrytyczne i Nosnosc Cienkosciennych Konstrukcji o Ortotropowych Scianach Plaskich, Monographs, TUL Press, Lodz, */in Polish/*.

[55] Teter A., Kolakowski Z. (2003) Natural Frequencies of a Thin-Walled Structures with Central Intermediate Stiffeners or/and Variable Thickness, Thin-Walled Structures, 41(4): 291–316.

[56] Kubiak T. (2011) Estimation of Dynamic Buckling for Composite Columns with Open Cross-Section, Computers and Structures, 89 (21-22): 2001–2009.

[57] Prince P.J., Dormand J.R. (1981) High Order Embedded Runge-Kutta Formulae, Journal of Computational and Applied Mathematics., 7(1): 67–75.

[58] Jones R. M. (1975) Mechanics of Composite Materials, International Student Edition, McGraw-Hill Kogakusha, Ltd., Tokyo.

[59] Kelly A. (Ed.), (1989) Concise Encyclopedia of Composite Materials, Pergamon Press.

Chapter 10

FSM Scenarios of Laminar-Turbulent Transition in Incompressible Fluids

N.M. Evstigneev and N.A. Magnitskii

Additional information is available at the end of the chapter

http://dx.doi.org/10.5772/48811

1. Introduction

The problem of turbulence arose more than hundred years ago to explain the nature of chaotic motion of the nonlinear continuous medium and to find ways for its description; so far it remains one of the most attractive and challenging problems of classical physics. Researchers of this problem have met with exclusive difficulties and there was an understanding of that the problem of turbulence always considered difficult, is actually extremely difficult. This problem is named by Clay Mathematics Institute as one of seven millennium mathematical problems [1] and it is also in the list of 18 most significant mathematical problems of XXI century formulated by S.Smale [2].

The nature of turbulence - the disordered chaotic motion of a nonlinear continuous medium, the causes and mechanisms of chaos generation remain the main issue in the turbulence problem. Several models trying to explain the mechanisms of turbulence generation in nonlinear solid media were suggested at different time. Among such models the most known are Landau-Hopf and Ruelle-Takens models, explaining generation of turbulence by the infinite cascade of Andronov-Hopf bifurcations and, accordingly, by destruction of three-dimensional torus with generation of strange attractor. However, these models have not been justified by experiments with hydrodynamic turbulence.

The universal unified mechanism of transition to dynamical chaos in all nonlinear dissipative systems of differential equations including autonomous and nonautonomous systems of ordinary and partial differential equations and differential equations with delay argument was theoretically and experimentally proven in number of recent papers by the authors [3–12]. The mechanism is developing by FSM (Feigenbaum-Sharkovskii-Magnitskii) scenario through subharmonic and homoclinic bifurcation cascades of stable cycles or stable two dimensional or many-dimensional tori.

In this chapter we are presenting a consistent numerical solution method for 3D evolutionary Navier-Stokes equations with an arbitrary initial-boundary value problem posed. Then we

INTECH
open science | open minds

© 2012 The Author(s). Licensee InTech. This chapter is distributed under the terms of the Creative Commons Attribution License http://creativecommons.org/licenses/by/3.0), which permits unrestricted use, distribution, and reproduction in any medium, provided the original work is properly cited.

consider two well studied problems for incompressible Navier-Stokes equations, namely flow over a backward facing step and Rayleigh-Benard convection in cubic cavity. Numerical solutions of these problems for transitional regimes indicated existence of complicated scenarios formed by theory FSM. Thus, it seems reasonable, that there is no unified laminar-turbulent transition scenario, it can be a cascade of stable limit cycles or stable two dimensional or many dimensional tori, but all these scenarios lay in the frameworks of the FSM-theory.

2. Construction of high order numerical method for Navier-Stokes equations

2.1. Some theorems and assertions

Here we are talking only about three dimensional evolutionary incompressible Navier-Stokes equations:

$$\begin{aligned} &\nabla \cdot \mathbf{V} = 0, \\ &\tfrac{\partial \mathbf{V}}{\partial t} + (\mathbf{V} \cdot \nabla)\mathbf{V} + \rho_0^{-1} \nabla P = \nu \nabla^2 \mathbf{V}, \end{aligned} \tag{1}$$

and suppose that laminar-turbulent transition is well described by these equations. Here ρ_0 is a constant fluid density, ν is the kinematic fluid viscosity, $\mathbf{V}$ is the velocity vector-function and P is the scalar pressure function. Main questions of numerical solution trustworthiness of Navier-Stokes equations and its attraction to real system (1) attractor are:

1. Does the change of infinite dimensional system for finite-dimensional alters the attractor and solution?

2. Do numerical approximation errors have crucial affect on the attractor trajectory?

3. Does a numerical solution convergence to a real solution and how close are they?

Some results of attractor approximation and numerical attractors for nonlinear PDEs were obtained in papers [13, 14], but none of them covers such a complicated topic as existence of and numerical trajectories attraction for Navier-Stokes attractors. However in papers [15–19] it is shown that for a well-posed initial-boundary problem there exists an attractor and its dimension and volume is limited from above. This allows us to pose some assumptions on numerical methods that must be used to get a numerical system attractor and that it is close to the attractor of Navier-Stokes equations. To do so we will be following closely to work of R.Temam [16, 20].

Existence of a global attractor for 3D Navier-Stokes equations is proven [15, 16]. Let that attractor A exist for a well-posed initial-boundary problem in domain $\Omega \in \mathbb{R}$ with local Lipschitz-contonous boundary $\partial\Omega_i$ and velocity vector-function $\mathbf{V} \in \mathbb{R}^3$. Let k_0 be the macroscopic wave number that corresponds with the characteristic length scale L and k_k is the Kolmogorov wave number defined as:

$$k_k = \left(\frac{\epsilon}{\nu^3}\right)^{1/4} \tag{2}$$

ν - kinematic fluid viscosity and ϵ - rate of turbulent energy dissipation. whole spectra of wave numbers can be described as:

$$k_i = (k_0, k_1, k_2, ..., k_k), i \in \mathbb{N} \tag{3}$$

Hence the minimum space dimension that is required to hold all possible flow scales in bounded domain $\mathbf{V} \in \mathbb{R}^3$ is defined as:

$$Y = \left(\frac{k_k}{k_0}\right)^3 \tag{4}$$

The value of Y gives us number of degrees of freedom for the given turbulent flow. We can use the value for ϵ from [16] where it is shown that ϵ is limited by the supremum:

$$\epsilon = \nu \cdot \limsup_{t \to \infty} \left\{ \sup_{\mathbf{V}(x,0) \subset A} \frac{1}{t} \int_0^t \sup \left| \nabla^2 V(x,t) \right|^2_{x \in \Omega} dx \right\}, \tag{5}$$

and we can find the following:

Theorem 01. *[16] the Hausdorff-Besicovitch dimension of any attractor A in the Navier-Stokes equations for a well-posed boundary value problem is limited by:*

$$\dim A \leq C \cdot \frac{k_k}{k_0}, \tag{6}$$

where C is an arbitrary constant. One can derive a more explicit estimate. Since:

$$\epsilon \leq \overline{\epsilon} = \nu \left| \nabla^2 \mathbf{V}(x,t) \right|^2 \tag{7}$$

and

$$k_k \leq \overline{k_k} = \left(\frac{\overline{\epsilon}}{\nu^3}\right)^{1/4} = \sqrt{\left(\frac{|\nabla^2 \mathbf{V}(x,t)|}{\nu}\right)} \tag{8}$$

so taking (4) into account:

$$\dim A \leq L^3 \cdot \left(\frac{|\nabla^2 \mathbf{V}(x,t)|}{\nu}\right)^{3/2}, \tag{9}$$

where L is a macroscopic problem scale. Since all functions in (9) have finite limits then the dimension of an attractor is always limited from above and its dimension is linked with number of degrees of freedom (4). So we have the following:

Assertion 01. *If a well-posed initial-boundary value problem for Navier-Stokes equations in bounded domain $\Omega \in \mathbb{R}^3$ with local Lipschitz-continuous boundary $\partial\Omega \in \mathbb{R}^2$ has a suitable numerical approximation with degrees of freedom $Y^* \geq Y$ that convergence to the given problem on every time step of approximation then the approximated numerical attractor A^* converges to the real attractor A.*

This assertion can be easily verified using estimates for finite-difference and finite-element approximations of Roger Temam [20]. This assertion answers the first question. Now we consider the convergence of the numerical method with certain properties for an arbitrary given well posed initial-boundary problem to the real solution of the problem. We are using the following theorem by Roger Temam [20], page 281:

Theorem 02. *Let space dimension is 3 and we have some numerical approximation for Navier-Stokes equations with some conditions on which its stable and discrete elements are h for space and k for time. Then there exists the sequence for h and $k \to 0$, that:*

$$\left.\begin{array}{r}\mathbf{V}_h \to \mathbf{V}\\ P_h \to P\end{array}\right\};\tag{10}$$

strong in $L^2(Q)$, week in $L^\infty(Q,0,\Omega)$; $Q = \Omega \times [0,t]$.

Here: $\mathbf{V}$,P - velocity vector-function and pressure scalar-function that correspond to a solution for an initial-boundary value problem, $\mathbf{V}_h$, P_h - velocity vector-function and pressure scalar-function on sequence h. C is an arbitrary constant. Here we consider one of possible solutions in attractor A, since uniqueness of a solution for a problem is not proven in 3D case. Prove of this theorem is given in [20], p.282, we are using only some conditions for the theorem validity:

Condition 1. *If a function $x \mapsto \mathbf{V}(x); x \subset \mathbb{R}^3$ has the property $\nabla \cdot \mathbf{V} = 0$, then a discrete function $h \mapsto \mathbf{V}_h$ must maintain the property $\nabla_h \cdot \mathbf{V}_h = 0$.*

Condition 2. *For any sequence $\mathbf{V}_h$ the condition:*

$$\sup\left[\sum_h |\mathbf{V}_{h+1} - \mathbf{V}_h|^{k+1} - \sum_h |\mathbf{V}_{h+1} - \mathbf{V}_h|^k\right] \leq 0,\tag{11}$$

has to hold.

Appropriate approximations and numerical procedures where used in [16, 20] . A screw-symmetric numerical operator ∇_h was used for $(\mathbf{V} \cdot \nabla)\,\mathbf{V}$, linear high order operators where used for linear diffusion parts, etc. Time integration was conducted by implicit Crank-Nicolson method for linear part of Navier-Stokes system and explicit second order for nonlinear part. Pressure correction was used so that Condition 1 is true on each time step. However the theorem is true only for h and $k \to 0$ [19], i.e. when the numerical system phase space dimension goes to ∞ . It is obvious that h and k are finite for a real numerical method that is applied on computers. Since the attractor dimension is limited from above (9) and number of degrees of freedom is very large but finite (4), one can stipulate that h and k can be finite if condition (4) and hence (9) are satisfied. It is hard to give a precise estimate but one can approximately evaluate those values by using Kolmogorov turbulent theory for invariant scales.

The least motion scale of turbulence can be defined as $l_k = k_k^{-1}$, where k_k is given by (2). Since ϵ is bounded (7), the dissipation rate for a developed turbulent flow is given approximately as:

$$\epsilon \propto \frac{\mathbf{V}^3}{L},\tag{12}$$

One immediately derives:

$$l_k = \left(\frac{\nu^3 L}{\mathbf{V}^3}\right)^{1/4} = \left(\frac{\nu^3 L^4}{\mathbf{V}^3 L^3}\right)^{1/4} = L \cdot R^{-3/4}, \tag{13}$$

Hence Y for a developed turbulent regime can be given as:

$$Y = \left(\frac{L}{L \cdot R^{3/4}}\right)^3 = R^{9/4}, \tag{14}$$

that corresponds with the maximum number of degrees of freedom for Direct Numerical Simulation. For smaller values of Reynolds (R) number where the flow is in transitional regime the value for (14) is not perfectly true. For the purpose of finding Y in this case we are using a spatially constructed numerical procedure which is given below. For now we assume that appropriate values of h and k for the given R are defined and conditions 1 and 2 are true for some selected numerical solution procedure. Taking into account that the attractor is a phase space trajectories attraction manifold we have the following

Assertion 02. *For a well-posed initial boundary value problem the numerical solution method with correct values of h and k and true Conditions 1 and 2 approximates an attractor of 3D evolutionary Navier-Stokes equations up to the precision order of the numerical method.*

2.2. Numerical method description and consistent mesh adaptation

So we can construct the numerical method with the above mentioned properties and it can be used to analyze laminar-turbulent transition as a nonlinear dynamic system. However we should point out that bifurcation parameters do depend on the numerical method. So the values of these parameters can vary from one method to another for a given bifurcation. We are considering advection initial boundary problems with no external force acting for dimensionless Navier-Stokes equations. The general problem can be described as: Let Ω be a bounded domain with local Lipschiz-continuous boundaries $\partial\Omega_i$. One must find velocity vector-function $\mathbf{V} : \Omega \times [0,t] \to \mathbb{R}^3$ and scalar pressure function $P : \Omega \times [0,t] \to \mathbb{R}$ such as:

$$\begin{gathered} Sh \cdot \tfrac{\partial \mathbf{V}}{\partial t} + (\mathbf{V} \cdot \nabla)\,\mathbf{V} + Eu \cdot \nabla P = \tfrac{1}{R}\nabla^2 \mathbf{V} \text{ in } Q = \Omega \times (0,t); \\ \nabla \cdot \mathbf{V} = 0 \text{ in } Q; \\ \mathbf{V} = f(\vec{x}), \text{ on } \partial\Omega_0 \times (0,t), \mathbf{V} = 0, \text{ on } \partial\Omega_1 \times (0,t), \tfrac{\partial \mathbf{V}}{\partial \vec{n}} = 0, \text{ on } \partial\Omega_2 \times (0,t); \\ \mathbf{V}(\vec{x},0) = \mathbf{V}_0(x,y,z), \text{ in } \Omega \text{ with } \nabla \cdot \mathbf{V}_0(x,y,z) = 0. \end{gathered} \tag{15}$$

Similarity criteria are constructed using characteristic macroscopic scales: $Sh = fL/V$ - Strouhal number; $Eu = 2P/(\rho V^2)$ - Euler number; $R = VL/\nu$ - Reynolds number; L - macroscopic characteristic scale, f - frequency, V - characteristic velocity, P_0 - reference pressure, $\rho = const$ - fluid density, ν - fluid kinematic viscosity. Various boundary conditions are given as: $\partial\Omega_0$ - inflow condition, $\partial\Omega_1$- solid wall condition, $\partial\Omega_2$ - outflow condition and the problem is initialized with initial conditions. Most important similarity criterion is Reynolds number for laminar-turbulent transition. The rest criteria can be used to scale some real problems using π -theorem [21] and are not used since we are interested only in nonlinear dynamics of equations with no regard to a real specific problem. In order to determine necessary number of discrete elements one should take into account two factors; the first is to use the upper bound (14) as a start and derive ϵ from the averaged equations and the second is to use modified wave number analysis to calculate how many elements are needed to represent certain wave number and thus necessary number of degrease of freedom. Using this method, see [22] one can numerically determine anisotropic element density in Ω and optimize this number. This approach also satisfies all conditions of theorems and assertions described above. Since the rate of energy dissipation is a function of Reynolds number and particular initial-boundary conditions we are going to compute required Y using Reynolds averaging. Let us rewrite momentum equation in (15) in coordinates as:

$$\frac{\partial V_i}{\partial t} + \sum_j \left(V_j \frac{\partial V_i}{\partial x_j} \right) + \frac{1}{\rho} \frac{\partial P}{\partial x_i} = \nu \frac{\partial^2 V_i}{\partial x_i^2};\ i,j = \{1,2,3\} \tag{16}$$

and introduce averaging:

$$\mathbf{V} = \overline{\mathbf{V}} + \mathbf{V}', \tag{17}$$

thus:

$$\overline{V_i + V_j} = \overline{V_i} + \overline{V_j};\ \overline{\partial_\alpha V} = \partial_\alpha \overline{V};\ \overline{\overline{V_i} V_j} = \overline{V_i}\,\overline{V_j};\ \alpha = \{t;x\}. \tag{18}$$

Here $\overline{V}$ - averaged and V' instantaneous functions of velocity vector-function. By applying (17) to (16) one gets:

$$\frac{\partial \overline{V_i}}{\partial t} + \sum_j \left(\overline{V_j} \frac{\partial \overline{V_i}}{\partial x_j} + \overline{V_j{}' \frac{\partial V_i{}'}{\partial x_j}} \right) + \frac{1}{\rho} \frac{\partial \overline{P}}{\partial x_i} = \sum_j \frac{\partial \overline{\tau_{ij}}}{\partial x_j}, \tag{19}$$

where second rank tensor $\tau_{ij} = \nu \left(\frac{\partial V_i}{\partial x_j} + \frac{\partial V_j}{\partial x_i} \right)$ corresponds to the Newtonian fluid. Multiplying (16) on $\overline{V_i}$ and applying (17) one gets:

$$\overline{\frac{\partial V_i}{\partial t} V_i} + \sum_j \left(\overline{V_j \frac{\partial V_i}{\partial x_j} V_i} \right) + \overline{\frac{1}{\rho} \frac{\partial P}{\partial x_i} V_i} = \sum_j \overline{\frac{\partial \tau_{ij}}{\partial x_j} V_i}. \tag{20}$$

Multiplying (19) on $\overline{V_i}$:

$$\frac{\partial \overline{V_i}}{\partial t} \overline{V_i} + \sum_j \left(\overline{V_i}\,\overline{V_j} \frac{\partial \overline{V_i}}{\partial x_j} \right) + \frac{1}{\rho} \frac{\partial \overline{P}}{\partial x_i} \overline{V_i} = \sum_j \left[\left(\frac{\partial \overline{\tau_{ij}}}{\partial x_j} + \frac{\partial T_{ij}}{\partial x_j} \right) \overline{V_i} \right], \tag{21}$$

one gets the stress equation, where $T_{ij} = -\overline{V_i{}' V_j{}'}$ are the components of virtual (Reynolds) stress tensor. Subtraction of (21) from (20) reads:

$$\overline{\frac{\partial V_i{}'}{\partial t} V_i{}'} + \sum_j \left(\overline{V_j \frac{\partial V_i}{\partial x_j} V_i} - \overline{V_j} \frac{\partial \overline{V_i}}{\partial x_j} \overline{V_i} \right) + \overline{\frac{1}{\rho} \frac{\partial P'}{\partial x_i} V_i{}'} = \sum_j \left(\overline{\frac{\partial \tau_{ij}{}'}{\partial x_j} V_i{}'} - \frac{\partial T_{ij}}{\partial x_j} \overline{V_i} \right), \tag{22}$$

where summation is performed in accordance with (18). The value $\overline{\tau_{ij}{}'V_i{}'}$ is the second infinitesimal order to the other parts of (22) and then the latter can be simplified by using averaging rules (18) as:

$$\frac{1}{2}\left(\frac{\overline{(V_i')^2}}{\partial t}+\sum_j \frac{\partial\overline{(V_i')^2}}{\partial x_j}\overline{V_j}\right)+\frac{1}{\rho}\frac{\overline{\partial P'}}{\partial x}V_i'+\frac{1}{2}\sum_j \frac{\partial\overline{V_i'^2 V_j'}}{\partial x_j}+ \tag{23}$$

$$+\sum_j\left(\overline{\frac{\partial V_i'}{\partial x_j}\tau_{ij}'}+\overline{V_j'V_i'}\frac{\partial\overline{V_i'}}{\partial x_j}\right)=0.$$

One can get the equation for perturbation energy balance applying a summation of (23) by index i:

$$\frac{\partial k}{\partial t}+\sum_j\frac{\partial k}{\partial x_j}\overline{V_j}=-\sum_j\left(\frac{\partial}{\partial x_j}\left(\rho^{-1}\overline{P'V_j'}+\sum_i\left[\overline{V_i'^2V_j'}\right]\right)\right)+\sum_{i,j}\left(T_{ij}\frac{\partial\overline{V_i'}}{\partial x_j}\right)-\epsilon, \tag{24}$$

where perturbation kinetic energy is written as:

$$k=\frac{1}{2}\sum_i\overline{(V_i')^2}, \tag{25}$$

and perturbation rate of dissipation is described by:

$$\epsilon=\sum_{i,j}\left(\overline{\frac{\partial V_i'}{\partial x_j}\tau_{ij}'}\right)=\frac{\nu}{2}\sum_{i,j}\overline{\left(\frac{\partial V_i'}{\partial x_j}+\frac{\partial V_j'}{\partial x_i}\right)^2}. \tag{26}$$

The latter expression (26) is the exact value of (7) for a given Reynolds number. We are using the following correlation to define the Kolmogorov wave number (2):

$$k_k=\left(\frac{R^2}{2}\sum_{i,j}\overline{\left(\frac{\partial V_i'}{\partial x_j}+\frac{\partial V_j'}{\partial x_i}\right)^2}\right)^{1/4}, \tag{27}$$

where the sum expression is numerically calculated by a test simulation with the maximum possible number of discrete elements, defined by (14). After that we are constructing the isolines of NY, calculated by (4), where the dissipation wave number is applied through (27) and N is defined by modified wave number analysis, described bellow. Then the mesh that we are using is adopted to satisfy all calculated values of Y. Only after all these procedures we can say that dynamic nonlinear analysis results we obtained are true and can be considered trustworthy.

In order to solve an initial-value problem for Navier-stokes numerically we are introducing the following semi discrete scheme based on the fractional step method using high order TVD

Runge-Kutta forth order method with each step of RK as:

$$
\begin{aligned}
&1.\mathbf{V}' - \mathbf{V}^n = -\Delta t(\mathbf{V}^n \cdot \nabla)\mathbf{V}^n;\\
&2.\mathbf{V}'' - \mathbf{V}' = \Delta t \Theta \nabla^2 \mathbf{V}'^{('')}\\
&\text{while } \nabla \cdot \mathbf{V}^{\beta+1} \neq 0; \begin{cases} 3.\nabla^2 P = -\nabla \cdot \mathbf{V}^{\beta}/\Delta t \\ 4.\mathbf{V}^{\beta+1} = \mathbf{V}'' - \Delta t \nabla P \end{cases}\\
&5.\mathbf{V}^{n+1} = \mathbf{V}^{\beta+1}
\end{aligned}
\tag{28}
$$

Here $\mathbf{V}^n$ and $\mathbf{V}^{n+1}$ are the previous timestep and next timestep values of the velocity vector function; P is the pressure, Δt - timestep for the given Runge-Kutta stage and Θ is the diffusion parameter and for a forced advection problems equals R^{-1}. All other superscripts on $\mathbf{V}$ are intermediate values of velocity vector function inside the stage. More details on this numerical procedure can be found in [9, 23, 24]. Each step here is just shortly described. On step 1 in (28) advection equations are solved with the condition (11) satisfied. We are using fifth order WENO-type scheme. Several problems were solved for pure advection equation and Burgers equation (in 1D and 2D) before the final variant of WENO scheme was selected. Since the timestep is limited by the accuracy requirements the numerical scheme was explicit. On the second step of (28) diffusion equations are solved by using large stencil approximation with the 6-th order of accuracy. It is possible to apply implicit method here (in (28) in brackets) for natural advection problem described bellow, since values of Θ are of unity magnitude. But for forced advection where $\Theta = R^{-1}$ (for $R > 100$) it is possible to apply explicit time method. Academician Belotserkovsky O.M. [25] suggested a physical interpretation of this fractional step method. Step 1 and 2 are calculating not solenoidal vector field that breaks the mass conservation equation. But if we apply the operator $(\nabla\times)$ on step 1 and 2 for both (28) and Navier-Stokes equations (15), we get the curl transport equations, since $\nabla \times \nabla P = 0$. Curl properties are correctly simulated though $\nabla \cdot \mathbf{V} = 0$ is not satisfied even for steps 1 and 2. The latter corrected by applying the pressure correction on the third step, where the Poisson equation is solved for pressure scalar function until the solenoidal criteria is met up to the machine accuracy. After that the velocity field is corrected and he velocity vector function is solenoidal on step 4 up to machine accuracy, so condition 1 for correct approximation is satisfied.

Spatial discretization is a combined finite volume for $\mathbf{V}$ and finite element for P discretization. Since we are focusing on fundamental problems (with simple geometry) the discrete elements are rectangular cuboids, thus it allows us adjusting its dimensions in accordance with the calculated values of Y. Finite elements are the same cuboids but variables are stored on vertexes rather than centers of mass as for $\mathbf{V}$. So the described system of equations is rewritten for arbitrary convex element i with volume W_i and $\Omega = \bigcup_{i=1}^{N} W_i; W_i \cap W_j =$ for $\forall i \neq j$ as:

$$
\begin{aligned}
&\textstyle\iiint_{W_i} \nabla \cdot \mathbf{V} dW = \frac{1}{W_i} \oint_{S_i} \mathbf{V}^f \cdot \vec{n} dS = 0,\\
&\textstyle\frac{\partial}{\partial t} \iiint_{W_i} \mathbf{V} dW + \oint_{S_i} [\mathbf{V}^f \mathbf{V}^f \cdot \vec{n}] dS + \oint_{S_i} / P^f / \vec{n} dS - \Theta \oint_{S_i} [\nabla \mathbf{V}]^f ds = 0,
\end{aligned}
\tag{29}
$$

here S - is one of element side square, subscript f refers to the face value and $\vec{n}$ is the unit vector on the side, total number of sides is B. Using discrete elements and applying summation instead of integration

$$\oint_{S_i} f(U)\cdot\vec{n}dS \approx \sum_{j=1}^{B} f(U_j)\cdot\vec{n}\Delta S_j$$

one immediately reads:

$$\begin{gathered}\Sigma_{j=1}^{B}\left[\mathbf{V}_j^f\cdot\vec{n}_j\right]\Delta S_j = 0,\\ W_i\tfrac{d}{dt}\mathbf{V}_i + \Sigma_{j=1}^{B}\left[\mathbf{V}_j^f\mathbf{V}_i^f\vec{n}_j\right]\Delta S_j + /\nabla_h P/ - \Theta\Sigma_{j=1}^{B}\left[\nabla_h\mathbf{V}\right]_j^f\cdot\vec{n}\Delta S_j = 0.\end{gathered} \tag{30}$$

Here / / operator is treated by finite element method. Then discrete system (30) is applied in (28) by described numerical methods. Please note that $(\mathbf{V}\cdot\nabla)\mathbf{V} = \nabla\cdot(\mathbf{V}\mathbf{V}) + \mathbf{V}\nabla\cdot\mathbf{V} = \nabla\cdot(\mathbf{V}\mathbf{V})$ in (29) and (30) since $\nabla\cdot\mathbf{V} = 0$ on every timestep.

In order to complete the analysis of numerical scheme and answer all questions positively one must perform modified wave number analysis for used discrete schemes. We are considering first order PDE like on step 1, in (28). Let the differential operator ∂_x be approximated by central differences as $\partial_x = \delta_x + \mathcal{O}(\Delta x^2)$ using N discrete segments each with the length Δx and let the function be $u(x) = c_k\cdot\exp(ikx)$. One immediately gets analytical solution as $\partial_x u(x) = ikc_k\exp(ikx) = iku(x)$. For the given central differences approximation the real part $u(x) = \cos(kx)$ becomes:

$$\begin{aligned}\delta_x(\cos(kx)) = \frac{1}{2\Delta x}\cos(k(x+\Delta x)) - \frac{1}{2\Delta x}\cos(k(x-\Delta x)) =\\ = -\left(\frac{\sin(kx)}{\Delta x}\right)\sin(kx)\end{aligned} \tag{31}$$

Here we can see that the difference of analytical and numerical solutions is in wave number k that changes to $k' = \sin(kx)/\Delta x$ and if $k\Delta x << 1$ then $k' = k - k^3\Delta x^2 + ...$ For small k the result of approximate solution (31) is close to the analytical. But when wave number k increases i.e. discretization scale λ decreases ($\lambda = 2\pi/k$ and if $\Delta x = \lambda/N$ then $k\Delta x = 2\pi/N$), the error grows. So we call k' a modified wave number by the numerical scheme. Thus by applying this analysis one can get the minimal undisturbed scale representation of the given numerical discrete scheme. This allows us getting minimal number of finite volumes necessary for the given k say calculated k_k. In general for the given complex function one can write:

$$k'\Delta x = -i\sum_{j=0}^{N-1} c_j\exp\left(\frac{i2\pi jk}{N}\right) = -i\sum_{m=-s}^{m=s} a_m\exp(imk\Delta x), \tag{32}$$

where s is half length of the stencil in discrete space and a_m - are the interpolation coefficients of the considered numerical scheme. One can immediately see from (32) that for symmetric schemes only real part presents in the modified wave number since $a_m = -a_{-m}$ and $a_0 = 0$. That explains why symmetric approximation cannot be used for approximation of advection operator in (30). Numerical wave number analysis for WENO scheme was first made in [26]], but here we use analytical analysis. By applying WENO weights, since WENO scheme is fifth order everywhere (even for discontinuous functions, see [24, 27] one can get interpolation coefficients:

$$a_m^{WENO} = (-1/30; 1/4; 11/5; -1/2; -1/20; 0)\,; s = 3; \mathcal{O}(\Delta x^5). \tag{33}$$

By inserting (33) into (32) one can get modified wave number (real and imaginary parts) and derive minimal necessary numbers of elements to represent a given wave number, see fig.1. So one can see that for the 2-nd order central differences scheme at least 8 elements are needed

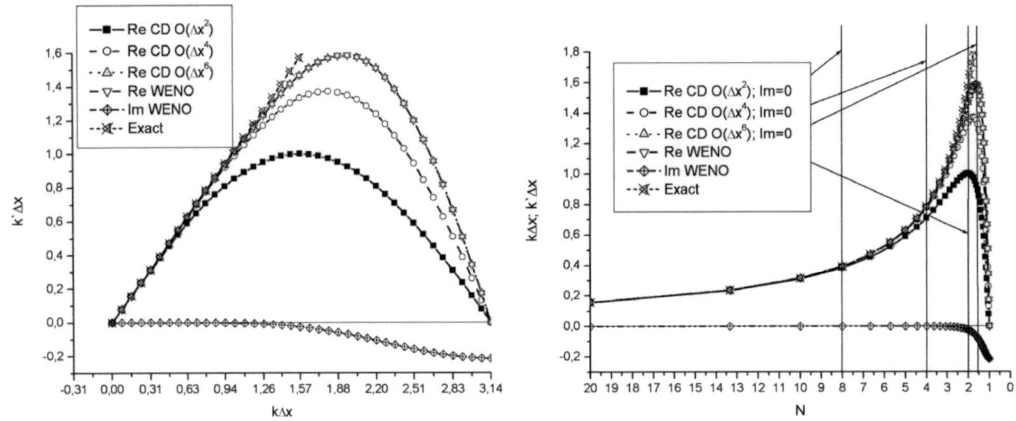

Figure 1. Modified wave number for different schemes to $k\Delta x \in [0, \pi]$(left) and minimal number of elements N for the correct given wave number representation (right). CD - central differences.

to describe a given wave number correctly, and 2 elements for the 6-th order scheme as well as for the WENO scheme. So now we know that the values that are found in accordance with (27) must be multiplied by two. Applying the same method for other parts of discretization one can point out that the given estimate for two elements is enough and here is omitted for the sake of brevity.

Time integration also requires some care since the condition (11) is in spatio-temporal condition and if CFD literature is known as TVD (TVB for equals sign) condition for nonlinear PDEs [28, 29]. Since WENO approximation is TVB, we are going to give a TVD condition (11) for time integration:

$$\frac{\partial V_i}{\partial t} = L(V_i), \tag{34}$$

where L is the given complex nonlinear TVB differential operator consisting of all steps (28). Simple TVD method for (34) is the Euler's method $V^{n+1} = V^n + \Delta t L(V^{n+s})$. If (34) is the explicit method (i.e. $s = 0$) then the stability criterion is a Courant number $\Delta t \leq C$. One can give the following

Lemma 01. *If a direct Euler's method (34) is applied for a TVB or TVD spatial approximation operator for $\Delta t \leq C$ and by applying m-stage Runge-Kutta scheme*

$$\begin{aligned} V^{n+1} &= \textstyle\sum_{i=1}^{m} \left(\alpha_i V^{n+1-i} + \Delta t \beta_i L(V^{n+1-i})\right); \\ &\textstyle\sum_{i=1}^{m} \alpha_i = 1, \forall \alpha_i, \beta_i \geq 0, \text{on } L(V), \end{aligned} \tag{35}$$

then the solution is stable in any norm

$$\left\|V^{n+1}\right\| \leq \max\left\{\left\|V^{n}\right\|; \left\|V^{n-1}\right\|;..;\left\|V^{n-m+1}\right\|\right\}$$

with time step limited by:

$$\Delta t \leq \min_{i} \frac{\alpha_i}{\beta_i} \Delta t(Euler); \tag{36}$$

if $\forall \beta_i = 0$, then $\frac{\alpha_i}{\beta_i} \to \infty$.

The same idea is given in [30]. Proving the lemma is done by considering the combinations of Euler's steps. This lemma gives us constants α_i and β_i that were used to construct the forth order TVD Runge-Kutta method that fulfills the condition (11) and has a maximum stable timestep possible. One should point out that the standard RK4 method with constants $(1/6;1/3;1/3;1/6)$ is not TVD and can't be used for time integration. So it is shown that the presented numerical method is guaranteed to satisfy all given theorems, conditions and assertions and, hence, can be used to describe nonlinear dynamics of laminar-turbulent transition.

The whole work described here took three and half years and 80% of the time was used on calculation. New CUDA technology [31] was applied lately and now all described numerical methods are calculated using NVIDIA GPUs that greatly accelerate the research.

3. Laminar-turbulent transition for the flow over a backward facing step

One of the best studied problems is the problem of the flow over a backward facing step. It was simulated by many different authors and has lots of benchmark results and even few results about nonlinear dynamics in the problem.

3.1. Initial-boundary value problem, mesh adaptation and benchmark verification

The geometry and boundary conditions are taken from [32] with small adaptation, since laminar-turbulent transition is investigated as a dynamic system so we compare not only benchmark results but bifurcation sequences as well.

Domain Ω with local Lipschiz-continuous boundaries $\partial\Omega_i$ is represented by a rectangular channel divided by two unequal parts in Cartesian coordinates. The length of the domain spans in X axis direction. The first part of the channel has length $L_1 = 2.0$, the second part of the channel length is $L_2 = 10.0$, so the whole length of the domain is $L = L_1 + L_2 = 12.0$. Height spans in Y axis and the second part of the domain is higher than the first one by the size of the step $h = 0.6$ with the first part height of $H = 0.9$. So the second part of the domain height is $H + h = 1.5$. Width of the domain $W = 3.5$ spans in Z axis direction and is the same for both part. The geometry is almost identical to [32] with $L_1 = 10.0$. The formed step causes the flow to create recirculation zones inside Ω and transit to turbulence with the growth of Reynolds number.

There are many different boundary conditions available in papers for this particular problem [33–35, 38, 39]. All boundaries in step Y axis direction $\partial\Omega_1 \in \min_y(\Omega)$ are given by solid walls with no slip condition. The upper boundary $\partial\Omega_2 \in \max_y(\Omega)$ is given as either wall or symmetry. Boundaries $\partial\Omega_3 \in \max_z(\Omega)$ and $\partial\Omega_4 \in \min_z(\Omega)$ vary from one work to another. The boundary $\partial\Omega_3$ is given as a solid wall but $\partial\Omega_4$ as symmetry plane in [33]. Boundaries in

Z direction in some other papers, i.e. [35, 40], are given by symmetry planes thus modeling a semi-2D problem and decreasing number of mesh elements due to the lack of side boundary layers. We choose boundary conditions like in [32] where boundaries in Z direction are no slip walls. Boundary $\partial\Omega_5 \in \min_x(\Omega)$ is the inflow boundary where Poiseuille laminar solution for R=200 in the channel with the same width and height. Boundary $\partial\Omega_6 \in \max_x(\Omega)$ is the outflow boundary. We introduce smooth outflow boundary conditions [41] with the outflow buffer so that unphysical outflow conditions are neglected. Pressure is set on the outflow boundary as a reference value. Initial conditions in whole domain Ω for velocity vector-function and scalar pressure function obtained by laminar stationary solution of this problem with $R = 200$ so that $\nabla \cdot \mathbf{V} = 0$ at t=0. Reynolds number is defined as:

$$R = V^0 L/\nu, L = 2WH/(W + H). \tag{37}$$

Here V^0 is the inflow velocity normalized by the flow rate, like in [32]. Due to the fact that DNS is very computational demanding we are limiting the maximum R by analyzing literature results. It is shown in [32] that a "chaotic attractor" that corresponds to noise frequencies band appear at $R > 1500$ so we set maximum $R = 1500$. For the purpose of minimizing computational efforts we are considering three meshes adopted for $R = 700, R = 1000$ and $R = 1500$. The rest Reynolds number meshes are lying inside the segment. Using an upper bound estimate (14) and taking modified wave number analysis in account one gets total mesh numbers as: 5 040 813(R=700), 11 246 827(R=1000) and 24 004 984 (R=1500). We set $L = 1\frac{3}{7}$ from the given geometry and since we know R, we can evaluate minimal Kolmogorov linear scale of one element l_k:

$$l_k = \{0.024495281; 0.018745885; 0.013830488\}, \tag{38}$$

and, hence, in Cartesian coordinates we have $M = \{491; 641; 869\}$, $N = \{62; 81; 109\}$, $K = \{144; 188; 254\}$, elements for the upper bound estimate. Time step is selected 0.005 for all calculations [**?**]. We performed these three control calculations on fine grids and calculated isosurfaces of minimal scales l_k by (27). Example of these isosurfaces for $R = 1000$ is shown in fig.2.

It is clearly visible that we must use fines mesh with element size of 0.01577 of boundary layers and recirculation zone, but the rest of the flow has s much lighter demand for small scales. So final meshes where chosen to be:

$$\begin{aligned}
G1 &= \{245X54X83\}, R = 700; \Delta x_\alpha = \{0.0677; 0.01957; 0.03964\} \approx 103^3;\\
G2 &= \{339X72X119\}, R = 1000; \Delta x_\alpha = \{0.04696; 0.01577; 0.028548\} \approx 143^3;\\
G3 &= \{560X100X177\}, R = 1500; \Delta x_\alpha = \{0.04696; 0.01577; 0.028548\} \approx 215^3;
\end{aligned} \tag{39}$$

where cubic values are shown for required memory estimate. After that all other calculations for the given problem are performed on these (39) three grids that satisfy necessary conditions of theorems. In order to perform benchmark tests boundaries $\partial\Omega_{3,4}$ where changed accordingly. Values of are the same l_k for inside flow region and near periodic boundaries so our grid resolution (39) is overdensed for test calculations with periodic boundaries. So new mesh was generated in the same manner through calculation of l_k but it contained less number of l_k elements, so benchmark calculations took less time to solve.

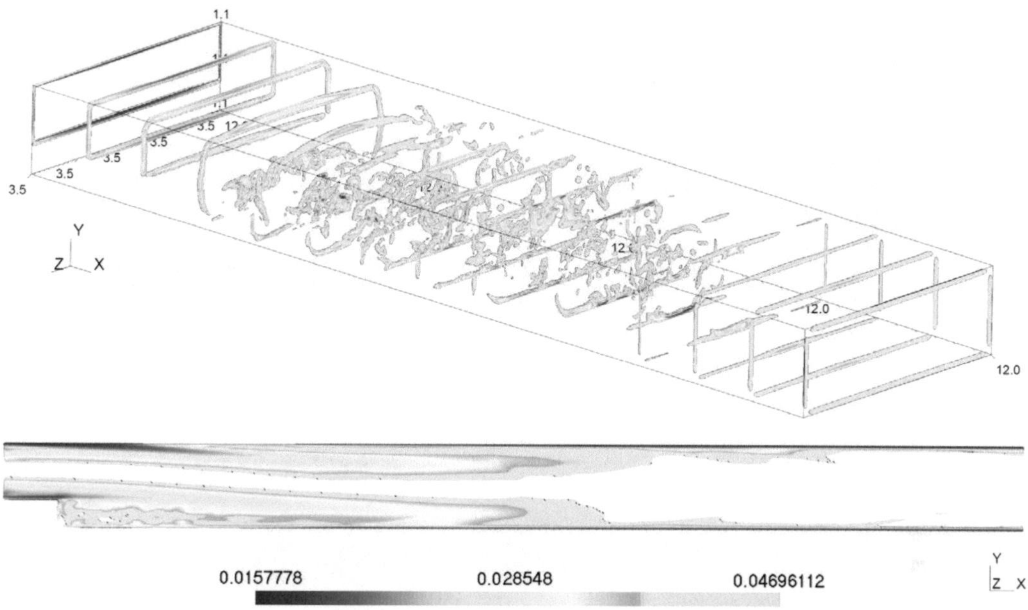

Figure 2. Calculated values of l_k for $R = 1000$. Central section and literal 3D sections are shown.

While solving various modifications of the problem we changed height of the step in accordance with the geometry presented for particular benchmark tests and here all detail data of these geometries is omitted due to brevity. Integral and statistical data of $\mathbf{V}$ and P functions is considered as well as friction wall coefficients.

First we overseen qualitative comparison of results for R=1500 with side walls with DNS [35], [42], [37] and physical model [43] results. Velocity isolines sections are presented in fig.3. One can clearly see the 3D structure of the flow plus the recirculation vertexes that appear near all walls. This agrees well with the presented papers. To compare qualitatively periodic boundary results we used papers [44] and [35]. The flow is much more simple for periodic boundaries in Z direction.

For the quantitative analysis we considered some papers: [44] for $R = \{100; 389; 1000\}$, [34] for $R = 150 - 800$, [35] for $R = 1500$, [45] in C_f comparison for $R = 5500$ (using LES model) and [38, 46] for virtual stress tensor comparison for $R = 1800$ (using LES model). All data agreed well for benchmarks with maximum difference of 7%. Data was extracted from papers with high precision using shareware GetData Graph Digitizer 2.24.

Detail results are omitted for the sake of brevity, but some results are presented here. Reattachment vortex length ration to the height of the step is presented in fig.4 and compared with results from various sources, i.e. [34]. One can clearly see that the results are well agreed with reference data.

More quantitative results are presented in fig. 5, where a mean wall friction coefficient is presented. The dynamic eddy viscosity model was used to solve the flow for $R = 5500$, and was compared with other LES and DNS results. One can see that the result agree well with presented data, especially with DNS data (rectangles) from [45].

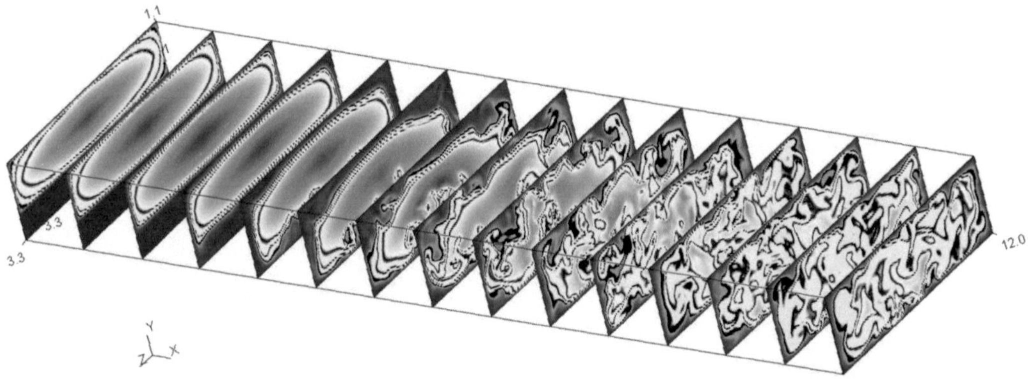

Figure 3. Instantaneous velocity vector-function modulus isolines in sections for R=1500.

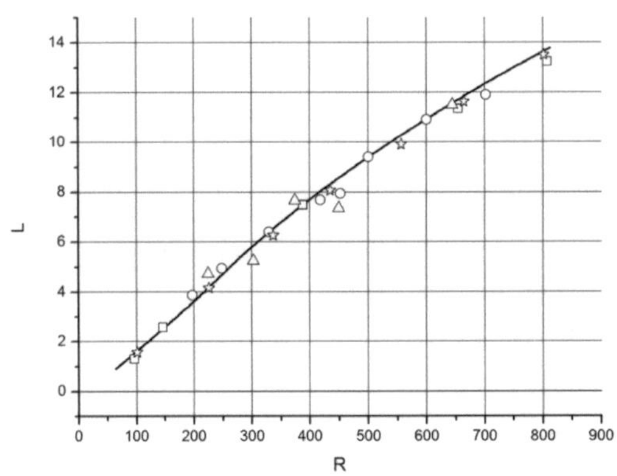

Figure 4. Mean reattachment length to the step height, compared with reference data, line - current method

Comparison for Reynolds stress correlations at $R = 1800$ with data from [38] presented in fig.6. One can see a very good agreement between DNS and experimental data.

More benchmark details are omitted for the sake of brevity, see [23, 47].

3.2. Nonlinear dynamics of laminar-turbulent transition

The idea of analysis was used many times in our research [7, 8, 23, 48] and corresponds with [32]. We saved data for velocity function in format V_x, V_y, V_z. Saving data for all domain was impossible since there's not enough disk space (it requires about 6,43 terabytes for one fixed Reynolds number), so we choose several points and saved data from them. Points in Cartesian coordinates normalized by length in each direction are: $p1 = \{0.1, 0.5, 0.5\}$; $p2 = \{0.2, 0.5, 0.5\}$; $p3 = \{0.5, 0.5, 0.5\}$; $p4 = \{0.7, 0.5, 0.5\}$; $p5 = \{0.8, 0.1, 0.1\}$. We are using Reynolds number as the bifurcation parameter for the problem and forming sets of three dimensional phase subspaces of the whole infinite dimensional phase space. The subspaces

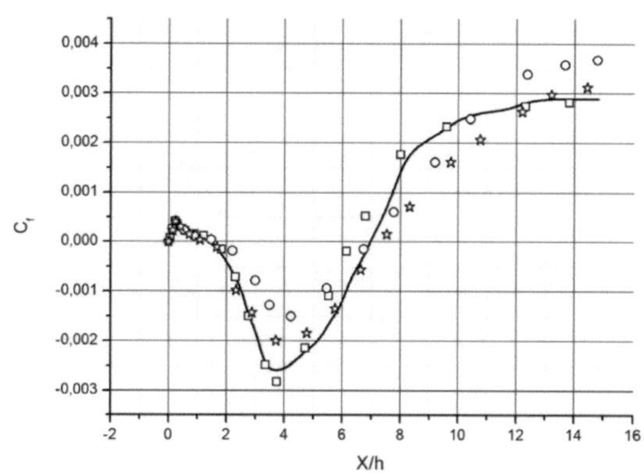

Figure 5. Mean wall friction coefficient C_f for $R = 5500$, line - current method, pints - reference data

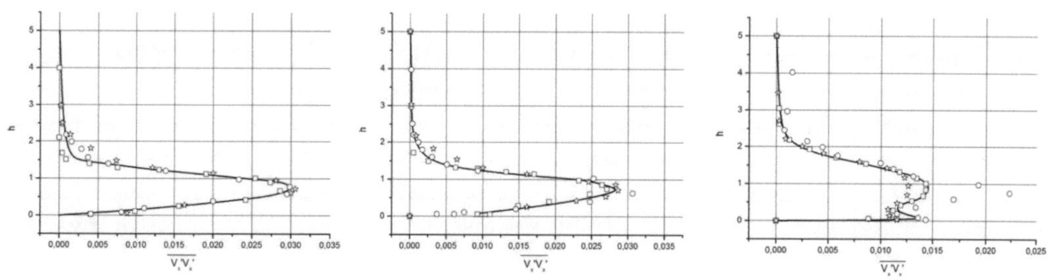

Figure 6. Reynolds stress correlations comparison for $R = 1800$ with experimental data at length $x/h = \{4;6;15\}$ in the center (W/2).

are formed by velocity vector components. Under infinite dimensional phase space here we consider finite dimensional phase space produced by the numerical system (30) whose space dimension is greater than the attractor dimension of the system (15) for the given initial-boundary value problem with a finite preset Reynolds number. It is important to notice that there exists a hysteresis of a solution if one approaches a fixed R from different sides, say solution G exists for R_0 if we travel to it by $R_1 \to R_0$, $R_1 < R_0$ but does not exist for R_0 if $R_1 > R_0$, i.e. see [49]. So we are only considering the following cascade of bifurcation parameters $R_1 < R_2 < ... < R_n = 1500$.

The solution is laminar for R from 100 up to 736 but the time of stationary solution formulation increases as R grows. For the stationary laminar solution we can monitor a fixed point in infinite dimensional phase space and in all five three dimensional subspaces of velocity vector function. Starting from a system exhibits one frequency mode regime. A limited cycle C_1 is formed from every stationary point in phase space and has projections in all subspaces. One can see a projection of the cycle for the point P_1 in fig. 7.

At the point of $R = 850$, the cycle looses stability and forms a stable two dimensional invariant torus $T_2 = C_1 \otimes C_2$ as the result of Andronov-Hopf bifurcation. This torus is located in all infinite dimensional phase space and can be found in all subspaces for points P_i. Presumably, this is due to the fact of incompressibility what expresses in elliptic operator for pressure

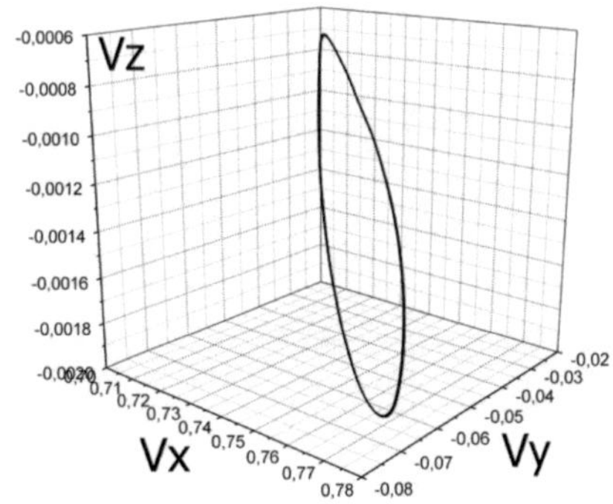

Figure 7. Point P_1, stable limit cycle, R=740

and hence the acoustic speed is infinite. An example of this torus is shown in fig.8. It is clearly seen that the system is very sensible for the values of bifurcation parameters. It is shown in fig.8, that a change of R by 1.0 changes the attractor from C_1 to T_2. Comparing these results with the mentioned work [32] we can say that the formation of the first cycle appears almost at the same Reynolds number, here we have $R = 737$ and in [32] the value is R=735. The formation of two-frequent mode in [32] appears at R=855, but there's no stable phase space trajectory available, since numerical errors started dominating. And further investigation in [32] is performed using frequency analysis that indicated creation of the other independent frequency. The recent report in TsAGI [50] of Sibgatullin I.N. indicated that other initial-boundary problems exhibit the resembling scenario by which an invariant torus is formatted in phase space. But all reports and papers indicate that a chaotic behavior follows the formation of two dimensional invariant tori. We suppose that this is due to the fact that numerical methods, used in the papers and reports, don't meet necessary conditions for theorems and assertions outlined here and so we continued the numerical analysis.

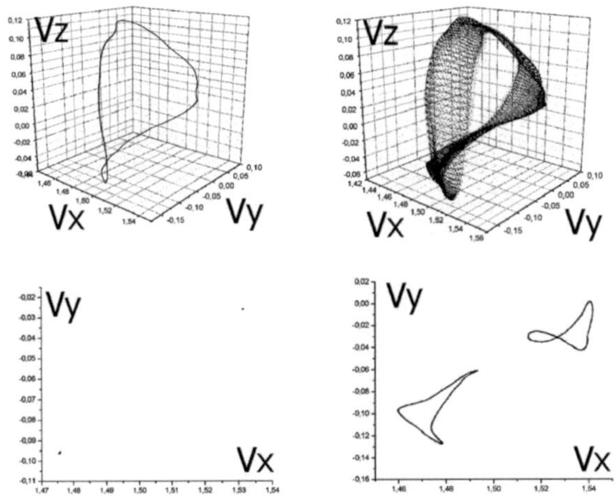

Figure 8. $R = 849$ Phase subspace cycle projection and section, point P4(left); $R = 850$ Phase subspace torus projection and section, point P4(right)

Further increase of R leads to more complex topology of cycles that are forming the torus. The process of this complication is shown in fig.9 at point P_3 and in fig.10 at point P_5.

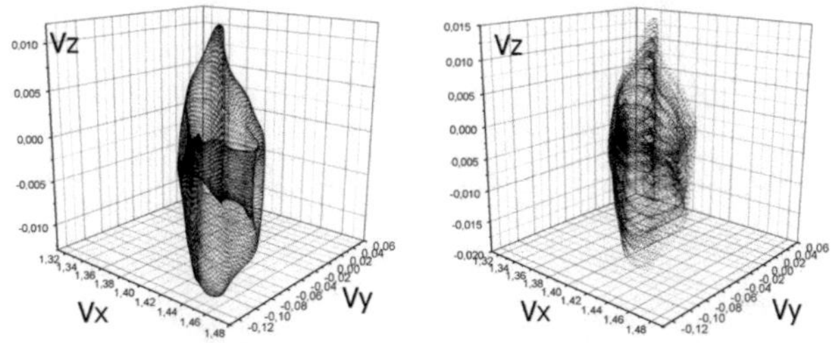

Figure 9. Torus complication in subspace projection at point P_3. $R = \{851; 883\}$

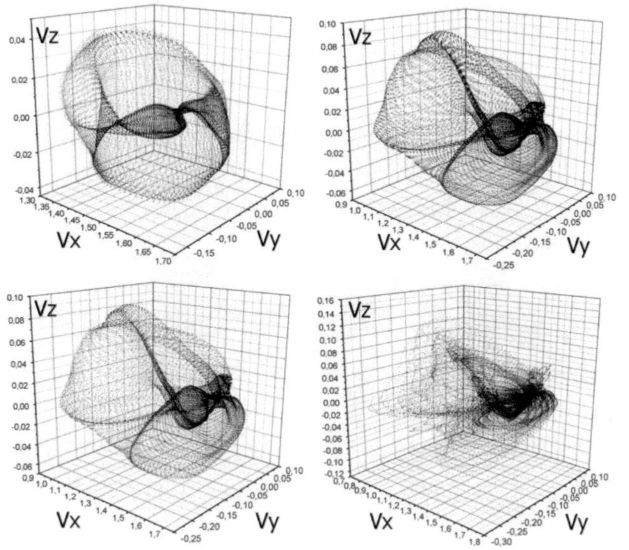

Figure 10. Torus complication in subspace projection at point P_5, $R = \{851; 882; 883; 885\}$

This process continues up to $R = 883$. It was found that at this point the two dimensional torus looses stability in the whole infinite dimensional space and forms a three dimensional invariant torus through the Andronov-Hopf bifurcation. This torus is formed by the topological multiplication of three irrational frequency cycles $T_3 = C_1 \otimes C_2 \otimes C_3$. Its projections in different subspaces are shown in fig.9, 10 for $R \geq 883$.

Since the space dimension in one point of the phase subspace is the same as the formed torus dimension one should increase the phase subspace dimension. It can be clearly seen on sections fig.11 where a new "coiling" can be monitored around the old cycle in plane section.

To do so we are taking data from another point thus forming a four dimensional subspace and then we are performing two sections by two dimensional planes. The first section in P_5 (plus data from P_2 marked with asterisk) is shown in fig.12 on the left. The torus structure can be seen by performing another section by the plane $W = 0.005$. One can see the three dimensional torus structure in fig.12 in sections for R=883.

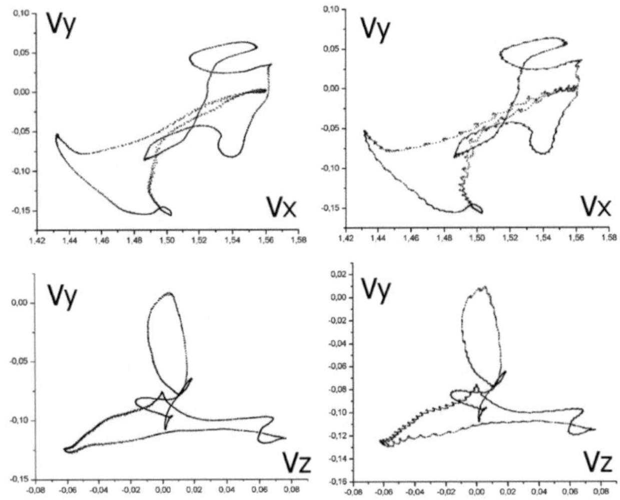

Figure 11. Plane sections of phase subspaces in P_4 and P_5 for $R = (882.5 \rightarrow 883)$ from left to right

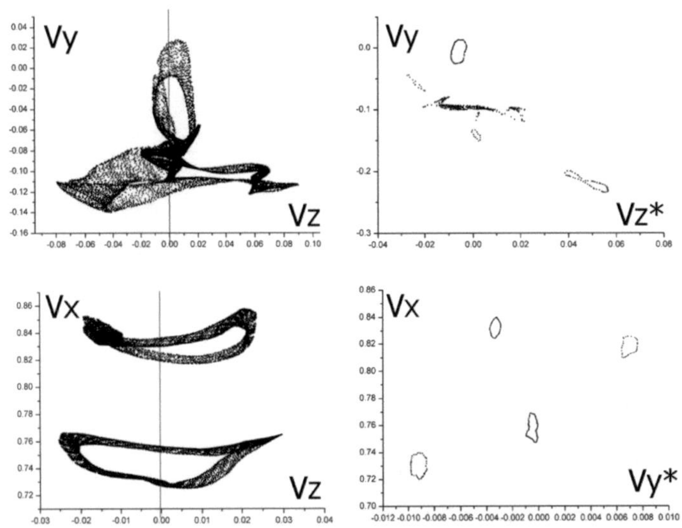

Figure 12. Section of the phase subspace in Point P_3 (top) and P_5 (bottom) plus P_2 data (marked asterisk), $R = 883$, on the left. Section of the four dimensional phase subspace by the additional plane, on the right

Three dimensional invariant torus losses its stability and through the doubling period bifurcation forms a double period three dimensional torus for $R = 883.8$ It requires huge amount of data to get results shown in fig.13. Depicted results required about 375Gb of data per one point and could only be completed by numerical methods on GPUs or on massive parallel clusters. All different sections are presented in fig.13 for four dimensional phase subspaces at $P_3 + P_2$ and $P_5 + P_2$.

It can be clearly seen that the torus topology is more complicated, i.e. see fig.13 at point P_5, and fig.12 - second section, although the bifurcation parameter only changed by 0.091%. These

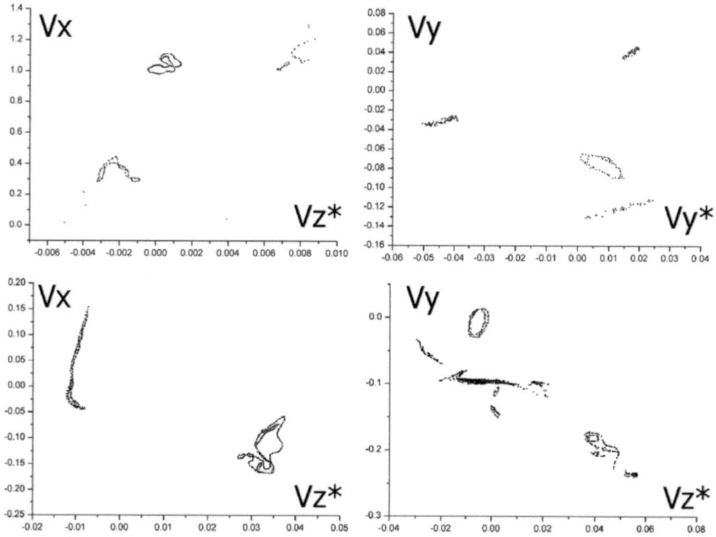

Figure 13. Section of the four dimensional phase subspaces by additional planes at point $P_3 + P_2(*)$ (top) and $P_5 + P_2(*)$ (bottom)

results show that a double period three dimensional torus is formed in all infinite dimensional phase space for $R = 883.8$. Further investigation is impossible due to the exponential growth of bifurcations. The second section view is already next to incomprehensible for $R = 890$ and consists of a filled black squire. In conclusion we found the following scenario of laminar-turbulent transition:

$$C_1 \rightarrow T_2 \rightarrow T_3 \rightarrow T_{3\otimes 2} \rightarrow, \tag{40}$$

It can be seen in (40) that the scenario exhibits initial stage of Landau-Hopf scenario (Hopf bifurcation cascades) and initial stage of FSM scenario, since the Feigenbaum scenario is progressing after the three dimensional torus. It corresponds with the initial stage of FSM scenario. One should point out that a three dimensional torus is stable at least for the time of numerical simulation for $1,5 \cdot 10^7$ timesteps at $R = 883$.

4. Laminar-turbulent transition for Rayleigh-Benard convection

One of the best studied and widely analyzed problems of fluid mechanics is the Rayleigh-Benard natural convection problem. The problem has been considered by many and has lots of results in the field of nonlinear bifurcation analysis, analytical, numerical and experimental.

4.1. The Oberbeck-Boussinesq approximation of Rayleigh-Benard convection, dimensionless form and benchmarks

One of the possible mathematical models for this problem is the Oberbeck-Boussinesq approximation. Here we are closely following [51] and assume, that fluid physical properties (ν,β) are only linear functions of temperature perturbations. The fluid density can be given as a function of temperature perturbation as:

$$\rho = \rho_0 \left(1 - \beta(T - T_0)\right), \tag{41}$$

where ρ - fluid density, T - fluid temperature, β - fluid thermal expansion coefficient, ρ_0,T_0 - mean values of fluid density and temperature. It is assumed under Oberbeck-Boussinesq approximation that density only changes due to temperature difference and, thus, causing buoyancy, yet fluid is considered incompressible. Temperature emission due to friction is also neglected. Introducing (41) to Navier-Stokes equations (1), assuming temperature passive advection-diffusion and taking gravity vector in Cartesian coordinates as $\vec{g} = \{0;0;-1\}$ in account, one gets:

$$\begin{gathered} \nabla \cdot \mathbf{V} = 0, \\ \frac{\partial \mathbf{V}}{\partial t} + (\mathbf{V} \cdot \nabla)\mathbf{V} + \rho_0^{-1} \nabla P = \nu \nabla^2 \mathbf{V} + \vec{g}\beta(T - T_0), \\ \frac{\partial T}{\partial t} + \mathbf{V}\nabla T = \chi \nabla^2 T, \end{gathered} \tag{42}$$

here χ is a fluid thermal conductivity coefficient. There are many scales can be chosen that make (42) dimensionless for the Rayleigh-Benard convection problem. One of the most common ways [51] is to use time scale τ as $\tau = h^2/\nu$, where h is the length between two planes with given temperature difference. Another way is to associate τ with the momentum transport by viscous terms and then $\tau = h^2/\nu$. Introducing additional dimensionless similarity criteria one can formulate a Rayleigh-Benard convection problem: find vector-function $\mathbf{V} : \Omega \times [0,t] \to \mathbb{R}^3$, scalar pressure function $P : \Omega \times [0,t] \to \mathbb{R}$ and and scalar temperature function $T : \Omega \times [0,t] \to \mathbb{R}$ that satisfy the following initial-boundary value problem:

$$\begin{gathered} \frac{\partial \mathbf{V}}{\partial t} + (\mathbf{V} \cdot \nabla)\,\mathbf{V} + \nabla P = \nabla^2 \mathbf{V} + RaPr^{-1}(T - T_0) \cdot (0;0;-1)^T \text{ in } Q; \\ \nabla \cdot \mathbf{V} = 0 \text{ in } Q = \Omega \times (0,t); \\ \frac{\partial T}{\partial t} + \mathbf{V}\nabla T = Pr^{-1}\nabla^2 T \text{ in } Q; \\ \mathbf{V} = 0, \partial T/\partial \vec{n} = 0, \text{ on } \partial\Omega_0 \times (0,t); \mathbf{V} = 0, T = T_\alpha \text{ on } \partial\Omega_1; \\ \mathbf{V}(\vec{x},0) = \mathbf{V}_0(\vec{x}), \nabla \cdot \mathbf{V}_0 = 0; T(\vec{x},0) = T_0(\vec{x}) \text{ in } \Omega. \end{gathered} \tag{43}$$

Here: Ω is a bounded domain with local Lipschiz-continuous boundary $\partial\Omega_i$; t is time; T is a fluid temperature; T_0 is a reference fluid temperature; $Pr = \nu/\chi$ is the dimensionless Prandtl number; $Ra = g\beta h^3 \Delta T/(\nu\chi)$ is the Rayleigh number; h is height in Ω between $\partial\Omega_1$; ΔT is a temperature difference between $\partial\Omega_1$.

There are two main types of boundary conditions. First consider boundary conditions whose plane is parallel to the temperature gradient, i.e. Ω_0. One usually chooses either periodic boundary conditions for temperature and velocity or wall boundary conditions with Neumann type for temperature. On other planes, namely Ω_1, temperature gradient by Dirichlet boundary conditions is set with wall no slip boundary for velocity.

Numerical solution method differs from (28) only in temperature equation and in diffusion part of Navier-Stokes equations. We applied implicit five diagonal matrix solution method [52] for all diffusion parts of (43) thus accuracy drops from 6-th order to 4-th order in space. It can be shown by the modified wave number analysis that number of elements remains the same just with a little lost of accuracy. In order to solve matrix equation $[A]\,[X] = [B]$ that arises from implicit method for diffusion parts of equation a five diagonal fast factorization solution routine is adopted for CPU calculations [53] and Geometric Multigrid method for GPU [54].

In order to satisfy conditions of the theorems we must adopt mesh by introducing Reynolds number through Prandtl and Payleigh numbers. We are using paper results [55, 56] that indicate the following relation is true:

$$R \simeq Ra^{0.44} Pr^{-0.76}, \tag{44}$$

for the range of $0.9 < Pr < 2$ and $1 \cdot 10^5 < Ra < 1 \cdot 10^9$. It is also known [57, 58] that transition form "soft" turbulence (where some frequencies can be determined in frequency analysis) to "hard" turbulence (where frequency-amplitude response becomes a coloured noise) occurs at $Ra_{cr} \approx 4 \cdot 10^7$ for $Pr > 0.9$. Assuming that it is impossible to make any quantitative analysis after similarity criteria greater than critical values we take relation (44) and find Reynolds number as:

$$R_{max} \simeq Ra_{cr}^{0.44} \min(Pr_{cr})^{-0.76} = 2856. \tag{45}$$

Using (45) and applying it to (27) one can get the following mesh adopted variants for cubic and cylindrical domains after calibration simulations:

$$\begin{aligned}
&G1 = \{250X250X250\} \text{ for } Pr = 0.9; \Delta x_{max} = 6.1\Delta x_{wall};\\
&G2 = \{216X216X216\} \text{ for } Pr > 1.0; \Delta x_{max} = 5.2\Delta x_{wall};\\
&G3 = \{185X185X185\} \text{ for } Pr > 1.25; \Delta x_{max} = 5.0\Delta x_{wall};\\
&G4 = \{250X250X100\} \text{ for } Pr > 4.0; \Delta x_{max} = 4.5\Delta x_{wall}; r/h = 4;\\
&G5 = \{400X400X47\} \text{ for } Pr > 5.0; \Delta x_{max} = 3.5\Delta x_{wall}; r/h = 14;
\end{aligned} \tag{46}$$

where r is the cylinder radius and h is the cylinder height. Immersed boundary is used for cylindrical approximation, for more information see [59]. For other geometry domains grid is specified in the same manner and omitted here for the sake of brevity.

Benchmarking the problem requires many different domain configurations and different dimensionless variations of the equations (43). We skip that for the sake of brevity and can recommend book by professor Getling A.V. [51] for more information on dimensionless forms, physical background and analytical analysis. It is known from the linear minimodal approximation of the problem [51] that the flow with wall boundary conditions on Ω_0 is more stable than that with periodic conditions or two dimensional problems. For the infinite Pr number the critical value of Rayleigh criterion for the first instability is $Ra_{cr} = 1707.762$ for the wave number $k = 3.117$ in case of wall boundary conditions and $Ra_{cr} = 27\pi^4/4 \approx 657.511$ with $k = 2.221$ for periodic boundary conditions. For benchmark verification several papers where considered [60].

Solution in cylindric domain for $Ra = 2000 - 31000$ and $r/h = 4$ with zero initial conditions is shown in fig.14, top and with initial perturbation $V_z = \cos(2\pi z/(2r))/100$ – in fig.14 bottom, where one can see the appearance of absolutely different solution for perturbed initial conditions with the prototype function $\cos(kx)$, see [51].

Some results in rectangular domain where also considered and compared with [61] with very good agreement. More results where compared with [51, 60] for $r/h = 14$ for range of

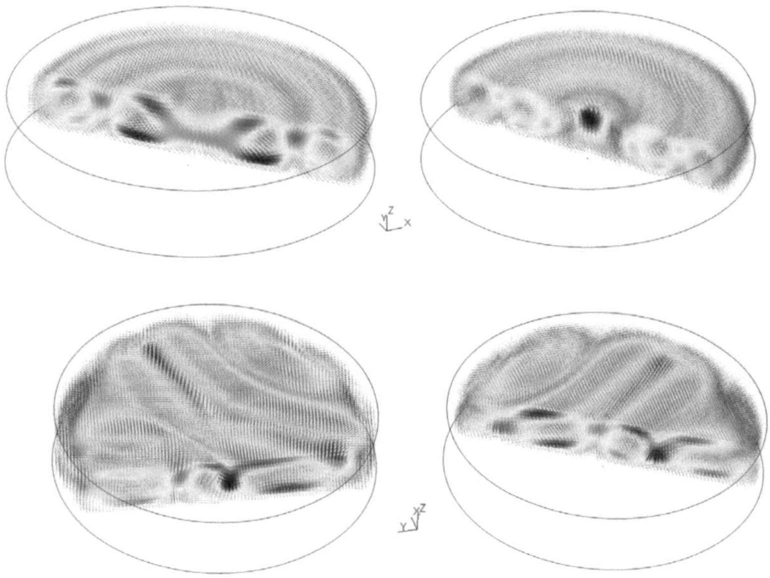

Figure 14. Velocity vector function section $Ra = 2000 - 31000$, $r/h = 4$. Zero initial conditions on the top and perturbed initial conditions on the bottom.

$Ra = \{2565, 4500, 6500, 16850\}$. We could clearly see the process of different pattern formation (round, circular, "vetruvian man", hexagon, irregular). For the analysis of nonlinear dynamics we considered a cubic domain absolutely identical to the physical experiment, presented in the book by P.G. Frik [62]. Experiment was made in a copper cubic domain with the height of 40mm. Horizontal boundaries where thermally stabilized and vertical boundaries formed a constant gradient. The experiment was aimed to investigate frequencies and attractors using thermal differential pares. Numerical simulation for $Ra = 2 \cdot 10^5$ and $Pr = 7$ is shown in fig.15.

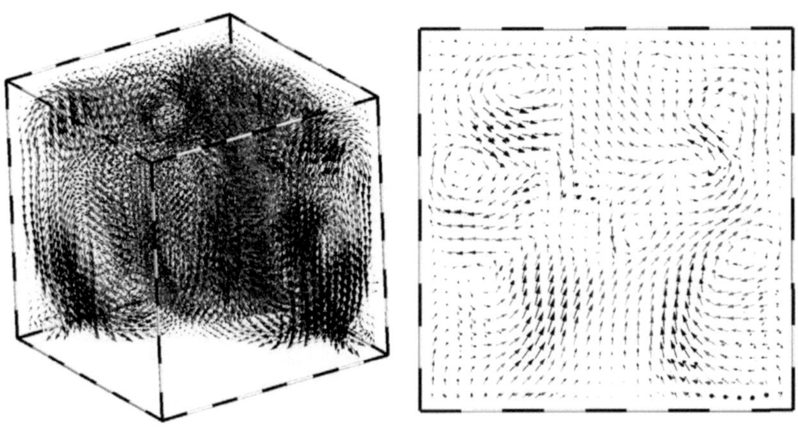

Figure 15. Velocity vectors in 3D and plane section (1;0;0), for $Ra = 2 \cdot 10^5$ and $Pr = 7$. Every six vector is shown.

4.2. Nonlinear dynamics of laminar-turbulent transition.

The analysis is absolutely identical to the one used for the problem of the flow over backward facing step. For this point we select five points with relative Cartesian coordinates $p_1 = \{0.5; 0.5; 0.5\}$; $p_2 = \{0.4; 0.5; 0.5\}$; $p_3 = \{0.5; 0.8; 0.5\}$; $p_4 = \{0.8; 0.5; 0.667\}$; $p_5 = \{0.16; 0.167; 0.1\}$. The subspaces of infinite dimensional phase space are constructed by velocity vector functions in various points and their combination. Similarity criteria of Ra and Pr numbers are used as bifurcation parameters. We took fixed values for Prandtl number and increased Ra number for every Pr number, thus we considered five various series of calculations.

4.2.1. First calculation series.

Prandtl number is set as 1.866666666. While we change Ra from 1 to 14100 we could see the formation of various stationary solutions with various recirculation zones formation in the domain with the complete correspondence to [62]. Laminar solution can be seen for $Ra < 2.5 \cdot 10^5$ in the whole domain, that corresponded to the fixed stationary point in the infinite dimensional phase space and in all subspaces. However this fixed point jumped from one position to another in the phase space as a function of Ra number. This stationary point looses stability for $Ra = 2.5 \cdot 10^5$ and a limited cycle is formed in the phase space and in each phase subspace. This cycle projection for point p_1 is shown in fig16.

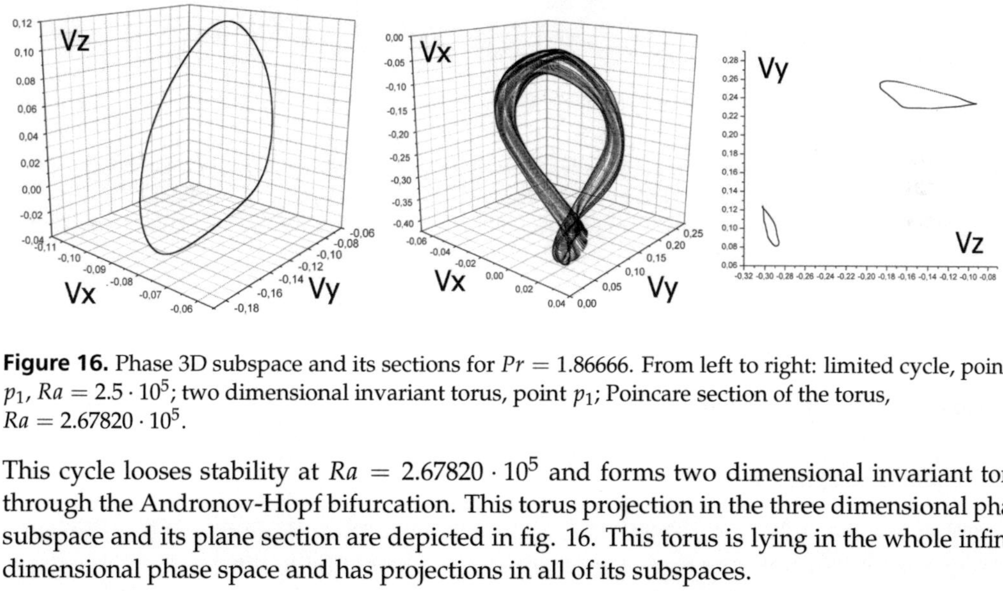

Figure 16. Phase 3D subspace and its sections for $Pr = 1.86666$. From left to right: limited cycle, point p_1, $Ra = 2.5 \cdot 10^5$; two dimensional invariant torus, point p_1; Poincare section of the torus, $Ra = 2.67820 \cdot 10^5$.

This cycle looses stability at $Ra = 2.67820 \cdot 10^5$ and forms two dimensional invariant torus through the Andronov-Hopf bifurcation. This torus projection in the three dimensional phase subspace and its plane section are depicted in fig. 16. This torus is lying in the whole infinite dimensional phase space and has projections in all of its subspaces.

The two-dimensional tori of double and quadruple periods are formed with the further increasing of Rayleigh number (fig. 17 and fig.18). Point p_5 in fig.18 depicts the torus projection near wall boundaries for $Ra = 2.9133225210 \cdot 10^5$.

Further increasing of Ra number led to chaotic instability. This can be related to the limited accuracy of the numerical method and solution trajectories slipped from one to another in such sensible dynamic system. The cascade of bifurcations agreed well with experimental

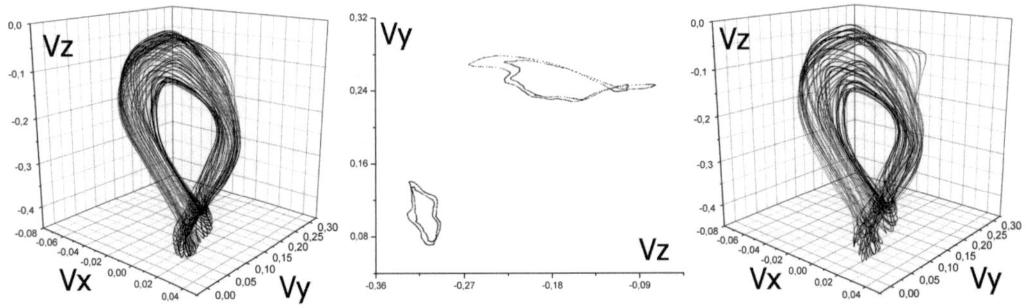

Figure 17. Phase 3D subspace and its sections for $Pr = 1.86666$. From left to right: double period two-dimensional invariant torus, point p_1, $Ra = 2.767858170 \cdot 10^5$; Poincare section of the double period torus; quadruple period two-dimensional invariant torus, point p_1, $Ra = 2.9133225210 \cdot 10^5$.

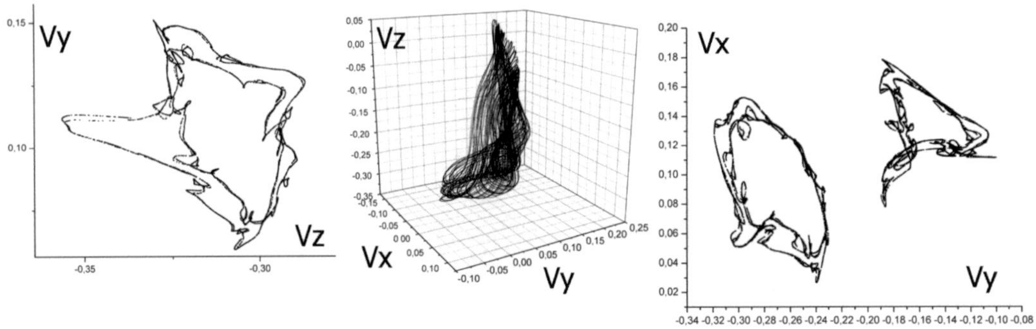

Figure 18. Phase 3D subspace and its sections for $Pr = 1.86666$. From left to right: 1/2 (zoomed) Poincare section of the quadruple period two-dimensional torus, point p_1, $Ra = 2.9133225210 \cdot 10^5$; quadruple period two-dimensional invariant torus, point p_5, $Ra = 2.9133225210 \cdot 10^5$; Poincare section of the quadruple period torus, point p_1.

results from [62] with only difference in exact values for Ra number. In [62] it could be clearly seen the formation of limit cycle and torus (two irrational frequencies), but further investigations led to chaos so it was impossible to tell from the experimental data of further bifurcations for the formed torus.

4.2.2. Second calculation series.

Pr number is fixed as 1.61290. The fluid motion is stationary up to $Ra = 1.361 \cdot 10^6$ when a limit cycle is formed from the stationary point in the whole infinite dimension phase space. Further increasing of Ra up to $1.365 \cdot 10^6$ led to the doubling period bifurcation with formation of double period limit cycle and, immediately, formation of the two-dimensional invariant torus through the Andronov-Hopf bifurcation.

This torus can be clearly seen for $Ra = 1.366 \cdot 10^6$ on the 3D subspace projection to the (V_x, V_y) plane in fig.19. Another doubling period bifurcation occurs at $Ra = 1.36905 \cdot 10^6$ forming a

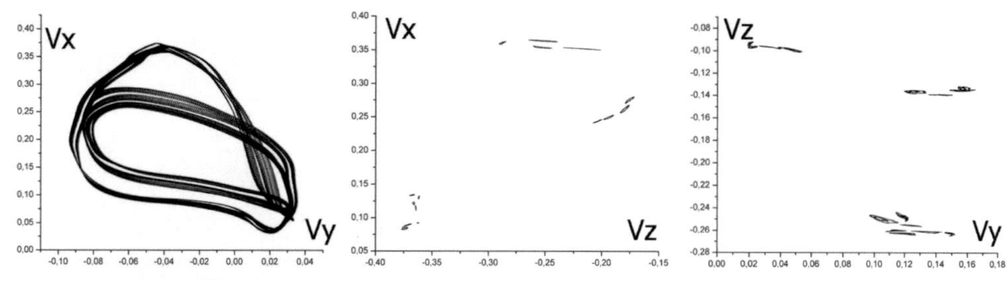

Figure 19. Phase projection and its sections for $Pr = 1.6129$. From left to right: the two dimensional invariant torus projection on (V_x, V_y) plane for point p_1, $Ra = 1.366 \cdot 10^6$; Poincare section of the torus at point p_1; Poincare section of the double period torus at point p_1, $Ra = 1.36905 \cdot 10^6$.

double period two-dimensional invariant torus that can be seen in fig.19. Further analysis was limited due to the numerical noise.

4.2.3. *Third calculation series.*

Prundtl number is set 1.354839. The limit cycle in system phase space is formed from the stationary point at $Ra = 1.286 \cdot 10^6$ and its projection on (V_x, V_y) plane is shown in fig.20.

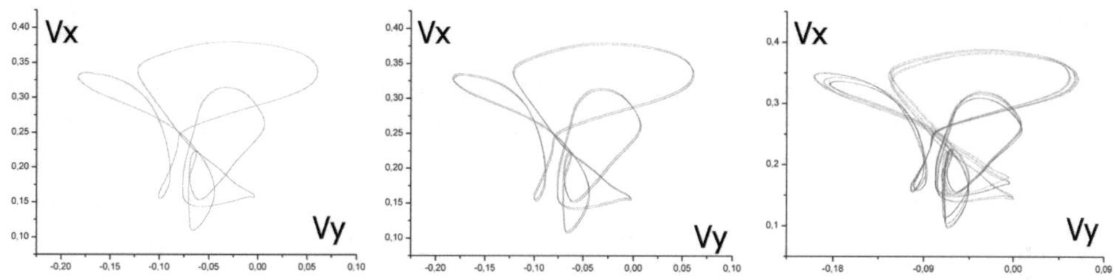

Figure 20. Phase subspace projection to (V_x, V_y) plane for $Pr = 1.354839$ at point p_1. From left to right: stable cycle, $Ra = 1.286 \cdot 10^6$; double period stable cycle, $Ra = 1.296 \cdot 10^6$; quintuple period stable cycle, $Ra = 1.306 \cdot 10^6$.

Next doubling period bifurcation occurs at $Ra = 1.296 \cdot 10^6$ with the formation of double period limit cycle, see fig.20. This cycle is lying in all infinite dimensional phase space and in all subspaces. With the increasing of Ra number the cycle suffers cascades of bifurcations in accordance with the FSM scenario. The quintuple period cycle is formed for $Ra = 1.306 \cdot 10^6$ who's projection on (V_x, V_y) plane is shown in fig.20.

A triple period cycle can be seen at $Ra = 1.308 \cdot 10^6$ in all subspaces and its projection on (V_x, V_y) plane is shown in fig.21. It means that the system suffered the whole subharmonic cascade of bifurcations and now there are all other unstable cycles exist in the system in accordance with the theory FSM. In conclusion one can say that there exist multiple scenarios of laminar-turbulent transition in Rayleigh-Benard convection as functions of Pr/Ra ratio:

1. Landau-Hopf bifurcation scenario that forms sets of many-dimensional tori T^n.

2. Landau-Hopf bifurcation scenario with doubling period bifurcation on n-dimensional invariant torus (Landau-Hopf scenario + FSM scenario).

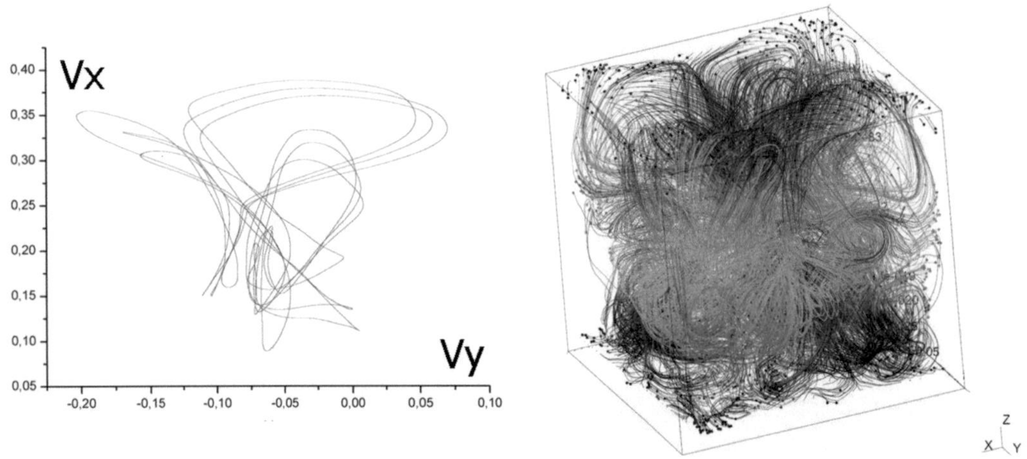

Figure 21. Triple period stable cycle at point, $Pr = 1.354839$, $Ra = 1.308 \cdot 10^6$; Stream lines in Ω for the same Pr and Ra.

3. FSM scenario with subharmonic cascade of bifurcations of stable limit cycles.

5. Conclusion

The following chapter covers five year work that has been conducted in the Chaotic Dynamics Laboratory in the Institute for Systems Analysis of Russian Academy of Sci., lead by professor N.A. Magnitskii. Our attempt in using standard open source or commercial software for this kind of analysis failed so we had to consider a specially constructed accurate and trustworthy numerical solution code for Navier-Stokes equations, partly described here. The results of numerical solution for initial-boundary value problems considered confirmed that laminar-turbulent transition undergoing the bifurcation process and has different scenarios. It is interesting to point out that similar scenarios with classical Feigenbaum scenario and Sharkovskiy windows of periodicity where recently found in [66] for initial-boundary value problems in continuous mechanical systems such as flexible plates and shallow shells. However, the universal FSM scenario [7] is found in all problems considered, despite the difference between problems. Recently we found that Boltzmann equations in hydrodynamic limit with BGK collision integral [63–65] also exhibit FSM scenario for laminar-turbulent transition process. It is likely that all hydrodynamic type chaotic solutions for PDEs are undergoing the FSM scenario in various modifications. The work continues and now we are considering compressible fluid dynamics (transonic and supersonic turbulence) and magnetohydrodynamics as well as other initial-boundary value problems for (1).

Author details

Nikolai Magnitskii and Nikolay Evstigneev
Laboratory 11-3, Chaotic Dynamical Systems, Institute for Systems Analysis of RAS, Russian Federation

6. References

[1] Carlson A., Jaffe A. and Wiles A. (Editors). (2006) The Millennium Prize Problems. Cambridge: Clay Mathematics Institute.

[2] Sadovnichii V. and Simo S.(Editors). (2002) Modern Problems of Chaos and Nonlinearity. - Moscow-Igevsk: Institute for Computational Studies.

[3] Magnitskii N.A. (2008) Universal theory of dynamical chaos in nonlinear dissipative systems of differential equations. // Commun. Nonlinear Sci. Numer. Simul., 13, pp. 416-33.

[4] Magnitskii N.A. (2007) Universal theory of dynamical and spatio-temporal chaos in complex systems. // Dynamics of Complex Systems 1, 1, pp. 18-39. (in Russian)

[5] Magnitskii N.A. (2008) New approach to analysis of Hamiltonian and conservative systems. //Differential Equations, 44, 12, pp. 1618-27.

[6] Magnitskii N.A. (2010) On topological structure of singular attractors of nonlinear systems of differential equations. //Differential Equations, 46, 11, pp.1551-1560.

[7] Magnitskii N.A., Sidorov S.V. (2006) New Methods for Chaotic Dynamics. - Singapure: World Scientific.

[8] Magnitskii N.A., Sidorov S.V. (2005) On transition to diffusion chaos through a subharmonic cascade of bifurcations of two-dimensional tori. // Differential Equations , 41, 11, pp. 1550-58.

[9] Evstigneev N.M., Magnitskii N.A., Sidorov S.V. (2009) On nature of laminar-turbulent flow in backward facing step problem. //Differential equations, v.45, 1, pp.69-73.

[10] Evstigneev N.M., Magnitskii N.A., Sidorov S.V. (2009) On the nature of turbulence in Rayleigh-Benard convection.//Differential Equations V.45, N.6, pp.909-912.

[11] Evstigneev N.M, Magnitskii N.A. and Sidorov S.V. (2010) Nonlinear dynamics of laminarturbulent transition in three dimensional Rayleigh-Benard convection.// Commun. Nonlinear Sci. Numer. Simul., 15, p. 2851-2859.

[12] Evstigneev N.M., Magnitskii N.A. (2010) On possible scenarios of the transition to turbulence in Rayleigh-Benard convection.// Doklady Akademii Nauk, Vol. 433, No. 3, pp. 318-322.

[13] Filatov A.N., Ipatova V.M. (1996) On globally stable difference schemes for the barotropic vorticity equation on a sphere // Russian J. Numer. Anal. Math. Modelling. - V. 11, N 1. - P. 1-26.

[14] Ipatova V.M. (1997) Attractors of approximations to non-autonomous evolution equations.//Mat. Sb., Volume 188, Number 6, Pages 47-56.

[15] Foias C., Temam R.(1987) The connection between the Navier-Stokes equations, dynamical systems, and turbulence theory, in Directions in Partial Differential Equations, Academic Press, NY, 55-73

[16] Temam R. (1991) Approximation of Attractors, Large Eddy Simulations and Multiscale Methods.// Proc. R. Soc. Lond.v.434, pp23-39.

[17] Ladyzhenskaya O.A. (1985) On the finiteness of the dimention of bounded invarian sets for the Navier-Stokes equations and other related dissipative systems.// J.Soviet Math. 28, pp714-725.

[18] Chepyzhov V., Vishik M. (2002) Attractors for Equations of Mathematical Physics.// Amer. Math. Soc. Colloq. Publ., Vol. 49, Amer. Math. Soc., Providence, RI.

[19] Vishik M.I., Titi E.S., Chepyzhov V.V. (2007) On convergence of trajectory attractors of the 3D Navier-Stokes-α model as α approaches 0.// SB MATH, 198 (12), pp 1703-1736.
[20] Temam R. (1983) Navier-Stokes Equations and Nonlinear Functional Analysis, SIAM.
[21] Buckingham E. (1914) On physically similar systems; illustrations of the use of dimensional equations. // Physical Review 4 pp.345-376.
[22] Evstigneev N.M. (2010) About one averaging method of the equations of a compressible and incompressible fluid. // Proc. GROU, PhM series, N2, pp.47-52. (in Russian)
[23] Evstigneev N.M., Magnitskii N.A., Sidorov S.V. (2008) New method for turbulent analysis in incompressible viscous fluid flow. // Proc. ISA RAS,v. 33, ed. 12. pp.49-65. (in Russian)
[24] Evstigneev N.M. (2008) Integration of 3D incompressible free surface Navier-Stokes equations on unstructured tetrahedral grid using distributed computation on TCP/IP networks. // Proc. Of the VII International conf. "Advances in Fluid Mechanics", pp. 194-208.
[25] Belotserkovskii O.M., Oparin A.M., Chechetkin V.M. (2005) Turbulence: New Approaches. - CISP, Science.
[26] Martin M.P. , Taylor E.M. , Wu M. , Weirs V.G. (2006) A bandwidth-optimized WENO scheme for the effective direct numerical simulation of compressible turbulence.// Journal of Computational Physics 220 (2006) 270-289.
[27] Evstigneev N.M. (2010) Numerical solution for Navier-Stokes equations on unstructured grid using semi-Lagrangian method.// Scientific bulletin of SpBSPolitechUniv., 1 (93). pp. 163-170. (in Russian)
[28] Cockburn B. and Shu C.-W. (1998) The Runge-Kutta discontinuous Galerkin method for conservation laws V: Multidimensional systems.// J. Comput. Phys., 141, pp. 199-224.
[29] Shu C.-W. (1987) TVB uniformly high-order schemes for conservation laws.// Math. Comp., 49, pp. 105-121.
[30] Shu C.-W. and Osher S. (1988) Efficient implementation of essentially non-oscillatory shock-capturing schemes.// J. Comput. Phys., v77, pp.439-471.
[31] CUDA C Programming Guide (2011)

http://developer.nvidia.com/category/zone/cuda-zone

[32] Rani H.P., Tony W.H. Sheu (2006) Nonlinear dynamics in a backward-facing step flow.// Physics of Fluids, Vol. 18, pp. 084101-14.
[33] Chiang T. P. and Tony W. H. Sheu. (2002) A numerical revisit of backward-facing step flow problem.// PHYSICS OF FLUIDS, VOLUME 11, NUMBER 4, 862-874.
[34] Rani H. P., Tony W. H. Sheu and Tsai E.S.F. (2007) Eddy structures in a transitional backward-facing step flow.// J. Fluid Mech. , vol. 588, pp. 43-58.
[35] Moin, Lee, T., Mateescu, D., (1997) Direct numerical simulation of turbulent flow over a backward-facing step.//J. of Fluids and Structures, vol. 330, pp. 349-374.
[42] Kim J., Moin P. (1985) Application of a fractional-step method to incompressible Navier-Stokes equations. //J Comp Phys, 59:308-23.
[37] Moin P., Mahesh K. (1998) Direct Numerical Simulation: A Tool in Turbulence Research.//Annu. Rev. Fluid Mech. 30:539-78.
[38] Driver, D.M., and Jovic, S. (1998) Backward facing step - experiment. Agard advisory report, agard-ar-345, a selection of test cases for the validation of large-eddy simulations of turbulent flows.// Report CMP31. 195-196.

[39] Adnan Meri and Wengle Hans (2004) DNS and LES of Turbulent Backward-Facing Step Flow Using 2ND-and 4TH-Order Discretization. //Fluid Mechanics and Its Applications, Volume 65, 2, 99-114.
[40] Dwight Barkley, Gabriela M., Gomes M., Henderson Ronald D. (2002) Three-dimensional instability in flow over a backward-facing step.// J. Fluid Mech., vol. 473, pp. 167-190.
[41] Varapaev, V. N. and Yagodkin, V. I. (1969) Flow stability in a channel with porous walls.// Izv. Akad. Nauk SSSR, Mekh. Zhid. i Gaza 4, 91-95.
[42] Kim J., Moin P. (1985) Application of a fractional-step method to incompressible Navier-Stokes equations. //J Comp Phys, 59:308-23.
[43] Mouza A.A., Pantzali M.N. ,Paras S.V., Tihon J. (2005) EXPERIMENTAL and NUMERICAL STUDY OF BACKWARD-FACING STEP FLOW.// Proc. Of 5th National Chemical Engineering Conference, Thessaloniki, Greece.
[44] Perez Guerrero J. S. and Cotta R. M. (1996) Benchmark integral transform results for flow over a backward-facing step.// Comput. Fluids 25, 527.
[45] Aider J.L. and Danet A. (2006) Large-eddy simulation study of upstream boundary conditions influence upon a backward-facing step flow.// Comptes Rendus M'ecanique, vol.334(7), 447 - 453.
[46] Jovic S., Driver D.M. (1994) Backward-facing step measurements at low Reynolds number, $Re_h = 5000$.// NASA Tech. Memo 108807.
[47] Evstigneev N.M, Magnitskii N.A. and Sidorov S.V. (2009) Nonlinear dynamics in Rayleigh-Benard convection problem.// Proc. 3d international conf. "System analysis and information technology", SAIT-2009, pp. 412-415. (in Russian)
[48] Evstigneev N.M., Ryabkov O.I. (2010) On the visualization qualitative analysis in nonlinear dynamic systems // Scientific Visualization. Electronic Journal of National Research Nuclear University "MEPhI".c.4 v.2 N.4 pp. 61-71.
[49] Volkov V. F., Tarnavskii G. A. (2001) Broken symmetry and hysteresis in steady-state and quasi-steady solutions to Euler and Navier-Stokes equations.// Zh. Vychisl. Mat. Mat. Fiz., 41:11, 1742-1750.
[50] TSAGI Web Seminar (2011)

`http://www.tsagi.ru/cgi-bin/jet/viewnews.cgi?id=`
`20110506121862279081`

[51] Getling A.V. (1998) Rayleigh-Benard Convection. Structures and Dynamics. - Advanced Series In Nonlinear Dynamics, vol.11, World Scientific.
[52] Van Daele M., Vanden Berghe G. and De Meyer H. (1992) Five-diagonal finite difference methods based on mixed-type interpolation for a certain fourth-order two-point boundary-value problem.// Computers and Mathematics with Applications. V24, I10, pp55-76.
[53] Diele F. (1998) The use of the factorization of five-diagonal matrices by tridiagonal Toeplitz matrices. // Applied Mathematics Letters Volume 11, Issue 3, Pages 61-69.
[54] Evstigneev N.M. (2009) Numerical integration of Poisson's equation using a graphics processing unit with CUDA-technology.// J. Numerical Methods and Programming, V.10, pp.268-274.

[55] Siegfried Grossmann, Detlef Lohse. (2002) Prandtl and Rayleigh number dependence of the Reynolds number in turbulent thermal convection.// PHYSICAL REVIEW E 66, pp16305-16305-6.
[56] Siegfried Grossmann, Detlef Lohse. (2000) //J. Fluid Mech. 407, pp27-35.
[57] Castaing, B., Gunaratne G. , Heslot F., Kadanoff L., Libchaber A., et al., (1989) Scaling of hard thermal turbulence in Rayleigh- Benard convection. // J. Fluid Mech. 204, 1-30.
[58] Niemela, J.J., Skrbek L. , Sreenivasan K.R., and Donnelly R.J.. (2000) Turbulent convection at very high Rayleigh numbers.// Nature, 398, 307-310.
[59] Evstigneev N.M. (2010) On the Lattice Boltzmann Method stabilization for turbulent flow regimes with extremely high Reynolds numbers.// Proc. GROU, PhM series, N2, pp. 53-62. (in Russian)
[60] Katarzyna Boronska, Laurette S. Tuckerman. (2010) Extreme multiplicity in cylindrical Rayleigh-Benard convection. I. Time dependence and oscillations.//PHYSICAL REVIEW E 81, 036320 pp 1-13.
[61] Watanabe T. (2004) Numerical Evidence for Coexistence of Roll and Square Patterns in Rayleigh-Benard Convetion.// Phys. Lett. A, Vol.321, pp.185-189.
[62] Frik P. G. (2003) Turbulence: Approaches and Models. - Moscow, IKI, (in Russian).
[63] Evstigneev N. M.; Magnitskii, N. A. (2010) Nonlinear Dynamics of Laminar-Turbulent Transition in Back Facing Step Problem for Bolzmann Equations in Hydrodynamic Limit.// Int. Conf. ICNAAM 2010. AIP Conference Proceedings, Volume 1281, pp. 896-900.
[64] Evstigneev N. M., Magnitskii, N. A. (2010) Nonlinear Dynamics in the Initial-Boundary Value Problem on the Fluid Flow from a Ledge for the Hydrodynamic Approximation to the Boltzmann Equations.// Differential Equations, V.46, N.12, pp.1794-1798.
[65] Evstigneev N.M. (2010) Lattice Boltzman method with entropy stabilization on GPU. // Proc. ISA RAS,v. 53, ed. 14. pp.49-65. (in Russian)
[66] Awrejcewicz J., Krysko V.A., Papkova I.V., Krysko A.V. (to appear 2012) Routes to chaos in continuous mechanical systems. Part 1,2,3. // Chaos Solitons and Fractals. Nonlinear Science, Non-equilibrium and Complex Phenomena.

Chapter 11

Numerical Algorithms of Finding the Branching Lines and Bifurcation Points of Solutions for One Class of Nonlinear Integral Equations

B. M. Podlevskyi

Additional information is available at the end of the chapter

http://dx.doi.org/10.5772/48735

1. Introduction

When investigating the nonlinear equations of the form

$$A(\lambda, f) = f,$$

where the operator $A(\lambda, f)$ nonlinearly depends both on the parameter λ and the function f, the formalistic approach, which is based on linearization, is applied. The application of this approach shows, that the branching points of equation can be only those values of parameter λ, for which unit ($\nu = 1$) is the eigenvalue of the corresponding linearized equation (see, eg, [20])

$$A(\lambda) f = f$$

with the operator-valued function $A: C \to X(H)$ ($X(H)$ is a set of linear operators, $\lambda \in C$ is the spectral parameter), nonlinearly depending on the parameter λ. If the linearized equation linearly depends on the parameter λ, i.e. $Af = \lambda f$, then its eigenvalues will be the branching points of initial equation. In a general case the curves of eigenvalues $\nu(\lambda)$ appears and then the branching points will be those values of parameter λ of the problem

$$A(\lambda) f = \nu(\lambda) f,$$

for which $\nu(\lambda) = 1$.

The theory of branching solutions of nonlinear equations arose in close connection with applied problems and development of its ever-regulated by the new applied problems.

INTECH
open science | open minds

© 2012 The Author(s). Licensee InTech. This chapter is distributed under the terms of the Creative Commons Attribution License http://creativecommons.org/licenses/by/3.0), which permits unrestricted use, distribution, and reproduction in any medium, provided the original work is properly cited.

Some of these problems is reflected in the monographs [3, 4, 20], as well as in several articles (see, eg, [5, 6] and references therein)

Application of the cited above approach to the nonlinear integral operator arising at synthesis of the antenna systems according to the given amplitude directivity pattern, brings to the nonlinear two-parameter eigenvalue problem

$$T(\lambda,\mu)f = f$$

with an integral operator $T(\lambda,\mu)$ analytically depending on two spectral parameters λ and μ.

The essential difference of the two-parameter problems from the one-parameter ones is that the two-parameter problem can not have at all the solutions or, on the contrary, to have them as a continuum set, which in the case of real parameters are the curves of eigenvalues.

Such problems are still not investigated because there are still many open questions connected with this problem such as, for example, the existence of solutions and their number, and also the development of numerical methods of solving such spectral problems for algebraic, differential and integral equations.

In the given work an algorithm of finding the branching lines of the integral equation arising in the variational statement of the synthesis problem of antenna array according to the given amplitude directivity pattern as, for example, in [2] is proposed.

2. Preliminary. Nonlinear synthesis problem

We consider the radiating system, which consists of identical and identically oriented radiators of the same for all radiators directivity pattern (DP), in which the phase centers are located on the plane XOY (grid plane) of Cartesian coordinate system. We believe that the coordinates of radiators (x_n, y_m) form a rectangular equidistant grid, focused on the axes and symmetric with respect to these axes. Then the function that describes the DP (plane array factor) of equidistant plane system of radiators (plane array) has the form [2].

$$f(\vartheta,\varphi) = \sum_{n=-M_1}^{M_1} \sum_{m=-M_2(n)}^{M_2(n)} I_{nm} e^{ik(x_n \sin\vartheta\cos\varphi + y_m \sin\vartheta\sin\varphi)}, \tag{1}$$

where I_{nm} are the complex currents on the radiators, ϑ,φ are the angular coordinates of a spherical coordinate system (R,ϑ,φ) whose center coincides with the center of the Cartesian coordinate system XOY, $M_2(n)$ is the integer function that sets the number of elements $N_2(n) = 2M_2(n)+1$ in the $n-$th row of the array. Thus, the number of elements N in this array is equal to $\sum_{n=-M_1}^{M_1} (2M_2(n)+1)$.

We introduce the generalized variables

$$\tilde{\xi}_1 = \sin\vartheta\cos\varphi,\ \tilde{\xi}_2 = \sin\vartheta\sin\varphi$$

and denote by d_1 and d_2, respectively, the distance between adjacent radiators along the axes Ox and Oy. Then the coordinates of the radiators are calculated as

$$x_n = d_1 n,\ y_m = d_2 m,$$

and the plane array factor (1) can be represented as

$$f(\tilde{\xi}_1, \tilde{\xi}_2) = \sum_{n=-M_1}^{M_1} \sum_{m=-M_2(n)}^{M_2(n)} I_{nm} e^{i(\tilde{c}_1 n \tilde{\xi}_1 + \tilde{c}_2 m \tilde{\xi}_2)}, \tag{2}$$

where

$$\tilde{c}_1 = kd_1,\ \tilde{c}_2 = kd_2.$$

Note that the function $f(\tilde{\xi}_1, \tilde{\xi}_2)$ is periodic with a period $2\pi / \tilde{c}_1$ for the variable $\tilde{\xi}_1$ and with a period $2\pi / \tilde{c}_2$ for the variable $\tilde{\xi}_2$. Denote by R_2 the region that corresponds to one period $R_2 : \left\{ \left|\tilde{\xi}_1\right| \le \pi / \tilde{c}_1, \left|\tilde{\xi}_2\right| \le \pi / \tilde{c}_2 \right\}$ and assume that the required amplitude directivity pattern $F(\tilde{\xi}_1, \tilde{\xi}_2)$ is given in some region $\Omega \subset R_2$ and is described by the function that is continuous and nonnegative in Ω and is equal to zero outside.

We must find such currents I_{nm} on radiators that created by them directivity pattern will approach by the amplitude to the given directivity pattern $F(\tilde{\xi}_1, \tilde{\xi}_2)$ in the best way. To this end, we consider the variational statement of the problem as, for example, in [2] or [18].

2.1. Variational statement of the synthesis problem

Thus, the synthesis problem we formulate as a problem of minimizing the functional [18]

$$\sigma(I) = \iint_{\Omega} \left[F(\tilde{\xi}_1, \tilde{\xi}_2) - \left| f(\tilde{\xi}_1, \tilde{\xi}_2) \right| \right]^2 d\tilde{\xi}_1 d\tilde{\xi}_2 + \iint_{R_2 \setminus \Omega} \left| f(\tilde{\xi}_1, \tilde{\xi}_2) \right|^2 d\tilde{\xi}_1 d\tilde{\xi}_2 \tag{3}$$

on the space $H_I = C^N$, i.e.

$$\sigma(I) \to \min_{I \in H_I},\ I_{nm} \in H_I,$$

which characterizes the magnitude of mean-square deviation of modules of the given directivity pattern and the synthesized one in the region Ω.

From the necessary condition of the functional $\sigma(I)$ minimum, we obtain a nonlinear system of equations for the optimum currents on radiators

$$I_{nm} = \frac{\tilde{c}_1 \tilde{c}_2}{(2\pi)^2} \iint_{\Omega} F(\tilde{\xi}_1, \tilde{\xi}_2) e^{-i(\tilde{c}_1 n \tilde{\xi}_1 + \tilde{c}_2 m \tilde{\xi}_2)} \times \exp\left\{ i \arg \sum_{n=-M_1}^{M_1} \sum_{m=-M_2(n)}^{M_2(n)} I_{nm} e^{i(\tilde{c}_1 n \tilde{\xi}_1 + \tilde{c}_2 m \tilde{\xi}_2)} \right\} d\tilde{\xi}_1 d\tilde{\xi}_2,$$

$$(n=-M_1 \div M_1,\ m=-M_2 \div M_2) \tag{4}$$

or the equation for the optimum directivity pattern, which is equivalent to (4)

$$f(\tilde{\xi}_1,\tilde{\xi}_2)=\frac{\tilde{c}_1\tilde{c}_2}{(2\pi)^2}\iint_\Omega F(\tilde{\xi}_1',\tilde{\xi}_2')K(\tilde{\xi}_1,\tilde{\xi}_2,\tilde{\xi}_1',\tilde{\xi}_2',\tilde{c}_1,\tilde{c}_2)e^{i\arg f(\tilde{\xi}_1',\tilde{\xi}_2')}d\tilde{\xi}_1'd\tilde{\xi}_2', \tag{5}$$

where

$$K(\tilde{\xi}_1,\tilde{\xi}_2,\tilde{\xi}_1',\tilde{\xi}_2',\tilde{c}_1,\tilde{c}_2)=\sum_{n=-M_1}^{M_1}\sum_{m=-M_2(n)}^{M_2(n)} e^{i[\tilde{c}_1 n(\tilde{\xi}_1-\tilde{\xi}_1')+\tilde{c}_2 m(\tilde{\xi}_2-\tilde{\xi}_2')]}$$

is the kernel, which essentially depends on the coordinates of antenna array.

Next, consider the rectangular grid with geometric center at the origin, which consists of $N=N_1\times N_2=(2M_1+1)(2M_2+1)$ elements. Here $M_2=M_2(n)=\text{const}$. We believe also that the amplitude directivity pattern $F(\tilde{\xi}_1,\tilde{\xi}_2)$ is given in the region $\Omega:\{|\tilde{\xi}_1|\le b_1, |\tilde{\xi}_2|\le b_2\}$. Denote by $2\alpha_1$ and $2\alpha_2$ the intervals of change of the angle ϑ in the region Ω at $\varphi=0$ and $\varphi=\pi/2$, respectively, and introduce new variables

$$\xi_1=\tilde{\xi}_1/\sin\alpha_1,\ \xi_2=\tilde{\xi}_2/\sin\alpha_2.$$

Then $\Omega:\{|\xi_1|\le 1, |\xi_2|\le 1\}$, and the kernel in equation (5) is real and takes the form [2]

$$K(\xi_1,\xi_2,\xi_1',\xi_2',c_1,c_2)=\sum_{n=-M_1}^{M_1}\sum_{m=-M_2}^{M_2} e^{i[c_1 n(\xi_1-\xi_1')+c_2 m(\xi_2-\xi_2')]}=\frac{\sin N_1\frac{c_1}{2}(\xi_1-\xi_1')}{\sin\frac{c_1}{2}(\xi_1-\xi_1')}\cdot\frac{\sin N_2\frac{c_2}{2}(\xi_2-\xi_2')}{\sin\frac{c_2}{2}(\xi_2-\xi_2')}, \tag{6}$$

where $c_1=kd_1\sin\alpha_1$, $c_2=kd_2\sin\alpha_2$, N_1 and N_2 are the main parameters of the problem. Thus, equation (5) for optimal DP takes the form

$$f(\xi_1,\xi_2)=\frac{c_1c_2}{(2\pi)^2}\iint_\Omega F(\xi_1,\xi_2)K(\xi_1,\xi_2,\xi_1',\xi_2',c_1,c_2)e^{i\arg f(\xi_1',\xi_2')}d\xi_1'd\xi_2', \tag{7}$$

and equation (4) for optimal currents takes the form

$$I_{nm}=\frac{c_1c_2}{(2\pi)^2}\iint_\Omega F(\xi_1,\xi_2)e^{-i(c_1 n\xi_1+c_2 m\xi_2)}\times\exp\left\{i\arg\sum_{n=-M_1}^{M_1}\sum_{m=-M_2}^{M_2} I_{nm}e^{i(c_1 n\xi_1+c_2 m\xi_2)}\right\}d\xi_1 d\xi_2,$$

$$(n=-M_1\div M_1,\ m=-M_2\div M_2). \tag{8}$$

Equivalence of equations (7) and (8) means that between the solutions of these equations one-to-one correspondence exists, i.e., to each solution of equation (7) corresponds the solution of equation (8) and vice versa. This means that if I_{nm}, $n=\overline{-M_1,M_1}$, $m=\overline{-M_2,M_2}$, is a solution of equation (8), the corresponding to it solution of equation (7) is determined by the formula

$$f(\xi_1,\xi_2) = \sum_{n=-M_1}^{M_1} \sum_{m=-M_2}^{M_2} I_{nm} e^{i(c_1 n\xi_1 + c_2 m\xi_2)}, \tag{9}$$

and if $f(\xi_1,\xi_2)$ is a solution of equation (7), the corresponding to it solution of equation (8) is determined by the relation

$$I_{nm} = \frac{c_1 c_2}{(2\pi)^2} \iint_{\Omega} F(\xi_1,\xi_2) e^{i\left[\arg f(\xi_1,\xi_2) - (c_1 n\xi_1 + c_2 m\xi_2)\right]} d\xi_1 d\xi_2. \tag{10}$$

Since equations (7) and (8) are nonlinear equations (Hammerstein type), they may have nonunique solutions, the number and properties of which depend on the number of elements in the antenna array and their placement, and also on the properties of the given amplitude directivity pattern $F(\xi_1,\xi_2)$.

It is easy to see that one of possible solutions of equation (7) (call it trivial) is

$$f_0(\xi_1,\xi_2,c_1,c_2) = \iint_{\Omega} F(\xi_1',\xi_2') K(\xi_1,\xi_2,\xi_1',\xi_2',c_1,c_2) d\xi_1' d\xi_2'. \tag{11}$$

Experimental results of numerical synthesis of the directivity pattern for different values of parameters c_1 and c_2 show that with growth of parameters c_1 and c_2 there are other solutions that branch off from a trivial solution and they are more effective in terms of the values of functional (3), from 0% to ≈ 75 %

In particular, for the given directivity pattern $F(\xi_1,\xi_2) = 1$ the values of functional (3), which it takes on the optimal solution $f(\xi_1,\xi_2,c_1,c_2)$ for different values of the main parameters c_1 and c_2 ($\sigma_1(I) = 0.739543$, correspond, to $c_1 = c_2 = 0.57$, $\sigma_2(I) = 0.719989$ - $c_1 = c_2 = 0.60$, $\sigma_3(I) = 0.644291$ - $c_1 = c_2 = 0.65$, $\sigma_4(I) = 0.559552$ - $c_1 = c_2 = 0.70$, $\sigma_5(I) = 0.493709$ - $c_1 = c_2 =$ 0.75) is smaller than the values of functional (3) for the trivial solution $f_0(\xi_1,\xi_2,c_1,c_2)$ at the same parameter values c_1 and c_2 ($\sigma_1^0(I) = 0.739769$, $\sigma_2^0(I) = 0.741211$, $\sigma_3^0(I) = 0.734128$, $\sigma_4^0(I) = 0.707903$, $\sigma_5^0(I) = 0.661929$), respectively by 0.03%, 2.86%, 12.24%, 20.95% and 25.41%.

Numerical examples of the trivial and branching solutions for the given directivity pattern $F(\xi_1,\xi_2) = 1$ and the basic parameters of $c_1 = c_2 = 0.75$ are shown in Fig. 1 -- Fig. 4.

In Fig. 1 shows the trivial solution, which creates a symmetrical inphase current distribution on the radiators of array (Fig. 2). The amplitude of the synthesized directivity pattern, which branches off from $f_0(\xi_1,\xi_2,c_1,c_2)$, is shown in Fig. 3, and the optimal current on the radiators that it creates, is asymmetric and is shifted to the first quadrant relatively of the center of array (Fig. 4).

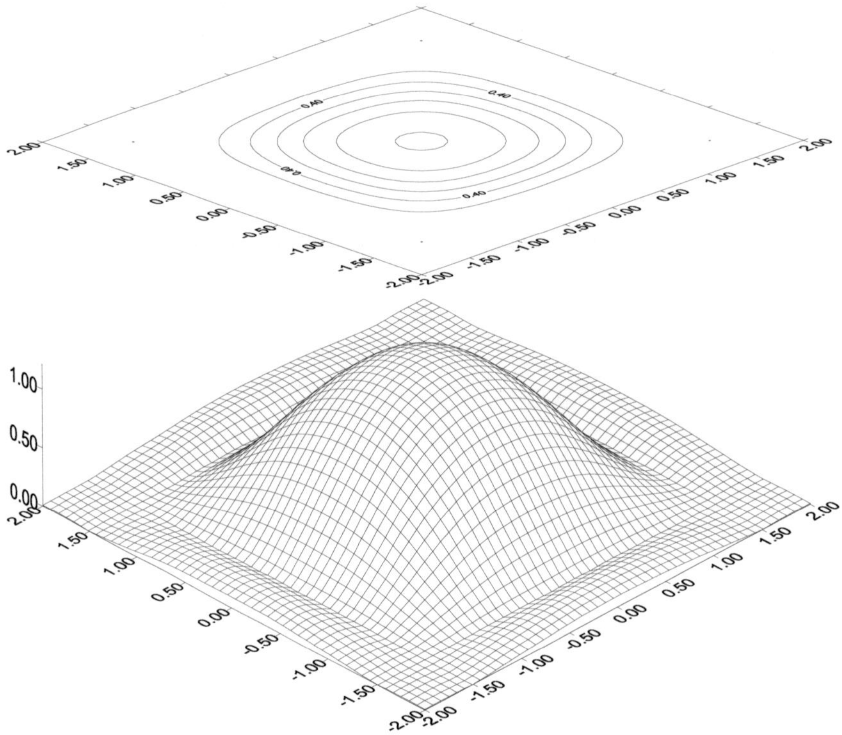

Figure 1. Amplitude directivity pattern of a trivial solution $f_0(\xi_1, \xi_2, c_1, c_2)$

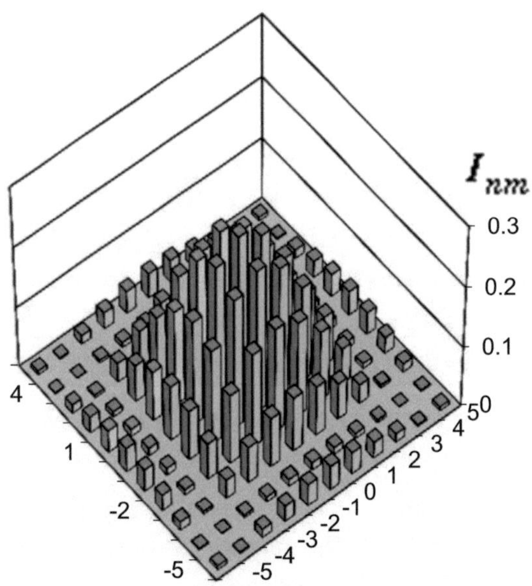

Figure 2. Current cophasal distribution on the radiators, which creates the diagram $f_0(\xi_1, \xi_2, c_1, c_2)$

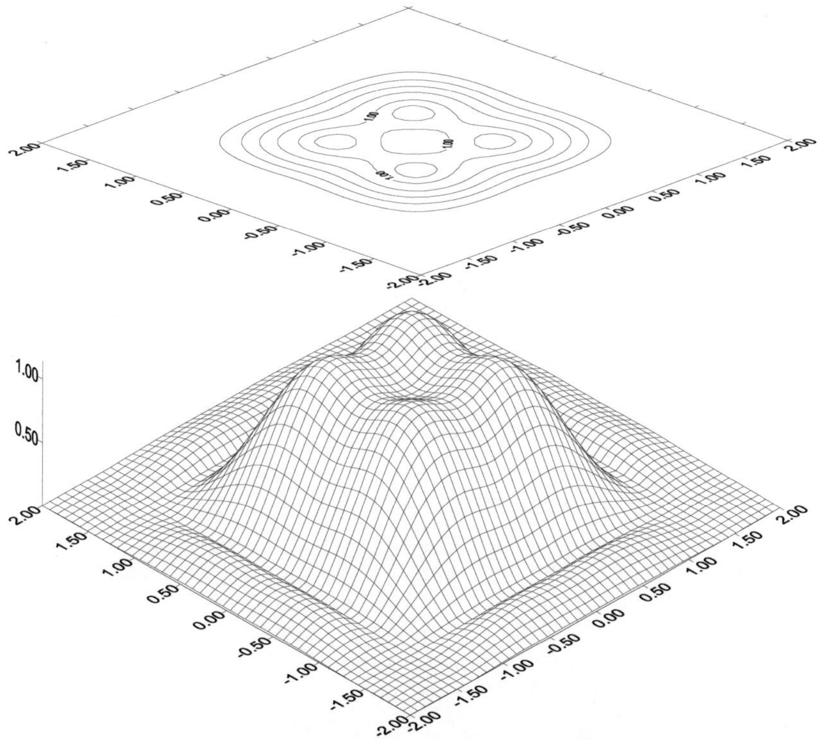

Figure 3. Amplitude directivity pattern of the solution branched from $f_0(\xi_1, \xi_2, c_1, c_2)$

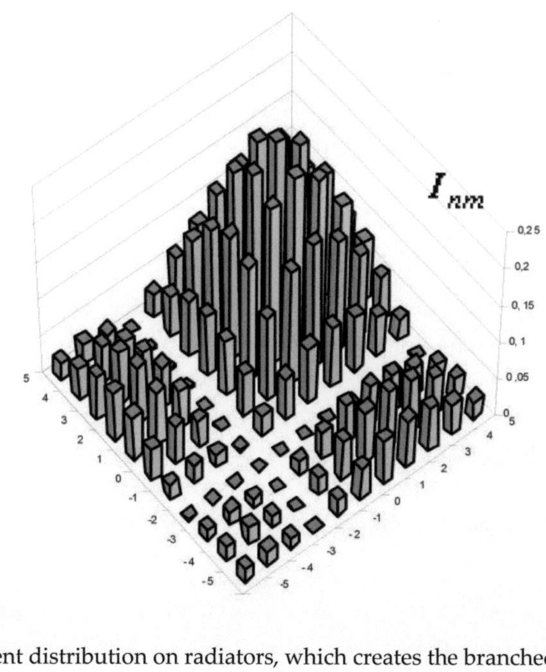

Figure 4. Optimal current distribution on radiators, which creates the branched solution

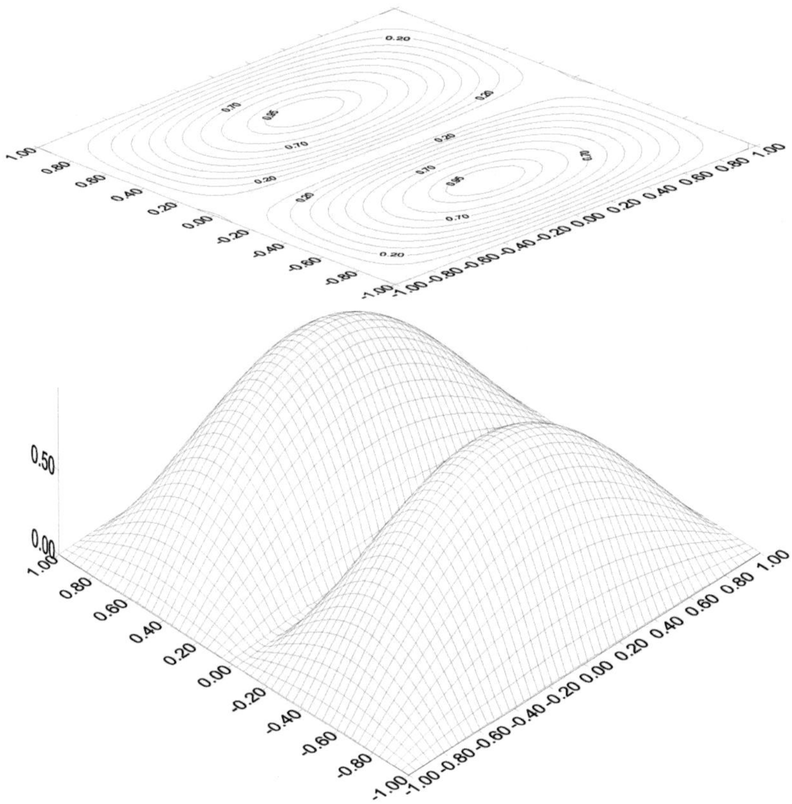

Figure 5. Given amplitude directivity pattern $F(\xi) = \cos\frac{\pi\xi_1}{2} | \sin \pi\xi_2 |$

The branching solutions are still more effective for directivity patterns that do not have central symmetry. For example, for the given directivity pattern $F(\xi) = \cos\frac{\pi\xi_1}{2} | \sin \pi\xi_2 |$ and the main parameters $c_1 = 1.0, c_2 = 0.85$, which is shown in Fig. 5, numerical examples of the trivial and the branching solutions are shown in Fig. 6 - Fig. 9. From these figures we see that the branching solution (the amplitude directivity pattern of which is shown in Fig. 8, and the optimal distribution of the current on the radiators that it generates is shown in Fig. 9) more accurately than the trivial solution (11) (amplitude directivity pattern is shown in Fig. 6, which is created by symmetric inphase current (Fig. 7)) approximates the given directivity pattern not only in the mean square approximation (in terms of the values of functional (3) - 0.050109 in comparison with 0.185960) to about 73%, but also with respect to the form.

Thus, in most cases from the practical point of view the nontrivial solution, that branches off from $f_0(\xi_1, \xi_2, c_1, c_2)$ with growth of parameters c_1 and c_2 is interesting.

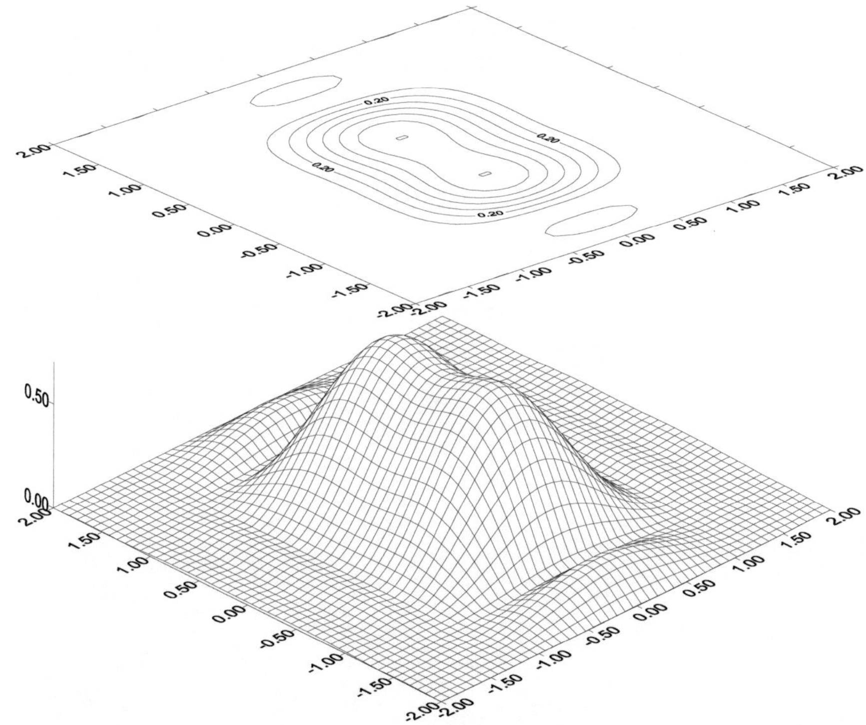

Figure 6. Amplitude directivity pattern of trivial solution $f_0(\xi_1, \xi_2, c_1, c_2)$

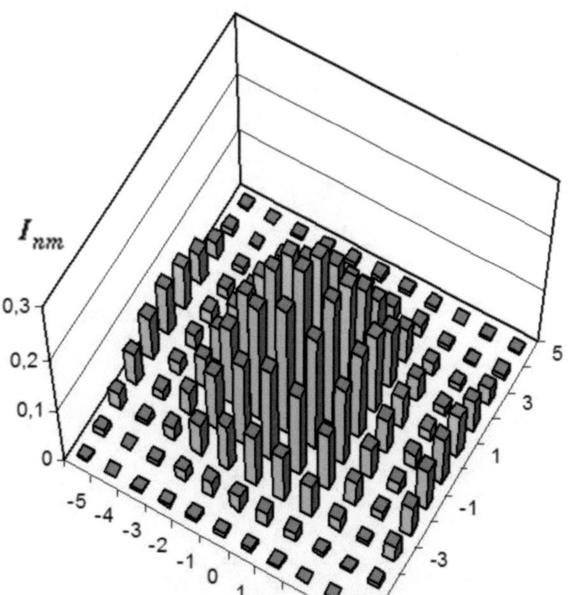

Figure 7. Inphase current distribution on the radiators, which creates the diagram $f_0(\xi_1, \xi_2, c_1, c_2)$

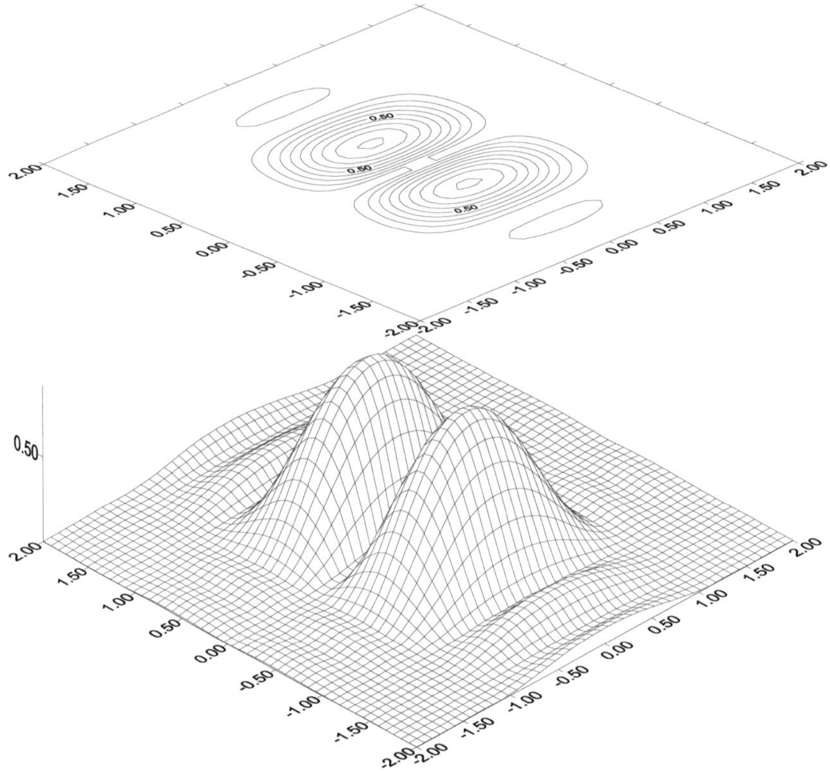

Figure 8. Amplitude directivity pattern of solution branched off from $f_0(\xi_1, \xi_2, c_1, c_2)$

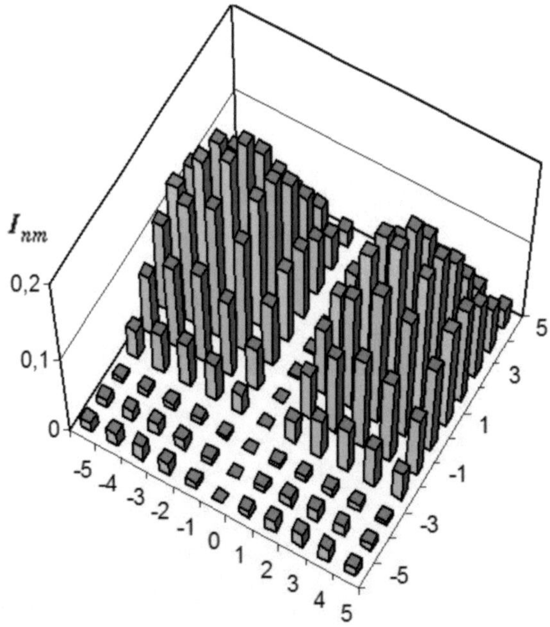

Figure 9. Optimal distribution of current on the radiators, which creates the branching solution

2.2. Problem of finding the branching lines

The points of possible branching of solutions of integral equation (7) are such values of real physical parameters $(c_1,c_2) \in R^2$, in which homogeneous integral equation [18]

$$u(\xi_1,\xi_2,c_1,c_2) = \tilde{T}(c_1,c_2)u(\xi_1,\xi_2,c_1,c_2) \equiv \iint_{\Omega} F(\xi_1',\xi_2')K(\xi_1,\xi_2,\xi_1',\xi_2',c_1,c_2)\frac{u(\xi_1',\xi_2',c_1,c_2)}{f_0(\xi_1',\xi_2',c_1,c_2)}d\xi_1'd\xi_2', \quad (12)$$

obtained by linearization of equation (7), has solutions distinct from identical zero [20]. Thus, we have obtained the nonlinear (with respect to parameters c_1 and c_2) two-parameter eigenvalue problem

$$\left(\tilde{T}(\lambda_1,\lambda_2) - I\right)u(\xi_1,\xi_2,\lambda_1,\lambda_2) = 0,\ \lambda_1 = c_1,\ \lambda_2 = c_2. \quad (13)$$

It is easy to be convinced, that at arbitrary finite values $c_1 > 0$, $c_2 > 0$, the function $f_0(\xi_1,\xi_2,c_1,c_2)$ is the eigenfunction of equation (12). From this it follows, that the operator $T(c_1,c_2)$ has a spectrum, which coincides with the first quadrant of the plane R^2.

The problem consists in finding such range of real parameters $\lambda_1 = c_1$ and $\lambda_2 = c_2$ of the problem (13), for which there appear the solutions different from $f_0(\xi_1,\xi_2,c_1,c_2)$.

It should be noted that in a special case, when it is possible to separate variables in the function $F(\xi_1,\xi_2)$, i.e. $F(\xi_1,\xi_2)$ to present as $F(\xi_1,\xi_2) = F_1(\xi_1)\cdot F(\xi_2)$, the equation (12), provided that function $u(\xi_1,\xi_2,c_1,c_2)$ can also be presented as $u(\xi_1,\xi_2,c_1,c_2) = u(\xi_1,c_1)\cdot u(\xi_2,c_2)$, decomposes in two independent one-parameter equations, i.e.

$$u_j(\xi_j,c_j) = T(c_j)u_j(\xi_j,c_j),\ j = 1,2,$$

with operators

$$T_j(c_j)u(\xi_j,c_j) = \int_{-1}^{1} \frac{F_j(\xi_j)}{f_0(\xi_j,c_j)} \frac{\sin N_j \frac{c_j}{2}(\xi_j - \xi_j')}{\sin\frac{c_j}{2}(\xi_j - \xi_j')} u(\xi_j',c_j)d\xi_j',\ j = 1, 2..$$

The study of such equations is carried out in [2, 18], and it is possible to apply , for example, the algorithms of the work [11, 13, 15] to solution of such equations.

In the given work the numerical algorithms to solve more complicated problem when the variables are not separated, are proposed.

3. Basic equations

First we will show that the kernel $K(\xi_1,\xi_2,\xi_1',\xi_2',c_1,c_2)$ (as in one-dimensional case [18]) in the region $\Omega\!:\!\{|\,\xi_1\,|\le 1, |\,\xi_2\,|\le 1\}$ for arbitrary $c_1 > 0$ and $c_2 > 0$ is a positive kernel of the integral operator

$$Af(\xi_1,\xi_2)=\iint\limits_{\Omega} K(\xi_1,\xi_2,\xi_1',\xi_2',c_1,c_2)f(\xi_1',\xi_2')d\xi_1'd\xi_2'.$$

To this end, consider the scalar product

$$(Af,f)=\iint\limits_{\Omega}\iint\limits_{\Omega} K(\xi_1,\xi_2,\xi_1',\xi_2',c_1,c_2)f(\xi_1',\xi_2')\overline{f(\xi_1',\xi_2')}d\xi_1'd\xi_2'\,d\xi_1 d\xi_2.$$

Substituting the expression for $K(\xi_1,\xi_2,\xi_1',\xi_2',c_1,c_2)$ (6), we obtain

$$(Af,f)=\iint\limits_{\Omega}\iint\limits_{\Omega}\left(\sum_{n=-M_1}^{M_1}\sum_{m=-M_2}^{M_2} e^{i\left[nc_1(\xi_1-\xi_1')+mc_2(\xi_2-\xi_2')\right]}\right)f(\xi_1',\xi_2')\overline{f(\xi_1',\xi_2')}d\xi_1'd\xi_2'\,d\xi_1 d\xi_2=$$
$$=\sum_{n=-M_1}^{M_1}\sum_{m=-M_2}^{M_2}\left\{\iint\limits_{\Omega}\iint\limits_{\Omega} f(\xi_1',\xi_2')e^{-i(nc_1\xi_1'+mc_2\xi_2')}\overline{f(\xi_1,\xi_2)e^{i(nc_1\xi_1+mc_2\xi_2)}}d\xi_1'd\xi_2'\,d\xi_1 d\xi\right\}=$$
$$=\sum_{n=-M_1}^{M_1}\sum_{m=-M_2}^{M_2}\left\|\iint\limits_{\Omega} f(\xi_1',\xi_2')e^{-i(nc_1\xi_1'+mc_2\xi_2')}d\xi_1'd\xi_2'\right\|^2=\frac{4\pi^2}{c_1c_2}\sum_{n=-M}^{M_1}\sum_{m=-M_2}^{M_2}|I_{nm}|^2\geq 0.$$

Obviously, the last inequality transforms into equality only when $I_{nm}=0,\ n=\overline{-M_1,M_1},\ m=\overline{-M_2,M_2}$. From this it follows that $K(\xi_1,\xi_2,\xi_1',\xi_2',c_1,c_2)$ is positive, and positive operator A leaves invariant a cone $\mathbf{K}$ $(A\mathbf{K}\subset\mathbf{K})$ of the continuous nonnegative functions on Ω. As a result, we obtain that $f_0(\xi_1,\xi_2,c_1,c_2)$ is positive on Ω function. Taking it into account, we shall reduce the operator (12) to a selfajoint form by a standard method. Introducing a new function

$$\varphi(\xi_1,\xi_2,c_1,c_2)=\sqrt{w(\xi_1,\xi_2,c_1,c_2)}\,u(\xi_1,\xi_2,c_1,c_2), \tag{14}$$

where $w(\xi_1,\xi_2,c_1,c_2)=F(\xi_1,\xi_2)/f_0(\xi_1,\xi_2,c_1,c_2)$, we obtain the integral equation

$$\varphi(\xi_1,\xi_2,c_1,c_2)=\iint\limits_{\Omega}\Phi(\xi_1,\xi_2,\xi_1',\xi_2',c_1,c_2)\varphi(\xi_1',\xi_2',c_1,c_2)d\xi_1'd\xi_2' \tag{15}$$

with a symmetric kernel

$$\Phi(\xi_1,\xi_2,\xi_1',\xi_2',c_1,c_2)=K(\xi_1,\xi_2,\xi_1',\xi_2',c_1,c_2)\sqrt{w(\xi_1,\xi_2,c_1,c_2)w(\xi_1',\xi_2',c_1,c_2)}.$$

Since at arbitrary $c_1>0$ and $c_2>0$ the function $f_0(\xi_1,\xi_2,c_1,c_2)$ is the eigenfunction of the equation (12), then with regard for (14), the eigenfunction of the equation (15) at arbitrary $c_1>0$ and $c_2>0$ will be the function

$$\varphi_0(\xi_1,\xi_2,c_1,c_2)=\sqrt{F(\xi_1,\xi_2)f_0(\xi_1,\xi_2,c_1,c_2)},$$

which corresponds to a spectrum of the operator (15), coinciding with the first quadrant of the plane R^2.

To find the solutions distinct from $\varphi_0(\xi_1,\xi_2,c_1,c_2)$, we shall eliminate this function from the kernel $\Phi(\xi_1,\xi_2,\xi_1',\xi_2',c_1,c_2)$, then the equation (15) will be reduced to the integral equation

$$\varphi(\xi_1,\xi_2,c_1,c_2)=T(c_1,c_2)\varphi(\xi_1,\xi_2,c_1,c_2)\equiv\int_{-1}^{1}\int_{-1}^{1}E(\xi_1,\xi_2,\xi_1',\xi_2',c_1,c_2)\varphi(\xi_1',\xi_2',c_1,c_2)d\xi_1'd\xi_2' \quad (16)$$

with a symmetric kernel

$$E(\xi_1,\xi_2,\xi_1',\xi_2',c_1,c_2)=\sqrt{w(\xi_1,\xi_2,c_1,c_2)w(\xi_1',\xi_2',c_1,c_2)}\times$$

$$\times\left[K(\xi_1,\xi_2,\xi_1',\xi_2',c_1,c_2)-\frac{f_0(\xi_1,\xi_2,c_1,c_2)f_0(\xi_1',\xi_2',c_1,c_2)}{||\varphi_0(\xi_1,\xi_2,c_1,c_2)||^2}\right]. \quad (17)$$

From Schmidt's lemma [20] it follows, that $\varphi_0(\xi_1,\xi_2,c_1,c_2)$ will not be the eigenfunction of equation (16) anymore. That is, we have eliminated a continuum set of eigenvalues from a spectrum of the operator (16), which coincides with the first quadrant of the plane R^2 that corresponds to the function $\varphi_0(\xi_1,\xi_2,c_1,c_2)$. So, we obtain a self-adjoint generalized eigenvalue problem

$$L(\lambda_1,\lambda_2)\varphi\equiv\left(T(\lambda_1,\lambda_2)-I\right)\varphi=0,\ \lambda_1=c_1,\ \lambda_2=c_2, \quad (18)$$

with the operator $T(c_1,c_2)$ which is a continuously differentiable with respect to the parameters c_1 and c_2. The existence of partial Frechet derivatives $\frac{\partial T(c_1,c_2)}{\partial c_i}$, $i=1,2$, and $\frac{\partial^2 T(c_1,c_2)}{\partial c_i \partial c_j}$, $i,j=1,2$ at arbitrary points $c_1=\mu$, $c_2=\nu$ follows from the continuity of the kernel $E(\xi_1,\xi_2,\xi_1',\xi_2',c_1,c_2)$ on the set of its variables in the region $\Omega\times\Omega$ and the existence and continuity in $\Omega\times\Omega$ the derivatives $\frac{\partial E(\xi_1,\xi_2,\xi_1',\xi_2',c_1,c_2)}{\partial c_i}$, $i=1,2$ and $\frac{\partial^2 E(\xi_1,\xi_2,\xi_1',\xi_2',c_1,c_2)}{\partial c_i \partial c_j}$, $i,j=1,2$ which because of their bulky form, are not presented.

Using the property of degeneracy of the kernel $K(\xi_1,\xi_2,\xi_1',\xi_2',c_1,c_2)$, we will reduce equation (16) to an equivalent system of algebraic equations.

Using the formula (6), we write the kernel $E(\xi_1,\xi_2,\xi_1',\xi_2',c_1,c_2)$ as

$$E(\xi_1,\xi_2,\xi_1',\xi_2',c_1,c_2)=\sum_{n=-M_1}^{M_1}\sum_{m=-M_2}^{M_2}K_{nm}^1(\xi_1,\xi_2,c_1,c_2)\cdot K_{nm}^2(\xi_1',\xi_2',c_1,c_2)-$$

$$-\left(\sum_{n=-M_1}^{M_1}\sum_{m=-M_2}^{M_2}K_{nm}^1(\xi_1,\xi_2,c_1,c_2)q_{nm}^1\right)\left(\sum_{n=-M_1}^{M_1}\sum_{m=-M_2}^{M_2}K_{nm}^2(\xi_1',\xi_2',c_1,c_2)q_{nm}^2\right),$$

where

$$K_{nm}^1(\xi_1,\xi_2,c_1,c_2)=\frac{\sqrt{c_1c_2}}{2\pi}\sqrt{\frac{F(\xi_1,\xi_2)}{f_0(\xi_1,\xi_2,c_1,c_2)}}\cdot e^{i(c_1n\xi_1+c_2m\xi_2)},$$

$$K_{nm}^2(\xi_1',\xi_2',c_1,c_2)=\frac{\sqrt{c_1c_2}}{2\pi}\sqrt{\frac{F(\xi_1',\xi_2')}{f_0(\xi_1',\xi_2',c_1,c_2)}}\cdot e^{-i(c_1n\xi_1'+c_2m\xi_2')},$$

$$q_{nm}^1=\iint_\Omega F(\xi_1',\xi_2')e^{-i(c_1n\xi_1'+c_2m\xi_2')}d\xi_1'd\xi_2',$$

$$q_{nm}^2=\iint_\Omega F(\xi_1,\xi_2)e^{i(c_1n\xi_1+c_2m\xi_2)}d\xi_1d\xi_2.$$

Then equation (16) takes the form

$$\varphi(\xi_1,\xi_2,c_1,c_2)=\sum_{n=-M_1}^{M_1}\sum_{m=-M_2}^{M_2}a_{nm}^1(\xi_1,\xi_2,c_1,c_2)b_{nm},$$

where

$$b_{nm}=\iint_\Omega K_{nm}^2(\xi_1',\xi_2',c_1,c_2)\varphi(\xi_1',\xi_2',c_1,c_2)d\xi_1'd\xi_2',$$

$$a_{nm}^1(\xi_1,\xi_2,c_1,c_2)=K_{nm}^1(\xi_1,\xi_2,c_1,c_2)-\frac{1}{\gamma}\left(\sum_{s=-M_1}^{M_1}\sum_{t=-M_2}^{M_2}K_{st}^1(\xi_1,\xi_2,c_1,c_2)q_{st}^1\right)q_{nm}^2,$$

$$\gamma=\sum_{n=-M_1}^{M_1}\sum_{m=-M_2}^{M_2}q_{nm}^1q_{nm}^2,$$

and the unknown coefficients b_{nm} are determined as solutions of a homogeneous system of linear algebraic equations

$$b_{kl}=\sum_{n=-M_1}^{M_1}\sum_{m=-M_2}^{M_2}\alpha_{nm}^{(kl)}(c_1,c_2)b_{nm},\quad k=\overline{-M_1,M_1},\ l=\overline{-M_2,M_2},$$

where

$$\alpha_{nm}^{(kl)}(c_1,c_2)=\iint_\Omega a_{kl}^2(\xi_1,\xi_2,c_1,c_2)a_{nm}^1(\xi_1,\xi_2,c_1,c_2)d\xi_1d\xi_2,$$

$$a_{nm}^2(\xi_1,\xi_2,c_1,c_2)=K_{nm}^2(\xi_1,\xi_2,c_1,c_2)-\frac{1}{\gamma}\left(\sum_{s=-M_1}^{M_1}\sum_{t=-M_2}^{M_2}K_{st}^2(\xi_1,\xi_2,c_1,c_2)q_{st}^2\right)q_{nm}^1.$$

So, we have obtained the two-parameter nonlinear (with respect to the spectral parameters) matrix eigenvalue problem equivalent to (18)

$$\mathbf{D}_N(\lambda,\mu)\mathbf{b}_N\equiv(\mathbf{A}_N(\lambda,\mu)-\mathbf{I}_N)\mathbf{b}_N=0 \tag{19}$$

with symmetric matrix $\mathbf{A}_N(\lambda,\mu)$ of dimension $N\times N$, $\mathbf{I}_N$ is the identity matrix of dimension $N\times N$, $\mathbf{b}_N\in R^N$, $\lambda=c_1$, $\mu=c_2$.

Thus, the problem of finding lines the branching of solutions of equation (7) is reduced to finding the eigenvalues curves of nonlinear two-parameter spectral problem (19). Obviously, in order the problem (19) to have a nonzero solution it is necessary that

$$\psi(\lambda,\mu)\equiv\det\mathbf{D}_N(\lambda,\mu)=0, \tag{20}$$

i.e. the eigenvalues of problem (19) are zeros of function $\psi(\lambda,\mu)$.

4. Algorithm of finding the eigenvalue curves

The main calculational part of algorithm proposed is the implementation method proposed in [14, 15] to compute all eigenvalues of the nonlinear matrix spectral problem

$$\mathbf{T}_n(\lambda,\mu)u_n=0, \tag{21}$$

belonging to some given range of the spectral parameter λ at the given value of parameter μ. In the problem (21) $u_n\in\mathbb{R}^n$, and $\mathbf{T}_n(\lambda,\mu)$ is the real $(n\times n)$ matrix whose elements depend nonlinearly on the parameters λ and μ. In order to detail how the method [15] is applied to the problem under consideration in this paper, we present the necessary results from [15].

Thus, we replace in the problem (21), for example, the parameter μ by the expression $\mu=\alpha\lambda+\beta$ and consider the appropriate one-parameter problem

$$\mathbf{T_n}(\lambda)u_n\equiv\mathbf{T_n}(\lambda,\alpha,\beta)u_n=0, \tag{22}$$

at the given fixed values α and β. Then, obviously, the eigenvalues of problem (21) are zeros of function

$$f(\lambda)\equiv\det\mathbf{T}_n(\lambda)=0,$$

where $\mathbf{T}_n(\lambda)$ is a real $(n\times n)$ matrix whose elements depend nonlinearly on the parameter λ.

One should determine how many zeros of the function $f(\lambda)$, and, therefore, the eigenvalues of the problem are in some given range of change of parameter $\lambda\in[\lambda_{c_k},\lambda_{d_k}]\subset\mathbb{R}$ and calculate each of them.

4.1. The argument principle of meromorphic function

In the basis of algorithm of finding number of zeros and their approximations, which are in some areas G, is the statement that follows from the argument principle of meromorphic functions.

Statement. *Let the meromorphic function* $f(\lambda)$ *have in the region* G m *zeros* $\lambda_1, \lambda_2, \ldots, \lambda_m$ *(with regard for their multiplicity) and no zeros on the boundary* Γ *of region* G*, then the number* m *is determined in accordance with the principle of the argument*

$$m = s_0 = \frac{1}{2\pi i}\int_{\Gamma} \frac{f'(\lambda)}{f(\lambda)} d\lambda \tag{23}$$

and relations

$$\sum_{j=1}^{m} (\lambda_j)^k = s_k, \quad k = 1, \ldots, m \tag{24}$$

are true, where

$$s_k = \frac{1}{2\pi i}\int_{\Gamma} \lambda^k \frac{f'(\lambda)}{f(\lambda)} d\lambda, \quad k = 0, 1, \ldots. \tag{25}$$

Thus, knowing s_k, $k = 1, 2, \ldots, m$, from the system (24) we can find the zeros of functions $f(\lambda)$ that are in the region G.

By putting the interval $[\lambda_{c_t}, \lambda_{d_t}]$ in the region G, such as a circle with center at $r_{0_t} = (\lambda_{c_t} + \lambda_{d_t})/2$ and radius $\rho_t = (\lambda_{d_t} - \lambda_{c_t})/2$, and applying the above statement to the meromorphic function $f(\lambda) = \det \mathbf{T}_n(\lambda)$, you can find all the eigenvalues of problem (22), belonging to the given region G, i.e. to the given interval $[\lambda_{c_t}, \lambda_{d_t}]$. The integrals in (23) and (25) we can replace by some approximate quadrature formulas, such as rectangles at N points on Γ, and since Γ is a circle, then to calculate quantities $s_k, k = 0, 1, 2, \ldots$, we obtain the relation

$$s_k = \frac{1}{N}\sum_{j=1}^{N} (\lambda_j)^k \rho_t \exp\left(i\frac{2\pi j}{N}\right)\frac{f'(\lambda_j)}{f(\lambda_j)}, \tag{26}$$

where $\lambda_j = r_{0_t} + \rho_t \exp\left(i\frac{2\pi j}{N}\right)$. The system itself (24) we solve using Newton's method, by choosing the initial approximation on the border Γ of the region G:

$$\lambda_j^{(0)} = r_{0_t} + \rho_t \exp\left(i\frac{2\pi j}{s_0}\right), \quad j = 1, 2, \ldots, s_0. \tag{27}$$

The found eigenvalues can be refined, using them as initial approximations for Newton's method

$$\lambda_{l+1} = \lambda_l - \frac{f(\lambda_l)}{f'(\lambda_l)}, \qquad l = 0,1,2,\ldots \tag{28}$$

or for one of bilateral analogies of Newton's method [15], for example,

$$\lambda_{2l+1} = \lambda_{2l} - \frac{f(\lambda_{2l})f'(\lambda_{2l})}{f'(\lambda_{2l})^2 - f(\lambda_{2l})f''(\lambda_{2l})}$$

$$\lambda_{2l+2} = \lambda_{2l+1} - \frac{f(\lambda_{2l+1})}{f'(\lambda_{2l+1})}, \qquad l = 0,1,2,\ldots \tag{29}$$

The argument principle (23) and formula of the argument principle type (24), (25) were repeatedly applied in solving various spectral problems (see, for example, [1, 7, 9] and references therein), but the peculiarity of the proposed algorithm is to compute the values of function $f(\lambda)$ and its derivatives basing on *LU*-decomposition of the matrix $\mathbf{T}_n(\lambda)$.

4.2. Numerical procedure of calculating the derivatives (the first and the second) for the matrix determinant

Theorem. *If the elements of square matrix* $\mathbf{D}(\lambda)$ *are differentiable functions with respect to parameter* λ, *then for any* λ *for derivatives of determinant* $\det\mathbf{D}(\lambda) \equiv f(\lambda)$ *of matrix* $\mathbf{D}(\lambda)$ *the relations*

$$f'(\lambda) \equiv \left[\det\mathbf{D}(\lambda)\right]' = \sum_{k=1}^{n} v_{kk}(\lambda) \prod_{i=1, i\neq k}^{n} u_{ii}(\lambda), \tag{30}$$

$$f''(\lambda) \equiv \left[\det\mathbf{D}(\lambda)\right]'' = \sum_{k=1}^{n} w_{kk}(\lambda) \prod_{i=1, i\neq k}^{n} u_{ii}(\lambda) + \sum_{k=1}^{n} v_{kk}(\lambda) \left(\sum_{j=1, j\neq k}^{n} v_{jj}(\lambda) \prod_{i=1, i\neq k, i\neq j}^{n} u_{ii}(\lambda) \right), \tag{31}$$

are true, where $u_{ii}(\lambda), v_{ii}(\lambda)$ *and* $w_{ii}(\lambda)$ *are, respectively, the elements of the upper triangular matrix* $\mathbf{U}(\lambda)$, $\mathbf{V}(\lambda)$ *and* $\mathbf{W}(\lambda)$ *in decompositions*

$$\mathbf{D}(\lambda) = \mathbf{L}(\lambda)\mathbf{U}(\lambda), \tag{32}$$

$$\mathbf{B}(\lambda) = \mathbf{M}(\lambda)\mathbf{U}(\lambda) + \mathbf{L}(\lambda)\mathbf{V}(\lambda), \tag{33}$$

$$\mathbf{C}(\lambda) = \mathbf{N}(\lambda)\mathbf{U}(\lambda) + 2\mathbf{M}(\lambda)\mathbf{V}(\lambda) + \mathbf{L}(\lambda)\mathbf{W}(\lambda), \tag{34}$$

and $\mathbf{L}(\lambda)$ *is the lower triangular matrix with single diagonal elements.*

Proof. It is known that matrix $\mathbf{D}(\lambda)$ of order n, which for any λ has the major minorities of all orders from 1 to $(n-1)$, differen from zero, by using the *LU*-decomposition can be written as (32), where $\mathbf{L}(\lambda)$ is the lower triangular matrix with single diagonal elements, and $\mathbf{U}(\lambda)$ is the upper triangular matrix. Then

$$f(\lambda) = \det \mathbf{L}(\lambda) \det \mathbf{U}(\lambda) = \prod_{i=1}^{n} u_{ii}(\lambda).$$

Since the elements of a square matrix $\mathbf{D}(\lambda)$ (and hence the matrix $\mathbf{U}(\lambda)$) are differentiable functions with respect to λ, then for any λ we obtain that

$$f'(\lambda) = \sum_{k=1}^{n} u'_{kk}(\lambda) \prod_{i=1, i\neq k}^{n} u_{ii}(\lambda),, \tag{35}$$

$$f''(\lambda) = \sum_{k=1}^{n} u''_{kk}(\lambda) \prod_{i=1, i\neq k}^{n} u_{ii}(\lambda) + \sum_{k=1}^{n} u'_{kk}(\lambda) \left(\sum_{j=1, j\neq k}^{n} u'_{jj}(\lambda) \prod_{i=1, i\neq k, i\neq j}^{n} u_{ii}(\lambda) \right).. \tag{36}$$

To find the values $u'_{ii}(\lambda)$ we differentiate (32) with respect to λ. We obtaine (33), i.e.

$$\mathbf{B}(\lambda) = \mathbf{M}(\lambda)\mathbf{U}(\lambda) + \mathbf{L}(\lambda)\mathbf{V}(\lambda),$$

where $\mathbf{B}(\lambda) = \mathbf{D}'(\lambda)$, $\mathbf{M}(\lambda) = \mathbf{L}'(\lambda)$, $\mathbf{V}(\lambda) = \mathbf{U}'(\lambda)$, and $v_{ii}(\lambda) = u'_{ii}(\lambda)$ are the elements of matrix $\mathbf{V}(\lambda)$. Now, differentiating the last equality with respect to λ, we obtain (34), namely:

$$\mathbf{C}(\lambda) = \mathbf{N}(\lambda)\mathbf{U}(\lambda) + 2\mathbf{M}(\lambda)\mathbf{V}(\lambda) + \mathbf{L}(\lambda)\mathbf{W}(\lambda),$$

where $\mathbf{C}(\lambda) = \mathbf{B}'(\lambda) = \mathbf{D}''(\lambda)$, $\mathbf{N}(\lambda) = \mathbf{M}'(\lambda)$, $\mathbf{W}(\lambda) = \mathbf{V}'(\lambda) = \mathbf{U}''(\lambda)$, and a $w_{ii}(\lambda) = v'_{ii}(\lambda) = u''_{ii}(\lambda)$ are the elements of matrix $\mathbf{W}(\lambda)$. Thus, from (33) and (34) we obtain (35) and (36), i.e. (30) and (31). Theorem is proved.

Therefore, to calculate, $f(\lambda_m)$, $f'(\lambda_m)$ and $f''(\lambda_m)$ it is necessary to calculate

$$\mathbf{D} = \mathbf{LU}$$

$$\mathbf{B} = \mathbf{MU} + \mathbf{LV} \tag{37}$$

$$\mathbf{C} = \mathbf{NU} + 2\mathbf{MV} + \mathbf{LW},$$

at a fixed $\lambda = \lambda_m$, from which

$$f(\lambda_m) = \prod_{i=1}^{n} u_{ii}, \; f'(\lambda_m) = \sum_{k=1}^{n} v_{kk} \prod_{i=1, i\neq k}^{n} u_{ii}, \tag{38}$$

$$f''(\lambda_m) = \sum_{k=1}^{n} w_{kk} \prod_{i=1, i\neq k}^{n} u_{ii} + \sum_{k=1}^{n} v_{kk} \left(\sum_{j=1, j\neq k}^{n} v_{jj} \prod_{i=1, i\neq k, i\neq j}^{n} u_{ii} \right).$$

Elements of matrices in the decompositions (37) can be calculated using the recurrent relations

$$r = 1, 2, \dots, n,,$$

$$u_{rk} = d_{rk} - \sum_{j=1}^{r-1} l_{rj} u_{jk}, \quad k = r, \dots, n,$$

$$l_{ir} = \left(d_{ir} - \sum_{j=1}^{r-1} l_{ij} u_{jr} \right) / u_{rr}, \quad i = r+1, \dots, n,$$

$$v_{rk} = b_{rk} - \sum_{j=1}^{r-1} (m_{rj} u_{jk} + l_{rj} v_{jk}), \quad k = r, \dots, n,$$

$$m_{ir} = \left[b_{ir} - \sum_{j=1}^{r-1} (m_{ij} u_{jr} + l_{ij} v_{jr}) - l_{ir} v_{rr} \right] / u_{rr}, \quad i = r+1, \dots, n,$$

$$w_{rk} = c_{rk} - \sum_{j=1}^{r-1} (n_{rj} u_{jk} + 2 m_{rj} v_{jk} + l_{rj} w_{jk}), \quad k = r, \dots, n,$$

$$n_{ir} = \left[c_{ir} - \sum_{j=1}^{r-1} (n_{ij} u_{jr} + 2 m_{ij} v_{jr} + l_{ij} w_{jr}) - 2 m_{ir} v_{rr} - l_{ir} w_{rr} \right] / u_{rr}, \quad i = r+1, \dots, n.$$

If some of the principal minors of the matrix of order $j \leq n-1$ are zero, then the decomposition (32) may not exist or, if it exists, it is ambiguous.

In practice, the best way to establish the possibility of *LU*-decomposition is to try to calculate it. It may happen that $u_{rr} = 0$ (r is the order of the main minor of the matrix, which is zero). To avoid this, in the process of decomposition one may use a series of permutations of rows (and/or columns) of matrix $\mathbf{D}$ with a choice of principal element. In this case the decomposition (37) can be written as

$$\mathbf{PD} = \mathbf{LU} \tag{39}$$

$$\mathbf{PB} = \mathbf{MU} + \mathbf{LV} \tag{40}$$

$$\mathbf{PC} = \mathbf{NU} + 2\mathbf{MV} + \mathbf{LW}$$

where $\mathbf{P}$ is a permutation matrix, moreover $\det\mathbf{P}=(-1)^q$, where q is a number of permutations (for example, rows). In this case the relations (38) take the form

$$f(\lambda_m)=(-1)^q\prod_{i=1}^{n}u_{ii},\ f'(\lambda_m)=(-1)^q\sum_{k=1}^{n}v_{kk}\prod_{i=1,i\neq k}^{n}u_{ii}, \tag{41}$$

$$f''(\lambda_m)=(-1)^q\sum_{k=1}^{n}w_{kk}\prod_{i=1,i\neq k}^{n}u_{ii}+(-1)^q\sum_{k=1}^{n}v_{kk}\left(\sum_{j=1,j\neq k}^{n}v_{jj}\prod_{i=1,i\neq k,i\neq j}^{n}u_{ii}\right).$$

Since in the relations (38) the value of function and its derivative is calculated only on the boundary region G, i.e. at the given points $\lambda_j,\ j=1,\ldots,N$, then for their calculation we use the same numerical procedure of decomposition of matrices (37). As a result, to calculate the values $s_k,\ k=0,1,2,\ldots,$ we obtain the relation

$$s_k=\frac{1}{N}\sum_{j=1}^{N}\left(\left(\lambda_j\right)^k\rho_t\exp(i\tfrac{2\pi j}{N})\sum_{r=1}^{n}\frac{v_{rr}}{u_{rr}}\right), \tag{42}$$

where $u_{kk},\ v_{kk}$ are the elements of matrices $\mathbf{U},\mathbf{V}$ in decomposition (37) at fixed $\lambda=\lambda_j$.

Now, if we know some approximation to the eigenvalue, then the correction $\Delta\lambda_l=f(\lambda_l)/f'(\lambda_l)$ to construct the successive approximations for Newton's method (28) assumes the form

$$\Delta\lambda_l=1/\sum_{k=1}^{n}\frac{v_{kk}}{u_{kk}},\ l=0,1,\ldots, \tag{43}$$

and bilateral analogue of Newton's method (29) takes the form

$$\begin{cases}\lambda_{2l+1}=\lambda_{2l}-\left(\sum_{k=1}^{n}\frac{v_{kk}}{u_{kk}}\right)/\sum_{k=1}^{n}\left(\left(\frac{v_{kk}}{u_{kk}}\right)^2-\frac{w_{kk}}{u_{kk}}\right),\\ \lambda_{2l+2}=\lambda_{2l+1}-1/\sum_{k=1}^{n}\frac{\overline{v}_{kk}}{\overline{u}_{kk}},\end{cases}\quad l=0,1,2,\ldots,, \tag{44}$$

where $u_{kk},\ v_{kk},\ w_{kk}$ are the elements of matrices $\mathbf{U},\mathbf{V}$ and $\mathbf{W}$ in the decompositions (37) at fixed $\lambda=\lambda_{2l}$, and $\overline{u}_{kk},\ \overline{v}_{kk}$ are the elements of matrices $\mathbf{U},\mathbf{V}$ in the decompositions (37) at fixed $\lambda=\lambda_{2l+1}$.

Thus, the algorithm of finding the eigenvalue curves of the problem (19) consists of the following steps.

Algorithm 1.

Step 1. Determine the interval $\Lambda=[\lambda_c,\lambda_d]$, where we find the eigenvalues of problem (21). This can be a single-spaced interval or a sequence of intervals $\Lambda_t=[\lambda_{c_t},\lambda_{d_t}]$ such that

$\Lambda = \cup \Lambda_t$. For this purpose we put this interval Λ_t in a circle (area G), setting the center of the circle $r_{0_t} = (\lambda_{c_t} + \lambda_{d_t})/2$ and radius $\rho_t = (\lambda_{d_t} - \lambda_{c_t})/2$, and also the number of points of partition N of the boundary Γ of the region G, i.e. the circle.

Step 2. Determine the value of parameter $\mu = \alpha_k \lambda + \beta_k$ giving the next meaning for the values α_k and β_k .

Step 3. Using the decomposition (37) for complex values λ, we determine the number of eigenvalues that are in the selected area G, by the formula

$$m = s_0 = \frac{1}{N}\sum_{j=1}^{N} \rho_t \exp\left(i\frac{2\pi j}{N} \right) \sum_{r=1}^{n} \frac{v_{rr}}{u_{rr}},$$

and their approximate values we find by solving the system of equations (24), after calculating the right part of the formula (42).

Step 4. Using the decomposition (37) for real values λ, we refine all eigenvalues that fall in the area G, using the Newton method

$$\lambda_{\ell+1} = \lambda_\ell - 1 / \left(\sum_{r=1}^{n} \frac{v_{rr}}{u_{rr}} \right),$$

or bilateral analogue of Newton's method (44). As initial approximation we take the approximate values obtained in *Step 3*.

Step 5. Go to Step 2.

Step 6. If necessary, we correct the area G by changing it center and / or radius and go to Step 2, otherwise go to Step 7.

Step 7. The end.

Application of modification of algorithm for linear two-parameter problems was considered in [16].

5. Algorithm of finding the bifurcation points of eigenvalue curves

Note that if two eigenvalue curves intersect at some point, this point is called the point of bifurcation (or branch the point). Sufficient criterion for the existence of such points have been known long ago (see, for example, [8]) and consists in that the point (λ_b, μ_b) is a bifurcation point of equation

$$f(\lambda, \mu) \equiv \det D_n(\lambda, \mu) = 0, \tag{45}$$

if conditions $\frac{\partial f(\lambda,\mu)}{\partial \lambda} = 0$, $\frac{\partial f(\lambda,\mu)}{\partial \mu} = 0$ are satisfied, and the second order partial derivatives are different from zero. But this criterion was not often used in practical calculations, because it requires calculation of derivatives of the determinant of matrix.

Using the algorithm of computing derivatives of the determinant of the matrix proposed in section 2 this criterion can be effectively used to calculate the bifurcation points of equation (45).

Thus, the problem consists in determining such parameters λ and μ which are the solutions of two nonlinear algebraic equations

$$\begin{aligned}\frac{\partial f(\lambda,\mu)}{\partial\lambda} &\equiv \left[\det D_n(\lambda,\mu)\right]'_{\lambda}=0,\\ \frac{\partial f(\lambda,\mu)}{\partial\mu} &\equiv \left[\det D_n(\lambda,\mu)\right]'_{\mu}=0.\end{aligned} \tag{46}$$

Note that, with some approximation to solution of (46), for its solution the iterative process of Newton's method can be applied as in [12]

$$\begin{bmatrix}\lambda_{m+1}\\ \mu_{m+1}\end{bmatrix}=\begin{bmatrix}\lambda_{m}\\ \mu_{m}\end{bmatrix}-\left[J(\lambda_m,\mu_m)\right]^{-1}\begin{bmatrix}\left[f(\lambda_m,\mu_m)\right]'_{\lambda}\\ \left[f(\lambda_m,\mu_m)\right]'_{\mu}\end{bmatrix},\quad m=0,1,\ldots \tag{47}$$

where

$$J(\lambda_m,\mu_m)=\begin{bmatrix}\left[f(\lambda_m,\mu_m)\right]''_{\lambda\lambda} & \left[f(\lambda_m,\mu_m)\right]''_{\lambda\mu}\\ \left[f(\lambda_m,\mu_m)\right]''_{\mu\lambda} & \left[f(\lambda_m,\mu_m)\right]''_{\mu\mu}\end{bmatrix}, \tag{48}$$

Further we assume that the determinant of matrix of the second derivatives (48) whose elements are calculated at point different (λ_m,μ_m) from zero.

Thus, at each step iterative process to compute the function $f(\lambda,\mu)=\det\mathbf{T}(\lambda,\mu)$ and its partial derivatives (first and second) only for fixed values of the parameters λ and μ. This can be realized in a numerical procedure using the *LU*-decomposition of matrix $\mathbf{T}(\lambda,\mu)$, namely:

$$f'_{\lambda}(\lambda,\mu)=\sum_{k=1}^{n}v_{kk}^{1}(\lambda,\mu)\prod_{i=1,i\neq k}^{n}u_{ii}(\lambda,\mu),$$

$$f'_{\mu}(\lambda,\mu)=\sum_{k=1}^{n}v_{kk}^{2}(\lambda,\mu)\prod_{i=1,i\neq k}^{n}u_{ii}(\lambda,\mu),$$

$$f''_{\lambda\lambda}(\lambda,\mu)=\sum_{k=1}^{n}w_{kk}^{1,1}(\lambda,\mu)\prod_{i=1,i\neq k}^{n}u_{ii}(\lambda,\mu)+\sum_{k=1}^{n}v_{kk}^{1}(\lambda,\mu)\left(\sum_{j=1,j\neq k}^{n}v_{jj}^{1}(\lambda,\mu)\prod_{i=1,i\neq k,i\neq j}^{n}u_{ii}(\lambda,\mu)\right),$$

$$f''_{\mu\mu}(\lambda,\mu)=\sum_{k=1}^{n}w_{kk}^{2,2}(\lambda,\mu)\prod_{i=1,i\neq k}^{n}u_{ii}(\lambda,\mu)+\sum_{k=1}^{n}v_{kk}^{2}(\lambda,\mu)\left(\sum_{j=1,j\neq k}^{n}v_{jj}^{2}(\lambda,\mu)\prod_{i=1,i\neq k,i\neq j}^{n}u_{ii}(\lambda,\mu)\right),$$

$$f''_{\lambda\mu}(\lambda,\mu)=\sum_{k=1}^{n} w_{kk}^{1,2}(\lambda,\mu)\prod_{i=1,i\neq k}^{n} u_{ii}(\lambda,\mu)+\sum_{k=1}^{n} v_{kk}^{1}(\lambda,\mu)\left(\sum_{j=1,j\neq k}^{n} v_{jj}^{2}(\lambda,\mu)\prod_{i=1,i\neq k,i\neq j}^{n} u_{ii}(\lambda,\mu)\right),$$

$$f''_{\mu\lambda}(\lambda,\mu)=\sum_{k=1}^{n} w_{kk}^{2,1}(\lambda,\mu)\prod_{i=1,i\neq k}^{n} u_{ii}(\lambda,\mu)+\sum_{k=1}^{n} v_{kk}^{2}(\lambda,\mu)\left(\sum_{j=1,j\neq k}^{n} v_{jj}^{1}(\lambda,\mu)\prod_{i=1,i\neq k,i\neq j}^{n} u_{ii}(\lambda,\mu)\right),$$

where $v_{ii}^{1}(\lambda,\mu)=\left[u_{ii}(\lambda,\mu)\right]'_{\lambda}$, $v_{ii}^{2}(\lambda,\mu)=\left[u_{ii}(\lambda,\mu)\right]'_{\mu}$, $w_{ii}^{1,1}(\lambda,\mu)=\left[v_{ii}^{1}(\lambda,\mu)\right]'_{\lambda}$, $w_{ii}^{2,2}(\lambda,\mu)=\left[v_{ii}^{2}(\lambda,\mu)\right]'_{\mu}$, $w_{ii}^{1,2}(\lambda,\mu)=\left[v_{ii}^{1}(\lambda,\mu)\right]'_{\mu}$, $w_{ii}^{2,1}(\lambda,\mu)=\left[v_{ii}^{2}(\lambda,\mu)\right]'_{\lambda}$ are the diagonal elements of matrices $V^1(\lambda,\mu)$, $V^2(\lambda,\mu)$, $W^{1,1}(\lambda,\mu)$, $W^{2,2}(\lambda,\mu)$, $W^{1,2}(\lambda,\mu)$ and $W^{2,1}(\lambda,\mu)$ in decompositions

$$\left[D(\lambda,\mu)\right]'_{\lambda}\equiv B^1(\lambda,\mu)=M^1(\lambda,\mu)U(\lambda,\mu)+L(\lambda,\mu)V^1(\lambda,\mu),$$

$$\left[D(\lambda,\mu)\right]'_{\mu}\equiv B^2(\lambda,\mu)=M^2(\lambda,\mu)U(\lambda,\mu)+L(\lambda,\mu)V^2(\lambda,\mu),$$

$$\left[D(\lambda,\mu)\right]''_{\lambda\lambda}\equiv C^{1,1}(\lambda,\mu)=N^{1,1}(\lambda,\mu)U(\lambda,\mu)+2M^1(\lambda,\mu)V^1(\lambda,\mu)+L(\lambda,\mu)W^{1,1}(\lambda,\mu),$$

$$\left[D(\lambda,\mu)\right]''_{\lambda\lambda}\equiv C^{1,1}(\lambda,\mu)=N^{1,1}(\lambda,\mu)U(\lambda,\mu)+2M^1(\lambda,\mu)V^1(\lambda,\mu)+L(\lambda,\mu)W^{1,1}(\lambda,\mu),$$

$$\left[D(\lambda,\mu)\right]''_{\lambda\mu}\equiv C^{1,2}(\lambda,\mu)=N^{1,2}(\lambda,\mu)U(\lambda,\mu)+M^1(\lambda,\mu)V^2(\lambda,\mu)+M^2(\lambda,\mu)V^1(\lambda,\mu)+L(\lambda,\mu)W^{1,2}(\lambda,\mu),$$

$$\left[D(\lambda,\mu)\right]''_{\lambda\mu}\equiv C^{2,1}(\lambda,\mu)=N^{2,1}(\lambda,\mu)U(\lambda,\mu)+M^2(\lambda,\mu)V^1(\lambda,\mu)+M^1(\lambda,\mu)V^2(\lambda,\mu)+L(\lambda,\mu)W^{2,1}(\lambda,\mu),$$

which are obtained from decomposition

$$\mathbf{D}(\lambda,\mu)=\mathbf{L}(\lambda,\mu)\mathbf{U}(\lambda,\mu).$$

Here $M^1(\lambda,\mu)=\left[L(\lambda,\mu)\right]'_{\lambda}$, $M^2(\lambda,\mu)=\left[L(\lambda,\mu)\right]'_{\mu}$, $N^{1,1}(\lambda,\mu)=\left[M^1(\lambda,\mu)\right]'_{\lambda}$, $N^{2,2}(\lambda,\mu)=\left[M^2(\lambda,\mu)\right]'_{\mu}$, $N^{1,2}(\lambda,\mu)=\left[M^1(\lambda,\mu)\right]'_{\mu}$, $N^{2,1}(\lambda,\mu)=\left[M^2(\lambda,\mu)\right]'_{\lambda}$.

From this it follows that to calculate $f'_{\lambda}(\lambda_m,\mu_m)$, $f'_{\mu}(\lambda_m,\mu_m)$, $f''_{\lambda\lambda}(\lambda_m,\mu_m)$, $f''_{\mu\mu}(\lambda_m,\mu_m)$, $f''_{\lambda\mu}(\lambda_m,\mu_m)$ and $f''_{\mu\lambda}(\lambda_m,\mu_m)$ it is necessary for fixed $\lambda=\lambda_m$ and $\mu=\mu_m$ to calculate

$$\begin{aligned}
&D=LU,\\
&B^1=M^1U+LV^1,\\
&B^2=M^2U+LV^2,\\
&C^{1,1}=N^{1,1}U+2M^1V^1+LW^{1,1},\\
&C^{2,2}=N^{2,2}U+2M^2V^2+LW^{2,2},\\
&C^{1,2}=N^{1,2}U+M^1V^2+M^2V^1+LW^{1,2},\\
&C^{2,1}=N^{2,1}U+M^2V^1+M^1V^2+LW^{2,1},
\end{aligned}\tag{49}$$

from which

$$f'_{\lambda}(\lambda_m,\mu_m)=\sum_{k=1}^{n} v_{kk}^1 \prod_{i=1,i\neq k}^{n} u_{ii},\; f'_{\mu}(\lambda_m,\mu_m)=\sum_{k=1}^{n} v_{kk}^2 \prod_{i=1,i\neq k}^{n} u_{ii},$$

$$f''_{\lambda\lambda}(\lambda_m)=\sum_{k=1}^{n} w_{kk}^{1,1} \prod_{i=1,i\neq k}^{n} u_{ii}+\sum_{k=1}^{n} v_{kk}^1\left(\sum_{j=1,j\neq k}^{n} v_{jj}^1 \prod_{i=1,i\neq k,i\neq j}^{n} u_{ii}\right),$$

$$f''_{\mu\mu}(\lambda_m)=\sum_{k=1}^{n} w_{kk}^{2,2} \prod_{i=1,i\neq k}^{n} u_{ii}+\sum_{k=1}^{n} v_{kk}^2\left(\sum_{j=1,j\neq k}^{n} v_{jj}^2 \prod_{i=1,i\neq k,i\neq j}^{n} u_{ii}\right), \tag{50}$$

$$f''_{\lambda\mu}(\lambda_m)=\sum_{k=1}^{n} w_{kk}^{1,2} \prod_{i=1,i\neq k}^{n} u_{ii}+\sum_{k=1}^{n} v_{kk}^1\left(\sum_{j=1,j\neq k}^{n} v_{jj}^2 \prod_{i=1,i\neq k,i\neq j}^{n} u_{ii}\right),$$

$$f''_{\mu\lambda}(\lambda_m)=\sum_{k=1}^{n} w_{kk}^{2,1} \prod_{i=1,i\neq k}^{n} u_{ii}+\sum_{k=1}^{n} v_{kk}^2\left(\sum_{j=1,j\neq k}^{n} v_{jj}^1 \prod_{i=1,i\neq k,i\neq j}^{n} u_{ii}\right).$$

The elements of matrix from decomposition (49) can be calculated directly using the recurrent relations

$$r=1,2,\dots,n,$$

$$u_{rk}=d_{rk}-\sum_{j=1}^{r-1} l_{rj}u_{jk},\quad k=r,\dots,n,$$

$$l_{ir}=\left(d_{ir}-\sum_{j=1}^{r-1} l_{ij}u_{jr}\right)/u_{rr},\quad i=r+1,\dots,n,$$

$$v_{rk}^1=b_{rk}^1-\sum_{j=1}^{r-1}(m_{rj}^1u_{jk}+l_{rj}v_{jk}^1),\quad k=r,\dots,n,$$

$$m_{ir}^1=\left[b_{ir}^1-\sum_{j=1}^{r-1}(m_{ij}^1u_{jr}+l_{ij}v_{jr}^1)-l_{ir}v_{rr}^1\right]/u_{rr},\quad i=r+1,\dots,n,$$

$$v_{rk}^2=b_{rk}^2-\sum_{j=1}^{r-1}(m_{rj}^2u_{jk}+l_{rj}v_{jk}^2),\quad k=r,\dots,n,$$

$$m_{ir}^2=\left[b_{ir}^2-\sum_{j=1}^{r-1}(m_{ij}^2u_{jr}+l_{ij}v_{jr}^2)-l_{ir}v_{rr}^2\right]/u_{rr},\quad i=r+1,\dots,n,$$

$$w_{rk}^{1,1} = c_{rk}^{1,1} - \sum_{j=1}^{r-1} (n_{rj}^{1,1} u_{jk} + 2m_{rj}^{1} v_{jk}^{1} + l_{rj} w_{jk}^{1,1}) , \quad k = r, \dots, n,$$

$$n_{ir}^{1,1} = \left[c_{ir}^{1,1} - \sum_{j=1}^{r-1} (n_{ij}^{1,1} u_{jr} + 2m_{ij}^{1} v_{jr}^{1} + l_{ij} w_{jr}^{1,1}) - 2m_{ir}^{1} v_{rr}^{1} - l_{ir} w_{rr}^{1,1} \right] / u_{rr} , \quad i = r+1, \dots, n,$$

$$w_{rk}^{2,2} = c_{rk}^{2,2} - \sum_{j=1}^{r-1} (n_{rj}^{2,2} u_{jk} + 2m_{rj}^{2} v_{jk}^{2} + l_{rj} w_{jk}^{2,2}) , \quad k = r, \dots, n,$$

$$n_{ir}^{2,2} = \left[c_{ir}^{2,2} - \sum_{j=1}^{r-1} (n_{ij}^{2,2} u_{jr} + 2m_{ij}^{2} v_{jr}^{2} + l_{ij} w_{jr}^{2,2}) - 2m_{ir}^{2} v_{rr}^{2} - l_{ir} w_{rr}^{2,2} \right] / u_{rr} , \quad i = r+1, \dots, n,$$

$$w_{rk}^{1,2} = c_{rk}^{1,2} - \sum_{j=1}^{r-1} (n_{rj}^{1,2} u_{jk} + m_{rj}^{1} v_{jk}^{2} + m_{rj}^{2} v_{jk}^{1} + l_{rj} w_{jk}^{1,2}) , \quad k = r, \dots, n,$$

$$n_{ir}^{1,2} = \left[c_{ir}^{1,2} - \sum_{j=1}^{r-1} (n_{ij}^{1,2} u_{jr} + m_{ij}^{1} v_{jr}^{2} + m_{ij}^{2} v_{jr}^{1} + l_{ij} w_{jr}^{1,2}) - m_{ir}^{1} v_{rr}^{2} - m_{ir}^{2} v_{rr}^{1} - l_{ir} w_{rr}^{1,2} \right] / u_{rr} , \quad i = r+1, \dots, n,$$

$$w_{rk}^{2,1} = c_{rk}^{2,1} - \sum_{j=1}^{r-1} (n_{rj}^{2,1} u_{jk} + m_{rj}^{2} v_{jk}^{1} + m_{rj}^{1} v_{jk}^{2} + l_{rj} w_{jk}^{2,1}) , \quad k = r, \dots, n,$$

$$n_{ir}^{2,1} = \left[c_{ir}^{2,1} - \sum_{j=1}^{r-1} (n_{ij}^{2,1} u_{jr} + m_{ij}^{2} v_{jr}^{1} + m_{ij}^{1} v_{jr}^{2} + l_{ij} w_{jr}^{2,1}) - m_{ir}^{2} v_{rr}^{1} - m_{ir}^{1} v_{rr}^{2} - l_{ir} w_{rr}^{2,1} \right] / u_{rr}, \quad i = r+1, \dots, n, ,$$

which are generalization of recurent relations of Section 3.2.

Thus, the algorithm of finding of the bifurcation points of eigenvalue curves of two-parametric spectral problem consists of the following steps.

Algorithm 2.

Step 1. To set the accuracy of calculations: with respect of the parameters - ε_p and with respect of the function - ε_f

Step 2. Initialize λ_0, μ_0

Step 3. **for** m =1,2, ... up to achievement of accuracy ε_p **do**

Step 4. Calculate the matrix $D_n(\lambda,\mu) = T_n(\lambda,\mu) - I$, $B^1(\lambda,\mu) = \left[D_n(\lambda,\mu)\right]'_{\lambda}$, $B^2(\lambda,\mu) = \left[D_n(\lambda,\mu)\right]'_{\mu}$ and $C^{1,1}(\lambda,\mu) = \left[D_n(\lambda,\mu)\right]''_{\lambda\lambda}$, $C^{2,2}(\lambda,\mu) = \left[D_n(\lambda,\mu)\right]''_{\mu\mu}$, $C^{1,2}(\lambda,\mu) = \left[D_n(\lambda,\mu)\right]''_{\lambda\mu}$, $C^{2,1}(\lambda,\mu) = \left[D_n(\lambda,\mu)\right]''_{\mu\lambda}$ for $\lambda = \lambda_m$, $\mu = \mu_m$.

Step 5. Using the decomposition (49) and relations (50) we calculate $f'_\lambda(\lambda_m,\mu_m)$, $f'_\mu(\lambda_m,\mu_m)$, $f''_{\lambda\lambda}(\lambda_m,\mu_m)$, $f''_{\mu\mu}(\lambda_m,\mu_m)$, $f''_{\lambda\mu}(\lambda_m,\mu_m)$ and $f''_{\mu\lambda}(\lambda_m,\mu_m)$ and construct the matrix of the second derivatives of (48).

Step 6. Compute the next approximation to λ and μ by the formula (47)

Step 7. **end for** m

Step 8. **if** $|f(\lambda_m,\mu_m)| \le \varepsilon_f$ **then go to** *Step 10.*

Step 9. **else** Initialize a different initial approximation to the bifurcation point and **go to** *Step 3.*

Step 10. The end

6. Analysis of numerical results

In conducting a series of numerical experiments on the synthesis of antenna arrays, Algorithms 1 and 2 were used to find the curves of eigenvalues for two-parameter eigenvalue problem, which are the branching lines of solutions of nonlinear synthesis equation (7) and their bifurcation points. Numerical calculations were carried out as for the problems in which in the function $F(\xi_1,\xi_2)$ that describes the given directivity pattern of the array, the variables are separated and are not separated.

In Fig. 10 - Fig. 13 are shown the curves of eigenvalues for four problems, in which the given directivity patterns are defined by the formulas $F(\xi_1,\xi_2)=1$, $F(\xi_1,\xi_2)=\cos\frac{\pi}{2}\xi_1\cdot|\sin\pi\xi_2|$, $F(\xi_1,\xi_2)=\sqrt{1-(\xi_1^2+\xi_2^2)/2}$ and $F(\xi_1,\xi_2)=\cos\frac{\pi}{2}\left(\sqrt{\xi_1^2+\xi_2^2}\right)$, respectively.

In conducting numerical experiments the interval of changing of parameter λ is divided into sequence intervals, each of which is puted in a circle of corresponding radius with respective center. Number of points partition of boundary of each circle was constant and equal $N=512$. On the step 2 of algorithm the values of α was selected with the interval $[0,7 \div 1,5]$ with a step $\Delta\alpha = 0,05$ as well $\beta \equiv 0$.

Table. 1 presents the bifurcation points for four directivity patterns $F(\xi_1,\xi_2)$, when the variables are separated and three directivity patterns when the variables are not separated. For the first four diagrams a bifurcation point is shown also, which can be obtained by other methods, provided that in the function $f(\xi_1,\xi_2)$ the variables are separated.

The Table shows that when the variables in the functions $F(\xi_1,\xi_2)$ and $f(\xi_1,\xi_2)$ are separated, the results obtained by different approaches (reduction of one-parameter problems to the transcendental equations and solving them [2], by methods of descent [18] and also by bilateral methods proposed in [11, 13, 15]) are the same.

Note that the bifurcation points (at least their rough estimates) may be obtained graphically from Fig. 10 - Fig. 13, and may be clarified by the Algorithm 2.

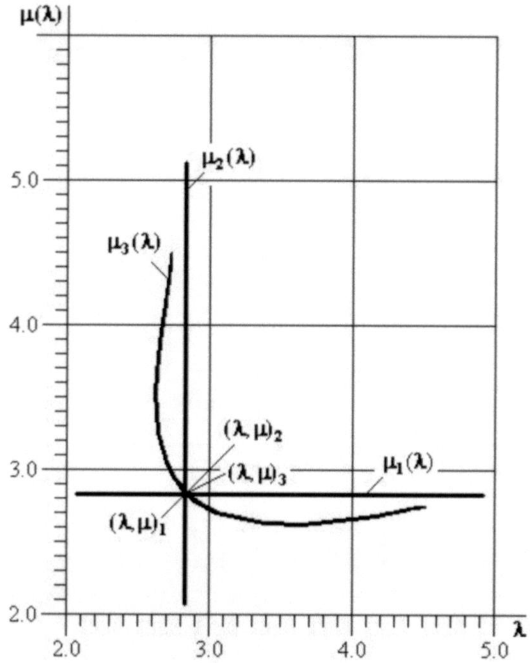

Figure 10. Eigenvalue curves for $F(\xi_1, \xi_2) = 1$

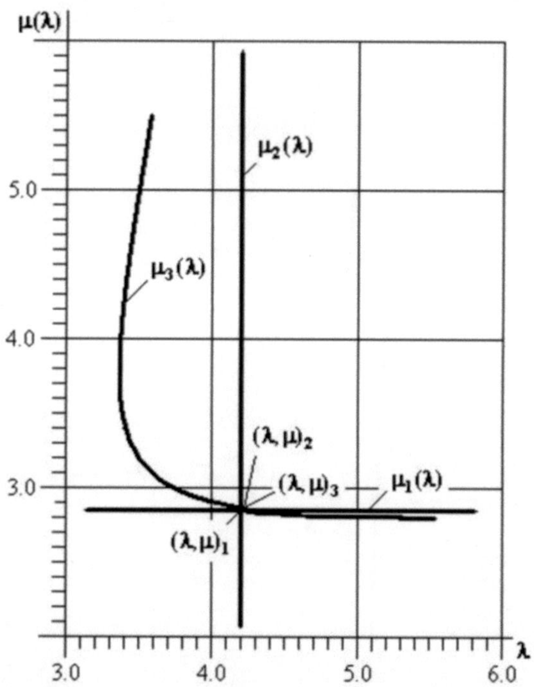

Figure 11. Eigenvalue curves for $F(\xi_1, \xi_2) = \cos\frac{\pi}{2}\xi_1 \cdot |\sin \pi\xi_2|$

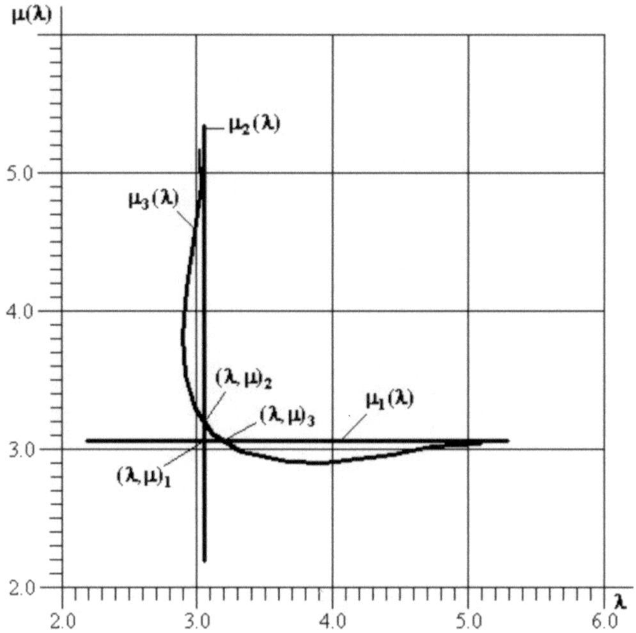

Figure 12. Eigenvalue curves for $F(\xi_1,\xi_2) = \sqrt{1-(\xi_1^2+\xi_2^2)/2}$

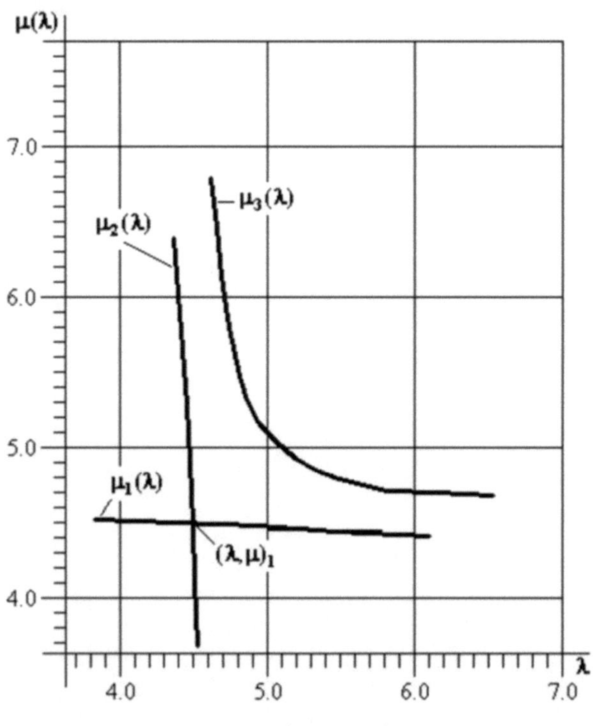

Figure 13. Eigenvalue curves for $F(\xi_1,\xi_2) = \cos\frac{\pi}{2}\left(\sqrt{\xi_1^2+\xi_2^2}\right)$

$F(\xi_1,\xi_2)$	Bifurcation point $(\lambda_b,\mu_b)_i$	Bifurcation point, obtained by other methods
const = 1	$(\lambda_b,\mu_b)_1=(\lambda_b,\mu_b)_2=$ $=(\lambda_b,\mu_b)_3=(2.832715,\ 2.832715)$	$(\lambda_b,\mu_b)_1=(2.832715,\ 2.832715)$
$\cos\frac{\pi\xi_1}{2}\cdot\cos\frac{\pi\xi_1}{2}$	$(\lambda_b,\mu_b)_1=(\lambda_b,\mu_b)_2=$ $=(\lambda_b,\mu_b)_3=(4.207065,\ 4.207065)$	$(\lambda_b,\mu_b)_1=(4.207065,\ 4.207065)$
$\lvert\sin\pi\xi_1\rvert\cdot\lvert\sin\pi\xi_2\rvert$	$(\lambda_b,\mu_b)_1=(\lambda_b,\mu_b)_2=$ $=(\lambda_b,\mu_b)_3=(2.855425,\ 2.855425)$	$(\lambda_b,\mu_b)_1=(2.855425,\ 2.855425)$
$\cos\frac{\pi\xi_1}{2}\cdot\lvert\sin\pi\xi_2\rvert$	$(\lambda_b,\mu_b)_1=(\lambda_b,\mu_b)_2=$ $=(\lambda_b,\mu_b)_3=(4.207065,\ 2.855425)$	$(\lambda_b,\mu_b)_1=(4.207065,\ 2.855425)$
$\sqrt{1-\frac{1}{2}(\xi_1^2+\xi_2^2)}$	$(\lambda_b,\mu_b)_1=(3.064250,\ 3.064250)$ $(\lambda_b,\mu_b)_2=(3.064250,\ 3.186696)$ $(\lambda_b,\mu_b)_3=(3.186696,\ 3.064250)$	-
$1-\frac{1}{2}(\xi_1^2+\xi_2^2)$	$(\lambda_b,\mu_b)_1=(3.302395,\ 3.302395)$ $(\lambda_b,\mu_b)_2=(3.302395,\ 3.565660)$ $(\lambda_b,\mu_b)_3=(3.565660,\ 3.302395)$	-
$\cos\frac{\pi}{2}\sqrt{\xi_1^2+\xi_2^2}$	$(\lambda_b,\mu_b)_1=(4.503957,\ 4.503957)$	-

Table 1. Bifurcation points for different types of the given directivity pattern

7. Concluding remarks

Numerical experiments with the calculation of eigenvalues and eigenvectors, realized for certain specified types of directivity pattern by the proposed algorithms, and comparison them with existing results obtained by other methods shows their efficiency (in terms of bilateral approximations and convergence rate). Developed and implemented algorithms for numerical finding of the branching lines of nonlinear integral equations, the kernels of which nonlinearly depend on two spectral parameters and their bifurcation points, yielded the new results, namely:

- We have found all valid solutions (the curves of eigenvalues) of the problem (19), which fall in the interval of modified parameters λ and μ, that we are interested in.
- For the problems in which the function $F(\xi_1,\xi_2)$ admits separation of variables, another solution to the problem (19) has been found (for example, for $F(\xi_1,\xi_2)=1$ and $F(\xi_1,\xi_2)=\cos\frac{\pi}{2}\xi_1\cdot\lvert\sin\pi\xi_2\rvert$ this is $\mu_3(\lambda)$, what is shown in Fig. 10 and Fig. 11, respectively), which corresponds to the synthesized directivity patterns in which the variables are not separated.

- For the problems in which the function $F(\xi_1,\xi_2)$ does not allow separation of variables, we have found the solutions to the problem (19) (for example, for $F(\xi_1,\xi_2)=\sqrt{1-(\xi_1^2+\xi_2^2)/2}$ and $F(\xi_1,\xi_2)=\cos\frac{\pi}{2}\left(\sqrt{\xi_1^2+\xi_2^2}\right)$ are $\mu_1(\lambda)$ and $\mu_2(\lambda)$, shown in Fig. 12 and Fig. 13, respectively), which are supposed to exist only for the diagrams, where the variables are separated.
- We have calculated the bifurcation point of eigenvalue curves for the problems in which the function $F(\xi_1,\xi_2)$ does not allow separation of variables (eg, $F(\xi_1,\xi_2)=1-(\xi_1^2+\xi_2^2)/2$, $F(\xi_1,\xi_2)=\sqrt{1-(\xi_1^2+\xi_2^2)/2}$ and $F(\xi_1,\xi_2)=\cos\frac{\pi}{2}\left(\sqrt{\xi_1^2+\xi_2^2}\right)$). For these diagrams there are no known results. The results have been obtained for the first time.

Since the spectral parameters are the geometric and electromagnetic characteristics of radiating systems, the solution of this problem makes it possible to obtain the necessary information at the design stage, choosing the optimal ones with respect to the size and electrodynamic characteristics of the radiating system.

Note that such two-dimensional problem was studied also in the works [10, 17, 19], but there numerical results obtained for some directivity patterns $F(\xi_1,\xi_2)$ are not reliable.

To complete we shall mark, that the offered algorithm of calculation of derivatives of matrix determinant can be used and in the approach in which basis the implicit function theorem is. In such approach it is necessary to solve the Cauchy problem

$$\begin{aligned} &\frac{d\lambda}{d\mu}=-\frac{\left[\det T_n(\lambda,\mu)\right]'_\mu}{\left[\det T_n(\lambda,\mu)\right]'_\lambda}, \\ &\lambda(\mu_1)=\lambda_1, \end{aligned} \tag{51}$$

for which the right part of equation (51) can be calculated by the algorithm of calculation of derivatives of matrix determinant. Besides by the algorithm, given in this paper, it is possible numerically to define a number of eigenvalues, and, therefore, the eigenvalue curves, which are in the given range of spectral parameters and to calculate the initial value for Cauchy problem for each curve.

Author details

B. M. Podlevskyi
Institute of Applied Problems of Mechanics and Mathematics of NASU, Ukraine

8. References

[1] Abramov, A. A., Ul'yanova, V. I. & Yukhno, L. F. (1998). The Argument Principle in a Spectral Problem for Systems of Ordinary Differential Equations with Singularities, *Comput. Math. Math. Phys*, Vol. 38: 57-63.

[2] Andriychuk, M. I., Voitovich, N. N., Savenko, P. O. & Tkachuk, V. P. (1993). *The antenna synthesis according to prescribed amplitude radiation pattern: numerical methods and algorithms*, Naukova Dumka, Kiev. [in Russian].

[3] Awrejcewicz, J. (1989). *Bifurcation and Chaos in Simple Dynamical Systems,* World Scientific, Singapore.

[4] Awrejcewicz, J. (1991). *Bifurcation and Chaos in Coupled Oscillators,* World Scientific, Singapore.

[5] Awrejcewicz, J. (1990). Numerical investigations of the constant and periodic motions of the human vocal cords including stability and bifurcation phenomena, *Dynamics and Stability of Systems Journal,* Vol. 5 (1): 11-28.

[6] Cliffe, K. A., Spence, A. and Tavener, S. J. (2000). The numerical analysis of bifurcation problems with application to fluid mechanics, *Acta Numerica,* Vol. 9: 39-131.

[7] Delves, L. M. & Lyness, J. N. (1967). A numerical method for locating the zeros of analytic function, *Math. Comput,* Vol. 21: 543-561.

[8] Goursat, E. (1930). *Cours d'analyse mathematique.Tome I.*, Gauthier-Villars, Paris.

[9] Kravanja, P., Sakurai, T. & M. van Barel, (1998). On locating clusters of zeros of analytic functions, *BIT,* Vol. 38 (2): 101-104.

[10] Kravchenko, V. F., Protsakh, L. P., Savenko, P. O., Tkach, M. D. (2010). The Mathematical Features of the Synthesis of Plane Equidistant Antenna Arrays According to the Prescribed Amplitude Directivity Pattern, *Antennas,* No. 3: 34-45. [in Russian].

[11] Podlevskyi, B. M. (2007). On certain two-sided analogues of Newton's method for solving nonlinear eigenvalue problems, *Comput. Math. Math. Physics,* Vol. 47 (11): 1745-1755.

[12] Podlevskyi, B. M. (2008). Newton's Method as a Tool for Finding the Eigenvalues of Certain Two-Parameter (Multiparameter) Spectral Problems, *Comput. Math. Math. Phys,* Vol. 48 (12): 2140-2145.

[13] Podlevskyi, B. M. (2009). Bilateral analog of the Newton method for determination of eigenvalues of nonlinear spectral problems, *J. Mathematical Sciences,* Vol. 160 (3): 357-367.

[14] Podlevskyi, B. M. (2010). On one approach to finding the branching lines and bifurcation points of solutions of nonlinear integral equations whose kernels depend analytically on two spectral parameters, *J. Mathematical Sciences,* Vol. 171 (4): 433-452.

[15] Podlevskyi, B. M. (2011). The numerical methods and algorithms of solution of the generalized spectral problems. The thesis for a Doctor's Degree in Physics and Mathematics, Institute of Mathematics NAS of Ukraine, Kyiv. [in Ukraine]

[16] Podlevskyi, B. M. & Khlobystov, V. V. (2010). On one approach to finding eigenvalue curves of linear two-parameter spectral problems, *J. Mathematical Sciences,* Vol. 167 (1): 96-106.

[17] Protsakh, L. P., Savenko, P. O., Tkach, M. D. (2006). Method of implicit function for solving eigenvalue problem with nonlinear two-dimensional spectral parameter, *Math. Methods and Physicomechanical Fields,* Vol. 49 (3): 41-46. [in Ukraine].

[18] Savenko, P. O. (2002). *Nonlinear problems of radiating systems synthesis (theory and methods of solution),* IAPMM NASU, Lviv. [in Ukraine].

[19] Savenko, Petro & Tkach, Myroslava (2010). Numerical Approximation of Real Finite Nonnegative Function by Modulus of Discrete Fourier Transform, *Applied Math.*, Vol. 1 (1): 65-75.

[20] Vainberg, M. M. & Trenogin, V. A. (1969). *Branching theory of the solution of nonlinear equations,* Nauka, Moscow. [in Russian].

Chapter 12

Memory and Asset Pricing Models with Heterogeneous Beliefs

Miroslav Verbič

Additional information is available at the end of the chapter

http://dx.doi.org/10.5772/48070

1. Introduction

Heterogeneous agent models are present in various fields of economic analysis, such as market maker models, exchange rate models, monetary policy models, overlapping generations models and models of socio-economic behaviour. Yet the field with the most systematic and perhaps most promising nonlinear dynamic approach seems to be asset price modelling. Contributions by Brock and Hommes (1998), LeBaron (2000), Hommes *et al.* (2002), Chiarella and He (2002), Chiarella *et al.* (2003), Gaunersdorfer *et al.* (2003), Brock *et al.* (2005), Hommes *et al.* (2005), and Hommes (2006) thoroughly demonstrate how a simple standard pricing model is able to lead to complex dynamics that makes it extremely hard to predict the evolution of prices in asset markets. The main framework of analysis of such asset pricing models constitutes a financial market application for the evolutionary selection of expectation rules, introduced by Brock and Hommes (1997a) and is called the adaptive belief system.

As a model in which different agents have the ability to switch beliefs, the adaptive belief system in a standard discounted value asset pricing set-up is derived from mean-variance maximization and extended to the case of heterogeneous beliefs (Hommes, 2006, p. 47). It can be formulated in terms of deviations from a benchmark fundamental and therefore used in experimental and empirical testing of deviations from the rational expectations benchmark. Agents are boundedly rational, act independently of each other and select a forecasting or investment strategy based upon its recent relative performance. The key feature of such systems, which often incorporate active learning and adaptation, is endogenous heterogeneity (*cf.* LeBaron, 2002), which means that markets can move through periods that support a diverse population of beliefs, and others in which these beliefs and strategies might collapse down to a very small set.

INTECH
open science | open minds

© 2012 The Author(s). Licensee InTech. This chapter is distributed under the terms of the Creative Commons Attribution License http://creativecommons.org/licenses/by/3.0), which permits unrestricted use, distribution, and reproduction in any medium, provided the original work is properly cited.

The mixture of different trader types leads to diverse dynamics exhibiting some stylized, qualitative features observed in practice on financial markets (*cf.* Campbell *et al.*, 1997; Johnson *et al.*, 2003), e.g. persistence in asset prices, unpredictability of returns at daily horizon, mean reversion at long horizons, excess volatility, clustered volatility, and leptokurtosis of asset returns. An important finding so far was that irregular and chaotic behaviour is caused by rational choice of prediction strategies in the bounded-rationality framework, and that this also exhibits quantitative features of asset price fluctuations, observed in financial markets. Namely, due to differences in beliefs these models generate a high and persistent trading volume, which is in sharp contrast to no trade theorems in rational expectations models. Fractions of different trading strategies fluctuate over time and simple technical trading rules can survive evolutionary competition. On average, technical analysts may even earn profits comparable to the profits earned by fundamentalists or value traders.

While recent literature on asset price modelling focuses mainly on impacts of heterogeneity of beliefs in the standard adaptive belief system as set up by Brock and Hommes (1997a) on market dynamics and stability on one hand, and the possibility of the survival of such 'irrational' and speculative traders in the market on the other, several crucial issues regarding the foundations of asset price modelling and its underlying theoretical findings remain open and indeterminate. One of those issues is related to heterogeneity in investors' time horizon; both their planning and their evaluation perspective. Namely, it has been scarcely addressed so far how memory in the fitness measure, i.e. the share of past information that boundedly rational economic agents take into account as decision makers, affects stability of evolutionary adaptive systems and survival of technical trading.

LeBaron (2002) was using simulated agent-based financial markets of individuals following relatively simple behavioural rules that are updated over time. Actually, time was an essential and critical feature of the model. It has been argued that someone believing that the world is stationary should use all available information in forming his or her beliefs, while if one views the world as constantly in a state of change, then it will be better to use time series reaching a shorter length into the past. The dilemma is thus seen as an evolutionary challenge where long-memory agents, using lots of past data, are pitted against short-memory agents to see who takes over the market. Agents with a short-term perspective appear to both influence the market in terms of increasing volatility and create an evolutionary space where they are able to prosper. Changing the population to more long-memory types has led to a reliable convergence in strategies. Memory or perhaps the lack of it therefore appeared to be an important aspect of the market that is likely to keep it from converging and prevent the elimination of 'irrational', speculative strategies from the market.

Honkapohja and Mitra (2003) provided basic analytical results for dynamics of adaptive learning when the learning rule had finite memory and the presence of random shocks precluded exact convergence to the rational expectations equilibrium. The authors focused on the case of learning a stochastic steady state. Even though their work is not done in the heterogeneous agent setting, the results they obtained are interesting for our analysis. Their fundamental outcome was that the expectational stability principle, which plays a central role in situations of complete learning, as discussed e.g. in Evans and Honkapohja (2001), retains its importance in the analysis of incomplete learning, though it takes a new form. In the models that were analyzed, expectational stability guaranteed stationary dynamics in the learning economy and unbiased forecasts.

Chiarella *et al.* (2006) proposed a dynamic financial market model in which demand for traded assets had both a fundamentalist and a chartist component in the boundedly rational framework. The chartist demand was governed by the difference between current price and a (long-run) moving average. By examining the price dynamics of the moving average rule they found out that an increase of the window length of the moving average rule can destabilize an otherwise stable system, leading to more complicated, even chaotic behaviour. The analysis of the corresponding stochastic model was able to explain various market price phenomena, including temporary bubbles, sudden market crashes, price resistance and price switching between different levels.

The objective of this chapter is to lay the foundations for a competent and critical theoretical analysis setting the memory assumption in a simple, analytically tractable asset pricing model with heterogeneous beliefs. We shall thus analyze the effects of additional memory in the fitness measure on evolutionary adaptive systems and the nature of consequences for survival of technical trading. In order to examine our research hypothesis adequately, both analytical and numerical analysis will have to be employed and complemented. Therefore, we shall first expand the asset pricing model to include more memory, and then solve it both analytically and numerically. Two cases are going to be analyzed, hopefully sufficiently general to cover some main aspects of financial markets; (1) a two-type case of fundamentalists versus contrarians and (2) a three-type case of fundamentalists versus opposite biased beliefs. Complementing the stability analysis with local bifurcation theory (*cf.* Awrejcewicz, 1991; Palis and Takens, 1993; Kuznetsov, 1995; Awrejcewicz and Lamarque, 2003), we will also be able to analyze numerically the effects of adding different amounts of additional memory to fitness measure on stability of the standard asset pricing model and survival of technical trading. Thus the analysis of both local and global stability can be performed for different combinations of trader types in the market.

2. The heterogeneous agents model

The adaptive belief system employs a mechanism dealing with interaction between fractions of market traders of different types, and the distance between the fundamental and the actual price. Financial markets are thus viewed as an evolutionary system, where price fluctuations are driven by an evolutionary dynamics between different expectation schemes. Pioneering work in this field has been done by Brock and Hommes (1997a), who attempted to conciliate the two main perspectives concerning economic fluctuations, i.e. the new classical and the Keynesian view (*cf.* Hommes, 2006, pp. 1-5), and the underlying rules relating to the formation of expectations. In order to get some insight into possible ways of theoretical analysis to follow, we shall describe a simple, analytically tractable version of the asset pricing model as constructed by Brock and Hommes (1998). The model can be viewed as composed of two simultaneous parts; present value asset pricing and the evolutionary selection of strategies, resulting in equilibrium pricing equation and fractions of belief types equation. We shall also make an indication of where memory in the fitness measure (and in expectation rules) enters the model and how it might affect the analysis.

2.1. Present value asset pricing

The model incorporates one risky asset and one risk free asset. The latter is perfectly elastically supplied at given gross return R, where $R = 1 + r$. Investors of different types h have different beliefs about the conditional expectation and the conditional variance of modelling variables based on a publicly available information set consisting of past prices and dividends. The present value asset pricing part of the adaptive demand system is used to model each investor type as a myopic mean variance maximizer of expected wealth demand, $E_{h,t}W_t$, for the risky asset:

$$E_{h,t}W_{t+1} = RE_{h,t}W_t + (p_{t+1} + y_{t+1} - Rp_t)z_{h,t}, \tag{1}$$

where p_t is the price (ex dividend) at time t per share of risky asset, y_t is an IID dividend process at time t of the risky asset, $z_{h,t}$ is number of shares purchased at date t by agent of type h, and $R_{t+1} = p_{t+1} + y_{t+1} - Rp_t$ is the excess return.

In order to perform myopic mean variance maximization of expected wealth demand for risky asset of type h, we seek for $z_{h,t}$ that solves:

$$\max_{z_{h,t}} \left\{ E_{h,t}W_{t+1} - \frac{1}{2}aV_{h,t}W_{t+1} \right\} \tag{2}$$

and thus:

$$z_{h,t} = \frac{E_{h,t}\left[p_{t+1} + y_{t+1} - Rp_t\right]}{aV_{h,t}\left[p_{t+1} + y_{t+1} - Rp_t\right]} = \frac{1}{a\sigma^2}E_{h,t}\left[p_{t+1} + y_{t+1} - Rp_t\right], \tag{3}$$

where the belief about expected value of wealth at time $t + 1$, conditional on all publicly available information at time t, for a trader of type h is $E_{h,t}W_{t+1}$, the belief about conditional variance is $V_{h,t}W_{t+1}$, and there is a risk factor $k = \frac{1}{a\sigma^2}$ present. Beliefs about the conditional variance of excess return are assumed to be constant and the same for all types of investors, i.e. $V_{h,t} = \sigma^2$. All traders are assumed to be equally risk averse with a given risk aversion parameter a, which is constant over time[1].

[1] Gaunersdorfer (2000) investigated the case of time varying variance and supported the assumption of a constant and homogeneous variance term.

Solving this optimization problem produces quantities of shares purchased by agents of different types, which enables us to seek for the equilibrium between the constant supply of the risky asset per trader z^s and the sum of demands:

$$\sum_{h=1}^{H} n_{h,t} k E_{h,t}\left[p_{t+1} + y_{t+1} - Rp_t\right] = z^s , \tag{4}$$

where the fraction of traders of type h out of altogether H types at time t is denoted by $n_{h,t}$, where $\sum_{h=1}^{H} n_{h,t} = 1$. The price of the risky asset is determined by market clearing, which can be seen by rewriting expression (4) in the form:

$$Rp_t = \sum_{h=1}^{H} n_{h,t} E_{h,t}\left[p_{t+1} + y_{t+1}\right] - a\sigma^2 z^s , \tag{5}$$

where $a\sigma^2 z^s$ is the risk premium. The latter is an extra amount of money that traders get for holding the risky asset. Traders will only purchase the risky asset if its expected value is equal or higher than the expected value of the risk-free asset. Since the outcome of the risky asset is uncertain, a risk premium is associated with it.

In the simplest case of IID dividends with mean $\bar{y}$ and with traders having correct beliefs about dividends, i.e. $E_{h,t}\left[y_{t+1}\right] = \bar{y}$, the market price of the risky asset p_t at time t is determined by:

$$Rp_t = \sum_{h=1}^{H} n_{h,t} E_{h,t}\left[p_{t+1}\right] + \bar{y} - a\sigma^2 z^s + \varepsilon_t , \tag{6}$$

where a noise term ε_t is included, which represents random fluctuations in the supply of risky shares. Considering a special case with a constant zero supply of outside shares, i.e. $z^s = 0$, we obtain:

$$Rp_t = \sum_{h=1}^{H} n_{h,t} E_{h,t}\left[p_{t+1}\right] + \bar{y} + \varepsilon_t .$$

If we instead consider for a moment the case of homogeneous beliefs with no noise and all traders being rational, the pricing equation simplifies to:

$$Rp_t = E_t\left[p_{t+1}\right] + \bar{y} - a\sigma^2 z^s . \tag{7}$$

In equilibrium the expectations of the price will be the same and equal to the fundamental price. The constant fundamental value of the price of the risky asset p^* in the case of homogeneous beliefs is derived from the expression:

$$Rp^* = p^* + \bar{y} - a\sigma^2 z^s . \tag{8}$$

By imposing a transversality condition on expression (7) with infinitely many solutions we exclude bubble solutions (*cf.* Cuthbertson, 1996) and expression (8) now has only one solution. We are thus able to derive the fundamental price as the discounted sum of expected future dividends:

$$p^* = \frac{1}{R-1}\left[\overline{y} - a\sigma^2 z^s\right]. \tag{9}$$

By simplification of the fundamental price equation for the case of the IID dividend process with constant conditional expectation we thus obtain the standard benchmark notion of the 'fundamental', i.e. $p_t^* = \frac{\overline{y}}{r}$, to be used in the model hereinafter.

Taking into account the appropriate form of heterogeneous beliefs of future prices, i.e. including some deterministic function $f_{h,t}$, which can differ across trader types:

$$E_{h,t}\left[p_{t+1}\right] = E_t\left[p_{t+1}^*\right] + E_{h,t}\left[x_{t+1}\right] = p_{t+1}^* + f_h(x_{t-1},...,x_{t-L}),$$

we restrict beliefs about the next deviation of the actual from the fundamental price, x_t, to deterministic functions of past deviations from the fundamental:

$$E_{h,t}\left[p_{t+1}\right] = p^* + f_h(x_{t-1},...,x_{t-L}), \tag{10}$$

where L is the number of lags of past information, taken into account. Since the deterministic function in the expectation rule depends on preceding price deviations, it can also be seen as including memory. However, due to rapidly increasing analytical complexity, *viz.* including more preceding price deviations rapidly increases the dimension of the system, this issue has so far mainly been neglected. In this chapter we are focusing on the memory in the fitness measure and will thus include only one lag in the memory in the expectation rule, i.e. $f_h(x_{t-1})$.

Taking into account that $p_t^* = \frac{\overline{y}}{r}$, the equilibrium pricing equation (5) can thus finally be rewritten in terms of deviations from the fundamental price, $x_t = p_t - p^*$:

$$Rx_t = \sum_{h=1}^{H} n_{h,t} E_{h,t}\left[x_{t+1}\right] = \sum_{h=1}^{H} n_{h,t} f_{h,t}. \tag{11}$$

The particular form of deterministic function in the forecasting or expectation rule is thus what determines different types of heterogeneous agents in an adaptive belief system. In general, we distinguish between two typical investor types; fundamentalists and 'noise traders' or technical analysts. Fundamentalists believe that the price of an asset is defined solely by its efficient market hypothesis fundamental value (Fama, 1991), i.e. the present value of the stream of future dividends. Since they have no knowledge about other beliefs and fractions, $f_{h,t} \equiv 0$. Actual financial data show that fundamentalists have a stabilizing effect on prices (De Grauwe and Grimaldi, 2006).

Technical analysts or chartists, on the other hand, believe that asset prices are not completely determined by fundamentals, but may be predicted by inferences on past prices. Depending on the purpose of analysis, it is possible to distinguish between (pure) trend chasers with expectation rule $f_{h,t} = g_h x_{t-1}; g_h > 0$, (pure) contrarians with expectation rule $f_{h,t} = g_h x_{t-1}; g_h < 0$, and (pure) biased beliefs with expectation rule $f_{h,t} = b_h$, where g_h is the trend and b_h is the bias (difference between p^* and trader's belief of p^*) of the trader of type h.

2.2. Evolutionary selection of strategies

In order to be able to understand the dynamics of fractions of different trader types, we consider the appropriate formulations of realized excess return R_t from expression (1), and demand of different types of market traders, $z_{h,t-1}$, defined by expression (3). Taking again into account the nature of the dividend process $y_t = \overline{y} + \delta_t$ with constant conditional expectation, $\overline{y} = E\left[y_{t+1}\right]$, and assumed distribution $\delta_t \sim \text{IIDN}(0, \vartheta^2)$, we are thus able to formulate profits for a particular type of traders in each period as the product of realized excess return and number of shares purchased by traders of that type:

$$\pi_{h,t} = R_t z_{h,t-1} - C_h = (p_t + y_t - Rp_{t-1})kE_{h,t-1}\left[p_t + y_t - Rp_{t-1}\right] - C_h, \tag{12}$$

where C_h represents the costs traders have to pay to use strategy h. Albeit introducing additional analytical complexity, we usually take into account the costs for predictor of particular trader type, since more information-intense predictors are evidently more costly. It is of course convenient to rewrite profits of different types of traders in terms of deviations from the benchmark fundamental:

$$\pi_{h,t} = (x_t - Rx_{t-1} + \delta_t)kE_{h,t-1}\left[x_t - Rx_{t-1}\right] - C_h. \tag{13}$$

The fitness function or performance measure of each trader type can now be defined in terms of its realized profits. In fact, it can be expressed as the weighted sum of realized profits, i.e. as the sum of current realized profits and a share of past fitness, which is in turn defined as past realized profits:

$$U_{h,t} = wU_{h,t-1} + (1-w)\pi_{h,t}, \tag{14}$$

where current realized profits are defined in the following final form:

$$\pi_{h,t} = k(x_t - Rx_{t-1})(f_{h,t-1} - Rx_{t-1}) - C_h. \tag{15}$$

The fitness function can for $U_{h,0} = 0$ also be rewritten in the following expanded form with exponentially declining weights:

$$U_{h,t} = w^{t-1}(1-w)\pi_{h,1} + w^{t-2}(1-w)\pi_{h,2} + \ldots + w(1-w)\pi_{h,t-1} + (1-w)\pi_{h,t}.$$

In case of the equilibrium pricing equation, herein formulated as the sum over trader types of products of a fraction of particular trader type and its deterministic function, the fitnesses enter the adaptive belief system before the equilibrium price is observed. This is suitable for analyzing the asset pricing model as an explicit nonlinear difference equation. Even though nonlinear asset pricing dynamics can be modelled either as a deterministic or a stochastic process, only the latter enables investigation of the effects of noise upon the asset pricing dynamics.

The share of past fitness in the performance measure is expressed by the parameter w; $0 \leq w \leq 1$, called memory strength. When the value of this parameter is zero ($w = 0$), the fitness is given by most recent net realized profit. Due to analytical tractability this is at present, y for the most part, the case in the existing literature on asset pricing models with heterogeneous agents, though not in this chapter. The main contribution of this chapter is that it analyzes the case of nonzero memory in the fitness measure. When the memory strength parameter takes a positive value, some share of current realized profits in any given period is taken into account when calculating the performance measure in the next time period. If the value of memory strength parameter amounts to one then of course the entire accumulated wealth is taken into account.

The expression (14) for the fitness function is somewhat different that the one used in Brock and Hommes (1998), where the coefficient of the current realized profits was fixed to 1. Namely, if we rewrite the memory strength parameter as $w = 1 - \frac{1}{T}$, where T is considered to be a specific number of time periods, we obtain the following expression for the fitness function:

$$U_{h,t} = \left(1 - \frac{1}{T}\right) U_{h,t-1} + \frac{1}{T} \pi_{h,t}, \tag{16}$$

which is equivalent to taking the last T observations into account with equal weight (as benchmark). When T approaches infinity, the memory parameter approaches 1 and the entire accumulated wealth is taken into account. We thus believe the expression (14) to be a more suitable formulation of the fitness measure than the one used in Brock and Hommes (1998), and in several other contributions.

Finally, we can express fractions of belief types, $n_{h,t}$, which are updated in each period, as a discrete choice probability by a multinomial logit model:

$$n_{h,t} = \frac{\exp\left[\beta U_{h,t-1}\right]}{\sum_{i=1}^{H} \exp\left[\beta U_{i,t-1}\right]}, \tag{17}$$

by using parameter β, determining the intensity of choice. The latter measures how fast economic agents switch between different prediction strategies; if the value of intensity of choice is zero, then all trader types have equal weight and the mass of traders distributes itself evenly across the set of available strategies, while on the other hand the entire mass of traders tends to use the best predictor, i.e. the strategy with the highest fitness, when the intensity of choice approaches infinity (the neoclassical limit).

Trader fractions are therefore determined by fitness and intensity of choice. Rationality in the asset pricing model is evidently bounded, since fractions are ranked according to fitness, but not all agents choose the best predictor. To ensure that fractions of belief types depend only upon observable deviations from the fundamental at any given time period, fitness function in the fractions of belief types equation may only depend on past fitness and past return. This indeed ensures that past realized profits are observable quantities that can be used in predictor selection.

One might wonder whether the traders' myopic mean-variance maximization is a reasonable assumption, especially when we allow for traders with a longer memory span. This assumption is widely used in modelling in economics and finance, though it would certainly be interesting to let traders plan longer ahead, even with an infinite planning horizon, as in the Lucas (1978) asset pricing model. However, in this kind of model one usually assumes perfect rationality to keep the analysis tractable. So far very little work has been done on infinite horizon models with bounded rationality and heterogeneous beliefs. Furthermore, one can also discuss whether individuals are really able to plan over a long horizon, or whether they might use simple heuristics over a short horizon and occasionally adapt them. After all, memory in the fitness measure is not equivalent to the planning horizon, but rather an "evaluation horizon" used to decide whether or not to switch strategies. There is empirical and experimental evidence that humans give more weight to the recent past than the far distant past, and this is formalized in our model.

3. Fundamentalists versus Contrarians

The first case we are going to examine is a two-type heterogeneous agents model with fundamentalists and contrarians as market participants. Fundamentalists exhibit deterministic function of the form:

$$f_{1,t} \equiv 0 \tag{18}$$

and have some positive information gathering costs C, i.e. $C > 0$. Contrarians exhibit a deterministic function:

$$f_{2,t} = g x_{t-1}; \quad g < 0 \tag{19}$$

and zero information gathering costs. It is thus a case of fundamentalists versus pure contrarians. We have the following fractions of belief types equation:

$$n_{h,t} = \frac{\exp\left[\beta U_{h,t-1}\right]}{\exp\left[\beta U_{1,t-1}\right] + \exp\left[\beta U_{2,t-1}\right]}; \quad h = 1,2\,. \tag{20}$$

For convenience we shall also introduce a difference in fractions m_t:

$$m_t = n_{1,t} - n_{2,t} = \frac{\exp\left[\beta U_{1,t-1}\right] - \exp\left[\beta U_{2,t-1}\right]}{\exp\left[\beta U_{1,t-1}\right] + \exp\left[\beta U_{2,t-1}\right]} = \tanh\left[\frac{\beta}{2}\left(U_{1,t-1} - U_{2,t-1}\right)\right]. \tag{21}$$

Finally, we have the fitness measure equation of each type:

$$U_{1,t} = wU_{1,t-1} + (1-w)\left[-kRx_{t-1}\left(x_t - Rx_{t-1}\right) - C\right], \tag{22}$$

$$U_{2,t} = wU_{2,t-1} + (1-w)\left[k\left(x_t - Rx_{t-1}\right)\left(gx_{t-2} - Rx_{t-1}\right)\right]. \tag{23}$$

In order to analyze memory in our heterogeneous asset pricing model, we shall first determine the position and stability of the steady state and the period two-cycle in relation to the memory strength parameter. We will also examine the possible qualitative changes in dynamics. Then we will perform some numerical simulations to combine global stability analysis with local stability analysis.

3.1. Position of the steady state

In our two-type heterogeneous agents model of fundamentalists versus contrarians the equilibrium pricing equation has the following form:

$$Rx_t = n_{2,t}gx_{t-1} = \frac{1-m_t}{2}gx_{t-1}, \tag{24}$$

where $n_{1,t} - n_{2,t} = m_t$ and $n_{1,t} + n_{2,t} = 1$. The difference in fractions of belief types equation, on the other hand, has the following form:

$$m_t = \tanh\left[\frac{\beta}{2}\left(w\left(U_{1,t-2} - U_{2,t-2}\right) - (1-w)\left(kgx_{t-3}\left(x_{t-1} - Rx_{t-2}\right) + C\right)\right)\right]. \tag{25}$$

A steady state price deviation x is a fixed point of the system, if it satisfies $x = f(x)$ for mapping $f(x)$. In our two-type heterogeneous agents model of fundamentalists versus contrarians we have:

$$Rx = \frac{1-m}{2}gx, \tag{26}$$

where either $x^{eq} = 0$, or $R = \frac{1-m^*}{2}g$ and thus $m^* = 1 - \frac{2R}{g}$. In the former case we get the fundamental steady state, where the price is equal to its fundamental value and the difference in fractions is:

$$m^{eq} = \tanh\left[\frac{\beta}{2}\left(w\left(U_1^{eq} - U_2^{eq}\right) - (1-w)C\right)\right].$$

Since it follows from expressions (22) and (23) that $U_1^{eq} = -C$ and $U_2^{eq} = 0$ when $w \neq 1$, the steady state difference in fractions simplifies:

$$m^{eq} = \tanh\left[\frac{\beta}{2}\left(-wC - (1-w)C\right)\right] = \tanh\left[-\frac{\beta C}{2}\right]. \tag{27}$$

Possible other (non-fundamental) steady states should satisfy:

$$m^* = \tanh\left[\frac{\beta}{2}\left(w(U_1^* - U_2^*) - (1-w)\left(kgx^*\left(x^* - Rx^*\right) + C\right)\right)\right]. \tag{28}$$

Since it can be derived that $U_1^* = -kRx^{*2}\left(1-R\right) - C$ and $U_2^* = kx^{*2}\left(1-R\right)\left(g-R\right)$, we finally obtain:

$$m^* = \tanh\left[-\frac{\beta}{2}\left(kgx^{*2}\left(1-R\right) + C\right)\right]. \tag{29}$$

Therefore we can state the following lemma.

Lemma 1: *The fundamental steady state in case of fundamentalists versus contrarians is a unique steady state of the system. Memory does not affect the position of this steady state.*

Proof of Lemma 1:

Since $g < 0$, $\frac{2R}{g} < 0$ holds and expression $m^* = 1 - \frac{2R}{g}$ is always greater than 1. On the other hand, the value of the hyperbolic tangent function is by definition between –1 and 1. In fact, since $k > 0$, $g < 0$, $R > 1$, $C > 0$ and the variable x is squared, the right-hand side of expression (29) is always between –1 and 0. Expression (29) thus never gives a solution and the fundamental steady state $(0,\ m^{eq})$ is a unique steady state of the system. Since there is no memory strength parameter in expression (27) and thus also in expression (26), memory does not affect the position of this steady state.

3.2. Stability of the steady state

In order to analyze stability of the steady state we shall rewrite our system as a difference equation:

$$X_t = F_1\left(X_{t-1}\right), \tag{30}$$

where $X_{t-1} = \left(x_{1,t-1}, x_{2,t-1}, x_{3,t-1}, u_{1,t-1}, u_{2,t-1}\right)$ is a vector of new variables, which are defined as: $x_{1,t-1} := x_{t-1}$, $x_{2,t-1} := x_{t-2}$, $x_{3,t-1} := x_{t-3}$, $u_{1,t-1} := U_{1,t-2}$ and $u_{2,t-1} := U_{2,t-2}$.

We therefore obtain the following 5-dimensional first-order difference equation:

$$x_{1,t} = x_t = \frac{1}{R} n_{2,t} g x_{1,t-1} =$$

$$= \frac{1}{R} g x_{1,t-1} \frac{\exp\left[\beta U_{2,t-1}\right]}{\exp\left[\beta U_{1,t-1}\right] + \exp\left[\beta U_{2,t-1}\right]} = \frac{1}{R} g x_{1,t-1} \frac{\exp\left[\beta u_{2,t}\right]}{\exp\left[\beta u_{1,t}\right] + \exp\left[\beta u_{2,t}\right]}, \tag{31}$$

$$x_{2,t} = x_{t-1} = x_{1,t-1}, \tag{32}$$

$$x_{3,t} = x_{t-2} = x_{2,t-1}, \tag{33}$$

$$u_{1,t} = U_{1,t-1} = wu_{1,t-1} + (1-w)\left[-kRx_{2,t-1}\left(x_{1,t-1} - Rx_{2,t-1}\right) - C\right], \tag{34}$$

$$u_{2,t} = U_{2,t-1} = wu_{2,t-1} + (1-w)\left[k\left(x_{1,t-1} - Rx_{2,t-1}\right)\left(gx_{3,t-1} - Rx_{2,t-1}\right)\right]. \tag{35}$$

The local stability of a steady state is determined by the eigenvalues of the Jacobian matrix, which we do not present here due to the spatial limitations. We then compute the Jacobian matrix of the 5-dimensional map. At the fundamental steady state X^{eq} = (0, 0, 0, –C, 0) we obtain the new Jacobian matrix. A straightforward computation shows that the characteristic equation is in our case given by:

$$g(\lambda) = \left(\frac{1}{R}n_2^{eq}g - \lambda\right)\lambda^2\left(w - \lambda\right)^2 = 0, \tag{36}$$

with solutions (eigenvalues): $\lambda_1 = \frac{1}{R}n_2^{eq}g$, $\lambda_{2,3} = 0$ and $\lambda_{4,5} = w$. The steady state X^{eq} is stable for $|\lambda| < 1$; therefore in cases $-R < gn_2^{eq} < R$ and $w < 1$.

Thus we can state the following lemma.

Lemma 2: *The fundamental steady state in case of fundamentalists versus contrarians is globally stable for –R < g < 0. Memory does not affect the stability of this steady state.*

Proof of Lemma 2:

From the characteristic equation (36) we can observe three eigenvalues, where two of them are in fact double eigenvalues. The first eigenvalue assures stability when $-R < gn_2^{eq} < R$, *while the second and third (double) eigenvalue always assure stability. The fundamental steady state is stable for* $-\frac{R}{n_2^{eq}} < g < \frac{R}{n_2^{eq}}$, *but since* n_2^{eq} *depends on other parameters of the system and g < 0, stability is (more conveniently) guaranteed at least for* $-R < g < 0$. *Since the memory strength parameter is represented (only) by the third (double) eigenvalue, memory does not affect the stability of the steady state, as has been shown by the reduced system.*

3.3. Bifurcations and the Period Two-cycle

A bifurcation is a qualitative change of the dynamical behaviour that occurs when parameters are varied (Brock and Hommes, 1998). A specific type of bifurcation that occurs when one parameter is varied is called a co-dimension one bifurcation. There are several types of such bifurcations, *viz.* period doubling, saddle-node and Hopf bifurcations. The first type has eigenvalue –1 of the Jacobian matrix, the second type has eigenvalue 1 and the third type has complex eigenvalues on the unit circle.

If we take a look at the eigenvalue λ_1, which we are in our case interested in, we can observe that a saddle-node bifurcation can never occur. Namely, the expression:

$$1 = \frac{1}{R} n_2^{eq} g \tag{37}$$

can never hold, since the left-hand side is a positive constant and the right-hand side is always negative for $g < 0$, $R > 0$ and $n_2^{eq} > 0$. On the other hand, the expression:

$$-1 = \frac{1}{R} n_2^{eq} g \tag{38}$$

may be satisfied for $n_2^{eq} \neq 0$, since both sides of the expression are then negative. Thus a (primary) period doubling bifurcation may occur in our model for the following β-value:

$$\beta^* = \frac{1}{C} \ln\left[-\frac{R}{R+g}\right], \tag{39}$$

which has been computed by plugging $n_2^{eq} = \dfrac{1}{\exp\left[-\beta C\right]+1}$ into expression (38) and solving for the memory strength parameter β.

Now we can check the existence of a period two-cycle $\left\{(x^*, m^*), (-x^*, m^*)\right\}$. Taking into account that $U_1^* = kRx^{*2}\left(1+R\right) - C$ and $U_2^* = kx^{*2}\left(1+R\right)\left(g+R\right)$, a period two-cycle occurs when $-R = \dfrac{1-m^*}{2} g$, and thus $m^* = 1 + \dfrac{2R}{g}$ satisfies:

$$m^* = \tanh\left[-\frac{\beta}{2}\left(kgx^{*2}\left(1+R\right)+C\right)\right]. \tag{40}$$

Therefore we can state the following lemma.

Lemma 3: *In case of fundamentalists versus contrarians the fundamental steady state (0, m^{eq}) is unstable for $g < -2R$ and there exists a period two-cycle $\left\{(x^*, m^*), (-x^*, m^*)\right\}$. For $-2R < g < -R$ there are two possibilities: (1) if $m^* = 1 + \dfrac{2R}{g} < m^{eq}$ then (0, m^{eq}) is the unique, globally stable steady state, while (2) if $m^* = 1 + \dfrac{2R}{g} > m^{eq}$ then the steady state (0, m^{eq}) is unstable and there exists a period two-cycle $\left\{(x^*, m^*), (-x^*, m^*)\right\}$. Memory does not affect the position of the period two-cycle.*

Proof of Lemma 3:

For $g < -2R$ it is clear from the expression for eigenvalue λ_1 of the characteristic equation (36) that the fundamental steady state is unstable. Furthermore, since $0 < m^* < 1$, the expression (40) has two *solutions, x^* and $-x^*$. If expression (38) is satisfied, it then follows from expressions $m^* = 1 + \frac{2R}{g}$ and (40) that $\{(x^*, m^*), (-x^*, m^*)\}$ is a period two-cycle. Finally, for $-2R < g < -R$, the fundamental steady state is unstable and expression (40) has solutions $\pm x^*$ if and only if $m^* > m^{eq} = \tanh\left[-\frac{\beta C}{2}\right]$. Since the memory strength parameter does not affect the difference in fractions of belief types, memory does not affect the position of the period two-cycle.*

As in the paper of Brock and Hommes (1998), very strong contrarians with $g < -2R$ may lead to the existence of a period two-cycle, even when there are no costs for fundamentalists ($C = 0$). When the fundamentalists' costs are positive ($C > 0$), strong contrarians with $-2R < g < -R$ may lead to a period two-cycle. As the intensity of choice increases to $\beta = \beta^*$, a period doubling bifurcation occurs in which the fundamental steady state becomes unstable and a (stable) period two-cycle is created, with one point above and the other one below the fundamental.

When the intensity of choice further increases, we are likely to find a value $\beta = \beta^{**}$, for which the period two-cycle becomes unstable and a Hopf bifurcation of this period two-cycle occurs, as in Brock and Hommes (1998). The model would then get an attractor consisting of two invariant circles around each of the two (unstable) period two-points, one lying above and the other one below the fundamental. Immediately after such a Hopf bifurcation, the price dynamics is either periodic or quasi-periodic, jumping back and forth between the two circles. The proof of this phenomenon is not straightforward due to the non-zero period points, although the 5-dimensional system (31) – (35) is still symmetric with respect to the origin. We shall thus demonstrate the occurrence of the Hopf bifurcation and the emergence of the attractor numerically in the next section.

3.4. Numerical analysis

Our numerical analysis in the case of fundamentalists and contrarians will be conducted for fixed values of parameters $R = 1.1$, $k = 1.0$, $C = 1.0$ and $g = -1.5$. We shall thus vary the intensity of choice parameter β and of course the memory strength parameter w. Four analytical tools will be used[2]; bifurcation diagrams, largest Lyapunov characteristic exponent (LCE) plots, phase plots, and time series plots.

The dynamic behaviour of the system can first and foremost be determined by investigating bifurcation diagrams. In Figure 1 the bifurcation diagrams for two different values of the memory strength parameter are presented. We can observe that for low values of β we have a stable steady state, i.e. the fundamental steady state. As has been proven in Lemma 1, the position of this steady state, i.e. $x^{eq} = 0$, is independent of the memory, which is clearly demonstrated by the simulations. For increasing β a (primary) period doubling bifurcation occurs at $\beta = \beta^*$; the steady state becomes unstable and a stable period two-cycle appears, as

[2] However, we will not discuss these tools here in more detail, since they are fairly well-known; instead we will direct the interested reader to more detailed discussions in Arrowsmith and Place (1990), Shone (1997), and Brock and Hommes (1998).

proven in Lemma 3. As can be seen from the simulations, this bifurcation value is also independent of the memory. The stability of the steady state is thus unaffected by the memory, as proven in Lemma 2.

If β increases further, indeed a (secondary) Hopf bifurcation occurs at $\beta = \beta^{**}$, as has been claimed in Section 3.3; the period two-cycle becomes unstable and an attractor appears consisting of two invariant circles around each of the two (unstable) period two-points, one lying above and the other one below the fundamental. It is a supercritical Hopf bifurcation, where the steady state gradually changes either into an unstable equilibrium or into an attractor (*cf.* Guckenheimer and Holmes, 1983; Frøyland, 1992; Kuznetsov, 1995). The position of the period two-cycle is independent of the memory, but it is not independent of the intensity of choice, as can be seen from expression (40). Numerical simulations suggest that the secondary bifurcation value also does not vary with changing memory strength parameter w. For $\beta > \beta^{**}$ chaotic dynamic behaviour appears, which is interspersed with many (mostly higher order) stable cycles. Such a bifurcation route to chaos was also called the rational route to randomness (Brock and Hommes, 1997a), while the last part of it has been referred to as the breaking of an invariant circle.

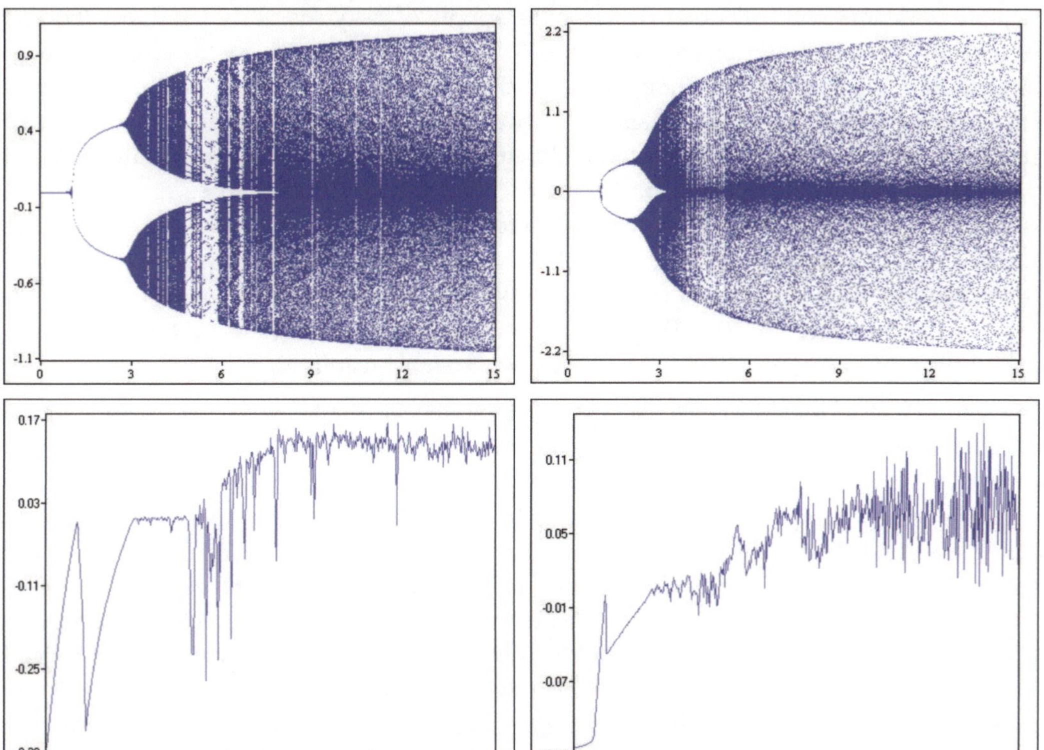

Notes: Horizontal axis represents the intensity of choice (β). Vertical axis represents deviations of the price from the fundamental value (x) in the upper two diagrams and the value of the largest LCE in the lower two diagrams, respectively. The diagrams differ with respect to the memory strength parameter w; the left one corresponds to $w = 0.3$, while the right one corresponds to $w = 0.9$.

Figure 1. Bifurcation diagrams and Largest LCE plots of β in case of fundamentalists versus contrarians

By examining largest Lyapunov characteristic exponent (LCE) plots of β we arrive at the same conclusions about the dynamic behaviour of the system. It can be seen from Figure 1 that the largest LCE is smaller than 0 and the system is thus stable until the primary bifurcation, which is independent of memory. At the bifurcation value, a qualitative change in dynamics occurs, i.e. a period doubling bifurcation and we obtain a stable period two-cycle. Largest LCE is again smaller than 0 and the system is thus stable until the secondary bifurcation. At this bifurcation value, again a qualitative change in dynamics occurs, i.e. a Hopf bifurcation, but the dynamics is more complicated.

For lower values of w the largest LCE after β^{**} is non-positive, but close to 0, which implies quasi-periodic dynamics. After some transient period the largest LCE becomes mainly positive with exceptions, which implies chaotic dynamics, interspersed with stable cycles. In fact, the largest LCE plot has a fractal structure (*cf.* Brock and Hommes, 1998, p. 1258). In the case of $w = 0.9$ the global dynamics after β^{**} immediately becomes chaotic. Memory thus certainly affects the dynamics after the secondary bifurcation. Since the latter is a period doubling bifurcation, we are talking about period doubling routes to chaos.

Next, we shall examine plots of the attractors in the (x_t, x_{t-1}) plane and in the $(x_t, n_{1,t})$ plane[3] without noise and with IID noise added to the supply of risky shares. In the upper left plot of each of the four parts of Figures 2 and 3 we can first observe the appearance of an attractor for the intensity of choice beyond the secondary bifurcation value. The orbits converge on such an attractor consisting of two invariant 'circles' around each of the two (unstable) period two-points[4], one lying above and the other one below the fundamental value. As the intensity of choice increases, the circles 'move' closer to each other. In the upper right and lower left plot of each of the four parts of Figures 2 and 3 we can observe that the system seems already to be close to having a homoclinic orbit. The stable manifold of the fundamental steady state, $W^s(0,m^{eq})$, contains the vertical segment, $x^{eq} = 0$, whereas the unstable manifold, $W^u(0,m^{eq})$, has two branches, one moving to the right and one to the left. Both of them are then 'folding back' close to the stable manifold.

For as Brock and Hommes (1998, p. 1254) have proven for the asset pricing model without additional memory, at infinite intensity of choice and strong contrarians, $g < -R$, that unstable manifold $W^u(0,-1)$ is bounded and all orbits converge on the saddle point (0, –1). In particular, all points of the unstable manifold converge on (0, –1) and are thus also on the stable manifold. Consequently, the system has homoclinic orbits for infinite intensity of choice. In the case of strong contrarians and high intensity of choice it is therefore reasonable to expect that we will obtain a system close to having a homoclinic intersection between the stable and unstable manifolds of the fundamental steady state. This is indeed what can be observed from the lower left plot of each of the two parts of Figures 2 and 3 and it suggests the occurrence of chaos for high intensity of choice. As can be seen from the lower right plot of each of the two parts of Figures 2 and 3, the addition of small dynamic noise to the system does not alter our findings.

---

[3] Attractors in the $(x_t, n_{2,t})$ plane are just flipped (rotated by 180 degrees) images of attractors in the $(x_t, n_{1,t})$ plane and will thus not be separately examined.

[4] Though we are topologically speaking about circles, the actual shape of such an attractor can be quite diverse, as seen from the figures.

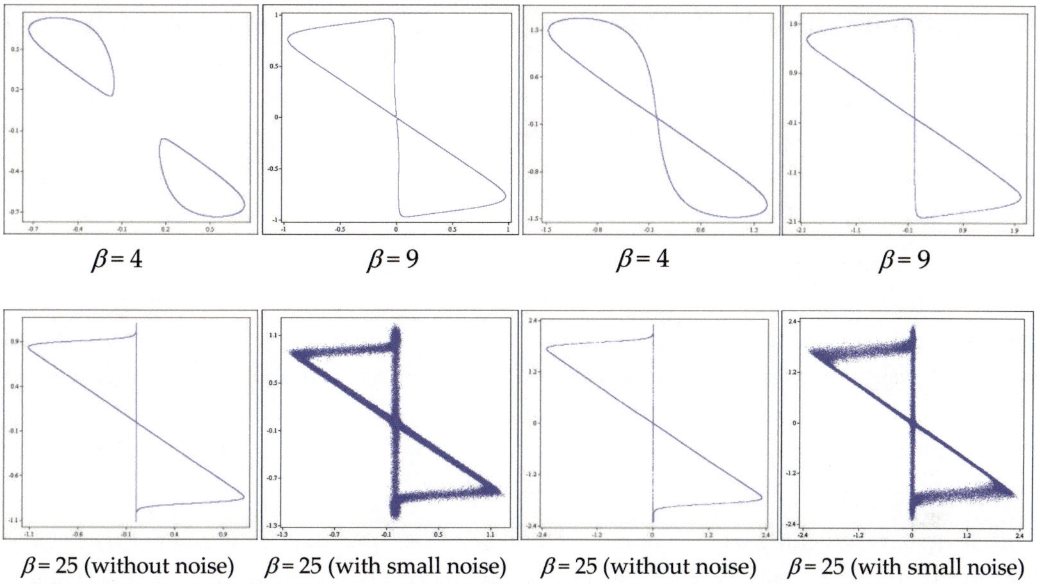

Notes: Horizontal axis represents deviations of the price from the fundamental value (x_t). Vertical axis represents lagged deviations of the price from the fundamental value (x_{t-1}). The groups of four diagrams differ with respect to the memory strength parameter w; the left group corresponds to $w = 0.3$, while the right group corresponds to $w = 0.9$.

Figure 2. Phase plots of (x_t, x_{t-1}) in case of fundamentalists versus contrarians

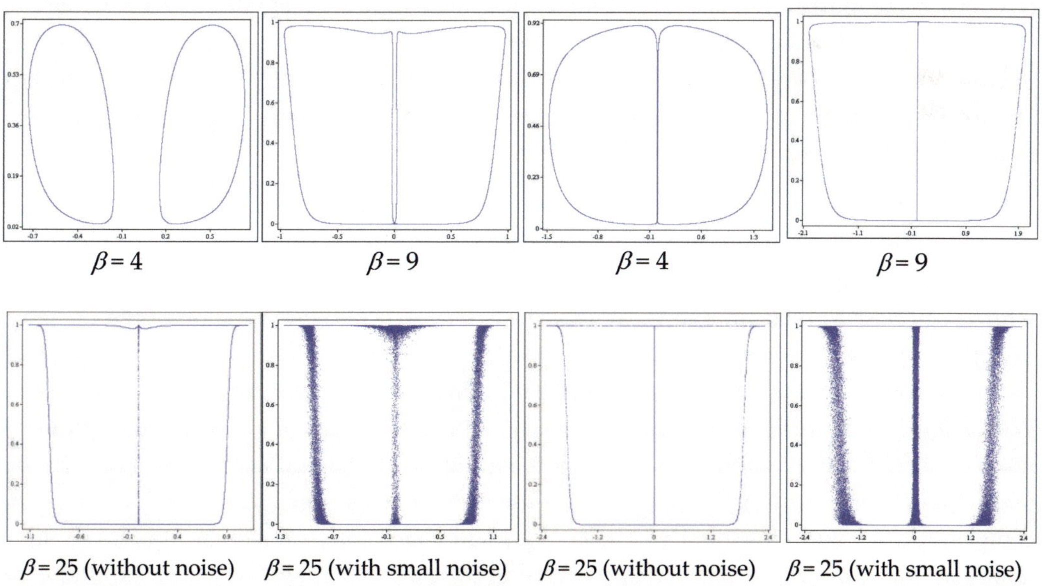

Notes: Horizontal axis represents deviations of the price from the fundamental value (x_t). Vertical axis represents the fraction of fundamentalists ($n_{1,t}$). The groups of four diagrams differ with respect to the memory strength parameter w; the left group corresponds to $w = 0.3$, while the right group corresponds to $w = 0.9$.

Figure 3. Phase plots of (x_t, $n_{1,t}$) in case of fundamentalists versus contrarians

Again, we can observe that memory has an impact on the global dynamics of the system. That is, both the convergence of the system on an attractor consisting of two invariant 'circles' around each of the two unstable period two-points and the 'moving' of the circles closer to each other seem to be happening faster (at lower intensity of choice) when more memory is present in the model. Moreover, at the same intensity of choice we seem to be closer to obtaining a system that has a homoclinic intersection between the stable and unstable manifolds of the fundamental steady state when the memory strength is higher.

Finally, we shall examine time series plots of deviations of the price from the fundamental value and of the fraction of fundamentalists[5]. Figure 4 shows some time series corresponding to the attractors in Figures 2 and 3, with and without noise added to the supply of risky shares. Similarly to the findings of Brock and Hommes (1998), we can observe that the asset prices are characterized by an irregular switching between a stable phase with prices close to their (unstable) fundamental value and an unstable phase of up and down price fluctuations with increasing amplitude.

This irregular switching is of course reflected in the fractions of fundamentalists and contrarians in the market. Namely, when the oscillations of the price around the unstable steady state gain sufficient momentum, it becomes profitable for the trader to follow efficient market hypothesis fundamental value despite the costs that are involved in this strategy. The fraction of fundamentalists approaches unity and the asset price stabilizes. But then the nonzero costs of fundamentalists bring them into position where they are unable to compete in the market; the fraction of fundamentalists rapidly decreases to zero, while the fraction of contrarians with no costs approaches unity with equal speed. The higher the intensity of choice, *ceteris paribus*, the faster this transition is complete; when β approaches the neoclassical limit, the entire mass of traders tends to use the best predictor with respect to costs, i.e. the strategy with the highest fitness.\

Additional memory does not change the pattern of asset prices *per se*, but it does affect its period. Namely, at the same intensity of choice and higher memory strength the period of this irregular cycle appears to be elongated on average, in such a way that the stable phase with prices close to their fundamental value lasts longer, while the duration of the unstable phase of up and down price fluctuations does not change significantly. The effect of including more memory thus mainly appears to be stabilizing with regard to asset prices. With regard to fractions of different trader types we could say that including additional memory affects the transition from the short period of fundamentalists' dominance to the longer period of contrarians' dominance in the market. This transition takes more time to complete at the same intensity of choice. More memory thus causes the traders to stick longer to the strategy that has been profitable in the past, but might not be so profitable in the recent periods.

[5] Since the fraction of contrarians is just the unity complement of the fraction of fundamentalists, i.e. $n_{1,t} + n_{2,t} = 1$, the former will thus not be separately graphically examined.

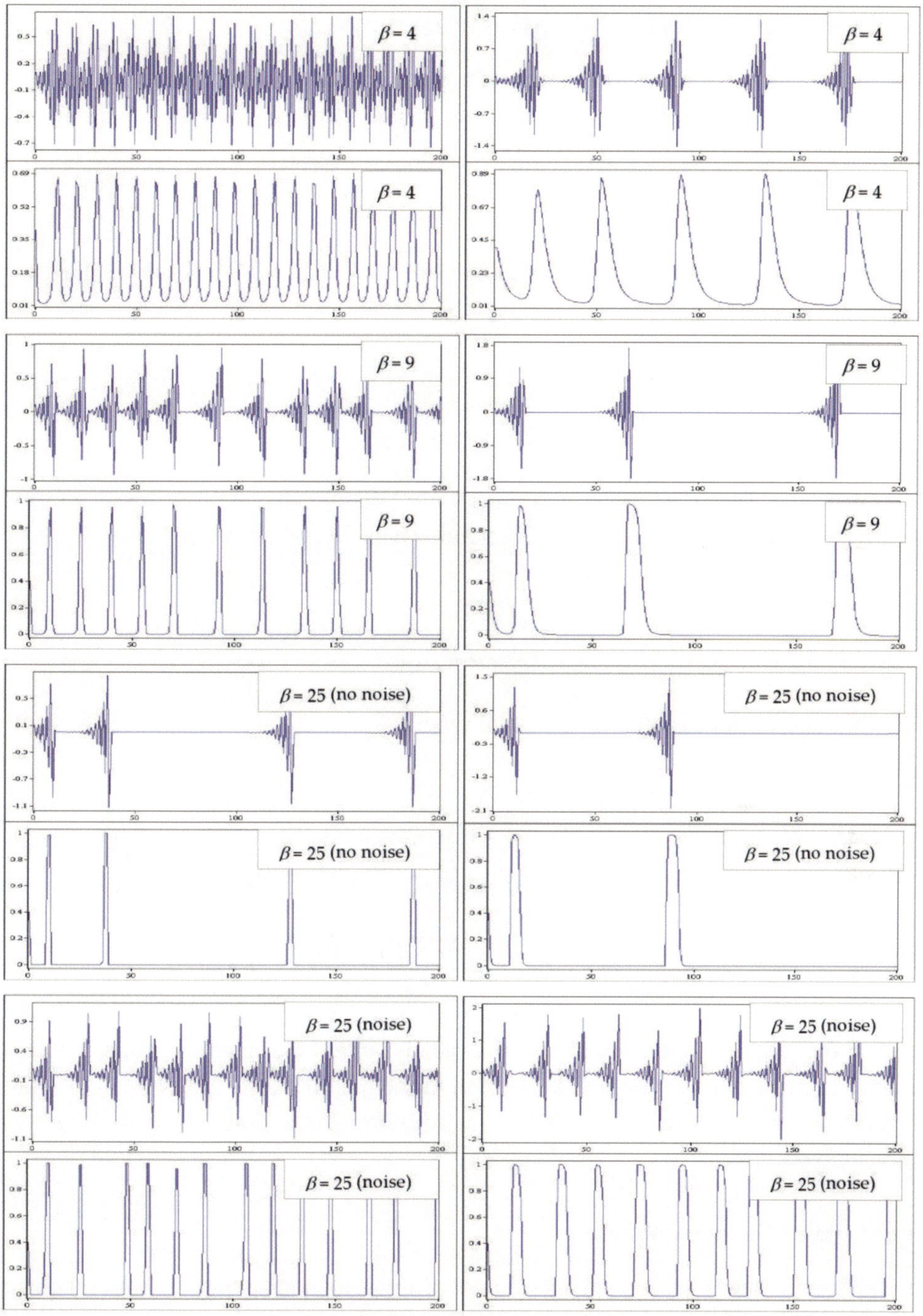

Notes: Horizontal axis represents the time (t). Vertical axis in each pair of time series plots first represents deviations of the price from the fundamental value (x_t), and then the fraction of fundamentalists ($n_{1,t}$). The plots on the left-hand side and the right-hand side of the figure differ with respect to the memory strength parameter w; the ones on the left correspond to $w = 0.3$, while the ones on the right to $w = 0.9$.

Figure 4. Time series of prices and fractions in case of fundamentalists versus contrarians

4. Fundamentalists versus opposite biased beliefs

The second case we are going to examine is a three-type heterogeneous agents model with fundamentalists and opposite biased beliefs as market participants. Fundamentalists again exhibit a deterministic function of the form:

$$f_{1,t} \equiv 0\,, \tag{41}$$

though this time with no information gathering costs, i.e. $C = 0$. Biased beliefs exhibit deterministic functions:

$$f_{2,t} = b_2; \quad b_2 > 0\,, \tag{42}$$

$$f_{3,t} = b_3; \quad b_3 < 0\,, \tag{43}$$

for optimist and pessimist biases, respectively[6]. Biases also exhibit zero information gathering costs. We have the following fractions of belief types equation:

$$n_{h,t} = \frac{\exp\left[\beta U_{h,t-1}\right]}{\sum_{i=1}^{3}\exp\left[\beta U_{i,t-1}\right]}; \quad h = 1,2,3\,. \tag{44}$$

Finally, we have the fitness measures of each type:

$$U_{1,t} = wU_{1,t-1} + (1-w)\left[-kRx_{t-1}\left(x_t - Rx_{t-1}\right)\right], \tag{45}$$

$$U_{2,t} = wU_{2,t-1} + (1-w)\left[k\left(x_t - Rx_{t-1}\right)\left(b_2 - Rx_{t-1}\right)\right], \tag{46}$$

$$U_{3,t} = wU_{3,t-1} + (1-w)\left[k\left(x_t - Rx_{t-1}\right)\left(b_3 - Rx_{t-1}\right)\right]. \tag{47}$$

In order to analyze memory in our heterogeneous asset pricing model, we shall first determine the position and stability of the steady state, and then examine the possible qualitative changes in dynamics; all in relation to the memory strength parameter. Then we shall perform some numerical simulations to combine global stability analysis with local stability analysis.

4.1. Position of the steady state

In our three-type heterogeneous agents model of fundamentalists versus biased beliefs, we shall again start by rewriting our system as a difference equation:

$$X_t = F_2\left(X_{t-1}\right), \tag{48}$$

---

[6] In this chapter we will mainly focus on the symmetric case.

where $X_{t-1} = \left(x_{1,t-1}, x_{2,t-1}, u_{1,t-1}, u_{2,t-1}, u_{3,t-1}\right)$ is a vector of new variables, defined as: $x_{1,t-1} := x_{t-1}$, $x_{2,t-1} := x_{t-2}$, $u_{1,t-1} := U_{1,t-2}$, $u_{2,t-1} := U_{2,t-2}$ and $u_{3,t-1} := U_{3,t-2}$.

We therefore obtain the following 5-dimensional first-order difference equation:

$$x_{1,t} = x_t = \frac{1}{R}\left(n_{2,t}b_2 + n_{3,t}b_3\right) = \frac{1}{R}\left(\frac{\exp\left[\beta U_{2,t-1}\right]}{\sum_{i=1}^{3}\exp\left[\beta U_{i,t-1}\right]}b_2 + \frac{\exp\left[\beta U_{3,t-1}\right]}{\sum_{i=1}^{3}\exp\left[\beta U_{i,t-1}\right]}b_3\right) =$$

$$= \frac{1}{R}\left(\frac{\exp\left[\beta u_{2,t}\right]}{\sum_{i=1}^{3}\exp\left[\beta u_{i,t}\right]}b_2 + \frac{\exp\left[\beta u_{3,t}\right]}{\sum_{i=1}^{3}\exp\left[\beta u_{i,t}\right]}b_3\right), \tag{49}$$

$$x_{2,t} = x_{t-1} = x_{1,t-1}, \tag{50}$$

$$u_{1,t} = U_{1,t-1} = wu_{1,t-1} + (1-w)\left[-kRx_{2,t-1}\left(x_{1,t-1} - Rx_{2,t-1}\right)\right], \tag{51}$$

$$u_{2,t} = U_{2,t-1} = wu_{2,t-1} + (1-w)\left[k\left(x_{1,t-1} - Rx_{2,t-1}\right)\left(b_2 - Rx_{2,t-1}\right)\right], \tag{52}$$

$$u_{3,t} = U_{3,t-1} = wu_{3,t-1} + (1-w)\left[k\left(x_{1,t-1} - Rx_{2,t-1}\right)\left(b_3 - Rx_{2,t-1}\right)\right]. \tag{53}$$

Our three-type heterogeneous agents model of fundamentalists versus biased beliefs in general can have the following steady state price deviations:

$$x = \frac{1}{R}\left(n_2 b_2 + n_3 b_3\right). \tag{54}$$

We obtain the fundamental steady state for $b_2 = -b_3 = b > 0$ (opposite biased beliefs), where $x^{eq} = 0$. This is implied by $u_1^{eq} = u_2^{eq} = u_3^{eq} = 0$ when $w \neq 1$ and consequently by $n_1^{eq} = n_2^{eq} = n_3^{eq} = \frac{1}{3}$, originating from the rewritten expression (44).

By performing a generalization we can state the following lemma.

Lemma 4: *The fundamental steady state in the case of fundamentalists versus opposite biased beliefs is a unique steady state of the system. Memory does not affect the position of this steady state.*

Proof of Lemma 4:

We will prove a more general result for the case with h = 1, …, H purely biased types b_h (including fundamentalists with $b_1 = 0$). Proceeding from the non-transformed variables the system is:

$$Rx_t = \sum_{h=1}^{H} n_{h,t} b_h, \tag{55}$$

$$n_{h,t} = \frac{\exp\left[\beta\left(wU_{h,t-2} + (1-w)\left[k\left(x_{t-1} - Rx_{t-2}\right)\left(b_h - Rx_{t-2}\right)\right]\right)\right]}{\sum_{i=1}^{H}\exp\left[\beta\left(wU_{i,t-2} + (1-w)\left[k\left(x_{t-1} - Rx_{t-2}\right)\left(b_i - Rx_{t-2}\right)\right]\right)\right]}; \quad 1 \le h \le H. \tag{56}$$

After subtracting off identical terms from the exponents of both numerator and denominator in expression (56) we obtain a new expression for the fractions:

$$n_{h,t} = \frac{\exp\left[\beta\left(wU^{\circ}_{h,t-2} + (1-w)k\left(x_{t-1} - Rx_{t-2}\right)b_h\right)\right]}{\sum_{i=1}^{H}\exp\left[\beta\left(wU^{\circ}_{i,t-2} + (1-w)k\left(x_{t-1} - Rx_{t-2}\right)b_i\right)\right]}; \quad 1 \le h \le H, \tag{57}$$

where $U^{\circ}_{h,t}$ is the fitness of trader type h, adjusted by subtracting off identical terms as above. The dynamic system defined by (55) and (57) is thus of the form:

$$Rx_t = V_{\beta k}(x_{t-1} - Rx_{t-2}), \tag{58}$$

where the right-hand side function is defined as:

$$V_{\beta k}(y_t) = \frac{\exp\left[\beta\left(wU^{\circ}_{h,t-2}(y_{t-1}) + (1-w)kb_h y_t\right)\right]}{\sum_{i=1}^{H}\exp\left[\beta\left(wU^{\circ}_{i,t-2}(y_{t-1}) + (1-w)kb_i y_t\right)\right]} = \sum_{h=1}^{H} b_h n_h = \langle b_h \rangle. \tag{59}$$

Since it follows from (52) and (53) that $U^*_h = kx^*(1-R)(b_h - Rx^*)$, steady states of expressions (55) and (57) or expression (58) are determined by:

$$Rx^* = V_{\beta k}(x^* - Rx^*) = V_{\beta k}(-rx^*) \tag{60}$$

where r = R – 1. Since a steady state has to satisfy expression (60), following Brock and Hommes (1998, p. 1271), a straightforward computation shows that:

$$\frac{\mathrm{d}}{\mathrm{d}y}V_{\beta k}(y) = \sum_{h=1}^{H}\left(\frac{\beta k b_h \exp[\beta k b_h y]}{\sum_{i=1}^{H}\exp[\beta k b_i y]} - \frac{\exp[\beta k b_h y]}{\left(\sum_{i=1}^{H}\exp[\beta k b_i y]\right)^2}\cdot\frac{\mathrm{d}}{\mathrm{d}y}\left(\sum_{i=1}^{H}\exp[\beta k b_i y]\right)\right)b_h =$$

$$= \sum_{h=1}^{H}\left(\beta k n_h b_h^2 - \beta k n_h b_h \sum_{h=1}^{H} n_h b_h\right) = \sum_{h=1}^{H}\left(\beta k n_h b_h^2 - \beta k n_h b_h \langle b_h \rangle\right) =$$

$$= \beta k\left[\langle b_h^2 \rangle - \langle b_h \rangle^2\right] > 0, \tag{61}$$

where the inequality follows from the fact that the term between square brackets can be interpreted as the variance of the stochastic process, where each b_h is drawn with probability n_h. Therefore, $V_{\beta k}(y)$ is increasing and $V_{\beta k}(-rx^*)$ decreasing in x. It then follows from expression (60) that the steady* *state x* has to be unique. From expression (59) we obtain* $V_{\beta k}(0)=\sum_{h=1}^{H}\frac{b_h}{H}=\bar{b}$ *, so that x* equals the fundamental steady state ifequation reference goes here and only if* $\bar{b}=0$ *, i.e. when all biases are exactly balanced. Since there is no memory strength parameter left in expressions (60) and* $V_{\beta k}(0)$ *, memory does not affect the position of this steady state. It has to be mentioned though, that our derivation holds for finite intensity of choice, since fractions are only then all positive.*

4.2. Stability of the steady state and bifurcations

The local stability of a steady state is again determined by the eigenvalues of the Jacobian matrix. At the fundamental steady state X^{eq} = (0, 0, 0, 0, 0) the Jacobian matrix exhibits the characteristic equation that is in our case given by:

$$g(\lambda)=-\left(\lambda^2-\left(w-\frac{2}{3R}k\beta b^2(w-1)\right)\lambda-\frac{2}{3}k\beta b^2(w-1)\right)\lambda\left(w-\lambda\right)^2=0, \qquad (62)$$

which has the following three solutions, two of them being double: $\lambda_1=0$, $\lambda_{2,3}=w$ and $\lambda_{4,5}=\frac{1}{6R}\left(2b^2\beta k(1-w)+3Rw\pm\sqrt{\left(2b^2\beta k(w-1)-3Rw\right)^2-24b^2\beta k(1-w)R^2}\right)$.

The fundamental steady state is stable for $|\lambda|<1$, which in our case is limited to the product of eigenvalues $\lambda_{4,5}$ being smaller than one, i.e. $-\frac{2}{3}k\beta b^2(w-1)<1$. In terms of the intensity of choice this happens for $\beta<-\frac{3}{2kb^2(w-1)}$, while in terms of the memory strength this is guaranteed for $w<1-\frac{3}{2k\beta b^2}$.

Thus we can state the following lemma.

Lemma 5: *The fundamental steady state in case fundamentalists versus opposite biased beliefs is globally stable for* $\beta<-\frac{3}{2kb^2(w-1)}$. *Memory affects the stability of this steady state by restricting it to the given interval of the parameter value.*

Proof of Lemma 5:

From the characteristic equation (62) we can observe five eigenvalues. The first three eigenvalues always assure stability, while the last two eigenvalues limit stability. Given k > 0, b > 0, β ≥ 0, R > 1 and 0 ≤ w ≤ 1, the condition for stability in terms of β implies $\beta<-\frac{3}{2kb^2(w-1)}$. Similarly, the condition for stability in terms of w indicates $w<1-\frac{3}{2k\beta b^2}$.
Memory therefore affects the stability of the steady state as shown.

If we now take a look at the eigenvalues $\lambda_{4,5}$ of the characteristic equation (62), which are of interest in our case, we can observe that a saddle-node bifurcation would occur for:

$$\beta = \frac{3R}{2b^2k(1-R)}. \tag{63}$$

This can never hold, since $\beta \geq 0$ and the left-hand side is always non-negative, while $R > 1$ and the right-hand side is always negative. On the other hand, a period doubling bifurcation would occur for:

$$\beta = \frac{3R(w+1)}{2b^2k(R+1)(w-1)}. \tag{64}$$

This can never hold either, since $\beta \geq 0$ and the left-hand side is again always non-negative, while $0 \leq w \leq 1$ and the right-hand side is either negative or not defined.

The remaining qualitative change of the three discussed in Section 4.3 is the Hopf bifurcation. For this to occur, a complex conjugate pair of eigenvalues has to cross the unit circle. Eigenvalues $\lambda_{4,5}$ are complex for $\left(2b^2\beta k(w-1) - 3Rw\right)^2 - 24b^2\beta k(1-w)R^2 < 0$, which produces the following interval of values:

$$\frac{R\left(3w - 6R - 2\sqrt{R(R-w)}\right)}{2b^2k(w-1)} < \beta < \frac{R\left(3w - 6R + 2\sqrt{R(R-w)}\right)}{2b^2k(w-1)}. \tag{65}$$

We therefore state the following lemma.

Lemma 6: There exists an intensity of choice value β* such that the fundamental steady state, which is stable for $0 \leq \beta < \beta^*$, becomes unstable and remains such for $\beta > \beta^*$. For $\beta^* = -\dfrac{3}{2kb^2(w-1)}$ the system exhibits a Hopf bifurcation. Memory affects the emergence of this bifurcation, viz. with more memory the bifurcation occurs later.

As we have just established, in the case of fundamentalists versus opposite biased beliefs increasing intensity of choice to switch predictors destabilizes the fundamental steady state. This happens through a Hopf bifurcation. We can thus conclude, as did Brock and Hommes (1998) for the simpler version of the model, that in the presence of biased agents the first step towards complicated price fluctuations is different from that in the presence of contrarians. This fact does not change when we take memory into account.

Proof of Lemma 6:

When β increases, terms with β in the expressions for the eigenvalues $\lambda_{4,5}$ increase as well, and one of the eigenvalues has to cross the unit circle at some critical $\beta = \beta^*$. The fundamental steady state thus becomes unstable. Since it is obvious from the characteristic equation (62) that for all $\beta \geq 0$ we have $g(1) > 0$ and $g(-1) < 0$, a bifurcation has to occur. At the moment of the bifurcation the *product of eigenvalues* $\lambda_{4,5}$ *has to be equal one, i.e.* $-\frac{2}{3}k\beta b^2(w-1)=1$. *This happens either when we have two real eigenvalues with product equal to one or a complex conjugate pair of eigenvalues. Since* $\beta^* = -\frac{3}{2kb^2(w-1)}$ *falls into the interval (65) for any given finite memory strength, we can conclude that for* $\beta = \beta^*$ *the eigenvalues have to be complex and thus a Hopf bifurcation occurs. Since the memory strength parameter is present in the expression for* β^**, memory affects the emergence of this bifurcation; the higher the value of this parameter, the higher the bifurcation value.*

4.3. Numerical analysis

Our numerical analysis in the case of fundamentalists and opposite biased beliefs will be conducted for fixed values of parameters $R = 1.1$, $k = 1.0$, $b_2 = 0.2$ and $b_3 = -0.2$. We shall thus vary the memory strength parameter w and the intensity of choice parameter β. The same four analytical tools will be used than in Section 3.4.

Dynamic behaviour of the system can again first and foremost be determined by investigating bifurcation diagrams. From Figure 5 we can observe that for low values of β we have a stable steady state, i.e. the fundamental steady state. As has been proven in Lemma 4, the position of this steady state, i.e. $x^{eq} = 0$, is independent of the memory, which is clearly demonstrated by the simulations. For increasing β a bifurcation occurs at $\beta = \beta^*$, which is a Hopf bifurcation; the steady state becomes unstable and an attractor appears, consisting of an invariant circle around the (unstable) steady state. It is again a supercritical Hopf bifurcation, where the steady state gradually changes either into an unstable equilibrium or into an attractor.

The bifurcation value varies with changing memory strength parameter, as given by expression in Lemma 6. As can also be seen from Figure 5 at higher memory strength the bifurcation occurs later. For $\beta > \beta^*$ complex dynamical behaviour appears, which is interspersed with stable cycles. As we have already discovered in Section 4.2, irrespective of the amount of additional memory that is taken into account such a (bifurcation) route to complicated dynamics is different from that in the presence of contrarians, where we observed period doubling route to chaos (rational route to randomness).

By examining largest Lyapunov characteristic exponent (LCE) plots of β we arrive at more precise conclusions about the dynamic behaviour of the system. It can be seen from Figure 5 that the largest LCE is smaller than 0 and the system is thus stable until the bifurcation. At the bifurcation value a qualitative change in dynamics occurs, i.e. a Hopf bifurcation. The dynamics is somewhat more complicated. Namely, we can observe that the largest LCE after $\beta = \beta^*$ is non-positive, but mainly close to 0, which implies periodic and quasi-periodic dynamics, i.e. for high values of the intensity of choice only regular (quasi-)periodic fluctuations around the unstable fundamental steady state occur. An important finding is that the predominating quasi-periodic dynamics does not seem to evolve to chaotic dynamics and the route to complex dynamics is indeed different from the routes examined so far.

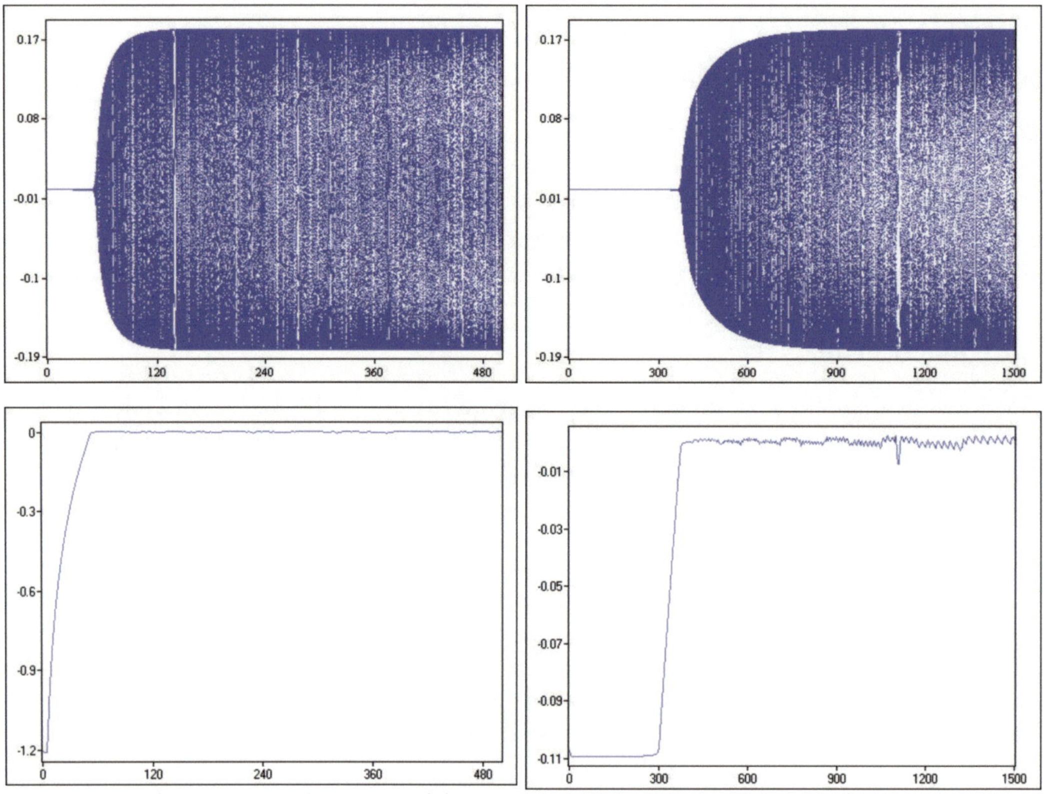

Notes: Horizontal axis represents the intensity of choice (β). Vertical axis represents deviations of the price from the fundamental value (x) in the upper two diagrams and the value of the largest LCE in the lower two diagrams, respectively. The diagrams differ with respect to the memory strength parameter w; the left one corresponds to $w = 0.3$, while the right one corresponds to $w = 0.9$.

Figure 5. Bifurcation diagrams and Largest LCE plots of β in case of fundamentalists versus opposite biased beliefs

Next, we shall examine plots of the attractors in the planes, determined by (x_t, x_{t-1}) and (x_t, $n_{1,t}$). In the upper left plot of each of the two parts of Figure 6 we can first observe the appearance of an attractor for the intensity of choice beyond the bifurcation value. The orbits converge to such an attractor consisting of an invariant 'circle' around the (unstable) fundamental steady state. The attractor obtained in the (x_t, $n_{1,t}$) plane is somewhat different. Namely, the unstable steady state dissipates into numerous points and evolves into a 'loop' shape, as shown in Figure 7.

As the intensity of choice increases, the dynamics remains periodic or quasi-periodic; in case of past deviations of prices from the fundamental value and fractions of biased beliefs the invariant circle slowly changes its shape into a '(full) square' (see Figure 6), while in case of fractions of fundamentalists the loop slowly changes into a 'three-sided square' (see Figure 7). For high values of intensity of choice we seem to obtain (stable) higher period cycles; in the case of past deviations of prices from the fundamental value and fractions of biased beliefs we seem to attain a stable period four-cycle, while in the case of fractions of fundamentalists it is difficult to obtain any solid indications based solely on numerical simulations due to

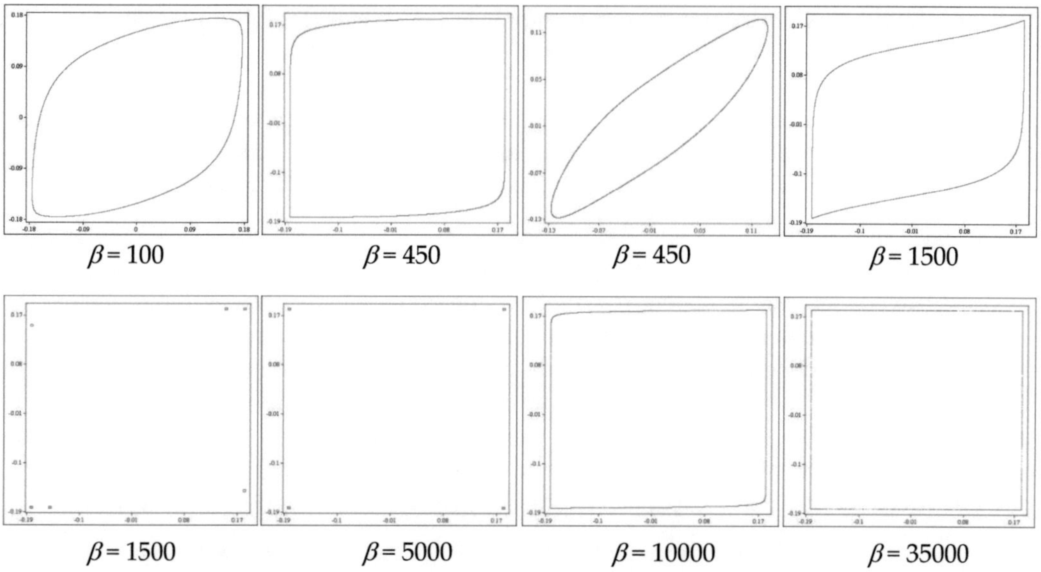

Notes: Horizontal axis represents deviations of the price from the fundamental value (x_t). Vertical axis represents lagged deviations of the price from the fundamental value (x_{t-1}). The groups of four diagrams differ with respect to the memory strength parameter w; the left group corresponds to $w = 0.3$, while the right group corresponds to $w = 0.9$.

Figure 6. Phase plots of (x_t, x_{t-1}) in case of fundamentalists versus opposite biases

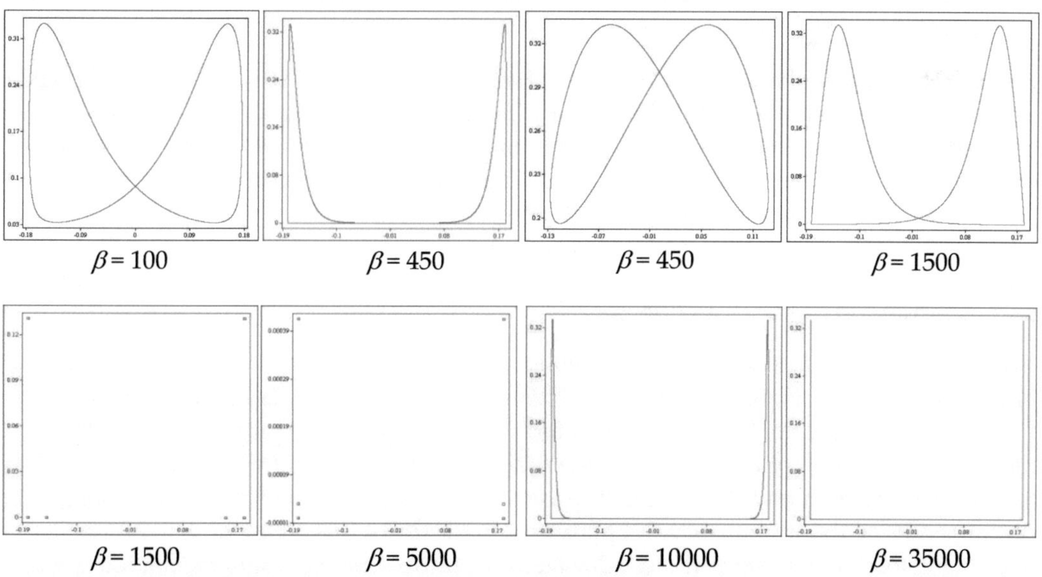

Notes: Horizontal axis represents deviations of the price from the fundamental value (x_t). Vertical axis represents the fraction of fundamentalists ($n_{1,t}$). The groups of four diagrams differ with respect to the memory strength parameter w; the left group corresponds to $w = 0.3$, while the right group corresponds to $w = 0.9$.

Figure 7. Phase plots of (x_t, $n_{1,t}$) in case of fundamentalists versus opposite biases convergence problems for very high values of intensity of choice. In the latter case we can observe stable period four- and six-cycles, however (see lower right plot of each of the two parts of Figure 7). Indeed, Brock and Hommes (1998) proved for the case of exactly opposite biased beliefs and infinite intensity of choice in their simpler version of the model without additional memory that the system has a stable four-cycle attracting all orbits, except for hairline cases converging on the unstable fundamental steady state. Additionally, they discovered that for all three trader types average profits along the four-cycle equal b^2.

Again, we can observe that the memory has an impact on the dynamics of the system. Namely, both the convergence of the system on an attractor and the further development of such an attractor seem to be dependent on the value of the memory strength parameter. The precise impact of memory is somewhat more difficult to establish due to the dependence of the bifurcation value on memory strength and the subsequent need to choose higher intensities of choice with higher memory strength in order to demonstrate different nature of attractors of the system. However, we can still establish that at the same intensity of choice (after the bifurcation value) the system apparently needs less additional memory in order to develop a specific stage of an attractor or even a (stable) higher period cycle.

Finally, we shall examine time series plots of deviations of the price from the fundamental value and of the fractions of all three types of traders. Figure 8 shows some time series corresponding to the attractors in Figures 6 and 7. We can observe that opposite biases may cause perpetual oscillations around the fundamental, even when there are no costs for fundamentalists, but can not lead to chaotic movements. Furthermore, as has already been indicated by the appearance of stable higher period cycles for high intensities of choice, in a three-type world, even when there are no costs and memory is infinite, fundamentalist beliefs can not drive out opposite purely biased beliefs, when the intensity of choice to switch strategies is high.

Hence, according to the argumentation of Brock and Hommes (1998, p. 1260), the market can protect a biased trader from his own folly if he is part of a group of traders whose biases are 'balanced' in the sense that they average out to zero over the set of types. Centralized market institutions can make it difficult for unbiased traders to prey on a set of biased traders provided they remain 'balanced' at zero. On the other hand, in a pit trading situation unbiased traders could learn which types are balanced and simply take the opposite side of the trade. In such situations biased traders would be eliminated, whereas a centralized trading institution could 'protect' them.

Additional memory does not change the pattern of asset prices and trader fractions *per se*, but it does affect its period. Namely, at the same intensity of choice and higher memory strength the period of these cycles appears to be elongated on average, in a way that both the negative and the positive deviation of the price from the fundamental value last longer. The same is valid for fractions ob biased traders, while in the case of fractions of fundamentalists the prolongation of the period of the irregular cycle appears in the form of less frequent 'spikes', which is understandable, since more persistent deviations of prices from the fundamental imply more space for biased traders and less chance for appearance of the fundamentalists. More memory causes the traders to stick longer to the strategy that has been profitable in the past, but might not be so profitable in the recent periods; therefore the system approaches purely quasi-periodic dynamics when the memory strength increases at given intensity of choice.

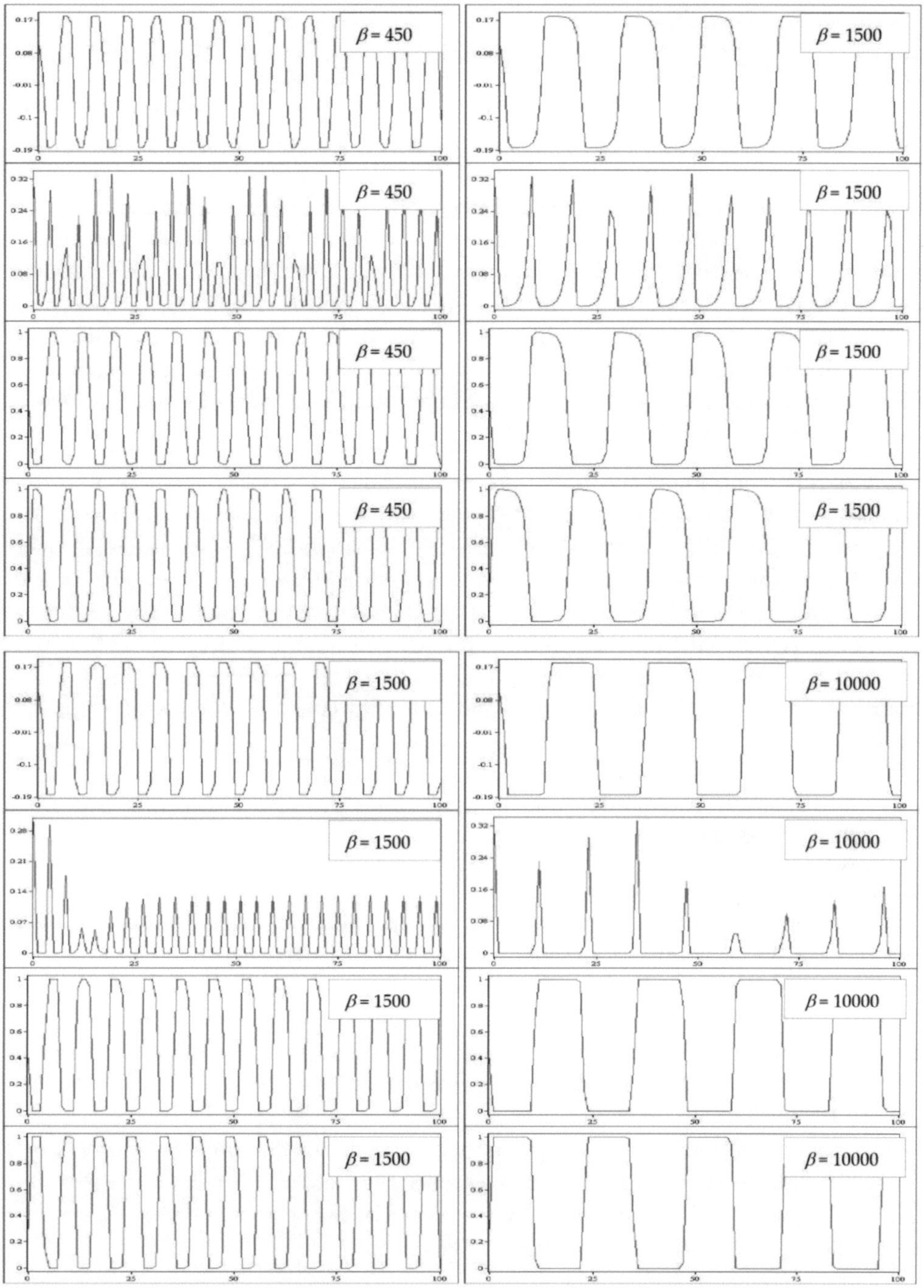

Notes: Horizontal axis represents the time (t). Vertical axis in each set of time series plots represents deviations of the price from the fundamental value (x_t), and the fractions of fundamentalists ($n_{1,t}$), optimistic biased beliefs ($n_{2,t}$) and pessimistic biased beliefs ($n_{3,t}$). The plots on the left-hand side and the right-hand side of the figure differ with respect to the memory strength parameter w; the ones on the left correspond to $w = 0.3$, while the ones on the right to $w = 0.9$.

Figure 8. Time series of prices and fractions in case of fundamentalists versus opposite biases

5. Concluding remarks

In a market with fundamentalists and contrarians the fundamental steady state is the unique steady state of the system, which arises for low values of intensity of choice. Memory affects neither the position of this steady state nor its stability. For increasing intensity of choice a primary bifurcation, i.e. a period doubling bifurcation occurs; the steady state becomes unstable and a stable period two-cycle appears. Both the primary bifurcation value and the position of the period two-cycle are independent of the memory. For further increasing intensity of choice a secondary bifurcation, i.e. a supercritical Hopf bifurcation, occurs; the period two-cycle becomes unstable and an attractor appears consisting of two invariant circles around each of the two (unstable) period two-points, one lying above and the other one below the fundamental. For high intensity of choice chaotic asset price dynamics occurs, interspersed with many stable period cycles. Such a bifurcation route to chaos is often called the rational route to randomness.

In case of strong contrarians and high intensity of choice it is reasonable to expect that we will obtain a system that is close to having a homoclinic intersection between the stable and unstable manifolds of the fundamental steady state, which indicates the occurrence of chaos. There exists a certain limited interval of memory strength values, for which at a given intensity of choice we are more likely to obtain such a system with more additional memory in the model. A rational choice between fundamentalists' and contrarians' beliefs triggers situations that do not reach fruition due to practical considerations and are thus unattainable, 'castles in the air', as Brock and Hommes (1998, p. 1258) would put it. As a consequence we obtain market instability, characterized by irregular up and down oscillations around the unstable efficient market hypothesis fundamental price. Additional memory lengthens on average the period of this irregular cycle and mainly appears to be stabilizing with regard to asset prices.

In a market with fundamentalists and opposite biases the fundamental steady state is also the unique steady state of the system, arising for low values of intensity of choice. Memory does not affect the position of this steady state, but does affect its stability. For increasing intensity of choice a supercritical Hopf bifurcation occurs; the steady state becomes unstable and an attractor appears. Memory affects the emergence of this bifurcation; the higher the memory strength, the higher the bifurcation value. More memory thus has a stabilizing effect on dynamics. For high intensity of choice the dynamic behaviour is more complex. However, irrespective of the amount of additional memory such a route to complicated dynamics is different from that in the presence of contrarians, for after the bifurcation value only regular (quasi-)periodic fluctuations around the unstable fundamental steady state occur. Consequently, an important finding is that the predominating quasi-periodic dynamics does not seem to evolve to chaotic dynamics.

After the incidence of the bifurcation the higher value of the memory strength parameter causes the dynamics to be less periodic and more quasi-periodic; the dynamics therefore converges on purely quasi-periodic behaviour with increasing memory strength. Opposite biases may cause perpetual oscillations around the fundamental, even without costs for fundamentalists, but can not lead to chaotic movements. Furthermore, in a three-type world, even when there are no costs and memory is infinite, fundamentalist beliefs can not drive out opposite purely biased beliefs, when the intensity of choice to switch strategies is high. Hence, following the argumentation of Brock and Hommes (1998, p. 1260), the market can protect a biased trader from his own folly if he is part of a group of traders whose biases are balanced.

In conclusion, both our analytical work and our numerical simulations suggest that biases alone do not trigger chaotic asset price fluctuations. Sensitivity to initial states and irregular switching between different phases seem to be triggered by trend extrapolators; in our case by contrarians. Apparently, some (strong) trend extrapolator beliefs are needed, such as strong trend followers or strong contrarians, in order to trigger chaotic asset price fluctuations. A key feature of our heterogeneous beliefs model is that the irregular fluctuations in asset prices are triggered by a rational choice in prediction strategies, based upon realized profits, *viz.* the observed deviations from the fundamentals are driven by short-run profit seeking. We can also talk about rational animal spirits that, according to Brock and Hommes (1997b), exhibit some qualitative features of asset price fluctuations in the actual financial markets, such as the autocorrelation structure of prices and returns.

Author details

Miroslav Verbič
Faculty of Economics, University of Ljubljana, Slovenia & Institute for Economic Research, Ljubljana, Slovenia

Acknowledgement

I am grateful for very helpful suggestions and comments from Cars H. Hommes, Valentyn Panchenko, Jan Tuinstra and Florian O. O. Wagener from the University of Amsterdam.

6. References

Arrowsmith, D. K. and Place, C. M., 1990. *An Introduction to Dynamical Systems*. Cambridge, UK: Cambridge University Press.

Awrejcewicz, J., 1991. *Bifurcation and Chaos in Coupled Oscillators*. Singapore: World Scientific Publishing.

Awrejcewicz, J. and Lamarque, C.-H., 2003. *Bifurcation and Chaos in Nonsmooth Mechanical Systems*. Singapore: World Scientific Publishing.

Brock, W. A. and Hommes, C. H., 1997a. »A Rational Route to Randomness«. *Econometrica* 65 (September), 1059-1095.

Brock, W. A. and Hommes, C. H., 1997b. »Models of Complexity in Economics and Finance« in: C. Heij [et al.]. *System Dynamics in Economic and Financial Models*. New York: John Wiley & Sons.

Brock, W. A. and Hommes, C. H., 1998. »Heterogeneous Beliefs and Routes to Chaos in a Simple Asset Pricing Model«. *Journal of Economic Dynamics and Control*, 22 (8-9), 1235-1274.

Brock, W. A., Hommes, C. H. and Wagener, F. O. O., 2005. »Evolutionary Dynamics in Markets with Many Trader Types«. *Journal of Mathematical Economics*, 41, 7-42.

Campbell, J. Y., Lo, A. W. and MacKinlay, A. C., 1997. *The Econometrics of Financial Markets.* Princeton, NJ: Princeton University Press.

Chiarella, C. and He, X.-Z., 2002. »Heterogeneous Beliefs, Risk and Learning in a Simple Asset Pricing Model«. *Computational Economics,* 19, 95-132.

Chiarella, C., He, X.-Z. and Hommes, C. H., 2006. »A Dynamic Analysis of Moving Average Rules«. *Journal of Economic Dynamics and Control,* 30, 1729-1753.

Chiarella, C., He, X.-Z. and Zhu, P., 2003. »Fading Memory Learning in the Cobweb Model with Risk Averse Heterogeneous Producers«. *Research Paper Series,* No. 108. Sydney: Quantitative Finance Research Centre, University of Technology.

Cuthbertson, K., 1996. *Quantitative Financial Economics: Stocks, Bonds and Foreign Exchange.* Chichester, UK: John Wiley & Sons.

De Grauwe, P. and Grimaldi, M., 2006. »Exchange Rate Puzzles: A Tale of Switching Attractors«. *European Economic Review,* 50 (1), 1-33.

Evans, G. W. and Honkapohja, S., 2001. *Learning and Expectations in Macroeconomics.* Princeton: Princeton University Press.

Fama, E. F., 1991. »Efficient Capital Markets: II«. *Journal of Finance,* 46 (5), 1575-1617.

Frøyland, J., 1992. *Introduction to Chaos and Coherence.* Bristol: Institute of Physics Publishing.

Gaunersdorfer, A., 2000. »Endogenous Fluctuations in a Simple Asset Pricing Model with Heterogeneous Agents«. *Journal of Economic Dynamics & Control,* 24 (5-7), 799-831.

Gaunersdorfer, A., Hommes, C. H. and Wagener, F. O. O., 2003. »Bifurcation Routes to Volatility Clustering under Evolutionary Learning«. *CeNDEF Working Paper,* No. 03-03. Amsterdam: University of Amsterdam.

Guckenheimer, J. and Holmes, P., 1983. *Nonlinear Oscillations, Dynamical Systems, and Bifurcations of Vector Fields.* New York: Springer-Verlag.

Hommes, C. H. [et al.], 2002. »Expectations and Bubbles in Asset Pricing Experiments«. *CeNDEF Working Paper,* No. 02-05. Amsterdam: University of Amsterdam.

Hommes, C. H., 2006. »Heterogeneous Agent Models in Economics and Finance« in: L. Tesfatsion and K. L. Judd, eds. *Handbook of Computational Economics, Volume 2: Agent-Based Computational Economics. Amsterdam:* Elsevier Science.

Hommes, C. H., Huang, H. and Wang, D., 2005. »A Robust Rational Route to Randomness in a Simple Asset Pricing Model«. *Journal of Economic Dynamics & Control,* 29 (6), 1043-1072.

Honkapohja, S. and Mitra, K., 2003. »Learning with Bounded Memory in Stochastic Models«. *Journal of Economic Dynamics & Control,* 27 (8), 1437-1457.

Johnson, N. F., Jefferies, P. and Hui, P. M., 2003. *Financial Market Complexity: What Physics Can Tell Us about Market Behaviour.* Oxford: Oxford University Press.

Kuznetsov, Y. A., 1995. *Elements of Applied Bifurcation Theory.* New York: Springer-Verlag.

LeBaron, B., 2000. »Agent Based Computational Finance: Suggested Readings and Early Research«. *Journal of Economic Dynamics and Control,* 24 (5-7), 679-702.

LeBaron, B., 2002. »Short-memory Traders and Their Impact on Group Learning in Financial Markets«, *Proceedings of the National Academy of Sciences (USA),* 99 (3), 7201-7206.

Lucas, R. E., 1978. »Asset Prices in an Exchange Economy«. *Econometrica,* 46 (6), 1429-1445.

Palis, J. and Takens, F., 1993. *Hyperbolicity and Sensitive Chaotic Dynamics at Homoclinic Bifurcations: Fractal Dimensions and Infinitely Many Attractors in Dynamics.* Cambridge: Cambridge University Press.

Shone, R., 1997. *Economic Dynamics: Phase Diagrams and their Economic Application.* Cambridge: Cambridge University Press.